# The Six Sigma Handbook

# About the Authors

**Thomas Pyzdek** has worked in continuous improvement and Lean Six Sigma for more than 50 years. His clients include Ford, McDonald's, Intuit, NASA, and many other organizations in industries ranging from manufacturing to services to healthcare. A Fellow of the American Society for Quality (ASQ), Mr. Pyzdek has received numerous medals and honors from professional organizations, including the ASQ Six Sigma Forum's first ever award for the Advancement of Six Sigma. Pyzdek is currently President and CEO of The Pyzdek Institute, which offers online Lean Six Sigma training and certification.

**Paul Keller** is President of Quality America, Inc., where he works with clients to successfully implement quality improvement concepts and techniques across a full spectrum of manufacturing and service sectors. His online courses and training materials are available through select colleges and universities worldwide, as well as through Quality America's website. In addition to a number of published articles in quality improvement, Keller has authored *Six Sigma Demystified* (McGraw-Hill, 2005) and *Statistical Process Control Demystified* (McGraw-Hill, 2011), and authored all the revisions in *The Handbook for Quality Management, Second Edition* (McGraw-Hill, 2011) and the third, fourth, and fifth editions of *The Six Sigma Handbook*.

# The Six Sigma Handbook

Fifth Edition

Thomas Pyzdek
Paul Keller

New York  Chicago  San Francisco
Athens  London  Madrid
Mexico City  Milan  New Delhi
Singapore  Sydney  Toronto

Library of Congress Control Number: 2018933610

McGraw-Hill Education books are available at special quantity discounts to use as premiums and sales promotions, or for use in corporate training programs. To contact a representative please visit the Contact Us page at www.mhprofessional.com.

The Six Sigma Handbook, Fifth Edition

Copyright © 2018 by McGraw-Hill Education. All rights reserved. Except as permitted under the United States Copyright Act of 1976, no part of this publication may be reproduced or distributed in any form or by any means, or stored in a database or retrieval system, without the prior written permission of the publisher.

1 2 3 4 5 6 7 8 9   LCR   23 22 21 20 19 18

ISBN   978-1-260-12182-7
MHID   1-260-12182-8

**Sponsoring Editor**
Robert Argentieri

**Editorial Supervisor**
Donna M. Martone

**Acquisition Coordinator**
Elizabeth Houde

**Project Manager**
Patricia Wallenburg

**Copy Editor**
James Madru

**Proofreader**
Claire Splan

**Indexer**
Claire Splan

**Production Supervisor**
Lynn M. Messina

**Composition**
TypeWriting

**Art Director, Cover**
Jeff Weeks

Information contained in this work has been obtained by McGraw-Hill Education from sources believed to be reliable. However, neither McGraw-Hill Education nor its authors guarantee the accuracy or completeness of any information published herein, and neither McGraw-Hill Education nor its authors shall be responsible for any errors, omissions, or damages arising out of use of this information. This work is published with the understanding that McGraw-Hill Education and its authors are supplying information but are not attempting to render engineering or other professional services. If such services are required, the assistance of an appropriate professional should be sought.

# CONTENTS

Preface .................................................................... xiii

**PART I**     **Six Sigma Implementation and Management**

**CHAPTER 1**    **Building the Responsive Six Sigma Organization** ................ 3
         What Is Six Sigma? ........................................................ 3
              *Why Six Sigma?* ....................................................... 4
              *The Six Sigma Philosophy* ........................................ 6
              *Six Sigma Versus Traditional Three Sigma Performance* ......... 8
              *The Change Imperative* .......................................... 12
         Implementing Six Sigma .............................................. 17
              *Timetable* ............................................................ 18
              *Infrastructure* ....................................................... 21
              *Integrating Six Sigma and Related Initiatives* ................ 38
              *Deployment to the Supply Chain* ............................. 52
              *Communications and Awareness* ............................ 54

**CHAPTER 2**    **Recognizing and Capitalizing on Opportunity** ................. 61
         Methods for Collecting Customer Data ........................ 61
              *Surveys* ............................................................... 62
              *Focus Groups* ...................................................... 71
              *Operational Feedback Systems* ............................... 72
         Cost of Poor Quality ................................................... 75
              *Cost of Quality Examples* ....................................... 78
              *Quality Cost Bases* ............................................... 81
         Benchmarking ............................................................ 82
              *The Benchmarking Process* ................................... 82
              *Getting Started with Benchmarking* ........................ 83
              *Why Benchmarking Efforts Fail* .............................. 85
              *The Benefits of Benchmarking* ............................... 86

>  *Some Dangers of Benchmarking*. . . . . . . . . . . . . . . . . . . . . . . . . . . . 87
> Innovation. . . . . . . . . . . . . . . . . . . . . . . . . . . . . . . . . . . . . . . . . . . . . . . . . . . 87
> > *Kano Model* . . . . . . . . . . . . . . . . . . . . . . . . . . . . . . . . . . . . . . . . . . . . 88
> > *Quality Function Deployment* . . . . . . . . . . . . . . . . . . . . . . . . . . . . . 89
> > *Translating Customer Demands*. . . . . . . . . . . . . . . . . . . . . . . . . . . . 93
> > *Creative Destruction* . . . . . . . . . . . . . . . . . . . . . . . . . . . . . . . . . . . 101
> Strategic Planning. . . . . . . . . . . . . . . . . . . . . . . . . . . . . . . . . . . . . . . . . . . 106
> > *Organizational Vision*. . . . . . . . . . . . . . . . . . . . . . . . . . . . . . . . . . . 107
> > *Strategy Development* . . . . . . . . . . . . . . . . . . . . . . . . . . . . . . . . . . 109
> > *Strategic Styles* . . . . . . . . . . . . . . . . . . . . . . . . . . . . . . . . . . . . . . . . 110
> > *Possibilities-Based Strategic Decisions*. . . . . . . . . . . . . . . . . . . . . . 111
> Strategic Development Using Constraint Theory . . . . . . . . . . . . . . . . 113
> > *The Systems Approach* . . . . . . . . . . . . . . . . . . . . . . . . . . . . . . . . . . 114
> > *Basic Constraint Management Principles and Concepts* . . . . . . . . 117
> > *Tools of Constraint Management* . . . . . . . . . . . . . . . . . . . . . . . . . 126
> > *Constraint Management Measurements* . . . . . . . . . . . . . . . . . . . . 138
> > *Summary and Conclusion* . . . . . . . . . . . . . . . . . . . . . . . . . . . . . . . 143

**CHAPTER 3** **Data-Driven Management**. . . . . . . . . . . . . . . . . . . . . . . . . . . . . . . . . . . . **145**

> Attributes of Good Metrics. . . . . . . . . . . . . . . . . . . . . . . . . . . . . . . . . . . 145
> > *Measuring Causes and Effects* . . . . . . . . . . . . . . . . . . . . . . . . . . . . 147
> The Balanced Scorecard. . . . . . . . . . . . . . . . . . . . . . . . . . . . . . . . . . . . . 149
> > *Translating the Vision*. . . . . . . . . . . . . . . . . . . . . . . . . . . . . . . . . . . 151
> > *Communicating and Linking* . . . . . . . . . . . . . . . . . . . . . . . . . . . . 159
> > *Business Planning* . . . . . . . . . . . . . . . . . . . . . . . . . . . . . . . . . . . . . 162
> > *Feedback and Learning*. . . . . . . . . . . . . . . . . . . . . . . . . . . . . . . . . 166

**CHAPTER 4** **Maximizing Resources**. . . . . . . . . . . . . . . . . . . . . . . . . . . . . . . . . . . . . . . 177

> Choosing the Right Projects . . . . . . . . . . . . . . . . . . . . . . . . . . . . . . . . . 177
> > *Types of Projects*. . . . . . . . . . . . . . . . . . . . . . . . . . . . . . . . . . . . . . . 178
> > *Analyzing Project Candidates*. . . . . . . . . . . . . . . . . . . . . . . . . . . . 179
> > *Using Pareto Analysis to Identify Six Sigma Project Candidates* . 187
> > *Throughput-Based Project Selection* . . . . . . . . . . . . . . . . . . . . . . . 189
> Ongoing Management Support . . . . . . . . . . . . . . . . . . . . . . . . . . . . . . 195
> > *Internal Roadblocks*. . . . . . . . . . . . . . . . . . . . . . . . . . . . . . . . . . . . 196
> > *External Roadblocks* . . . . . . . . . . . . . . . . . . . . . . . . . . . . . . . . . . . 197
> > *Individual Barriers to Change* . . . . . . . . . . . . . . . . . . . . . . . . . . . 197

*Ineffective Management Support Strategies*..................198
*Effective Management Support Strategies*...................199
*Cross-Functional Collaboration*...........................200
Tracking Six Sigma Project Results...........................201
*Financial Results Validation*.............................204
*Team Performance Evaluation*.............................204
*Team Recognition and Reward*............................205
*Lessons-Learned Capture and Replication*..................207

## PART II   Six Sigma Tools and Techniques

### CHAPTER 5   Project Management Using DMAIC and DMADV..............211
DMAIC and DMADV Deployment Models......................211
*Project Scheduling*......................................216
*Project Reporting*.......................................228
*Project Budgets*........................................230
*Project Records*........................................231
Six Sigma Teams..........................................232
*Team Membership*......................................233
*Team Dynamics Management, Including Conflict Resolution*...233
*Stages in Group Development*.............................234
*Member Roles and Responsibilities*........................236
*Management's Role*.....................................238
*Facilitation Techniques*..................................238

### CHAPTER 6   The Define Phase................................................243
Project Charters............................................243
Project Decomposition......................................246
*Work Breakdown Structures*..............................246
*Pareto Analysis*.........................................247
Deliverables................................................248
*Critical to Quality Metrics*................................250
*Critical to Schedule Metrics*..............................257
*Critical to Cost Metrics*..................................259
Top-Level Process Definition.................................265
*Process Maps*...........................................266
Assembling the Team........................................267

## CHAPTER 7  The Measure Phase .................................................. 269

Process Definition .................................................. 269
    *Flowcharts* .................................................. 270
    *SIPOC* .................................................. 271
Metric Definition .................................................. 275
    *Measurement Scales* .................................................. 276
    *Discrete and Continuous Data* .................................................. 278
Process Baseline Estimates .................................................. 278
    *Enumerative and Analytic Studies* .................................................. 280
    *Principles of Statistical Process Control* .................................................. 283
    *Estimating Process Baselines Using Process Capability Analysis* .................................................. 289

## CHAPTER 8  Process Behavior Charts .................................................. 291

Distributions .................................................. 291
    *Methods of Enumeration* .................................................. 291
    *Frequency and Cumulative Distributions* .................................................. 293
    *Sampling Distributions* .................................................. 294
    *Binomial Distribution* .................................................. 295
    *Poisson Distribution* .................................................. 297
    *Hypergeometric Distribution* .................................................. 298
    *Normal Distribution* .................................................. 299
    *Lognormal Distribution* .................................................. 306
    *Exponential Distribution* .................................................. 307
    *Weibull Distribution* .................................................. 309
Control Charts for Variables Data .................................................. 310
    *Averages and Ranges Control Charts* .................................................. 310
    *Averages and Standard Deviation (Sigma) Control Charts* .................................................. 314
    *Control Charts for Individual Measurements (X Charts)* .................................................. 317
Control Charts for Attributes Data .................................................. 323
    *Control Charts for Proportion (p Charts)* .................................................. 323
    *Control Charts for Count of Items (np Charts)* .................................................. 327
    *Control Charts for Average Occurrences-Per-Unit (u Charts)* .................................................. 329
    *Control Charts for Counts of Occurrences-Per-Unit (c Charts)* .................................................. 332
Control Chart Selection .................................................. 336
    *Rational Subgroup Sampling* .................................................. 336
Control Chart Interpretation .................................................. 341
    *Run Tests* .................................................. 347

Short-Run Statistical Process Control Techniques ............... 348
    *Variables Data* ............................................. 350
    *Attribute SPC for Small and Short Runs* .................... 361
    *Summary of Short-Run SPC* ................................. 368
SPC Techniques for Automated Manufacturing .................. 369
    *Problems with Traditional SPC Techniques* ................. 369
    *Special and Common Cause Charts* ......................... 370
    *EWMA Common Cause Charts* ............................... 371
    *EWMA Control Charts Versus Individuals Charts* ........... 378
Process Capability Indices ................................... 380
    *Example of Non-Normal Capability Analysis Using Minitab* ... 385

**CHAPTER 9**  **Measurement Systems Evaluation** ........................... 391

Definitions .................................................. 391
    *Measurement System Discrimination* ....................... 395
    *Stability* ................................................ 395
    *Bias* .................................................... 396
    *Repeatability* ............................................ 397
    *Reproducibility* .......................................... 400
    *Part-to-Part Variation* ................................... 403
    *Summary Reporting* ....................................... 403
    *Gage R&R Analysis Using Minitab* ......................... 404
    *Linearity* ................................................ 407
    *Linearity Analysis Using Minitab* ......................... 409
Attribute Measurement Error Analysis ......................... 410
    *Operational Definitions* .................................. 412
    *How to Conduct Attribute Inspection Studies* .............. 413
    *Minitab Attribute Gage R&R Example* ...................... 417

**CHAPTER 10**  **Analyze Phase** ............................................. 423

Value Stream Analysis ........................................ 423
    *Value Stream Mapping* ................................... 427
    *Spaghetti Charts* ........................................ 433
Analyzing the Sources of Variation ........................... 434
    *Cause and Effect Diagrams* ............................... 435
    *Boxplots* ................................................ 437
    *Statistical Inference* .................................... 439

    *Chi-Square, Student's* **t**, *and* **f** *Distributions* . . . . . . . . . . . . . . . . . . 440
    *Point and Interval Estimation* . . . . . . . . . . . . . . . . . . . . . . . . . . . . 444
    *Hypothesis Testing* . . . . . . . . . . . . . . . . . . . . . . . . . . . . . . . . . . . . . 452
    *Resampling (Bootstrapping)* . . . . . . . . . . . . . . . . . . . . . . . . . . . . . 459
  Regression and Correlation Analysis. . . . . . . . . . . . . . . . . . . . . . . . . . . . 460
    *Linear Models*. . . . . . . . . . . . . . . . . . . . . . . . . . . . . . . . . . . . . . . . . . 463
    *Least-Squares Fit* . . . . . . . . . . . . . . . . . . . . . . . . . . . . . . . . . . . . . . . 466
    *Correlation Analysis* . . . . . . . . . . . . . . . . . . . . . . . . . . . . . . . . . . . . 470
  Designed Experiments. . . . . . . . . . . . . . . . . . . . . . . . . . . . . . . . . . . . . . . 472
    *The Traditional Approach Versus*
      *Statistically Designed Experiments* . . . . . . . . . . . . . . . . . . . . . . 472
    *Terminology* . . . . . . . . . . . . . . . . . . . . . . . . . . . . . . . . . . . . . . . . . . . 472
    *Design Characteristics*. . . . . . . . . . . . . . . . . . . . . . . . . . . . . . . . . . . 474
    *Types of Design*. . . . . . . . . . . . . . . . . . . . . . . . . . . . . . . . . . . . . . . . 475
    *One-Factor ANOVA* . . . . . . . . . . . . . . . . . . . . . . . . . . . . . . . . . . . . 477
    *Two-Way ANOVA with No Replicates*. . . . . . . . . . . . . . . . . . . . . . 479
    *Two-Way ANOVA with Replicates*. . . . . . . . . . . . . . . . . . . . . . . . . 480
    *Full and Fractional Factorial* . . . . . . . . . . . . . . . . . . . . . . . . . . . . . 482
    *Power and Sample Size* . . . . . . . . . . . . . . . . . . . . . . . . . . . . . . . . . 491
    *Testing Common Assumptions* . . . . . . . . . . . . . . . . . . . . . . . . . . . 492
  Analysis of Categorical Data. . . . . . . . . . . . . . . . . . . . . . . . . . . . . . . . . . 499
    *Making Comparisons Using Chi-Square Tests* . . . . . . . . . . . . . . . 499
    *Logistic Regression*. . . . . . . . . . . . . . . . . . . . . . . . . . . . . . . . . . . . . 501
    *Binary Logistic Regression*. . . . . . . . . . . . . . . . . . . . . . . . . . . . . . . 503
    *Ordinal Logistic Regression*. . . . . . . . . . . . . . . . . . . . . . . . . . . . . . 506
    *Nominal Logistic Regression*. . . . . . . . . . . . . . . . . . . . . . . . . . . . . 510
  Non-Parametric Methods . . . . . . . . . . . . . . . . . . . . . . . . . . . . . . . . . . . . 513

**CHAPTER 11**  **The Improve/Design Phase** . . . . . . . . . . . . . . . . . . . . . . . . . . . . . . . . . . .517
  Using Customer Demands to Make
    Design and Improvement Decisions. . . . . . . . . . . . . . . . . . . . . . . . . . 517
    *Pugh Concept Selection Method*. . . . . . . . . . . . . . . . . . . . . . . . . . .517
  Lean Techniques for Optimizing Flow . . . . . . . . . . . . . . . . . . . . . . . . . . 518
    *Unnecessary Process Steps*. . . . . . . . . . . . . . . . . . . . . . . . . . . . . . . 519
    *Excessive Movement of Material or Personnel* . . . . . . . . . . . . . . . 519
    *Bottleneck or Constraint* . . . . . . . . . . . . . . . . . . . . . . . . . . . . . . . . 520
    *Process Errors Requiring Rework* . . . . . . . . . . . . . . . . . . . . . . . . . 521
    *Excess In-process Inventory*. . . . . . . . . . . . . . . . . . . . . . . . . . . . . . 521

    *Understanding Queues to Balance Processes*................526
  Using Empirical Model Building to Optimize....................529
    *Phase 0: Getting Your Bearings*............................531
    *Phase I: The Screening Experiment*.........................532
    *Phase II: Steepest Ascent (Descent)*........................536
    *Phase III: The Factorial Experiment*.......................538
    *Phase IV: The Composite Design*...........................540
    *Phase V: Robust Product and Process Design*.................544
  Data Mining, Artificial Neural Networks,
    and Virtual Process Mapping..............................548
    *Example of Neural Net Models*.............................549
  Optimization Using Simulation................................552
    *Predicting CTQ Performance*...............................552
    *Simulation Tools*........................................553
    *Random Number Generators*................................557
    *Model Development*.......................................560
    *Virtual DOE Using Simulation Software*....................570
  Risk Assessment Tools.......................................572
    *Design Review*..........................................573
    *Fault-Tree Analysis*....................................574
    *Safety Analysis*........................................575
    *Failure Mode and Effect Analysis*........................578
  Defining New Performance Standards Using Statistical Tolerancing. 581
    *Assumptions of Formula*..................................585
    *Tolerance Intervals*.....................................585

**CHAPTER 12**   **Control/Verify Phase**..........................................587

  Validating the New Process or Product Design...................587
  Business Process Control Planning............................588
    *Maintaining Gains*......................................588
    *Tools and Techniques Useful for Control Planning*...........590
    *Preparing the Process Control Plan*.......................591
    *Process Control Planning for Short and Small Runs*..........593
    *Process Audits*.........................................596
    *Selecting Process Control Elements*......................596
    *Other Elements of the Process Control Plan*...............599
  Multivariate Control Charts..................................599
  Principal Component Analysis.................................608

| | | |
|---|---|---|
| **APPENDIX 1** | Glossary of Basic Statistical Terms | 617 |
| **APPENDIX 2** | Area Under the Standard Normal Curve | 623 |
| **APPENDIX 3** | Critical Values of the $t$-Distribution | 627 |
| **APPENDIX 4** | Chi-Square Distribution | 629 |
| **APPENDIX 5** | $F$ Distribution ($\alpha = 1\%$) | 631 |
| **APPENDIX 6** | $F$ Distribution ($\alpha = 5\%$) | 633 |
| **APPENDIX 7** | Poisson Probability Sums | 635 |
| **APPENDIX 8** | Tolerance Interval Factors | 639 |
| **APPENDIX 9** | Control Chart Constants | 643 |
| **APPENDIX 10** | Control Chart Equations | 645 |
| **APPENDIX 11** | Table of $d_2^*$ Values | 647 |
| **APPENDIX 12** | Factors for Short Run Control Charts for Individuals, $\bar{X}$, and $R$ Charts | 649 |
| **APPENDIX 13** | Sample Customer Survey | 651 |
| **APPENDIX 14** | Process $\sigma$ Levels and Equivalent PPM Quality Levels | 653 |
| **APPENDIX 15** | Black Belt Effectiveness Certification | 655 |
| **APPENDIX 16** | Green Belt Effectiveness Certification | 667 |
| **APPENDIX 17** | AHP Using Microsoft Excel | 679 |
| | Notes | 683 |
| | References | 685 |
| | Index | 693 |

# PREFACE

Even a brief overview of Six Sigma literature will uncover more than a few approaches to implementing Six Sigma. Some have been remarkably unsuccessful, while others have provided the vision and means for the organization to prosper. Online chat groups provide a global sounding board for the discontented as well as the enthused. For some, Six Sigma is dead, and for those, perhaps it should be. They are ready to move on to the "next big thing."

What I tend to find most intriguing about these discussions are the details. What *specifically* failed in an implementation? Why didn't Six Sigma work? What alternative provides an improved approach? Invariably, the failures suffered from predictably flawed approaches, hampered by the usual suspects of implementation issues: lack of commitment and/or resources, and poor or misdirected focus. The fundamentals of the Six Sigma approach remain sound: Prioritize specific issues impacting customers, shareholders and/or employees, and problem-solve over a relatively rapid time frame using data-driven cross-functional project teams sponsored by the functional stakeholder groups. Effectively-managed projects will quickly identify organizational issues, highlighting the advantage of the project as the means to achieve manageable bites of improvement. Are there best practices that improve success rates? Absolutely! Best practices in Six Sigma are continuously evolving, just as Six Sigma itself evolved from earlier best practices in quality improvement. Those who truly "get it" know that Six Sigma will die only when organizations stop caring about their customers, employees and shareholders. Did Six Sigma really "kill innovation at XYZ Company?" Of course not! Can an overly bureaucratic approach to innovation stifle creativity? Yes, just as overzealous standardization emphasizes process consistency and internal productivity at the expense of an improved customer experience. These failures result from neglecting a key tenet of Six Sigma: customer focus.

This fifth edition addresses many of these challenges. If this is your first copy of *The Six Sigma Handbook*, you may find the implementation approach includes many of the practices advocated by "newer" disciplines, such as Business Process Management (BPM) or Business Process Improvement (BPI). The overall approach remains consistent with the earlier editions of the text, with elaboration and best practices added to more fully develop the approach in the reader's mind.

You'll notice many references to free online materials within the text, such as Excel file templates that can be used for analyzing projects, or videos that provide an in-depth narrative on specific topics. Additional links will be added over time to further extend the learning potential offered by the text, so be sure to regularly check back into the online site at www.mhprofessional.com/ssh5. We expect to offer additional learning options, some free and some at reasonable expense, based on reader feedback.

With well over 120,000 copies in print for its first four editions, *The Six Sigma Handbook* is well-established as a key reference guide for forward-thinking managers and customer-focused process improvement specialists alike. We'd like to thank our faithful readers who have made this work a lasting tribute to the concepts and techniques known as Six Sigma. We hope this fifth edition provides additional insight and direction to achieve ever-higher levels of value for your customers.

*Paul Keller*

# PART I

# Six Sigma Implementation and Management

## Chapter 1
Building the Responsive Six Sigma Organization

## Chapter 2
Recognizing and Capitalizing on Opportunity

## Chapter 3
Data-Driven Management

## Chapter 4
Maximizing Resources

# CHAPTER 1

# Building the Responsive Six Sigma Organization

## What Is Six Sigma?

Six Sigma is a rigorous, focused, and highly effective implementation of proven quality principles and techniques. Incorporating elements from the work of many quality pioneers, Six Sigma aims for virtually error-free business performance. Sigma, σ, is a letter in the Greek alphabet used by statisticians to measure the variability in any process. A company's performance is measured by the sigma level of their business processes. Traditionally, companies accepted three or four sigma performance levels as the norm, despite the fact that these processes created between 6,200 and 67,000 problems per million opportunities! The Six Sigma standard of 3.4 problems-per-million opportunities[1] is a response to the increasing expectations of customers and the increased complexity of modern products and processes.

Despite its name, Six Sigma's magic isn't in statistical or high-tech razzle-dazzle. Six Sigma relies on tried and true methods that have been used for decades. By some measures, Six Sigma discards a great deal of the complexity that characterized Total Quality Management (TQM). Six Sigma takes a handful of proven methods and trains a small cadre of in-house technical leaders, known as Six Sigma Black Belts, to a high level of proficiency in the application of these techniques. To be sure, some of the methods Black Belts use are highly advanced, including up-to-date computer technology. But the tools are applied within a simple performance improvement model known as Define-Measure-Analyze-Improve-Control, or DMAIC. DMAIC is described briefly as follows:

D  Define the goals of the improvement activity.
M  Measure the existing system.
A  Analyze the system to identify ways to eliminate the gap between the current performance of the system or process and the desired goal.
I  Improve the system.
C  Control the new system.

The DMAIC methodology is discussed in detail in Part II.

## *Why Six Sigma?*

When a Japanese firm took over a Motorola factory that manufactured Quasar television sets in the United States in the 1970s, they promptly set about making drastic changes in the way the factory operated. Under Japanese management, the factory was soon producing TV sets with 1/20th as many defects as they had produced under Motorola's management. They did this using the same workforce, technology, and designs, and did it while lowering costs, making it clear that the problem was Motorola's management. It took a while, but, eventually, even Motorola's own executives finally admitted "Our quality stinks" (Main, 1994).

It took until nearly the mid-1980s before Motorola figured out what to do about it. Bob Galvin, Motorola's CEO at the time, started the company on the quality path known as Six Sigma and became a business icon largely as a result of what he accomplished in quality at Motorola. Using Six Sigma, Motorola became known as a quality leader and a profit leader. After Motorola won the Malcolm Baldrige National Quality Award in 1988 the secret of their success became public knowledge and the Six Sigma revolution was on. Today it's hotter than ever. Even though Motorola has been struggling for the past few years, companies such as GE and AlliedSignal have taken up the Six Sigma banner and used it to lead themselves to new levels of customer service and productivity.

It would be a mistake to think that Six Sigma is about quality in the traditional sense. Quality, defined traditionally as conformance to internal requirements, has little to do with Six Sigma. Six Sigma focuses on helping the organization make more money by improving customer value and efficiency. To link this objective of Six Sigma with quality requires a new definition of quality: the value added by a productive endeavor. This quality may be expressed as potential quality and actual quality. Potential quality is the known maximum possible value added per unit of input. Actual quality is the current value added per unit of input. The difference between potential and actual quality is waste. Six Sigma focuses on improving quality (i.e., reducing waste) by helping organizations produce products and services better, faster, and cheaper. There is a direct correspondence between quality levels and "sigma levels" of performance. For example, a process operating at Six Sigma will fail to meet requirements about 3 times per million transactions. The typical company operates at roughly four sigma, equivalent to approximately 6,210 errors per million transactions. Six Sigma focuses on customer requirements, defect prevention, cycle time reduction, and cost savings. Thus, the benefits from Six Sigma go straight to the bottom line. Unlike mindless cost-cutting programs that also reduce value and quality, Six Sigma identifies and eliminates costs that provide no value to customers: waste costs.

For non–Six Sigma companies, these costs are often extremely high. Companies operating at three or four sigma typically spend between 25 and 40% of their revenues fixing problems. This is known as the cost of quality, or more accurately the cost of poor quality (COPQ). Companies operating at Six Sigma typically spend less than 5% of their revenues fixing problems (Fig. 1.1). COPQ values shown in Fig. 1.1 are at the lower end of the range of results reported in various studies. The dollar cost of this gap can be huge. General Electric estimated that the gap between three or four sigma and Six Sigma was costing them between $8 billion and $12 billion per year.

One reason why costs are directly related to sigma levels is very simple: sigma levels are a measure of error rates, and it costs money to correct errors. Figure 1.2 shows the relationship between errors and sigma levels. Note that the error rate drops exponentially as the sigma level goes up, and that this correlates well to the empirical cost data shown in Fig. 1.1. Also note that the errors are shown as errors per million opportunities, not as percentages. This is another convention introduced by Six Sigma. In the past we could tolerate percentage error rates (errors per hundred opportunities). In today's competitive, global business climate, we cannot.

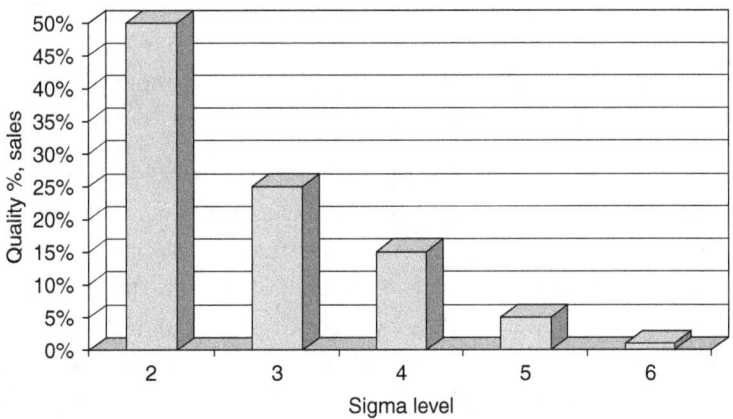

**Figure 1.1**  Cost of poor quality versus sigma level.

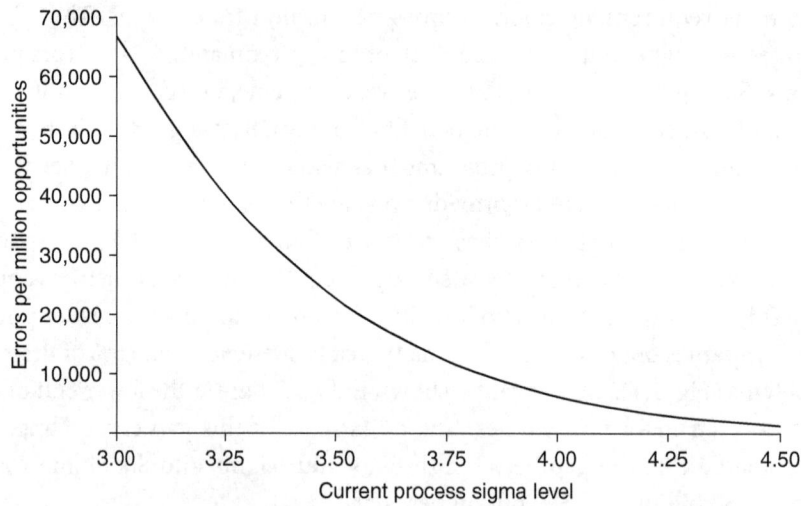

**Figure 1.2** Error rate versus sigma level.

## *The Six Sigma Philosophy*

Six Sigma is the application of the scientific method to the design and operation of management systems and business processes that enable employees to deliver the greatest value to customers and owners. The scientific method works as follows:

1. Observe some important aspect of the marketplace or your business.
2. Develop a tentative explanation, or hypothesis, consistent with your observations.
3. Based on your hypothesis, make predictions.
4. Test your predictions by conducting experiments or making further careful observations. Record your observations. Modify your hypothesis based on the new facts. If variation exists, use statistical tools to help you separate signal from noise.
5. Repeat steps 3 and 4 until there are no discrepancies between the hypothesis and the results from experiments or observations.

At this point you have a viable theory explaining an important relationship in your market or business. The theory is your crystal ball, which you can use to predict the future. As you can imagine, a crystal ball is very useful for any organization. Furthermore, it often happens that your theory will explain phenomena other than that you initially studied. Isaac Newton's theory of gravity may have begun with the observation that apples fell toward the earth, but Newton's laws of motion explained a great deal about the way planets moved about the sun. By applying the scientific method over a period of years, you will develop a deep understanding of what makes your customer and your business tick.

When this approach is applied across the organization, the political influence that stalls organizations is minimized and a "show me the data" attitude prevails. While corporate politics can never be eliminated where human beings interact, politics is much less an influence in Six Sigma organizations than in traditional organizations. People are often quite surprised at the results of this seemingly simple shift in attitude. The essence of these results is stated quite succinctly by "Pyzdek's law:"

> Most of what you know is wrong!

Like all such "laws," this is an overstatement. However, you'll be stunned by how often people are unable to provide data supporting positions on basic issues when challenged. For example, the manager of a technical support call center was challenged by the CEO to show that customers cared deeply about hold time. Upon investigation, the manager determined that customers cared more about the time it took to reach a technician and whether or not their issue was resolved. The call center's information system measured hold time to include both the time until the technician first answered the phone and the time the customer was on hold while the technician researched the answer. The customers cared much less about this "hold time," since they recognized the value it added in resolution of the issue. This fundamental change in focus made a great deal of difference in the way the call center operated.

### What We Know

We all know that there was a surge in births nine months after the November 1965 New York City power failure, right? After all, the *New York Times* said so in a story that ran August 8, 1966. If that's not prestigious enough for you, consider that the source quoted in the *Times* article was the city's Mt. Sinai Hospital, one of the best.

### What the Data Show

The newspaper compared the births on August 8, 1965 with those on August 8, 1966. This one-day comparison did indeed show an increase year-over-year. However, J. Richard Udry, director of the Carolina Population Center at the University of North Carolina, studied birthrates at several New York City hospitals between July 27 and August 14, 1966. His finding: the birthrate nine months after the blackout was slightly below the five-year average.

The Six Sigma philosophy focuses the attention on the stakeholders for whom the enterprise exists. It is a cause-and-effect mentality. Well-designed management systems and business processes operated by happy employees cause customers and owners to be satisfied or delighted. Of course, none of this is new. Most leaders of traditional organizations honestly believe that this is what they already do. What distinguishes the traditional approach from Six Sigma is the degree of rigor and commitment to the core principles.

## *Six Sigma Versus Traditional Three Sigma Performance*

The traditional quality model of process capability differed from Six Sigma in two fundamental respects:

- ▲ It was applied only to manufacturing processes, while Six Sigma is applied to all important business processes.
- ▲ It stipulated that a "capable" process was one that had a process standard deviation of no more than one-sixth of the total allowable spread, where Six Sigma requires the process standard deviation be no more than one-twelfth of the total allowable spread.

These differences are far more profound than one might realize. By addressing all business processes Six Sigma not only treats manufacturing as part of a larger system, it removes the narrow, inward focus of the traditional approach. Customers care about more than just how well a product is manufactured. Price, service, financing terms, style, availability, frequency of updates and enhancements, technical support, and a host of other items are also important. Also, Six Sigma benefits others besides customers. When operations become more cost-effective and the product design cycle shortens, owners or investors benefit too. When employees become more productive their pay can be increased. Six Sigma's broad scope means that it provides benefits to all stakeholders in the organization.

The second point also has implications that are not obvious. Six Sigma is, basically, a process quality goal, where sigma is a statistical measure of variability in a process. As such it falls into the category of a process capability technique. The traditional quality paradigm defined a process as capable if the process's natural spread, plus and minus three sigma, was less than the engineering tolerance. Under the assumption of normality, this three sigma quality level translates to a process yield of 99.73%. A later refinement considered the process location as well as its spread and tightened the minimum acceptance criterion so that the process mean was at least four sigma from the nearest engineering requirement. Six Sigma requires that processes operate such that the nearest engineering requirement is at least Six Sigma from the process mean.

One of Motorola's most significant contributions was to change the discussion of quality from one where quality levels were measured in percent (parts-per-hundred), to a discussion of parts-per-million (ppm) or even parts-per-billion. Motorola correctly pointed out that modern technology was so complex that old ideas about "acceptable quality levels" could no longer be tolerated. Modern business requires near perfect quality levels.

One puzzling aspect of the "official" Six Sigma literature is that it states that a process operating at Six Sigma will produce 3.4 parts-per-million nonconformances. However, if a special normal distribution table is consulted (very few go out to Six Sigma) one finds that the expected nonconformances are 0.002 PPM (2 parts-per-billion, or PPB). The difference occurs because Motorola presumes that the process mean can drift 1.5 sigma in either direction. (This assumption is further discussed in Chap. 7.) The area of a normal distribution beyond 4.5 sigma from the mean is indeed 3.4 PPM. Since control charts will easily detect any process shift of this magnitude in a single sample, the 3.4 PPM represents a very conservative upper bound on the nonconformance rate.

In contrast to Six Sigma quality, the old three sigma quality standard of 99.73% translates to 2,700 PPM failures, even if we assume zero drift. For processes with a series of steps, the overall yield is the product of the yields of the different steps. For example, if we had a simple two-step process where step #1 had a yield of 80% and step #2 had a yield of 90%, then the overall yield would be 0.8 × 0.9 = 0.72 = 72%. Note that the overall yield from processes involving a series of steps is always less than the yield of the step with the lowest yield. If three sigma quality levels (99.97% yield) are obtained from every step in a 10-step process, the quality level at the end of the process will contain 26,674 defects per million. (See rolled throughput yield calculations in the "Deliverables" section in Chap. 6.) Considering that the complexity of modern processes is usually far greater than 10 steps, it is easy to see that Six Sigma quality isn't optional, it's required if the organization is to remain viable.

The requirement of extremely high quality is not limited to multiple-stage manufacturing processes. Consider what three sigma quality would mean if applied to other processes:

- Virtually no modern computer would function.
- 10,800,000 mishandled healthcare claims each year.
- 18,900 lost U.S. savings bonds every month.
- 54,000 checks lost each night by a single large bank.
- 4,050 invoices sent out incorrectly each month by a modest-sized telecommunications company.
- 540,000 erroneous call detail records each day from a regional telecommunications company.
- 270,000,000 (270 million) erroneous credit card transactions each year in the United States.

With numbers like these, it's easy to see that the modern world demands extremely high levels of error-free performance. Six Sigma arose in response to this realization.

## Just Do It!

It's important to note that Six Sigma organizations are not academic institutions. They compete in the fast-paced world of business, and they don't have the luxury of taking years to study all aspects of a problem before deciding on a course of action. A valuable skill for the leader of a Six Sigma enterprise, or for the sponsor of a Six Sigma project, is to decide when enough information has been obtained to warrant taking a particular course of action. Six Sigma leadership should be conservative when spending the shareholders' dollars. As a result, project research tends to be tightly focused on delivering information useful for management decision-making. Once a level of confidence is achieved, management must direct the Black Belt to move the project from the Analyze phase to the Improve phase, or from the Improve phase to the Control phase. Projects are closed and resources moved to new projects as quickly as possible.

Six Sigma organizations are not infallible; they make their share of mistakes and miss opportunities. Yet, research has shown they make fewer mistakes than their traditional counterparts and perform significantly better in the long run. Their systems incorporate the ability to learn from these mistakes, with resulting systematic improvements.

## What's Important?

While working with an aerospace client, I was helping an executive set up a system for identifying potential Six Sigma projects in his area. I asked "What are your most important metrics? What do you focus on?" "That's easy," he responded. "We just completed our monthly ops review so I can show you."

He then called his secretary and asked that she bring the ops review copies. Soon the secretary came in lugging three large, loose-leaf binders filled with copies of PowerPoint slides. This executive and his staff spend one very long day each month reviewing all of these metrics, hoping to glean some direction to help them plan for the future. This is not focusing, it's torture!

Sadly, this is not an isolated case. Over the years I've worked with thousands of people in hundreds of companies and this measurement nightmare is commonplace, even typical. The human mind isn't designed to make sense of such vast amounts of data. We can only hold a limited number of facts in our minds at one time. We are simply overwhelmed when we try to retain too much information. One study of information overload found the following (Waddington, 1996):

- Two-thirds of managers report tension with work colleagues and loss of job satisfaction because of stress associated with information overload.

- One-third of managers suffer from ill health as a direct consequence of stress associated with information overload. This figure increases to 43% among senior managers.
- Almost two-thirds (62%) of managers testify that their personal relationships suffer as a direct result of information overload.
- Forty-three percent of managers think important decisions are delayed and the ability to make decisions is affected as a result of having too much information.
- Forty-four percent believe the cost of collating information exceeds its value to business.

Clearly, more information isn't always better.

When pressed, nearly every executive or manager will admit that there are a half-dozen or so measurements that really matter. The rest are either derivatives or window dressing. When asked what really interested him, my client immediately turned to a single slide in the middle of one of the binders. There were two "biggies" that he focused on. The second-level drill down involved a half-dozen major drivers. Tracking this number of metrics is well within the abilities of humans, if not crows! With this tighter focus the executive could put together a system for selecting good Six Sigma projects and team members.

Six Sigma activities focus on the few things that matter most to three key constituencies: customers, shareholders, and employees. The primary focus is on customers, but shareholder interests are not far behind. The requirements of these two groups are determined using scientific methods, of course. Yet the science of identifying customer and shareholder desires is not fully mature, so the data are supplemented with a great deal of personal contact at all levels of the organization. Employee requirements are also aggressively sought. Well-treated employees stay longer and do a better job.

Focus comes from two perspectives: down from the top-level goals and up from problems and opportunities. The opportunities meet the goals at the Six Sigma project, whose selection and development become critical aspects of meeting organizational objectives. Six Sigma projects link the activities of the enterprise to its improvement goals. The linkage is so tight that in a well-run enterprise people working on Six Sigma projects can tell you which enterprise objectives will be impacted by their project, and senior leaders are able to measure the impact of Six Sigma on the enterprise in clear and meaningful terms. The costs and benefits of Six Sigma are monitored using enterprise-wide tracking systems that can slice and dice the data in many different ways. At any point in time an executive can determine if Six Sigma is pulling its weight. In many TQM programs of the past people were unable to point to specific bottom-line benefits, so interest gradually waned and the programs were shelved when times got tough. Six Sigma organizations know precisely what they're getting for their investment.

Six Sigma also has an indirect and seldom measured benefit to an enterprise: its impact on human behavior. Six Sigma doesn't operate in a vacuum. When employees observe Six Sigma's dramatic results, they naturally modify how they approach their work. Seat-of-the-pants management doesn't sit well (pardon the pun!) in Six Sigma organizations that have reached "critical mass." Critical mass occurs when the organization's culture has changed as a result of Six Sigma's successful deployment across a large segment of the organization. The initial clash of cultures has worked itself out, and those opposed to the Six Sigma way have either left, converted, or learned to keep quiet.

When deploying Six Sigma, it's important not to stifle creativity for the sake of operational efficiencies. For example, successful research and development (R&D) involves a good deal of original creative thinking. Research may actually suffer from too much rigor and focus on error prevention. Cutting-edge research is necessarily trial and error and requires a high tolerance for failure. The chaos of exploring new ideas is not something to be managed out of the system; it is expected and encouraged. To the extent that it involves process design and product testing, including the concept of manufacturability, Six Sigma will certainly make a contribution to the development part of R&D. The objective is to selectively apply Six Sigma to those areas where it provides benefit.

Taking a broader view, a business is a complex undertaking, requiring creativity, innovation, and intuition for successful leadership. While it's good to be "data-driven," leaders need to question data effectively, especially since some of the most important components of success in business are unmeasured and perhaps immeasurable. Challenge counterintuitive data and subject it to a gut check. It may be that the counterintuitive result represents a startling breakthrough in knowledge, but it may simply be wrong.

Consider this example. A software client had a technical support call center to help their customers solve problems with the software. Customer surveys were collected and the statistician made an amazing discovery: hold time didn't matter! The data showed that customer satisfaction was the same for customers served immediately and for those on hold for an hour or more. Discussions began along the lines of how many fewer staff would be required due to this new information. Impressive savings were forecast.

Fortunately, the support center manager hadn't left his skepticism at the front door. He asked for additional data, which showed that the abandon rate increased steadily as people were kept on hold. The surveys were given only to those people who had waited for service. These people didn't mind waiting. Those who hung up the phone before being served apparently did. In fact, when a representative sample was obtained, excessive hold time was the number one complaint.

## *The Change Imperative*

In traditional organizations the role of management is to design systems to create and deliver value to customers and shareholders. Unfortunately, however, too many of these organizations fail to recognize that this is a never-ending task. Competitors constantly innovate in an attempt to steal your customers. Customers continuously change their minds about what they want. Capital markets offer investors new ways to earn a return on their investment. The result is an imperative to constantly change management systems.

Despite the change imperative, most enterprises resist change until there are obvious signs that current systems are failing one or more stakeholder groups. Perhaps declining market share makes it clear that your products or services are not as competitive as they once were. Customers may remain loyal, but complaints have reached epidemic proportions. Or share price, the perceived market value of your business, may be trending ominously downward. Traditional organizations watch for such signs and react to them. Change occurs, as it must, but it does so in an atmosphere of crisis and confusion. Substantial loss may result before the needed redesign is complete. People may lose their jobs or even their careers. Many organizations that employ these reactionary tactics don't survive the shock.

Sadly, as this page is written, the U.S. automobile industry is reeling from the combined effects of global competition, a worldwide credit crisis, and an extended period of high fuel costs. While arguments can be made as to the predictability of these events, it is clear that the strength of their competitors lies primarily in their ability to adapt. A recent poll found that more than 60% of global respondents agreed that the ability to change is an organization's main competitive advantage (Blauth, 2008). The ability to respond to customer demand, whether that demand is stagnant or dynamic, is a key focus of Six Sigma projects. Applied at a process level, the Lean principles deployed within these projects stress reduced inventories with decreased cycle times to quickly satisfy shifts in customer demand. As an organizational strategy, these principles result in agile organizations that invest in adaptability rather than volume efficiencies. Resources are deployed only when needed, so they can be constantly refocused to meet the current customer value definitions.

In this way, the Six Sigma enterprise proactively embraces change by explicitly incorporating change into their management systems. Full- and part-time change agent positions are created with a supporting infrastructure designed to integrate change into the routine. Systems are implemented to monitor changing customer, shareholder, and employee inputs, and to rapidly integrate the new information into revised business processes. The approach may employ sophisticated computer modeling, or more basic statistical analysis, to minimize unneeded tampering by separating signal from noise. These analytical techniques are applied to stakeholder inputs and to enterprise and process metrics at all levels.

The intended consequence of deploying Six Sigma is a change in behavior, as well as the more obvious organizational effectiveness and efficiencies. Conventional wisdom is respectfully questioned: the phrase "How do you know?" is heard repeatedly.

- "Nice report on on-time deliveries, Joan, but show me why you think this is important to the customer. If it is, I want to see a chart covering the last 52 weeks, and don't forget the control limits."

- "This budget variance report doesn't distinguish between expected variation and real changes to the system! I want to see performance across time, with control limits, so we know how to effectively respond."
- "Have these employee survey results been validated? What is the reliability of the questions? What are the main drivers of employee satisfaction? How do you know?"
- "How do these internal dashboards relate to the top-level dashboards that are important to shareholders?"

Yet, the act of challenging accepted practices poses risk. The challenger may feel isolated; those being challenged may feel threatened. These represent behavioral costs to the change effort. The net result of the challenge, ultimately, is the need for further information, which comes at a monetary cost and opportunity risk to the organization. These risks and costs must be effectively managed.

## Managing Change

Three goals of change may be summarized as follows:

- **Change the way people in the organization think**—Helping people modify their perspective is a fundamental activity of the change agent. All change begins with the individual, at a personal level. Unless the individual is willing to change his behavior, no real change is possible. Changing behavior requires a change in thinking. In an organization where people are expected to use their minds, people's actions are guided by their thoughts and conclusions. The change agent's job starts here.
- **Change the norms**—Norms consist of standards, models, or patterns that guide behavior in a group. All organizations have norms or expectations of their members. Change cannot occur until the organization's norms change. In effective Six Sigma organizations, the desired norm is data-driven decision-making focused on providing maximum value to key stakeholders.
- **Change the organization's systems or processes**—This is the "meat" of the change. Ultimately, all work is a process and quality improvement requires change at the process and system level. However, this cannot occur on a sustained basis until individuals change their behavior and organizational norms are changed.

Change agents fundamentally accomplish these goals by building buy-in within the key stakeholder groups affected by the change. While this is challenging at the process level, it is considerably more so at the organizational level, as is discussed in the next section.

The press of day-to-day business, combined with the inherent difficulties of change, makes it easy to let time slip by without significant progress. Keeping operations going is a full-time job, and current problems present themselves with an urgency that meeting a future goal can't match. Without the constant reminders from change agents that goals aren't being met, the leadership can simply forget about the transformation. It is the change agent's job to become the "conscience" of the leadership and to challenge them when progress falls short of goals.

## The Transformation Process

Based on considerable research over a number of years, Kotter (1995) established a set of eight steps required to achieve organizational transformation. Kotter believes that change initiatives fail when management treat transformation as an event rather than a process. The successful transformation process occurs over a period of years, and steps are invariably skipped when pressure is exerted to speed up the process. Just as toxic to success is declaring victory prematurely, which saps momentum and can destroy the progress to date. Kotter considers the following eight steps necessary for success:

1. **Establish a sense of urgency**—Use market data, competitive analysis, or a convenient crisis (i.e., don't let a good crisis go to waste) to convince the broad majority of management (Kotter recommends 75% or more) that business as usual is riskier than the unknowns associated with the change. The head of the impacted area (the CEO for a company transformation, or a unit head for a business unit, for example) must certainly be onboard and advocate convincingly for the change (i.e., actions and words).
2. **Form a powerful guiding coalition**—Assemble a team of powerbrokers within the organization to lead the effort *as a team*. The involvement of these senior managers will certainly be needed at some point in the process, so charging them with leading the effort builds their buy-in and ensures the transformation will not be undermined by uninvolved senior managers. The executive council suggested earlier for managing Six Sigma in an organization would meet this condition. The team approach is necessary to maintain their active participation and also to prevent turf wars. (See Chap. 5 for further information on team development.)
3. **Create a vision**—How will the new organization differ from the current organization? Kotter emphasizes the need to simplify the vision into a coherent message that can be delivered in five minutes or less, yet generate interest and understanding from the audience.
4. **Communicate the vision**—Communication of the vision must be consistent and persistent, in both word and deed. The transformation will require employees to change behavior, which does not come easily and has at least perceived risk if not actual risk. These risks can only be overcome when the message is credible. Communication is a key aspect of building organizational buy-in. A DMAIC approach to building organizational buy-in is presented later in this chapter.
5. **Empower others to act on the vision**—The vision will only be realized when it becomes the new normal. To progress to that point may require identifying and

removing systems (or individuals) that serve as barriers to the new approach, or creating new systems that embody the new approach. A Six Sigma deployment achieves its objectives through a series of focused cross-functional projects sponsored by the manager(s) functionally responsible for the impacted area(s). The projects are deployed by teams, led by a trained Black Belt, that consist of local experts who perform the process activities daily. The sponsoring of the project by the functional managers ensures the team is empowered to affect change to the process. The oversight of the program by top management, and the alignment of the program with the strategic objectives, encourages the local managers to support the teams completely to achieve the project objectives. This system, discussed more completely in the next section, is critical to the success of the deployment effort.

6. **Plan for and create short-term wins**—A rational person does not decide to begin exercising as a New Year's resolution, then immediately embark on a 26-mile marathon or hike K2. Rather, success comes from setting and succeeding at smaller challenges, which builds expertise and confidence in the approach to apply for larger endeavors. It's the best of human nature to learn by doing. Typically, a successful Six Sigma deployment will seek to develop initial training projects that allow students to learn the techniques and apply them to familiar processes, with reasonable goals for improvement. This builds confidence for the teams as well as for the organization as whole. It is equally important for managers to become comfortable with their oversight responsibility. They must balance the empowerment given to the teams to affect change with the accountability of the teams to produce a meaningful, workable solution in a reasonable time frame.

7. **Consolidate improvements and produce more change**—For a Six Sigma deployment, as projects begin to impact change and produce results, management should celebrate the gains and congratulate the teams responsible. Use the early successes to build awareness throughout the organization, revise systems or policies that block effective change, and promote and develop employees with the skills needed to further affect change in the organization. Don't declare victory too soon. Instead, ramp up the efforts and build on the early success to gain critical mass and maintain momentum.

8. **Institutionalize new approaches**—Develop the leadership team and their practices to fully incorporate the change initiative into the organization's lifeblood. Constantly and consistently communicate the link between organization success and program success, and actively promote those responsible for the gains (at the expense of others who thought they could wait it out).

The approaches summarized above are more fully developed in the following section.

## Implementing Six Sigma

After nearly three decades of Six Sigma experience, there is now a solid body of scientific research that successful deployment involves focusing on a small number of high-leverage items. The activities and systems required to successfully implement Six Sigma are well documented.

- **Leadership**—Leadership's primary role is to create a clear vision for Six Sigma success and to communicate their vision clearly, consistently, and repeatedly throughout the organization. In other words, leadership must lead the effort. Their primary responsibility is to ensure that Six Sigma goals, objectives, and progress are properly aligned with those of the enterprise as a whole. This is done by modifying the organization such that personnel naturally pursue Six Sigma as part of their normal routine. This requires the creation of new positions and departments, and modified reward, recognition, incentive, and compensation systems. These key issues are discussed throughout this chapter. The Six Sigma deployment will begin with senior leadership training in the philosophy, principles, and tools they need to prepare their organization for success.
- **Infrastructure**—Using their newly acquired knowledge, senior leaders direct the development and training of an infrastructure to manage and support Six Sigma.
- **Communication and awareness**—Simultaneously, steps are taken to "soft-wire" the organization and to cultivate a change-capable environment where innovation and creativity can flourish. A top-level DMAIC project is focused on the change initiative and the communication required to build buy-in of the initiative, as outlined later in this chapter.
- **Stakeholder feedback systems**—Systems are developed for establishing close communication with customers, employees, and suppliers. This includes developing rigorous methods of obtaining and evaluating customer, owner, employee, and supplier input. Baseline studies are conducted to determine the starting point and to identify cultural, policy, and procedural obstacles to success. These systems are discussed in greater detail in Chap. 2.
- **Process feedback systems**—A framework for continuous process improvement is developed, along with a system of indicators for monitoring progress and success. Six Sigma metrics focus on the organization's strategic goals, drivers, and key business processes, as discussed in Chap. 3.
- **Project selection**—Six Sigma projects are proposed for improving business processes by people with process knowledge at various levels of the organization. Six Sigma projects are selected based on established protocol by senior management to achieve

business performance objectives linked to measurable financial results, as discussed in Chap. 4.

▲ **Project deployment**—Six Sigma projects are conducted by project teams lead by Black Belts (or by Green Belts with the technical assistance of Black Belts). Project deployment is discussed in detail in Part II of this book.

## Timetable

Figure 1.3 shows a typical set of deployment activities to reach system maturity within two years. The resulting benefits are dependent on the rate of project deployment and the organization's initial quality levels. A typical goal is an improvement rate of approximately 10 times every two years, measured in terms of errors (or defects) per million opportunities (DPMO).[2] For example, an organization starting at a typical sigma level of 3.0 would seek to reduce their overall error rate from approximately 67,000 to about 6,700 (or about 4.0 sigma level) in two years time. Figure 1.4 provides a rough guideline for determining when you will reach Six Sigma based on the initial quality level, assuming the 10 times

### Six Sigma Deployment Timeline

**Six Sigma initial planning**
- Identify 6σ leader
- Six Sigma plan drafted and approved
- Identify core team members

**Six Sigma phase 2A program development**
- Tailor BB training
- BB retention plan
- ID Black Belt candidates
- Train Black Belts
- Review management process documentation
- Train leadership

**Six Sigma phase 3 implementation**
- Certify BB
- Issue Six Sigma practice
- Train Green Belts
- ID future projects
- Train the trainers
- Issue Mgmt process doc.
- Recognition/reward

Q1  Q2  Q3  Q4  Q5  Q6–Q8

**Six Sigma phase 1 alignment**
- Establish exec. council
- Hire consultant
- Issue 6σ policy
- Detailed planning
- ID management process
- Knowledge discovery
- Conduct readiness survey

**Six Sigma phase 2B program development**
- Establish project validation criteria
- Continuing BB training
- ID master BBs
- Begin to establish Six Sigma practice

**Six Sigma phase 4 implementation maturity**
- Prepare detailed road map of next steps
- Train employees
- Train 2nd wave Black Belts
- Train additional Green Belts
- ID additional projects

**Figure 1.3** Typical deployment activities and timeline.

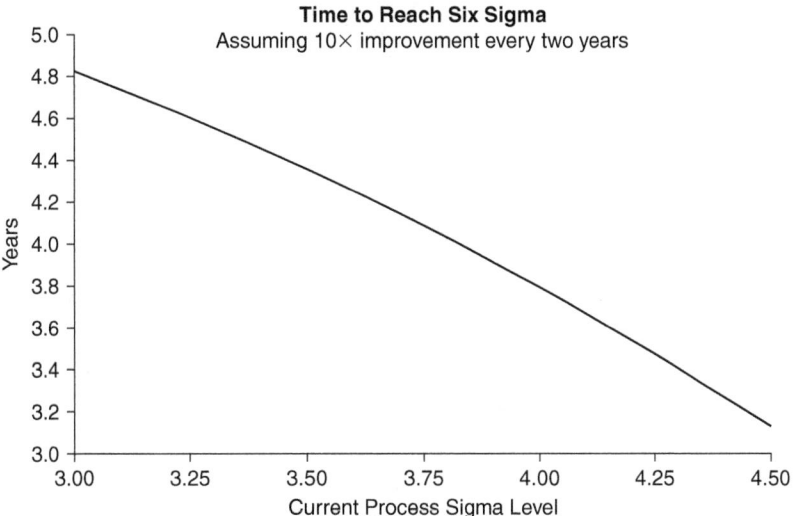

**Figure 1.4** Time to reach Six Sigma performance levels.

improvement every two years. For the typical company starting at three sigma, Fig. 1.4 indicates they will reach Six Sigma levels of performance after approximately five years from the time they have deployed Six Sigma. Given the deployment timeline shown in Fig. 1.3, it will be approximately seven years from date of program initiation. Of course, results will begin to appear within a year of starting the deployment.

Yet, even when the enterprise reaches a performance level of five or Six Sigma overall, there may still be processes operating at poor sigma levels, demonstrating the fallibility of the DPMO metric, especially when interpreted across an entire organization. Individual customers judge your organization based on their individual experiences, and customer expectations are a moving target, as previously discussed.

Figure 1.5 shows General Electric's published data on their Six Sigma program. Note there was sufficient savings to cover costs during the first year. In the second and subsequent years the benefits outpaced the costs, with the benefit-to-cost ratio improving steadily as costs level out. These results are consistent with those reported by academic research for companies that successfully implemented TQM.

The annual savings achieved by a given organization is largely dependent on their initial quality, as well as their resource commitment. The number of full-time personnel devoted to Six Sigma is a relatively small percentage of the total work force. Mature Six Sigma programs, such as those of General Electric, Johnson & Johnson, AlliedSignal, and others, average about 1% of their workforce as Black Belts, with considerable variation in that number. There is usually about one Master Black Belt for every 10 Black Belts, or about one Master Black Belt per 1,000 employees. A Black Belt will typically complete

**20** | The Six Sigma Handbook

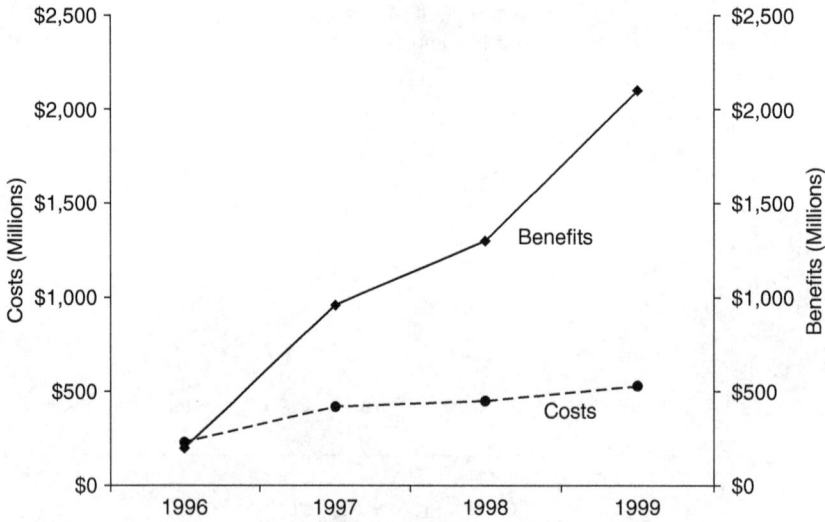

**Figure 1.5** GE's reported cost of Six Sigma versus benefits.

five to seven projects per year, usually working with teams. Project teams are led either by Black Belts or in some cases Green Belts, who, unlike Black Belts and Master Black Belts, are not engaged full time in the Six Sigma program. Green Belts usually devote between 5 and 10% of their time to Six Sigma project work.

Estimated savings per project vary from organization to organization, but average about $150,000 to $243,000 according to published figures. Some industries just starting their Six Sigma programs average as high as $700,000 savings per project, although these projects usually take longer. Note that these are not the huge megaprojects such as pursued by reengineering. Yet, by completing five to seven projects per year per Black Belt the company will add in excess of $1 million per year per Black Belt to its bottom line. For a company with 1,000 employees the resource requirement and estimated savings are shown in the following table:

| Master Black Belts: | 1 |
|---|---|
| Black Belts: | 10 |
| Projects: | 50 to 70 (5 to 7 per Black Belt) |
| Estimated saving: | $9 million to $14.6 million (i.e., $14,580 savings per employee) |

Savings for your organization can be easily estimated the same way. Recall from Fig. 1.1 the potential savings (about 25% of revenue) that exists in a typical three sigma organization, and it's easy to see there are many potential projects available within a typical organization. Since Six Sigma savings—unlike traditional slash and burn cost cutting—impact only non-value-added costs, they flow directly to your company's bottom line. Traditional, income-statement-based cost cutting inevitably hurts value-adding activities. As a result, the savings seldom measure up to expectations, and revenues often suffer as well. The predicted bottom-line impact is not actually realized. Firms engaging in these activities hurt their prospects for future success and delay their recovery.

## *Infrastructure*

A successful Six Sigma deployment demands an organizational infrastructure to manage and support the various activities summarized earlier in this chapter. Six Sigma is the primary strategy for enterprise-wide business process improvement; to ensure success it is necessary to institutionalize it as a way of doing business. It is not enough to train resources to act outside of the normal business functions. To the contrary, such a plan virtually guarantees failure by placing the Six Sigma activities somewhere other than the mainstream. Instead, process improvement must become an ongoing part of the business to meet the ever-changing market conditions and customer value definitions.

It's interesting to note that companies institutionalizing the principles of TQM obtained excellent results, which are comparable to the results reported by companies implementing Six Sigma. Those that didn't invariably failed to achieve lasting results. Six Sigma provides a quasi-standardized set of guidelines for deployment, resulting in a much higher success rate. Although each organization will develop its own unique approach to Six Sigma, it is helpful to review the practices of successful companies.

Most importantly, successful Six Sigma deployment is always a top-down affair. For Six Sigma to have a major impact on overall enterprise performance, it must be fully embraced and actively led by top management. Isolated efforts at division or department levels are doomed from the outset. Like flower gardens in a desert, they may flourish and produce a few beautiful results for a time, but sustaining the results requires immense effort by local heroes in constant conflict with the mainstream culture, placing themselves at risk. Sooner or later, the desert will reclaim the garden. Six Sigma shouldn't require heroic effort—there are never enough heroes to go around. Once top management has accepted its leadership responsibility the organizational transformation process can begin.

A key decision is whether Black Belts will report to a central Six Sigma organization or to managers located elsewhere in the organization. The experience of most successful Six Sigma enterprises is that centralized reporting is best. Internal studies by one company that experimented with both types of reporting revealed the results shown in Table 1.1. The major reason for problems with the decentralized approach was disengaging people from routine work and firefighting. Six Sigma is devoted to change, and it seems change tends to take a back seat to current problems. To be sure, the Black Belt possesses a skill set that can be very useful in putting out fires. Black Belts also tend to excel at whatever they do. This combination makes it difficult to resist the urge to pull the Black Belt off of his or her projects "just for a while." In fact, some organizations have trouble getting the Black Belts out of their current department and into the central organization. In one case the CEO intervened personally on behalf of the Black Belts to break them loose. Such stories are testimony to the difficulties encountered in making drastic cultural changes.

**Table 1.1** Black Belt Certification Versus Reporting Arrangement

| Where Black Belt Reported | Black Belts Successfully Certified |
| --- | --- |
| Local organization | 40% |
| Centralized Six Sigma organization | 80% |

The transformation process involves new roles and responsibilities on the part of many individuals in the organization. In addition, new change agent positions must be created. Table 1.2 lists some typical roles and responsibilities. In a Six Sigma organization, improvement and change are the full-time job of a small but critical percentage of the organization's personnel. These full-time change agents are the catalyst that institutionalizes change.

Education and training are important means of changing individual perceptions and behaviors. In this discussion, a distinction is made between training and education. *Training* refers to instruction and practice designed to teach a person how to perform one or more tasks. Training focuses on concrete tasks to be completed. *Education* refers to instruction in thinking. Education focuses on integrating abstract concepts into one's knowledge of the world. An educated person will view the world differently after being educated. This is an essential part of the process of change.

Six Sigma training is a subproject of the Six Sigma deployment plan, whose timetables must be tightly linked. Training provided too early or too late is a mistake. When training is provided too early, the recipient will forget much of what he has learned before it is needed. When it is provided too late, the quality of the employee's work will suffer. When it comes to training, just-in-time delivery is the goal.

Table 1.2  Six Sigma Roles and Responsibilities

| Responsible Entity | Roles | Responsibilities |
|---|---|---|
| Executive Six Sigma Council | Strategic leadership | • Ensures Six Sigma goals are linked to enterprise goals<br>• Develops new policies as required<br>• Aligns process excellence efforts across the organization<br>• Suggests high-impact projects<br>• Approves project selection strategy |
|  | Ensures progress | • Provides resources<br>• Tracks and controls progress toward goals<br>• Reviews improvement teams' results (BB, GB, Lean, Supply Chain, other)<br>• Reviews effectiveness of Six Sigma deployment: systems, processes, infrastructure, etc. |
|  | Cultural transformation | • Communicates vision<br>• Removes formal and informal barriers<br>• Commissions modification of compensation, incentive, reward, and recognition systems |
| Director, Six Sigma | Manages Six Sigma infrastructure and resources | • Six Sigma champion for ACME<br>• Develops Enterprise Six Sigma deployment<br>• Owns the Six Sigma project selection and prioritization process for ACME<br>• Ensures Six Sigma strategies and projects are linked through quality function deployment to business plans<br>• Achieves defect reduction and cost take-out targets through Six Sigma activities<br>• Member of Executive Six Sigma Council<br>• Leads and evaluates the performance of Black Belts and Master Black Belts<br>• Communicates Six Sigma progress with customers, suppliers, and the enterprise<br>• Champions Six Sigma reward and recognition, as appropriate |
| Six Sigma Certification Board | Certifies Black Belts Board representatives include Master Black Belts and key Six Sigma leaders | • Works with local units to customize Black Belt and Green Belt requirements to fit business needs<br>• Develops and implements systems for certifying Black Belts and Green Belts<br>• Certifies Black Belts |

*(continued on next page)*

Table 1.2  Six Sigma Roles and Responsibilities (*continued*)

| Responsible Entity | Roles | Responsibilities |
|---|---|---|
| Six Sigma Core Team | Cross-functional Six Sigma team Part-time change agent | • Provides input into policies and procedures for successful implementation of Six Sigma across ACME<br>• Facilitates Six Sigma activities such as training, special recognition events, Black Belt testing, etc. |
| Master Black Belt | Enterprise Six Sigma expert Permanent full-time change agent Certified Black Belt with additional specialized skills or experience especially useful in deployment of Six Sigma across the enterprise | • Highly proficient in using Six Sigma methodology to achieve tangible business results<br>• Technical expert beyond Black Belt level on one or more aspects of process improvement (e.g., advanced statistical analysis, project management, communications, program administration, teaching, project coaching)<br>• Identifies high-leverage opportunities for applying the Six Sigma approach across the enterprise<br>• Basic Black Belt training<br>• Green Belt training<br>• Coach/Mentor Black Belts<br>• Participates on ACME Six Sigma Certification Board to certify Black Belts and Green Belts |
| Black Belt | Six Sigma technical expert Temporary, full-time change agent (will return to other duties after completing a two- to three-year tour of duty as a Black Belt) | • Leads business process improvement projects where Six Sigma approach is indicated<br>• Successfully completes high-impact projects that result in tangible benefits to the enterprise<br>• Demonstrated mastery of Black Belt body of knowledge<br>• Demonstrated proficiency at achieving results through the application of the Six Sigma approach<br>• Internal Process Improvement Consultant for functional areas<br>• Coach/Mentor Green Belts<br>• Recommends Green Belts for Certification |

| Responsible Entity | Roles | Responsibilities |
|---|---|---|
| Green Belt | Six Sigma project originator<br>Six Sigma project leader<br>Part-time Six Sigma change agent. Continues to perform normal duties while participating on Six Sigma project teams<br>Six Sigma champion in local area | • Demonstrated mastery of Green Belt body of knowledge<br>• Demonstrated proficiency at achieving results through the application of the Six Sigma approach<br>• Recommends Six Sigma projects<br>• Participates on Six Sigma project teams<br>• Leads Six Sigma teams in local improvement projects<br>• Works closely with other continuous improvement leaders to apply formal data analysis approaches to projects<br>• Teaches local teams, shares knowledge of Six Sigma<br>• Successful completion of at least one Six Sigma project every 12 months to maintain their Green Belt certification |
| Six Sigma Improvement Team | Primary ACME vehicle for achieving Six Sigma improvements | • Completes chartered Six Sigma projects that deliver tangible results<br>• Identifies Six Sigma project candidates |
| ACME Leaders and Managers | Champions for Six Sigma | • Ensures flow-down and follow-through on goals and strategies within their organizations<br>• Plans improvement projects<br>• Charters or champions chartering process<br>• Identifies teams or individuals required to facilitate Six Sigma deployment<br>• Integrates Six Sigma with performance appraisal process by identifying measurable Six Sigma goals/objectives/results<br>• Identifies, sponsors, and directs Six Sigma projects<br>• Holds regular project reviews in accordance with project charters<br>• Includes Six Sigma requirements in expense and capital budgets<br>• Identifies and removes organizational and cultural barriers to Six Sigma success<br>• Rewards and recognizes team and individual accomplishments (formally and informally)<br>• Communicates leadership vision<br>• Monitors and reports Six Sigma progress<br>• Validates Six Sigma project results<br>• Nominates highly qualified Black Belt and/or Green Belt candidates |

*(continued on next page)*

**Table 1.2** Six Sigma Roles and Responsibilities (*continued*)

| Responsible Entity | Roles | Responsibilities |
|---|---|---|
| Project Sponsor | Charters and supports Six Sigma project teams | • Sponsor is ultimately responsible for the success of sponsored projects<br>• Actively participates in projects<br>• Ensures adequate resources are provided for project<br>• Personal review of progress<br>• Identifies and overcomes barriers and issues<br>• Evaluates and accepts deliverable |
| "Matrixed" Project Manager | Manages Six Sigma resources dedicated to a particular area (e.g., teams of Black Belts on special assignment) Champions Six Sigma Black Belt team | • Provides day-to-day direction for Six Sigma project Black Belt and team activities<br>• Provides local administrative support, facilities, and materials<br>• Conducts periodic reviews of projects<br>• Provides input on Black Belt performance appraisals<br>• Makes/implements decisions based on recommendations of Six Sigma Black Belts |
| Six Sigma Improvement Team Member | Learns and applies Six Sigma tools to projects | • Actively participates in team tasks<br>• Communicates well with other team members<br>• Demonstrates basic improvement tool knowledge<br>• Accepts and executes assignments as determined by team |

The cost of Six Sigma training should be included in the previously discussed estimates of Six Sigma cost-benefit ratios and include

- Trainer salaries
- Consulting fees
- Classroom space and materials
- Lost time from the job
- Staff salaries
- Office space of training staff

The estimated benefits of the training include the subsequent project deliverables, often on an annualized basis. Since trained Black Belts and Green Belts will often work on multiple projects during the year, it's best to consider these costs and benefits on a program-wide basis, rather than a per-class or per-project basis.

## Champions and Sponsors

Six Sigma champions are high-level individuals who understand Six Sigma and are committed to its success. In larger organizations Six Sigma will be led by a full-time, high-level champion, such as an executive vice president. In all organizations, champions also include informal leaders who use Six Sigma in their day-to-day work and communicate the Six Sigma message at every opportunity. Sponsors are owners of processes and systems that help initiate and coordinate Six Sigma improvement activities in their areas of responsibilities.

Leaders should receive guidance in the art of "visioning." Visioning involves the ability to develop a mental image of the organization at a future time; without a vision, there can be no strategy.

Leaders need to be masters of communication. Fortunately, most leaders already possess outstanding communication skills; few rise to the top without them. However, training in effective communication is still wise, even if it is only refresher training. When large organizations are involved, communications training should include mass communication media, such as video, radio broadcasts, and print media. Communicating with customers, investors, and suppliers differs from communicating with employees and colleagues, and special training is often required.

Finally, leaders should demonstrate strict adherence to ethical principles. Leadership involves trust, and trust isn't granted to one who violates a moral code that allows people to live and work together. Honesty, integrity, and other moral virtues should be second nature to the leader.

## Black Belts

Candidates for Black Belt status are technically oriented individuals held in high regard by their peers. They should be actively involved in the process of organizational change and development. Candidates may come from a wide range of disciplines and need not be formally trained statisticians or analysts. However, because they are expected to master a wide variety of technical tools in a relatively short period of time, Black Belt candidates will probably possess a background in college-level mathematics, the basic tool of quantitative analysis. Coursework in statistical methods should be considered a strong plus or even a prerequisite. Black Belts receive from three to six weeks of training in the technical tools of Six Sigma. Three-week curricula are usually given to Black Belts working in service or transaction-based businesses, administrative areas, or finance. Four-week programs are common for manufacturing environments. Six weeks of training are provided for Black Belts working in R&D or similar environments. Figure 1.6 shows the curriculum used for courses in General Electric for personnel with finance backgrounds who will be applying Six Sigma to financial, general business, and e-commerce processes. Figure 1.7 shows GE's curriculum for the more traditional manufacturing areas.

Although some training companies offer highly compressed two-week training courses, these are not recommended. Even in a six-week course, students receive the equivalent of two semesters of college-level applied statistics in just a few days. Humans require a certain "gestation period" to grasp challenging new concepts; providing too much material in too short a time period is counterproductive. Successful candidates will be comfortable with computers. At a minimum, they should be proficient with one or more operating systems, spreadsheets, database managers, presentation programs, and word processors. As part of their training they will also be required to become proficient in the use of one or more advanced statistical analysis software packages and probably simulation software. Six Sigma Black Belts work to extract actionable knowledge from an organization's information warehouse. To ensure access to the needed information, Six Sigma activities should be closely integrated with the information systems of the organization. Obviously, the skills and training of Six Sigma Black Belts must be enabled by an investment in software and hardware. It makes no sense to hamstring these experts by saving a few dollars on computers or software.

As a full-time change agent, the Black Belt needs excellent interpersonal skills. In addition to mastering a body of technical knowledge, Black Belts must

- Communicate effectively verbally and in writing
- Communicate effectively in both public and private forums
- Work effectively in small group settings as both a participant and a leader
- Work effectively in one-on-one settings
- Understand and carry out instructions from leaders and sponsors

**Week 1**

The DMAIC and DFSS (design for Six Sigma) improvement strategies
Project selection and "scoping" (define)
QFD (quality function deployment)
Sampling principles (quality and quantity)
Measurement system analysis (also called "Gage R&R")
Process capability
Basic graphs
Hypothesis testing
Regression

**Week 2**

Design of experiments (DOE) (focus on two-level factorials)
Design for Six Sigma tools
Requirements flowdown
Capability flowup (prediction)
Piloting
Simulation
FMEA (failure mode and effects analysis)
Developing control plans
Control charts

**Week 3**

Power (impact of sample size)
Impact of process instability on capability analysis
Confidence intervals (vs. hypothesis tests)
Implications of the Central Limit Theorem
Transformations
How to detect "lying with statistics"
General linear models
Fractional factorial DOEs

**Figure 1.6** Sample curriculum for finance Black Belts.
(From Hoerl, 2001, p. 395. Reprinted by permission of ASQ.)

**Context**[1]
- Why Six Sigma
- DMAIC and DFSS processes (sequential case studies)
- Project management fundamentals
- Team effectiveness fundamentals

**Define**[1]
- Project selection
- Scoping projects
- Developing a project plan
- Multigenerational projects
- Process identification (SIPOC)

**Measure**[1]
- QFD
- Developing measurable CTQs
- Sampling (data quantity and data quality)
- Measurement system analysis (not just gage R&R)
- SPC Part I
  - The concept of statistical control (process stability)
  - The implications of instability on capability measures
- Capability analysis

**Analyze**[2]
- Basic graphical improvement tools ("Magnificent 7")
- Management and planning tools (Affinity, ID, etc.)
- Confidence intervals (emphasized)
- Hypothesis testing (de-emphasized)
- ANOVA (de-emphasized)
- Regression
- Developing conceptual designs in DFSS

**Improve**[3,4]
- DOE (focus on two-level factorials, screening designs, and RSM)
- Piloting (of DMAIC improvements)
- FMEA
- Mistake-proofing
- DFSS design tools
  - CTQ flowdown
  - Capability flowup
  - Simulation

**Control**[4]
- Developing control plans
- SPC Part II
  - Control charts
- Piloting new designs in DFSS

**Figure 1.7** Sample curriculum for manufacturing Black Belts. (The week in which the material appears is noted as a superscript. From Hoerl, 2001, p. 399. Reprinted by permission of ASQ.)

A change agent deficient in these soft skills will nearly always be ineffective. They are usually frustrated and unhappy souls who don't understand why their technically brilliant case for change doesn't cause instantaneous compliance by all parties. The good news is that if the person is willing to apply as much time and effort to soft-skill acquisition and mastery as they applied to honing their technical skills, they will be able to develop proficiency.

In general, Black Belts are hands-on oriented people selected primarily for their ability to get things done. Tools and techniques are provided to help them do this. The training emphasis is on application, not theory. In addition, many Black Belts will work on projects in an area where they possess a high degree of subject-matter expertise. Therefore, Black Belt training is designed around projects related to their specific work areas. This requires Master Black Belts or trainers with very broad project experience to answer application-specific questions. When these personnel aren't available, examples are selected to match the Black Belt's work as closely as possible. For example, if no trainer with human resource experience is available, the examples might be from another service environment; manufacturing examples would be avoided. Another common alternative is to use consultants to conduct the training. Consultants with broad experience within the enterprise as well as with other organizations can sometimes offer insights.

Black Belts must work on projects while they are being trained. Typically, the training classes are conducted at monthly intervals and project work is pursued between classes. One of the critical differences between Six Sigma and other initiatives is the emphasis on using the new skills to get tangible results. It is relatively easy to sit in a classroom and absorb the concepts well enough to pass an exam. It's another thing entirely to apply the new approach to a real-world problem. The Black Belt has to be able to use change agent skills to recruit sponsors and team members and to get these people to work together on a project with a challenging goal and a tight timetable. While the instructors can provide coaching and project-specific training and advice, there's no better time to initiate the process than during the training.

The process for selecting Black Belts should be clearly defined. This ensures consistency and minimizes the possibility of bias and favoritism. Figure 1.8 provides a list of seven success factors, with their relative importance weights, that can be used to compare Black Belt candidates.

The weights are, of course, subjective and only approximate, and are based on an exercise with a group of consultants and Master Black Belts. Organizations can easily identify their own set of criteria and weights, such as shown by Keller (2005). The important thing is to determine the criteria and then develop a method of evaluating candidates on each criterion. The sum of the candidate's criterion score times the criterion weight will give you an overall numerical assessment for ranking the Black Belt candidates. Of course,

**Figure 1.8** Black Belt success factors and importance weights.

the numerical assessment is not the only input into the selection decision, but it is a very useful one.

Notice the relatively low weight given to math skills. The rationale is that Black Belts will receive 200 hours of training, much of it focused on the practical application of statistical techniques using computer software and requiring very little actual mathematics. Software automates the analysis, making math skills less necessary. The mathematical theory underlying a technique is not discussed beyond the level necessary to help the Black Belt properly apply the tool. Black Belts who need help with a particular tool have access to Master Black Belts, other Black Belts, consultants, professors, and a wealth of other resources. Most statistical techniques used in Six Sigma are relatively straightforward and often graphical; spotting obvious errors is usually not too difficult for trained Black Belts. Projects seldom fail due to a lack of mathematical expertise. In contrast, the Black Belts will often have to rely on their own abilities to deal with the obstacles they will inevitably encounter. Failure to overcome the obstacle will often spell failure of the entire project.

Figure 1.9 provides a sample of the selection process for a Black Belt candidate at a particular firm.

**Minimum Criteria**

*Education*—Bachelors degree, minimum.

*Work Experience*—At least three years of business, technical, or managerial experience plus technical application of education and experience as a member or leader of functional and cross-functional project teams.

*Technical Capability*—Project management experience is highly desired. Understanding of basic principles of process management. Basic college algebra proficiency as demonstrated by exam.

*Computer Proficiency*—MS Office Software Suite.

*Communication*—Demonstrate excellent oral and written communication skills.

*Team Skills*—Ability to conduct meetings, facilitate small groups, and successfully resolve conflicts. Ability to mentor and motivate people.

**Final Candidate Selection**

To ensure that the Black Belts will be able to address enterprise-wide issues and processes, the Director of Six Sigma and the Executive Six Sigma Council will determine the number of Black Belts to be trained in each functional area, division, department, etc. Black Belt candidates are ranked using a system of points assigned during the screening process. Rank-ordered lists of Black Belt candidates are prepared for designated areas and presented to the senior management of the area for final selection. Area management nominates candidates from their list in numbers sufficient to fill the spaces allocated by the Director of Six Sigma and the Executive Six Sigma Council.

**Commitment to Black Belt Assignment**

Selected candidates are required to attend 200 hours of Black Belt training (see Chap. 4 for the training content). Within one year of completing training, the Black Belt candidate is required to become certified by passing a written examination and successfully completing at least two major projects. (See Appendix 15 for detailed Black Belt certification process information.) The Black Belt is assigned to Six Sigma full time as a Black Belt for a minimum period of two full years, measured from the time he or she is certified as a Black Belt.

**Reintegration of Black Belts into the Organization**

Black Belts are employed in the Black Belt role for two or three years. After that time they leave the Six Sigma organization and return to other duties. Accomplishing this transition is the joint responsibility of the Black Belt, the Director of Six Sigma, and the management of the Black Belt's former department. Collectively this group comprises the "Transition Team" for the Black Belt. However, senior leadership must accept ultimate responsibility for assuring that Black Belts are not "homeless" after completing their Black Belt tour of duty.

The Director of Six Sigma will inform the Black Belt at least six months prior to the scheduled return. Black Belts should maintain contact with their "home" organization during their tenure in Six Sigma. If it appears that there will be a suitable position available at approximately the time the Black Belt is scheduled to return, arrangements should be made to complete or hand off the Black Belt's Six Sigma projects in preparation for his return. If no suitable openings will be available, the Transition Team needs to develop alternative plans. Alternatives might include extending the Black Belt's term of service in Six Sigma, looking for openings in other areas, or making temporary arrangements.

**Figure 1.9** Black Belt candidate selection process and criteria.

Past improvement initiatives, such as TQM, shared much in common with Six Sigma. TQM also had management champions, improvement projects, sponsors, etc. One of the main differences in the Six Sigma infrastructure is the creation of more formally defined change agent positions. Some observers criticize this practice as creating corps of "elites," especially Black Belts and Master Black Belts. Let's examine the commonly proposed alternatives to creating a relatively small group of highly trained technical experts:

- **Train the masses**—This is the "quality circles" approach, where people in the lowest level of the organizational hierarchy are trained in the use of basic tools and set loose to solve problems without explicit direction from leadership. When this approach was actually tried in America in the 1970s the results were disappointing. The originators of the quality circles idea, the Japanese, reported considerably greater success with the approach. This was no doubt due to the fact that Japanese circles were integrated into decades-old, company-wide process improvement activities, while American firms typically implemented circles by themselves. Indeed, when Six Sigma deployments reach a high level of maturity, more extensive training is often successful.
- **Train the managers**—This involves training senior and middle management in change agent skills. This isn't a bad idea in itself. However, if the basic structure of the organization doesn't change, there is no clear way to apply the newly acquired skills. Training in and of itself does nothing to change an organization's environment. Historically, trained managers return to pretty much the same job. As time goes by their skills atrophy and their self-confidence wanes. If opportunities to apply their knowledge do arise, they often fail to recognize them or, if they do recognize them, fail to correctly apply the approach. This is natural for a person trying to do something different for the first time. The full-time change agents in Six Sigma learn by doing. By the end of their tenure, they can confidently apply Six Sigma methodology to a wide variety of situations.
- **Use the experts in other areas**—The tools of Six Sigma are not new. In fact, Industrial Statisticians, ASQ Certified Quality Engineers, Certified Reliability Engineers, Certified Quality Technicians, Systems Engineers, Industrial Engineers, Manufacturing Engineers, and other specialists already possess a respectable level of expertise in many Six Sigma tools. Some have a level of mastery in some areas that exceeds that of Black Belts. However, being a successful change agent involves a great deal more than mastery of technical tools. Black Belts, Green Belts, and Master Black Belts learn tools and techniques in the context of following the DMAIC approach to drive organizational change. This is very different than using the same techniques in routine daily work. Quality analysts, for example, generally work in the quality department as permanent, full-time employees. They report to a single boss and have well-defined areas of responsibility. Black Belts, in contrast, go out and seek projects

rather than work on anything routine. They report to many different people, who use different criteria to evaluate the Black Belt's performance. They are accountable for delivering measurable, bottom-line results. Obviously, the type of person who is good at one job may not be suitable for the other.

- **Create permanent change agent positions**—Another option to the Black Belt position is to make the job permanent. After all, why not make maximum use of the training by keeping the person in the Black Belt job indefinitely? Furthermore, as Black Belts gain experience they become more proficient at completing projects. There are, however, arguments against this approach. Having temporary Black Belts allows more people to go through the position, thus increasing the number of people in management with Black Belt experience. Since Black Belts work on projects that impact many different areas of the enterprise, they have a broad, process-oriented perspective that is extremely valuable in top management positions. The continuous influx of new blood into Black Belt and Green Belt positions keeps the thinking fresh and prevents the "them-versus-us" mentality that often develops within functional units. New Black Belts have different networks of contacts throughout the organization, which leads to projects in areas that might otherwise be missed. Permanent Black Belts would almost certainly be more heavily influenced by their full-time boss than temporary Black Belts, thus leading to a more provincial focus.

## Green Belts

Green Belts are Six Sigma project leaders capable of forming and facilitating Six Sigma teams and managing Six Sigma projects from concept to completion. Green Belt training consists of five days of classroom training and is conducted in conjunction with Six Sigma projects. (In some cases a ten-day course is offered to increase the time allotted for software training and exercises.) Training covers project management, quality management tools, quality control tools, problem solving, and descriptive data analysis. Six Sigma champions should attend Green Belt training. Usually, Six Sigma Black Belts help Green Belts define their projects prior to the training, attend training with their Green Belts, and assist them with their projects after the training.

Green Belts are change agents who work part time on process improvement. The bulk of the Green Belt's time is spent performing their normal work duties. Although most experts advocate that the Green Belt spend 10 to 20% of their time on projects, in most cases it is only 2 to 5%. A Green Belt will usually complete one or two major projects per year, usually as a team member rather than a team leader. Since a Green Belt is not trained in all the tools needed in the DMAIC cycle, when they lead projects they must be actively supported by a Black Belt. Few Green Belt projects cover enterprise-wide processes. However, since there are usually more Green Belts than Black Belts (by a factor of

> **Minimum Criteria**
>
> *Education*—High school or equivalent.
>
> *Work Experience*—At least three years of business, technical, or managerial experience.
>
> *Technical Capability*—High school algebra proficiency as demonstrated by a passing grade in an algebra course.
>
> *Computer Proficiency*—Word processing, presentation, and spreadsheet software.
>
> *Team Skills*—Willingness to lead meetings, facilitate small groups, and successfully resolve conflicts. Ability to mentor and motivate people.
>
> **Final Candidate Selection**
>
> Based on the organizational need for Green Belts, as determined by the Director of Six Sigma and the Executive Six Sigma Council, Green Belt training allotments are provided to Master Black Belts, Black Belts, and/or General Managers. Green Belt candidacy requires the consent of the candidate's management.
>
> **Commitment**
>
> Each Green Belt candidate selected will be required to complete a 40-hour Green Belt training course, and to lead at least one successful Six Sigma project every 12 months, or participate on at least two successful Six Sigma projects every 12 months. Green Belt certification is accomplished as described in Appendix 16.

**Figure 1.10** Green Belt candidate selection process and criteria.

2 to 5), Green Belt projects can have a tremendous impact on the enterprise. Figure 1.10 provides an overview of a process for the selection of Green Belt candidates.

## Master Black Belts

This is the highest level of technical and organizational proficiency. Master Black Belts provide technical leadership of the Six Sigma program. They must be thoroughly familiar with the Black Belt Body of Knowledge, as well as additional skills including the mathematical theory that forms the basis of the statistical methods, project management, coaching, teaching, and program organization at the enterprise level. Master Black Belts must be able to assist Black Belts in applying the methods correctly in unusual situations. Whenever possible, statistical training should be conducted only by qualified Master Black Belts or equivalently skilled consultants. If it becomes necessary for Black Belts and Green Belts to provide training, they should only do so under the guidance of Master Black Belts. Otherwise the familiar "propagation of error" phenomenon will occur; that is, Black Belt trainers pass on errors to Black Belt trainees who pass them on to Green Belts, who pass on greater errors to team members. Because of the nature of the Master's duties, all Master Black Belts must possess excellent communication and teaching skills.

Master Black Belts are recruited from the ranks of Black Belts. The process is usually less formal and less well defined than that for Black Belts or Green Belts and there is a great deal of variability between companies. Master Black Belt candidates usually make their interest known to Six Sigma leadership. Leadership selects candidates based on the needs of the enterprise and Six Sigma's role in meeting those needs. For example, in the early stages of deployment Master Black Belt candidates with excellent organizational skills and the ability to communicate the leadership's Six Sigma vision may be preferred. Intermediate deployments might favor candidates who excel at project selection and Black Belt coaching. Mature Six Sigma programs might look for Master Black Belts with training ability and advanced statistical know-how. Master Black Belts often have advanced technical degrees and extensive Black Belt experience. Many organizations provide Master Black Belts with additional training. Certification requirements for Master Black Belts vary with the organization. Many organizations do not certify Master Black Belts.

### Change Agent Compensation and Retention

Experienced Black Belts and Master Black Belts are in great demand throughout the manufacturing and services sectors.[3] Given their proven talent for effecting meaningful change in a complex environment, this is no surprise. Since organizations exist in a competitive world, steps must be taken to protect the investment in these skilled change agents, or they will be lured away by other organizations, perhaps even competitors. The most common (and effective) actions involve compensation and other financial incentives, such as:

- Bonuses
- Stock options
- Results sharing
- Payment of dues to professional societies
- Pay increases

There are also numerous nonfinancial and quasi-financial rewards. For example, Black Belts reentering the workforce after their tour of duty often enter positions that pay significantly higher than the ones they left when becoming Black Belts. In fact, in some companies the Black Belt position is viewed as a step on the fast track to upper management positions. Also, change is "news" and it is only natural that the names of Master Black Belts and Black Belts involved in major change initiatives receive considerable publicity on company Web sites as well as in newsletters, recognition events, project fairs, etc. Even if they don't receive formal recognition, Six Sigma projects often generate a great deal of internal excitement and discussion. The successful Black Belt usually finds that his work has earned him a reputation that makes him a hot commodity when it's time to end his Black Belt career.

There are, of course, innumerable complexities and details to be decided and worked out. Usually these issues are worked out by a team of individuals with members from Human Resources, the Six Sigma Core Team, and other areas of the organization. The team will address such issues as:

- What pay grade is to be assigned to the Black Belt and Master Black Belt positions?
- Should the pay grade be determined by the pay grade of the candidate's job prior to becoming a Black Belt?
- Should the Black Belt pay grade be guaranteed when the Black Belt leaves the Black Belt position to return to the organization?
- How do we determine eligibility for the various rewards? For example, are there key events such as acceptance as a Black Belt candidate, completion of training, completion of first project, successful certification, and so forth?
- What about Black Belts who were certified by other organizations or third parties?
- Do we provide benefits to Green Belts as well? If so, what and how?
- Who will administer the benefits package?

The plan will be of great interest to Black Belt candidates. If not done properly, the organization will find it difficult to recruit the best people.

## *Integrating Six Sigma and Related Initiatives*

At any given time most companies have numerous activities underway to improve their operations. For example, the company may have functional areas devoted to Lean Implementation, Continuous Improvement, or Business Process Reengineering, as well as those tasked to more traditional quality functions of quality assurance and quality control. Collectively, these functions are often known as the *quality function* of an organization.

### The Quality Function[4]

Juran and Gryna (1988, p. 2.6) define the quality function as "the entire collection of activities through which we achieve fitness for use, no matter where these activities are performed." Quality is thus influenced by, if not the responsibility of, many different departments. In most cases, the quality department serves a secondary, supporting role. While the quality department is a specialized function, quality activities are dispersed throughout the organization. The term "quality function" applies to those activities, departmental and companywide, that collectively result in product or service quality. An analogy can be made with the finance department. Even though many specialized finance and accounting functions are managed by the finance department, every employee in the organization is expected to practice responsible management of his or her budgets and expenditures.

Juran and Gryna (1988) grouped quality activities into three categories, sometimes referred to as the Juran Trilogy: planning, control, and improvement. *Quality planning* is the activity of developing the products and processes required to meet customers' needs. It involves a number of universal steps (Juran and DeFeo, 2010):

- Define the customers.
- Determine the customer needs.
- Develop product and service features to meet customer needs.
- Develop processes to deliver the product and service features.
- Transfer the resulting plans to operational personnel.

*Quality control* is the process used by operational personnel to ensure that their processes meet the product and service requirements (defined during the planning stage). It is based on the feedback loop and consists of the following steps:

- Evaluate actual operating performance.
- Compare actual performance to goals.
- Act on the difference.

*Quality improvement* aims to attain levels of performance that are unprecedented—levels that are significantly better than any past level. The methodologies recommended for quality improvement efforts utilize Six Sigma project teams, as described in Chap. 4. Notably, whereas earlier versions of Juran's *Quality Handbook* did not specifically advocate cross-functional project-based teams for quality improvement efforts, the most recent sixth edition (2010) clearly prescribes their use.

The mission of the quality function is companywide quality management. Quality management is the process of identifying and administering the activities necessary to achieve the organization's quality objectives. These activities will fall into one of the three categories in Juran's Trilogy.

Since the quality function transcends any specialized quality department, extending to all the activities throughout the company that affect quality, the primary role in managing the quality function is exercised by senior leadership. Only senior leadership can effectively manage the necessary cross-functional activities.

Leadership must give careful thought as to how the various overlapping activities can best be organized to optimize their impact on performance and minimize confusion over jurisdiction, resources, and authority. An "umbrella concept" often provides the needed guidance to successfully integrate the different but related efforts, resulting in the Process Enterprise.

## The Six Sigma Process Enterprise

Organizations are typically designed along functional lines, with functions such as engineering, marketing, accounting, and manufacturing assigned responsibility for specific tasks often corresponding closely to university degree programs. Persons with higher education in a specific discipline specialize in the work assigned to that function. Resources are allocated to each function based on the needs of the enterprise.

If the enterprise is to be successful, the "needs of the enterprise" must be based on the needs of its customers. However, customers obtain value from products or services created by the cooperative efforts and resources of many different functional areas. Most customers couldn't care less about how the enterprise creates the values they are purchasing.[5] A similar discussion applies to owners and shareholders. There is a substantial body of opinion among management experts that focusing internally on functional concerns can be detrimental to the enterprise as a whole. Deming (1986) explained the risks of departmental improvements at the expense of the system as a whole. An alternative is a holistic focus on the process or value stream that creates and delivers value.

A process focus means that stakeholder values are determined and activities are classified as either relating to the creation of the final value (value-added activity) or not (non-value-added activity). Processes are evaluated on how effectively and efficiently they create value. *Effectiveness* is defined as delivering what the customer requires, or exceeding the requirements; it encompasses quality, price, delivery, timeliness, and everything else that goes into perceived value. *Efficiency* is defined as being effective using a minimum of resources; more of an owner's perspective. Excellent processes are those that are both effective and efficient.

***Processes Are the Fundamental Activities of a Business.*** There is a tendency to narrowly interpret the term "process" as a manufacturing operation to convert raw materials into finished products. Throughout this book it has a much broader meaning, referring to any activity or set of activities that transform inputs to create values for stakeholders. The inputs can be labor, expertise, raw materials, products, transactions, information, or services that someone is willing to pay more for than they cost to create. In other words, the process adds value to the inputs. Said another way, *the process is the act of creating value.* The value can be a cured disease, a tasty banana split, a great movie, a successfully completed credit card transaction, or a cold soda purchased at a convenience store.

Reengineering, the process redesign fad so popular in the early 1990s, has become associated in the minds of many with brutal downsizing. Many academics condemned it as heartless and cold in its implementation. Yet the problem wasn't caused by reengineering in itself. The proper implementation of reengineering (and Six Sigma) focuses attention on broken and inefficient processes, enabling companies to operate faster and more efficiently and to use information technology more productively. It gave employees greater authority and a clearer view of how their work fit into the broader scheme of things. Customers benefited from lower prices, higher quality, and better services, and investors enjoyed a higher rate of return. And, more germane to our discussion of processes, reengineering taught business leaders to see their organizations not as control structures, but as processes that deliver value to customers in a way that creates profits for shareholders.

Many business leaders think of their organizations as extremely complex. From a process perspective, this is seldom the case, at least at the high levels. For example, Texas Instruments was able to break its $4 billion semiconductor business into six core processes:

- Strategy development
- Product development
- Customer design and support
- Manufacturing capability development
- Customer communication
- Order fulfillment

A large financial software company described its four core processes in plain English:

- Provide good products at good prices.
- Acquire customers and maintain good relations with them.
- Make it easy to buy from us.
- Provide excellent service and support after the sale.

Both of these companies have thousands of employees and generate billions of dollars in sales. Yet what they do for customers is really very simple. Once the basic (core) processes have been identified, the relationship between them should be determined and drawn on a *process map*. (Process mapping is discussed in greater detail in Part II of this handbook.) The process map presents employees with a simple picture that illustrates how the enterprise serves its customers. It is the basis for identifying subprocesses and, eventually, Six Sigma projects. Table 1.3 gives some examples of high-level processes and subprocesses.

*The Source of Conflict.* The truth is, it's the organizational structure that's complicated, not the business itself. The belief that the business is complicated results from a misplaced internal perspective by its leaders and employees. In a traditional organization, tremendous effort is wasted trying to understand what needs to be done if goals are not well defined and people don't know how their work relates to the organization's purpose. A process focus is often the first real "focus" an employee experiences, other than pleasing one's superiors.

**Table 1.3** Examples of High-Level Processes and Subprocesses

| Core Process | Subprocess |
|---|---|
| Product development | • R&D<br>• Design creation<br>• Prototype development<br>• Design production support |
| Marketing | • Inspiration, concept discovery<br>• Customer identification<br>• Developing market strategies<br>• Concept production support<br>• Customer acquisition and maintenance |
| Product creation | • Manufacturing<br>• Procurement<br>• Installation |
| Sales and service | • Fulfillment (order through payment)<br>• Pre-sale customer support<br>• Installation and frontline service<br>• Usage |
| Meta-processes | • Process excellence (Six Sigma)<br>• Voice of customer<br>• Voice of shareholder<br>• Voice of employee |

Management structures, since the time of Alfred P. Sloan in the 1920s and 1930s, are designed to divide work into discrete units with clear lines of responsibility and authority. While this approach produced excellent results for a time, it has inherent flaws that became quite apparent by 1980. Organizations put leadership atop a pyramid-shaped control system designed to carry out their strategies. Control of the resources needed to accomplish this resided in the vertical pillars, known as "functions" or "divisions." This command-and-control approach is depicted in Fig.1.11.

This arrangement creates "turf" where, much like caste systems, activities within a given area are the exclusive domain of that area. Personnel in engineering, for example, are not allowed to engage in activities reserved to the finance group, nor is finance allowed to "meddle" in engineering activities. These turfs are jealously guarded. In such a structure employees look to the leadership to tell them what to do and to obtain the resources needed to do it. This upward-inward focus is the antithesis of an external-customer focus. As Fig. 1.11 also shows, customer value is created by processes that draw resources from several different parts of the organization and end at a customer contact point. If an organization wants to be customer-focused, then it must change the traditional structure so its employees look *across* the organization at processes. As you might expect, this calls for a radical rethinking of the way the enterprise operates.

**Figure 1.11** Traditional command-and-control organizational structure.

As long as control of resources and turf remain entirely with the functional units, the enterprise will remain focused inwardly. Goals will be unit-based, rather than process-based. In short, Six Sigma (or any other process-oriented initiative) will not work effectively and efficiently. Functional department leaders have both the incentive and the ability to thwart cross-functional process improvement efforts. This doesn't mean that these people are "bad." It's simply that their missions are defined in such a way that they are faced with a dilemma: pursue the mission assigned to my area to the best of my ability, or support an initiative that detracts from it but benefits the enterprise as a whole. Social scientists call this "the tragedy of the commons." It is in the best interest of all fishermen not to overharvest the fishing grounds, but it is in the interest of each individual fisherman to get all he can from this "common resource." Similarly, it is in the best interest of the enterprise as a whole to focus on customers, but it is in each functional leader's best interest to pursue his or her provincial self-interest. After all, if every other functional manager tries to maximize the resources devoted to their area and I don't, I will lose my department's share of the resources. Self-interest wins hands down.

The proper place of the customer in the organization's hierarchy is illustrated in Fig. 1.12. Note that this perspective is precisely the opposite of the traditional view of the organization. The difficulties involved in making such a radical change should not be underestimated.

Edosomwan (1993) defines a customer- and market-driven enterprise as one that is committed to providing excellent quality and competitive products and services to satisfy the needs and wants of a well-defined market segment. This approach is in contrast to that of the traditional organization, as shown in Table 1.4.

**Figure 1.12** The "correct" view of the company organization chart.
(From Kotler, 1991, p. 21. Copyright © 1991 by Prentice-Hall, Inc. Reprinted by permission.)

**Table 1.4**  Traditional Organizations Versus Customer-Driven Organizations

|  | **Traditional Organizations** | **Customer-Driven Organizations** |
|---|---|---|
| Product and service planning | • Short-term focus<br>• Reactionary management<br>• Management by objectives planning process | • Long-term focus<br>• Prevention-based management<br>• Customer-driven strategic planning process |
| Measures of performance | • Bottom-line financial results<br>• Quick return on investment | • Customer satisfaction<br>• Market share<br>• Long-term profitability<br>• Quality orientation<br>• Total productivity |
| Attitudes toward customers | • Customers are irrational and a pain<br>• Customers are a bottleneck to profitability<br>• Hostile and careless<br>• "Take it or leave it" attitude | • Voice of the customer is important<br>• Professional treatment and attention to customers is required<br>• Courteous and responsive<br>• Empathy and respectful attitude |
| Quality of products and services | • Provided according to organizational requirements | • Provided according to customer requirements and needs |
| Marketing focus | • Seller's market<br>• Careless about lost customers due to poor customer satisfaction | • Increased market share and financial growth achieved |
| Process management approach | • Focus on error and defect detection | • Focus on error and defect prevention |
| Product and service delivery attitude | • It is OK for customers to wait for products and services | • It is best to provide fast time to market products and services |
| People orientation | • People are the source of problems and are burdens on the organization | • People are an organization's greatest resource |
| Basis for decision-making | • Product-driven<br>• Management by opinion | • Customer-driven<br>• Management by data |
| Improvement strategy | • Crisis management<br>• Management by fear and intimidation | • Continuous process improvement<br>• Total process management |
| Mode of operation | • Career-driven and independent work<br>• Customers, suppliers, and process owners have nothing in common | • Management-supported improvement<br>• Teamwork between suppliers, process owners, and customers practiced |

*Source: From Johnson. Copyright © 1993 by ASQ. Reprinted by permission.*

The journey from a traditional to a customer-driven organization has been made by enough organizations to allow us to identify a number of distinct milestones that mark the path to success. Generally, the journey begins with recognition that a crisis is either upon the organization, or imminent. This wrenches the organization's leadership out of denial and forces them to abandon the status quo.

When the familiar ways of the past are no longer acceptable, the result is a feeling of confusion among the leaders. At this stage the leadership must answer some very basic questions:

- What is the organization's purpose?
- What are our values?
- What does an organization with these values look like?

A "value" is that which one acts to gain and/or keep. It presupposes an entity capable of acting to achieve a goal in the face of an alternative. Values are not simply nice-sounding platitudes, they represent *goals*. Pursuing the organization's values implies building an organization that embodies these values. This is the leadership's *vision*, to create a reality where their values have been achieved.

After this vision has been clearly developed, the next step is to develop a strategy for building the new organization. Customer-driven organizations share certain common features:

- **Flattened hierarchies**—Getting everyone closer to the customer involves reducing the number of bureaucratic "layers" in the organization structure. It also involves the "upside-down" perspective of the organization structure shown in Fig. 1.12. The customer comes first, not the boss. Everyone serves the customer.
- **Risk-taking**—Customers' demands tend to be unpredictable. Responsiveness requires that organizations be willing to change quickly, which involves uncertainty and risk. Customer-driven organizations encourage risk-taking in a variety of ways. One important aspect is to celebrate mistakes made by individuals who engage in risky behavior. Bureaucratic impediments such as excessive dependence on written procedures are minimized or eliminated. Employees are encouraged to act on their own best judgments and not to rely on formal approval mechanisms.
- **Effective communication**—During transformation the primary task of the leadership team is the clear, consistent, and unambiguous transmission of their vision to others in the organization. One way this is done is through "internal marketing," which involves using the principles of marketing to get the message to the target "market": the employees. It is vital that the leaders' actions are completely consistent with their words. The assistance of outside consultants may be helpful in identifying inconsistencies.

Leaders should realize that their behavior carries tremendous symbolic meaning. This can contribute to the failure of convincing employees; a single action that is inconsistent with the stated message is sufficient to destroy all credibility. On the plus side, an action that clearly shows a commitment to the vision can help spread the word that "They're serious this time." The leadership should seek out stories that capture the essence of the new organization and repeat these stories often. For example, Nordstrom employees all hear the story of the sales clerk who allowed the customer to return a tire (Nordstrom's doesn't sell tires). This story captures the essence of the Nordstrom "rule book," which states:

- **Rule 1**—Use your own best judgment.
- **Rule 2**—There are no other rules.

Leaders should expect to devote a minimum of 50% of their time to communication during the transition.

- **Supportive boards of directors**—It is vital to obtain the enthusiastic endorsement of the new strategy by the board. Management cannot focus their attention until this support has been received. This will require that management educate their board and ask them for their approval. However, boards are responsible for governance, not management. Don't ask the board to approve tactics. This bogs down the board, stifles creativity in the ranks, and slows the organization down.
- **Partnered trade unions**—In the transformed organization, everyone's job changes. If the organization's employees are unionized, changing jobs requires that the union become management's partner in the transformation process. In the flat organization union employees will have greater authority. Union representatives should be involved in all phases of the transformation, including planning and strategy development. By getting union input, the organization can be ensured that during collective bargaining the union won't undermine the company's ability to compete or sabotage the strategic plan. Unions also play a role in auditing the company's activities to ensure that they comply with contracts and labor laws.
- **Measured results**—It is important that the right things be measured. The "right things" are measurements that determine that you are delivering on your promises to customers, investors, employees, and other stakeholders. You must also measure for the right reasons. This means that measurements are used to learn about how to improve, not for judgment. Finally, you must measure the right way. Measurements should cover processes as well as outcomes. Data must be available quickly to the people who use them. Measurements must be easy to understand. These topics are discussed in more detail in Chap. 3.
- **Rewarded employees**—Care must be taken to avoid punishing with rewards. Rewarding individuals with financial incentives for simply doing their jobs well implies

that the employee wouldn't do the job without the reward. It is inherently manipulative. The result is to destroy the very behavior you seek to encourage (Kohn, 1993). The message is that rewards should not be used as control mechanisms. Employees should be treated like adults and provided with adequate and fair compensation for doing their jobs. Recognizing exceptional performance or effort should be done in a way that encourages cooperation and team spirit, such as parties and public expressions of appreciation. Leaders should ensure fairness; for example, management bonuses and worker pay cuts don't mix.

*The Process Enterprise Solution.* Some companies—such as IBM, Texas Instruments, Owens Corning, and Duke Power—have successfully made the transition from the traditional organizational structure to an alternative system called the "Process Enterprise" (Hammer and Stanton, 1999). In these companies the primary organizational unit is not the functional department, but the process development team. These cross-functional teams, like the reengineering teams of old, have full responsibility for a major business process. For example, a product development team would work together in the same location to build the product development process from concept to launch. They would produce the design, documentation, training materials, advertising, and so on. In a Process Enterprise authority and control of resources is redistributed in a manner that achieves a balance of power between the process-focused and structure-focused aspects of the enterprise.

The differences between Process Enterprises and traditional organizations are fundamental. In the Process Enterprise a new position is created, that of Process Owner or Business Process Executive (BPE). The BPE position is permanent. BPEs are assigned from the senior-most executive body and given responsibility for designing and deploying the process, as well as control over all expenditures and supporting technology. They establish performance metrics, set and distribute budgets, and train the frontline workers who perform the process work. However, the people who perform the process work report to unit heads, not BPEs. In the Process Enterprise process goals are emphasized over unit goals. Process performance is used as a basis for compensation and advancement.

In a Process Enterprise lines of authority are less well defined. BPEs and functional unit managers are expected to work together to resolve disagreements. The BPE doesn't exert direct control over the workers, but because he controls budgets and sets goals by which unit managers will be evaluated, he does have a good deal of influence. The unit managers have to see to it that the process designs are sound, the resource allocation sufficient, and the goals clear and fair. In short, managing in a Process Enterprise places a premium on collaboration and cooperation.

One tool that has been developed to help clarify the different roles and responsibilities is the Decision Rights Matrix (Hammer and Stanton, 1999). This matrix specifies the roles the different managers play for each major decision, such as process changes, personnel hiring, setting budgets, and so on. For example, on a given decision must a given manager:

1. Make the decision?
2. Be notified in advance?
3. Be consulted beforehand?
4. Be informed after the fact?

The Decision Rights Matrix serves as a road map for the management team, especially in the early stages of the transition from traditional organization to Process Enterprise. Eventually team members will internalize the matrix rules.

BPEs must also work together. Processes overlap and process handoffs are critical. Often the same worker works with different processes. To avoid "horizontal turf wars" senior leadership needs to set enterprise goals and develop compensation and incentive systems that promote teamwork and cooperation between process owners.

The need for inter-process cooperation highlights the fact that no process is an island. From the customer's perspective, it's all one process. Overall excellence requires that the entire business be viewed as the customer sees it. One way to accomplish this is to set up a separate core business process with a focus of overall Process Excellence (PEX). PEX is change-oriented and cross-functional, including Six Sigma, all of the initiatives listed earlier, and many more as well. By creating a top-level position for PEX, leadership assigns clear responsibility for this important work. The PEX leader, usually a Vice President, leads a Process Excellence Leadership Team (PELT), which includes functional leaders as well as full-time PEX personnel such as the Director of Six Sigma. The VP of PEX isn't responsible for particular processes, but she has the authority to identify key processes and nominate owners for approval by the CEO or the PELT.

The mission of PEX is to see to it that all business processes accomplish the enterprise goals as they relate to customers, shareholders, and employees. PEX is also concerned with helping BPEs improve their processes, both internally and across the process boundaries. In other words, PEX is a *meta-process*, a process of processes. BPEs, unit managers, and Process Excellence leaders work together through Process Excellence Leadership Teams to ensure that the needs of the major stakeholder groups are met (Fig.1.13).

*Six Sigma and the Process Enterprise.*   Once the decision is made to become a Six Sigma Process Enterprise, the question of how to integrate the Six Sigma infrastructure will arise. General recommendations include

```
                    ┌─────────────────────────────────────┐
                    │            PELTs /PEX               │
                    │ 6σ, collaborate, negotiate,         │
                    │      communicate, feedback,         │
                    │            teamwork                 │
                    └─────────────────────────────────────┘
```

|  | **Process Owners** | **Unit Managers** |  |
|---|---|---|---|
| Strategies<br>• Plans<br>• Metrics<br>• Goals | • Manage processes<br>• Design processes<br>• Set process goals<br>• Measure process results (Ys)<br>• Train, inspire frontline workers<br>• Control budgets | • Manage people<br>• Operate processes<br>• Set unit goals<br>• Measure process drivers (Xs)<br>• Coach front line workers<br>• Utilize resources | Happy<br>• Customers<br>• Shareholders<br>• Employees |

**Figure 1.13** Process enterprise roles and responsibilities.

1. Designate Process Excellence (PEX) as one of the enterprise's core processes and select a BPE.
2. Master Black Belts will report to PEX. The Master Black Belts will have an enterprise-wide perspective. They will be concerned with the internal processes in PEX, as well as the overall value creation and delivery produced by the cooperative efforts of the core processes.
3. Black Belts will report to the BPEs, but the budget for the Black Belts comes from Process Excellence. This gives PEX influence, which helps maintain the enterprise perspective, but leaves day-to-day management and control with the Black Belt's customers, the BPEs.
4. BPEs have PEX goals, tied to incentives. PEX incentives are in the PEX budget.
5. Unit managers have process-based incentives. Process incentives are in the BPE's budgets.
6. The PEX leader and BPEs should collaboratively create a Decision Rights Matrix identifying:
   - The roles and responsibilities of PEX, BPEs, and unit managers. For example, hiring, budgets, project selection.
   - Who makes the decision in the areas just described?
   - Who must be consulted in decision-making?
   - What is the communication plan?
7. PEX helps develop a BPE Coordination Plan addressing such inter-process issues as:
   - Where do the core processes overlap?
   - How will cross-process Six Sigma projects be chartered and coordinated?

▼ Who will ensure that overlapping activities and handoffs are coordinated? (PEX plays a facilitation role here.)
▼ When is standardization across processes best and when isn't it? The process intersections should be invisible to customers (e.g., customers shouldn't have to provide the same information more than once; single form information for ordering of products, support plans, and registration). However, diversity may be necessary to serve unique customer needs.

Having Black Belts report to BPEs instead of to PEX does not contradict recommendations provided earlier in the chapter for Black Belts to report centrally. There is a critical difference: The earlier discussion assumed a traditional organizational structure, so if the Black Belts didn't report to the Six Sigma organization (referred to here as PEX) they would have been reporting to the *unit managers*. Here, the recommendation is to report to BPEs who, as process owners, have a much different perspective than the unit manager. This perspective, unlike that of unit managers, meshes very well with the Six Sigma focus on process improvement.

Table 1.5 illustrates the contrast between the way that staff functions used to operate under the traditional system of management, and the way they can operate more effectively.

**Table 1.5** How Staff Functions Are Changing

|  | **From** | **To** |
| --- | --- | --- |
| Role | Customer—for information, evidence, and reports from others | Supplier—of information, expertise, and other services |
| Strategy | Control—by imposition of policies and procedures, and by audit and inspection | Support—by gearing efforts to the needs of others<br>Self-control by client |
| Goal | Departmental—achievement of departmental objectives | Collective achievement of the organization's objectives |
| Style of working with others | Competitive, adversarial | Integrating, collaborative |
| Focus of attention | Some aspects of outcomes; for example, product quality, financial results<br>Some pieces of the process; for example, adherence to policy and procedure | The relationship between the entire underlying process and the achievement of all the desired outcomes |
| Image | Regulator, inspector, policeman | Educator, helper, guide |

*Source: Hutton, 1994, p. 220. Reprinted with permission.*

## *Deployment to the Supply Chain*

In the early part of the twentieth century Henry Ford pursued a great vision by building the Ford River Rouge Complex. By 1927 the Rouge was handling all production of Ford automobiles. It was truly a marvel. The Rouge was the largest single manufacturing complex in the United States, with peak employment of about 120,000. Here Henry Ford achieved self-sufficiency and vertical integration in automobile production, a continuous work flow from iron ore and other raw materials to finished automobiles. The complex included dock facilities, blast furnaces, open-hearth steel mills, foundries, a rolling mill, metal stamping facilities, an engine plant, a glass manufacturing building, a tire plant, and its own power house supplying steam and electricity.

On June 2, 1978, the Rouge was listed a National Historic Landmark: From state-of-the-art wonder to historical curiosity in just 50 years.

A related historical artifact is the idea that a firm can produce quality products or services by themselves. This may have been the case in the heyday of the Rouge, when the entire "supply chain" was a single, vertically integrated behemoth entity, but it is certainly no longer true. In today's world fully 50 to 80% of the cost of a manufactured product is in purchased parts and materials. When the customer forks over her good money for your product, she doesn't differentiate between you and your suppliers.

You say you're not in manufacturing? The situation is likely the same, regardless of industry. Consider personal finance software. Your customers run your software on a computer you didn't design with an operating system you have no control over. They're using your software to access their account at their financial institution to complete a tax return, which they will file electronically with the Internal Revenue Service (IRS). When your customers click the icon to run your product, they consider all of these intermediaries to be part of the value they are paying to receive.

The service industry is no different. Consider a discount brokerage company, whose customers want to use your service to buy common stocks, fixed income instruments, derivatives, etc. They also want debit cards, check writing, bill paying, pension plans, and a variety of other services, including financial advice, investment portfolio analysis, and annuities. When your customers put their money into their account at your firm, they expect you to be responsible for making all of the "third parties" work together seamlessly.

In short, you'll never reach Six Sigma quality levels with three sigma suppliers.

A primary objective with regard to suppliers is to obtain Six Sigma levels of supplier quality with minimal costs through projects that involve suppliers. The organization responsible for supply chain management (SCM) will take the lead in developing the supplier Six Sigma program, including preparation of a Supplier Six Sigma Deployment Plan with the following attributes:

- Policies on supplier Six Sigma
- Goals and deliverables of the supplier Six Sigma program
- Supplier communication plan
- Timetable for deployment, including phases (e.g., accelerated deployment to most critical suppliers)
- Procedures defining supplier contact protocols, supplier project charter, supplier project reporting and tracking, etc.
- Training requirements and timeline
- Methods of assessing supplier Six Sigma effectiveness
- Integration of the supplier Six Sigma program and in-house activities

SCM receives guidance from the Executive Six Sigma Council and the Six Sigma organization. The Six Sigma organization often provides expertise and other resources to the supplier Six Sigma effort.

SCM should sponsor or cosponsor supplier Six Sigma projects. In some cases SCM will lead the projects, often with supplier personnel taking a coleadership role. In others they will assist Black Belts or Green Belts working on other projects that involve suppliers. Full SCM sponsorship is usually required when the project's primary focus is on the supplier's product or process, such as to reduce the number of late deliveries of a key product. Projects involving suppliers, but not focused on them, can be cosponsored by SCM, such as a project involving the redesign of an order fulfillment process requiring only minor changes to the supplier's web ordering form. SCM assistance can take a number of different forms, for example:

- Acting as a liaison between the internal team members and suppliers
- Negotiating funding and budget authority for supplier Six Sigma projects
- Estimating and reporting supplier project savings
- Renegotiating contract terms
- Resolving conflicts
- Defining responsibility for action items
- Scheduling supplier visits
- Defining procedures for handling of proprietary supplier information
- Responding to supplier requests for assistance with Six Sigma

In addition to SCM, other elements within your organization play important supporting roles. Usually, Black Belts will come from the Six Sigma organization, although some larger enterprises assign a team of Black Belts to work on SCM projects full time. Green Belts often come from organizations sponsoring supplier-related projects. Team members are assigned from various areas, as with any Six Sigma project.

The customer certainly has the final say in the process and project requirements, but ultimate responsibility for the process itself should remain with the supplier, who owns and controls their processes, and may have liability and warranty obligations. Six Sigma teams must be clear that only SCM has the authority to make official requests for change. It can be embarrassing if a Black Belt makes a suggestion that the supplier believes to be a formal requirement to change. SCM may receive a new bid, price change, complaint letter, etc. from the supplier over such misunderstandings. Supplier relationships are often quite fragile; "Handle with care" is a good motto for the entire Six Sigma team to follow.

In addition to accepting responsibility for their processes, suppliers must often take the lead role in Six Sigma teams operating in supplier facilities. Supplier leadership must support Six Sigma efforts within their organizations. Suppliers must agree to commit the resources necessary to successfully complete projects, including personnel and funding.

## *Communications and Awareness*

Top-level Six Sigma projects using the DMAIC methodology can be defined to build buy-in for the change initiative and build awareness through communication, as follows (Keller, 2005).

### Define

DEFINE the scope and objectives for the Six Sigma change initiative, which is usually an enterprise undertaking.

Define the key stakeholder groups that will be impacted by the change. The key stakeholders are those groups whose involvement is key to the success of the change initiative, which can include

- Key customers
- Shareholders or other owners
- Senior leadership
- Middle management
- Six Sigma change agents
- The general employee population
- Suppliers

Define one or more metrics that can be used to track the current organizational culture on quality, which is discussed in the Measure description in the following section.

## Measure

Measure the baseline level of buy-in for the change initiative among these key stakeholder groups, as well as the baseline quality culture.

Buy-in can be measured according to the following scale (Forum, 1996), from lowest to highest: Hostility, Dissent, Acceptance, Support, Buy-in. Note that the desired level of buy-in surpasses mere support; enthusiasm is required for complete buy-in. Surveys and focus groups, further discussed in Chap. 2, are often used to measure buy-in as well as perceptions on quality.

Juran and Gryna (1993) define the company quality culture as the opinions, beliefs, traditions, and practices concerning quality. While sometimes difficult to quantify, an organization's culture has a profound effect on the quality produced by that organization. Without an understanding of the cultural aspects of quality, significant and lasting improvements in quality levels are unlikely.

Two of the most common means of assessing organization culture are the focus group and the written questionnaire. These two techniques are discussed in greater detail below. The areas addressed generally cover attitudes, perceptions, and activities within the organization that impact quality. Because of the sensitive nature of cultural assessment, anonymity is usually necessary. The author believes that it is necessary for each organization to develop its own set of questions. The process of getting the questions is an education in itself. One method for getting the right questions that has produced favorable results in the past is known as the critical-incident technique. This involves selecting a small representative sample ($n \times 20$) from the group you wish to survey and asking open-ended questions, such as:

- "Which of our organization's beliefs, traditions, and practices have a beneficial impact on quality?"
- "Which of our organization's beliefs, traditions, and practices have a detrimental impact on quality?"

The questions are asked by interviewers who are unbiased, and the respondents are guaranteed anonymity. Although usually conducted in person or by phone, written responses are sometimes obtained. The order in which the questions are asked (beneficial/detrimental) is randomized to avoid bias in the answer. Interviewers are instructed not to prompt the respondent in any way. It is important that the responses be recorded verbatim, using the respondent's own words. Participants are urged to provide as many responses as they can; a group of 20 participants will typically produce 80 to 100 responses.

The responses themselves are of great interest and always provide a great deal of information. In addition, the responses can be grouped into categories and the categories examined to glean additional insight into the dimensions of the organization's quality culture. The responses and categories can be used to develop valid survey items and to prepare focus-group questions. The follow-up activity is why so few people are needed at this stage—statistical validity is obtained during the survey stage.

## Analyze

Analyze the primary causes of buy-in resistance, which can include issues and resolutions such as (Forum, 1996; resolutions by Keller, 2005):

- **Unclear goals**—Goals need to be clearly communicated throughout the stakeholder groups.
- **No personal benefit**—Goals should be stated in terms that provide a clear link to personal benefits for stakeholders, such as decreased hassles or improved working conditions.
- **Predetermined solutions**—When teams are given the solution without chance for analysis of alternatives, they will likely be skeptical of the result. The root cause of this practice is often management resistance to free thinking or experimentation by process personnel or a lack of customer focus, as further described in Chap. 2.
- **Lack of communication**—Analyses and results should be communicated throughout the stakeholder groups.
- **Too many priorities**—Teams need to be focused on achievable results.
- **Short-term focus**—Goals should provide clear benefits over short and longer terms.
- **No accountability**—Clearly defined Project Sponsors, stakeholders, and team members provide accountability.
- **Disagreement on the definition of customer**—Clearly defined stakeholder groups are needed for project success. This can also be associated with so-called turf wars between various functional areas within an organization.
- **Low probability of implementation**—Formal project sponsorship and approvals provide a clear implementation channel.
- **Insufficient resources**—Stakeholder groups need to understand that the project is sufficiently funded and resources allocated—training project teams is essential.
- **Midstream change in direction or scope**—Changes in project scope or direction provide a potential for a loss of buy-in. Changes must be properly communicated to stakeholder groups to prevent this reduction in buy-in.

## Improve

Improve buy-in by addressing the causes of resistance, such as suggested by the resolutions noted above. Communication is the primary method of building buy-in, and can be effectively improved by developing and managing a Six Sigma communication plan. Successful implementation of Six Sigma will only happen if the leadership's vision and implementation plans are clearly understood and embraced by employees, shareholders, customers, and suppliers. Because it involves cultural change, Six Sigma frightens many people, and good communication is an antidote to fear: without it rumors run rampant and morale suffers. The commitment to Six Sigma must be clearly and unambiguously understood throughout the organization. This doesn't happen by accident; it is the result of careful planning and execution.

Communicating the Six Sigma message is a multimedia undertaking. The modern organization has numerous communications technologies at its disposal. Keep in mind that communication is a two-way affair; be sure to provide numerous opportunities for upward and lateral as well as downward communication. Here are some suggestions to accomplish the communications mission:

- All-hands launch event, with suitable pomp and circumstance
- Mandatory staff meeting agenda item
- Internal newsletters, magazines, and Web sites, with high profile links to enterprise Six Sigma Web site on home page
- Six Sigma updates in annual report
- Stock analyst updates on publicly announced Six Sigma goals
- Intranet discussion forums
- Two-way mail communications
- Surveys
- Suggestion boxes
- Videotape or DVD presentations
- Closed-circuit satellite broadcasts by executives, with questions and answers
- All-hands discussion forums
- Posters
- Logo shirts, gear bags, keychains, coffee mugs, and other accessories
- Speeches and presentations
- Memoranda
- Recognition events
- Lobby displays
- Letters

Promoting Six Sigma awareness is, in fact, an internal marketing campaign. A marketing expert, perhaps from your company's marketing organization, should be consulted. If your organization is small, a good book on marketing can provide guidance (e.g., Levinson et al., 1995).

For each stakeholder group, the key concerns include

- Who is primarily responsible for communication with this group?

- ▲ What are the communication needs for this group? For example, key customers may need to know how Six Sigma will benefit them; employees may need to understand the process for applying for a change agent position such as Black Belt.
- ▲ What communication tools, techniques, and methods will be used? These include meetings, newsletters, email, one-on-one communications, and Web sites.
- ▲ What will be the frequency of communication? Remember, repetition will usually be necessary to be certain the message is received and understood.
- ▲ Who is accountable for meeting the communication requirement?
- ▲ How will we measure the degree of success? Who will do this?
- ▲ The requirements and responsibilities can be organized using tables, such as Table 1.6.

When buy-in is reduced because of a perceived lack of management support, action is necessary to increase the leadership involvement. Senior managers' time is in great demand from a large number of people inside and outside of the organization. It is all too easy to schedule a weekly meeting to discuss "Six Sigma" for an hour, and then think you've done your part. In fact, transforming an organization, large or small, requires a prodigious commitment of the time of senior leadership, not just a few hours a month for update meetings.

**Table 1.6** Six Sigma Communications Plan and Requirements Matrix

| Requirement | Method | Frequency | Accountability |
|---|---|---|---|
| **Senior Leadership** | | | |
| Program strategy, goals, and high-level program plan | • Senior staff meeting<br>• Senior leadership training | • At least monthly<br>• Start of program | • CEOs<br>• Six Sigma Director<br>• Training department |
| Metrics/status performance to program plan | • Senior staff meetings | • At least monthly | • Six Sigma Director |
| **Middle Management** | | | |
| Program strategy, goals, and management-level program plan | • Regular flow down of upper level staff meeting notes; newsletter<br>• Management training | • At least monthly for staff meetings; newsletter piece every two weeks during program rollout, as needed thereafter<br>• Prior to first wave of Six Sigma projects | • Senior Leadership for staff meeting flow down<br>• Internal communications via core team for company newsletter<br>• Training department |

Etc. for customers, owners, stock analysts, change agents, bargaining unit, exempt employees, suppliers, or other stakeholder group

One way to maximize the value of an executive's time investment is the use of symbolic events, or stories, that capture the essence of management's commitment (or lack of it) to the change effort. Stakeholders repeat and retain stories far better than proclamations and statements. For example, there's a story told by employees of a large U.S. automotive firm that goes as follows:

In the early 1980s the company was just starting their quality improvement effort. At a meeting between upper management and a famous quality consultant, someone casually mentioned that quality levels were seasonal—quality was worse in the summer months. The consultant asked why this should be so. Were different designs used? Were the machines different? How about the suppliers of raw materials? The answer to each of these questions was "No." An investigation revealed that the problem was vacations. When one worker went on vacation, someone else did her job, but not quite as well. And that "someone" also vacated a job, which was done by a replacement, etc. It turned out that the one person going on vacation led to six people doing jobs they did not do routinely. The solution was to have a vacation shutdown of two weeks. This greatly reduced the number of people on new jobs and brought summer quality levels up to the quality levels experienced the rest of the year. This worked fine for a couple of years, given an auto industry recession and excess capacity. One summer, however, the senior executives were asked by the finance department to reconsider their shutdown policy. Demand had picked up and the company could sell every car it could produce. The accountants pointed out that the shutdown would cost $100 million per day in lost sales.

When the vice president of the truck division asked if anything had been done to address the cause of the quality slippage in the summer, the answer was "No, nothing had been done." The president asked the staff "If we go back to the old policy, would quality levels fall like they did before?" Yes, he was told, they would. "Then we stay with our current policy and shut down the plants for vacations," the president announced.

The president was challenged by the vice president of finance. "I know we're committed to quality, but are you sure you want to lose $1.4 billion in sales just to demonstrate our commitment?" The president replied, "Frank, I'm not doing this to 'demonstrate' anything. We almost lost our company a few years back because our quality levels didn't match our overseas competition. Looking at this as a $1.4 billion loss is just the kind of short-term thinking that got us in trouble back then. I'm making this decision to save money."

This story had tremendous impact on the managers who heard it, and it spread like wildfire throughout the organization. It demonstrated many things simultaneously: senior leadership's commitment to quality, political parity between operations and finance, how seemingly harmless policies can have devastating effects, an illustration of how short-term thinking had damaged the company in the past, and how long-term thinking worked in a specific instance, etc. It is a story worth 100 speeches and mission statements.

## Control

Control the change effort with a plan to maintain buy-in. Personnel trained as change agents can be placed in strategic positions throughout the organization, as shown in Fig. 1.14. This makes it possible for them to assist in the development and implementation (including sponsorship) of future quality improvement projects. Quality improvement of any significance nearly always involves multiple departments and levels in the organization. Change agents help organize an assessment of the organization to identify its strengths and weaknesses. Change is usually undertaken to either reduce areas of weakness or exploit areas of strength. The assessment is part of the education process. Knowing one's specific strengths and weaknesses is useful in mapping the process for change.

**Figure 1.14** Cascading of sponsorship. (Copyright © 1994 by David W. Hutton.)

# CHAPTER 2

# Recognizing and Capitalizing on Opportunity

Chapter 1 focused on the need to become a responsive, change-capable organization, and proposed a basic implementation plan to meet that need. The infrastructure described in Chap. 1 allows the organization to define and deploy Six Sigma projects focused on specific needs of key stakeholder groups, including customers, shareholders, and employees. In this way, the Six Sigma projects serve as the means of achieving these broad organizational objectives. Chapter 2 describes the organizational mechanisms required to recognize the key stakeholder needs, which provide opportunities for organizational growth and, in some cases, survival.

A core principle of Six Sigma is the recognition of *customer value*, a term derived from Lean practices. Customer input is critical in defining value, as an organization exists fundamentally to meet the needs of its customers. Obtaining valid customer input is a science itself. Market research firms use scientific methods such as critical incident analysis, focus groups, content analysis, and surveys to identify the "voice of the customer."

This chapter describes a number of approaches for gathering customer input on their value definition. It then explores other means of identifying opportunities, including quantifying costs of poor quality, benchmarking best practices, and innovation. It concludes with the development of strategic plans to identify a vision for the future and define objectives to achieve the vision. Chapter 3 will then build on the strategic plan to ensure systems are in place to manage relative to the plan, with feedback to adapt to changing market conditions.

## Methods for Collecting Customer Data

There are several primary strategies commonly used to obtain information from or about internal or external customers:

- Sample surveys

- Focus groups
- Operational feedback systems, including experimentation

## *Surveys*

In sample surveys, data are collected from a sample of a universe to estimate the characteristics of the universe, such as their range or dispersion, the frequency of occurrence of events, or the expected values of important universe parameters. The reader should note that these terms are consistent with the definition of enumerative statistical studies as described in Chap. 7. This is the traditional approach to such surveys. However, if survey results are collected at regular intervals, the results can be analyzed using the statistical control charts as described in Chap. 8 to obtain information on the underlying process. The process excellence leader should not be reticent in recommending that survey budgets be allocated for conducting small, routine, periodic surveys rather than infrequent "big studies." Without the information available from time-ordered series of data, it will not be possible to learn about processes that produce changes in customer satisfaction or perceptions of quality, or to verify progress toward improvement.

Survey development consists of the following major tasks (GAO, 1986, p.15):

1. Initial planning of the questionnaire
2. Developing the measures
3. Designing the sample
4. Developing and testing the questionnaire
5. Producing the questionnaire
6. Preparing and distributing mailing materials
7. Collecting data
8. Reducing the data to forms that can be analyzed
9. Analyzing the data

### Guidelines for Developing Questions

The axiom that underlies the guidelines shown below is that the question-writer(s) must be thoroughly familiar with the respondent group and must understand the subject matter from the perspective of the respondent group.

This is often problematic for the employee when the respondent group is the customer; methods for dealing with this situation are discussed below. There are eight basic guidelines for writing good questions:

- Ask questions in a format that is appropriate to the questions' purpose and the information required.

- Make sure the questions are relevant, proper, and qualified as needed.
- Write clear, concise questions at the respondent's language level.
- Give the respondent a chance to answer by providing a comprehensive list of relevant, mutually exclusive responses from which to choose.
- Ask unbiased questions by using appropriate formats and item constructions and by presenting all important factors in the proper sequence.
- Get unbiased answers by anticipating and accounting for various respondent tendencies.
- Quantify the response measures where possible.
- Provide a logical and unbiased line of inquiry to keep the reader's attention and make the response task easier.

The above guidelines apply to the *form* of the question. Using the critical incident technique to develop good question *content* is described in the following section.

### Response Types

There are several commonly used types of survey responses.

- **Open-ended questions**—These are questions that allow the respondents to frame their own response without any restrictions placed on the response. The primary advantage is that such questions are easy to form and ask using natural language, even if the question writer has little knowledge of the subject matter. Unfortunately, there are many problems with analyzing the answers received to this type of question. This type of question is most useful in determining the scope and content of the survey, not in producing results for analysis or process improvement.
- **Fill-in-the-blank questions**—Here the respondent is provided with directions that specify the units in which the respondent is to answer. The instructions should be explicit and should specify the answer units. This type of question should be reserved for very specific requests, e.g., "What was your age on your last birthday?—(age in years)."
- **Yes/no questions**—Unfortunately, yes/no questions are very popular. Although they have some advantages, they have many problems and few uses. Yes/no questions are ideal for dichotomous variables, such as defective or not defective. However, too often this format is used when the measure spans a range of values and conditions, for example, "Were you satisfied with the quality of your new car (yes/no)?" A yes/no response to such questions contain little useful information.
- **Ranking questions**—The ranking format is used to rank options according to some criterion, for example, importance. Ranking formats are difficult to write and difficult to answer. They give very little real information and are very prone to errors that can

invalidate all the responses. They should be avoided whenever possible in favor of more powerful formats and formats less prone to error, such as rating. When used, the number of ranking categories should not exceed five.

- ▲ **Rating questions**—With this type of response, a rating is assigned on the basis of the score's absolute position within a range of possible values. Rating scales are easy to write, easy to answer, and provide a level of quantification that is adequate for most purposes. They tend to produce reasonably valid measures. Here is an example of a rating format:

    For the following statement, check the appropriate box: The workmanship standards provided by the purchaser are

    | Clear |
    | Marginally adequate |
    | Unclear |

- ▲ **Guttman format**—In the Guttman format, the alternatives increase in comprehensiveness; that is, the higher-valued alternatives include the lower-valued alternatives. For example:

    Regarding the benefit received from training in quality improvement:

    | No benefit identified |
    | Identified benefit |
    | Measured benefit |
    | Assessed benefit value in dollar terms |
    | Performed cost/benefit analysis |

- ▲ **Likert and other intensity scale formats**—These formats are usually used to measure the strength of an attitude or an opinion. For example:

    Please check the appropriate box in response to the following statement: "The customer service representative was knowledgeable."

    | Strongly disagree |
    | Disagree |
    | Neutral |
    | Agree |
    | Strongly agree |

Intensity scales are very easy to construct. They are best used when respondents can agree or disagree with a statement. A problem is that statements must be worded to present a single side of an argument. We know that the respondent agrees, but we must infer what he believes. To compensate for the natural tendency of people to agree, statements are usually presented using the converse as well, for example, "The customer service representative was not knowledgeable."

When using intensity scales, use an odd-numbered scale, preferably with five or seven categories. If there is a possibility of bias, order the scale in a way that favors the hypothesis you want to disprove and handicaps the hypothesis you want to confirm. In this way you will confirm the hypothesis with the bias against you—a stronger result. If there is no bias, put the most undesirable choices first.

▲ **Semantic differential format**—In this format, the values that span the range of possible choices are not completely identified; only the end points are labeled. For example:

Indicate the number of times you initiated communication with your customer in the past month.

| few | · | · | · | · | · | · | · | many |

The respondent must infer that the range is divided into equal intervals. The range seems to work well with seven categories.

Semantic differentials are very useful when we do not have enough information to anchor the intervals between the poles. However, they are very difficult to write well and if not written well the results are ambiguous.

### Case Study Using the Critical Incident Technique

This actual case study involves the development of a mail survey at a community hospital. The same process has been successfully used by the author to develop customer surveys for clientele in a variety of industries.

The study of service quality and patient satisfaction was performed at a 213-bed community hospital in the southwestern United States. The hospital is a nonprofit, publicly funded institution providing services to the adult community; pediatric services are not provided. The purpose of the study was to:

▲ Identify the determinants of patient quality judgments.
▲ Identify internal service delivery processes that impacted patient quality judgments.
▲ Determine the linkage between patient quality judgments and intent-to-patronize the hospital in the future or to recommend the hospital to others.

To conduct the study, the author worked closely with a core team of hospital employees, and with several ad hoc teams of hospital employees. The core team included the Nursing Administrator, the head of the Quality Management Department, and the head of Nutrition Services.[†]

The team decided to develop their criteria independently. It was agreed that the best method of getting information was directly from the target group, in-patients. Due to the nature of hospital care services, focus groups were not deemed feasible for this study. Frequently, patients must spend a considerable period of time convalescing after being released from a hospital, making it impossible for them to participate in a focus group soon after discharge. While the patients are in the hospital, they are usually too sick to participate. Some patients have communicable diseases, which makes their participation in focus groups inadvisable.

Since memories of events tend to fade quickly (Flanagan, 1954), the team decided that patients should be interviewed within 72 hours of discharge. The target patient population was, therefore, all adults treated as in-patients and discharged to their homes. The following groups were not part of the study: families of patients who died while in the hospital, patients discharged to nursing homes, patients admitted for psychiatric care.[‡]

The team used the Critical Incident Technique (CIT) to obtain patient comments. The CIT was first used to study procedures for selection and classification of pilot candidates in World War II (Flanagan, 1954). A bibliography assembled in 1980 listed over 700 studies about or using the CIT (Fivars, 1980). Given its popularity, it is not surprising that the CIT has also been used to evaluate service quality.

CIT consists of a set of specifically defined procedures for collecting observations of human behavior in such a way as to make them useful in addressing practical problems. Its strength lies in carefully structured data collection and data classification procedures that produce detailed information not available through other research methods. The technique, using either direct observation or recalled information collected via interviews, enables researchers to gather firsthand patient-perspective information. This kind of self-report preserves the richness of detail and the authenticity of personal experience of those closest to the activity being studied. Researchers have concluded that the CIT produces information that is both reliable and valid.

This study attempted to follow closely the five steps described by Flanagan as crucial to the CIT: (1) establishment of the general aim of the activity studied; (2) development of a plan for observers or interviewers; (3) collection of data; (4) analysis (classification) of data; and (5) interpretation of data.

## Establishment of the General Aim of the Activity Studied

The general aim is the purpose of the activity. In this case the activity involves the whole range of services provided to in-patients in the hospital. This includes every service activity between admission and discharge.[§] From the service provider's perspective the general aim is to create and manage service delivery processes in such a way as to produce a willingness by the patient to utilize the provider's services in the future. To do this, the service provider must know which particular aspects of the service are remembered by the patient.

Our general aim was to provide the service provider with information on what patients remembered about their hospital stay, both pleasant and unpleasant. This information was to be used to construct a new patient survey instrument that would be sent to recently discharged patients on a periodic basis. The information obtained would be used by the managers of the various service processes as feedback on their performance, from the patient's perspective.

## Interview Plan

Interviewers were provided with a list of patients discharged within the past three days. The discharge list included all patients. Nonpsychiatric patients who were discharged to "home" were candidates for the interview. Home was defined as any location other than the morgue or a nursing home. Interviewers were instructed to read a set of predetermined statements. Patients to be called were selected at random from the discharge list. If a patient could not be reached, the interviewer would try again later in the day. One interview form was prepared per patient. To avoid bias, 50% of the interview forms asked the patient to recall unpleasant incidents first and 50% asked for pleasant incidents first. Interviewers were instructed to record the patient responses using the patient's own words.

## Collection of Data

Four interviewers participated in the data collection activity; all were management level employees of the hospital. Three of the interviewers were female, one was male. The interviews were conducted when time permitted during the interviewer's normal busy work day. The interviews took place during the September 1993 time period. Interviewers were given the instructions recommended by Hayes (1992) for generating critical incidents.

A total of 36 telephone attempts were made and 23 patients were reached. Of those reached, three spoke only Spanish. In the case of one of the Spanish-speaking patients a family member was interviewed. Thus, 21 interviews were conducted, which is slightly greater than the 10 to 20 interviews recommended by Hayes (1992). The 21 interviews produced 93 critical incidents.

## Classification of Data

The Incident Classification System required by CIT is a rigorous, carefully designed procedure with the end goal being to make the data useful to the problem at hand while sacrificing as little detail as possible (Flanagan, 1954). There are three issues in doing so: (1) identification of a general framework of reference that will account for all incidents; (2) inductive development of major area and sub-area categories that will be useful in sorting the incidents; and (3) selection of the most appropriate level of specificity for reporting the data.

The critical incidents were classified as follows:

1. Each critical incident was written on a 3 × 5 card, using the patient's own words.
2. The cards were thoroughly shuffled.
3. Ten percent of the cards (ten cards) were selected at random, removed from the deck, and set aside.
4. Two of the four team members left the room while the other two grouped the remaining 83 cards and named the categories.
5. The ten cards originally set aside were placed into the categories found in step 4.
6. Finally, the two members not involved in the initial classification were told the names of the categories. They then took the reshuffled 93 cards and placed them into the previously determined categories.

The above process produced the following dimensions of critical incidents:

- Accommodations (5 critical incidents)
- Quality of physician (14 critical incidents)
- Care provided by staff (20 critical incidents)
- Food (26 critical incidents)
- Discharge process (1 critical incident)
- Attitude of staff (16 critical incidents)
- General (11 critical incidents)

## Interpretation of Data

Interjudge agreement, the percentage of critical incidents placed in the same category by both groups of judges, was 93.5%. This is well above the 80% cutoff value recommended by experts. The setting aside of a random sample and trying to place them in established categories is designed to test the comprehensiveness of the categories. If any of the withheld items were not classifiable it would be an indication that the categories do not adequately span the patient satisfaction space. However, the team experienced no problem in placing the withheld critical incidents into the categories.

Ideally, a critical incident has two characteristics: (1) it is specific and (2) it describes the service provider in behavioral terms or the service product with specific adjectives (Hayes, 1992). Upon reviewing the critical incidents in the General category, the team determined that these items failed to have one or both of these characteristics. Thus, the 11 critical incidents in the General category were dropped. The team also decided to merge the two categories "Care provided by staff" and "Attitude of staff" into the single category "Quality of staff care." Thus, the final result was a five dimension model of patient satisfaction judgments: Food, Quality of physician, Quality of staff care, Accommodations, and Discharge process.

A rather obvious omission in the above list is billing. This occurred because the patients had not yet received their bill within the 72 hour time frame. However, the patient's bill was explained to the patient prior to discharge. This item is included in the Discharge process dimension. The team discussed the billing issue and it was determined that billing complaints do arise after the bills are sent, suggesting that billing probably is a satisfaction dimension. However, the team decided not to include billing as a survey dimension because (1) the time lag was so long that waiting until bills had been received would significantly reduce the ability of the patient to recall the details of their stay; (2) fear that the patient's judgments would be overwhelmed by the recent receipt of the bill; and (3) a system already existed for identifying patient billing issues and adjusting the billing process accordingly.

## Survey Item Development

As stated earlier, the general aim was to provide the service provider with information on what patients remembered about their hospital stay, both pleasant and unpleasant. This information was then to be used to construct a new patient survey instrument that would be sent to recently discharged patients on a periodic basis. The information obtained would be used by the managers of the various service processes as feedback on their performance, from the patient's perspective.

The core team believed that accomplishing these goals required that the managers of key service processes be actively involved in the creation of the survey instrument. Thus, ad hoc teams were formed to develop survey items for each of the dimensions determined by the critical incident study. The teams were given brief instruction by the author in the characteristics of good survey items. Teams were required to develop items that, in the opinion of the core team, met five criteria: (1) relevance to the dimension being measured; (2) concise; (3) unambiguous; (4) one thought per item; and (5) no double negatives. Teams were also shown the specific patient comments that were used as the basis for the categories and informed that these comments could be used as the basis for developing survey items.

Writing items for the questionnaire can be difficult. The process of developing the survey items involved an average of three meetings per dimension, with each meeting lasting approximately two hours. Ad hoc teams ranged in size from four to eleven members. The process was often quite tedious, with considerable debate over the precise wording of each item.

The core team discussed the scale to be used with each ad hoc team. The core team's recommended response format was a five point Likert-type scale. The consensus was to use a five point agree-disagree continuum as the response format. Item wording was done in such a way that agreement represented better performance from the hospital's perspective.

In addition to the response items, it was felt that patients should have an opportunity to respond to open-ended questions. Thus, the survey also included general questions that invited patients to comment in their own words. The benefits of having such questions are well known.

In addition, it was felt that these questions might generate additional critical incidents that would be useful in validating the survey.

The resulting survey instrument contained 50 items and three open-ended questions and is included in Appendix 13.

## Survey Administration and Pilot Study

The survey was to be tested on a small sample. It was decided to use the total design method (TDM) to administer the survey (Dillman, 1983). Although the total design method is exacting and tedious, Dillman indicated that its use would ensure a high rate of response. Survey administration would be handled by the Nursing Department.

TDM involves rather onerous administrative processing. Each survey form is accompanied by a cover letter, which must be hand-signed in blue ink. Follow up mailings are done one, three, and sevenweeks after the initial mailing. The 3- and 7-week follow-ups are accompanied by another survey and another cover letter. No "bulk processing" is allowed, such as the use of computer-generated letters or mailing labels. Dillman's research emphasizes the importance of viewing the TDM as a completely integrated approach (Dillman, 1983).

Because the hospital in the study is small, the author was interested in obtaining maximum response rates. In addition to following the TDM guidelines, he recommended that a $1 incentive be included with each survey. However, the hospital administrator was not convinced that the additional $1 per survey was worthwhile. It was finally agreed that to test the effect of the incentive on the return rate $1 would be included in 50% of the mailings, randomly selected.

The hospital decided to perform a pilot study of 100 patients. The patients selected were the first 100 patients discharged to home starting April 1, 1994. The return information is shown in Table 2.1.

Although the overall return rate of 49% is excellent for normal mail-survey procedures, it is substantially below the 77% average and the 60% "minimum" reported by Dillman. As possible explanations, the author conjectures that there may be a large Spanish-speaking constituency for this hospital. As mentioned above, the hospital is planning a Spanish version of the survey for the future.

The survey respondent demographics were analyzed and compared to the demographics of the non-respondents to ensure that the sample group was representative. A sophisticated statistical analysis was performed on the responses to evaluate the reliability and validity of each item. Items with low reliability coefficients or questionable validity were reworded or dropped.

[†] *The nutrition services manager was very concerned that she gets sufficient detail on her particular service. Thus, the critical incident interview instrument she used included special questions relating to food service.*
[‡] *The team was unable to obtain a Spanish-speaking interviewer, which meant that some patients that were candidates were not able to participate in the survey.*
[§] *Billing was not covered in the CIT phase of the study because patients had not received their bills within 72 hours.*

Table 2.1  Pilot Patient Survey Return Information

|  | Delivered | Returned | Percent |
|---|---|---|---|
| Surveys | 92 | 45 | 49 |
| Surveys with $1 incentive | 47 | 26 | 55 |
| Surveys without $1 incentive | 46 | 10 | 42 |

## *Focus Groups*

The focus group is a special type of group in terms of purpose, size, composition, and procedures. A focus group is typically composed of seven to ten participants who are unfamiliar with each other. These participants are selected because they have certain characteristic(s) in common that relate to the topic of the focus group.

The researcher creates a permissive environment in the focus group that nurtures different perceptions and points of view, without pressuring participants to vote, plan, or reach consensus. The group discussion is conducted several times with similar types of participants to identify trends and patterns in perceptions. Careful and systematic analyses of the discussions provide clues and insights as to how a product, service, or opportunity is perceived.

A focus group can thus be defined as a carefully planned discussion designed to obtain perceptions on a defined area of interest in a permissive, nonthreatening environment. The discussion is relaxed, comfortable, and often enjoyable for participants as they share their ideas and perceptions. Group members influence each other by responding to ideas and comments in the discussion.

In Six Sigma, focus groups are useful in a variety of situations:

- Prior to starting the strategic planning process
- Generate information for survey questionnaires
- Needs assessment, for example, training needs
- Test new program ideas
- Determine customer decision criteria
- Recruit new customers

The focus group is a socially oriented research procedure. The advantage of this approach is that members stimulate one another, which may produce a greater number of comments than would individual interviews. If necessary, the researcher can probe for additional information or clarification. Focus groups produce results that have high face validity, that is, the results are in the participant's own words rather than in statistical jargon. The information is obtained at a relatively low cost, and can be obtained very quickly.

There is less control in a group setting than with individual interviews. When group members interact, it is often difficult to analyze the resulting dialogue. The quality of focus group research is highly dependent on the qualifications of the interviewer. Trained and skilled interviewers are hard to find. Group-to-group variation can be considerable, further complicating the analysis. Finally, focus groups are often difficult to schedule.

Another method somewhat related to focus groups is that of customer panels. Customer panels are composed of a representative group of customers who agree to communicate their attitudes periodically via phone calls or mail questionnaires. These panels are more representative of the range of customer attitudes than customer complaint and suggestion systems, yet more easily managed than focus groups. To be effective, the identity of customers on the panel must be withheld from the employees serving them.

## *Operational Feedback Systems*

Many organizations have complaint and suggestion systems, including chat or general customer support functions, which provide customers an easy-to-use avenue for both favorable and unfavorable feedback to management. Due to selection bias, these methods do not provide statistically valid information. However, because they are a census rather than a sample, they provide opportunities for individual customers to have their say. These are moments of truth that can be used to increase customer loyalty. They also provide anecdotes that have high face validity and are often a source of ideas for improvement.

Customer complaints and suggestions can be related to virtually any aspect of the customer-supplier relationship, including product defects, errors in accuracy of information, delivery issues, and inadequate delivery or communication processes, to name just a few. The likelihood of a client reporting an issue is often related to the seriousness of the deficiency, the cost of the deficiency to the customer, and perhaps also the customer's expectation of corrective action. With this in mind, the organizational systems should incorporate the recognition that the issue may be more widespread than reported, so that the true impact is investigated and all impacted customers are notified of the issue and its corrective action.

When a customer complaint has been received, it represents an opportunity to increase customer loyalty, as well as a risk of losing the customer. The importance of complaint handling is illustrated in Fig. 2.1. These data illustrate that the decision to repurchase is highly dependent on whether their complaint was handled to their satisfaction.

**Figure 2.1** Percent planning to repurchase versus how complaint was handled. (*Note: The large durable goods survey did not include a response category of "satisfied."*)

Considering that customers who complain are likely to tell as many as 14 others of their experience, the importance of complaint handling in customer relations becomes obvious.

Despite the impressive nature of Fig. 2.1, even these figures dramatically understate the true extent of the problem. Complaints represent people who were not only unhappy, they were unhappy enough to report that dissatisfaction to the company. Research indicates that up to 96% of unhappy customers never tell the company. This is especially unfortunate since it has been shown that customer loyalty is *increased* by proper resolution of complaints. Given the dramatic impact of a lost customer, it makes sense to maximize the opportunity of the customer to complain. Complaints should be actively sought, an activity decidedly against human nature. This suggests that a system must be developed and implemented to motivate employees to seek out customer complaints, ideas, and suggestions; to feel their pain. The system should also provide every conceivable way for an unhappy customers to contact the company on their own, including toll-free hotlines, email, comment cards, and so forth.

A customer-focused organization incorporates customer feedback regarding value into the organization's daily operations. In this way, the organization collects and disseminates information on value definition as part of its intrinsic function. Whereas surveys have a latency period associated with their development and implementation, customer contacts provide real-time feedback, so organizational systems must be defined to collect and disseminate the information. This visibility is a key aspect of the Lean techniques described throughout Part II. Each and every customer contact provides that opportunity, and at the heart of the responsive organization is the inherent motivation to capture that data as the first step in satisfying customer needs. It becomes the lifeblood of the organization, whose products and services exist only to satisfy the customers' need for value. As noted earlier, an organization's failure to meet the needs or expectations of any given customer will not necessarily result in a complaint, so systems that rely solely on customer-initiated complaints will not achieve optimal results. Instead, regular contact with a customer can seek to identify what is working, and what is not working, relative to the customer's expectations.

A practice recently popularized in sales departments is *Solution Selling*, the title of a best-selling book (Eades, 2004). In *Solution Selling*, Eades discusses the need to identify the customer's pain: his or her critical business issue, challenge, or source of dissatisfaction. In *Solution Selling*, pain that is identified and admitted by a prospective buyer provides the first level of qualification for a sale. The seller's product or service is then, where appropriate, positioned as the solution to that pain. The relevance of this technique is its simplicity: engaging in conversation with informed stakeholders is often sufficient to define the pain that exists in their system. The solutions to alleviate that pain provide value that the customer is willing to pay for.

An organization in the software sector has successfully deployed a similar model implemented (and improved) over the course of the last 15 years. Their sales and technical support areas have access to a common database that maintains every suggestion or complaint voiced by a customer. This Quality Management System (QMS) database is integrated with statistical process control analysis tools to provide trend analysis of issues, resolution times, etc. Most notable is their response to these Improvement Opportunities (IO, as they call them). As the IO is entered into the system, it is circulated throughout the technical support, sales, and development areas, up to the vice president level. This provides immediate visibility to all internal stakeholders. High-priority issues, such as errors that significantly impact customers or the potential for a significant sale based on a critical customer need, are immediately investigated and scheduled for deployment. Of course, this requires a number of back-and-forth discussions with the customer to fully understand her or his needs, sometimes at a fairly high level in the organization to ensure the proper information is received. While this takes resources and time, it serves the organization's long-term vision to build relationships with their customer base. Their relatively flat organizational structure is designed to provide customer access to nearly all aspects of the organization, including subject matter experts. Their ability to rapidly integrate these new ideas into their products is heavily influenced by deployment of Lean concepts for rapid turnaround of new products. They have effectively removed non-value-added bureaucracies, while focusing their limited resources on creating marketable value. Their business continues to grow based on new customers as well as the continued upgrades and growth opportunities from long-term customers.

In any organization, the leadership sets the vision and the strategy. A basketball coach can neither dribble down court for the team, nor provide constant and immediate instructions to players on shooting, passing, and defense. Instead, the coach needs to develop skills in the players, then empower them to exercise judgment in the use of these skills.

As on the basketball court, responsibility in a customer-focused organization must be delegated down the line. Rather than tightened oversight to achieve adherence to customer requirements, bureaucracy must be reduced and the organizational hierarchy flattened to increase the communication between organizational levels, particularly communication to and from the front-line personnel with access to the customer. Through empowerment, the responsibility, focus, and authority of frontline personnel are shifted toward doing what is necessary to win customer loyalty, without need for additional management approval. This empowerment demands flexibility, sharing of information, and an understanding of the business operations, which is afforded by cross-training. The ultimate result is what Carlzon (1987) called *moments of truth* in customer relations, where customers realize the commitment of the organization to solving their problems.

Once customer feedback has been obtained, it must be used to improve process and product quality. A system for utilizing customer feedback is described as follows:

1. Local managers and employees serve customers' needs daily, using locally modified procedures along with general corporate policies and procedures.
2. By means of a standardized and locally sensitive questionnaire, determine the needs and attitudes of customers on a regular basis.
3. Comparing financial data, expectations, and past attitude information, determine strengths and weaknesses and their probable causes.
4. Determine where and how effort should be applied to correct weaknesses and preserve strengths. Repeat the process by taking action—step 1—and maintain it to attain a steady state or to evolve in terms of customer changes.
5. A similar process can take place at higher levels, using aggregated data from the field and existing policy flows of the organization.

## Cost of Poor Quality

The history of quality costs dates back to the first edition of *Juran's QC Handbook* in 1951. Today, quality cost accounting systems are part of a modern organization's quality improvement strategy, as well as many quality standards. Quality cost systems identify internal opportunities for return on investment. As such, quality costs stress avoiding defects and other behaviors that cause customer dissatisfaction, yet provide little understanding of the product or service features that satisfy or delight customers. It is conceivable that a firm could drive quality costs to zero and still go out of business.

Cost of quality includes any cost that would not be expended if quality were perfect. This includes such obvious costs as scrap and rework, but it also includes less obvious costs, such as the cost to replace defective material, expedite shipments for replacement material, the staff and equipment to process the replacement order, etc. Service businesses also incur quality costs; for example, a hotel incurs a quality cost when room service delivers a missing item to a guest. Specifically, quality costs are a measure of the costs associated with the achievement or nonachievement of product or service quality—including all product or service requirements established by the company and its contracts with customers and society. Requirements include marketing specifications, end-product and process specifications, purchase orders, engineering drawings, company procedures, operating instructions, professional or industry standards, government regulations, and any other document or customer needs that can affect the definition of product or service. More specifically, quality costs are the total of the cost incurred by (1) investing in the *prevention* of nonconformances to requirements; (2) *appraising* a product or service for conformance to requirements; and (3) *failure* to meet requirements (Fig. 2.2).

For most organizations, quality costs are hidden costs. Unless specific quality cost identification efforts have been undertaken, few accounting systems include provision for identifying quality costs. Because of this, unmeasured quality costs tend to increase. Poor quality impacts companies in two ways: higher cost and lower customer satisfaction. The lower customer satisfaction creates price pressure and lost sales, which result in lower revenues. The combination of higher cost and lower revenues eventually brings on a crisis that may threaten the very existence of the company. Rigorous cost of quality measurement is one technique for preventing such a crisis from occurring. Figure 2.3 illustrates the hidden-cost concept.

As a general rule, quality costs increase as the detection point moves further up the production and distribution chain. The lowest cost is generally obtained when errors are prevented in the first place. If nonconformances occur, it is generally least expensive to detect them as soon as possible after their occurrence. Beyond that point there is loss incurred from additional work that may be lost. The most expensive quality costs are from nonconformances detected by customers. In addition to the replacement or repair loss, a company loses customer goodwill, and their reputation is damaged when the customer relates his experience to others. In extreme cases, litigation may result, adding even more cost and loss of goodwill.

**PREVENTION COSTS**

The costs of all activities specifically designed to prevent poor quality in products or services. Examples are the costs of new product review, quality planning, supplier capability surveys, process capability evaluations, quality improvement team meetings, quality improvement projects, quality education and training.

**APPRAISAL COSTS**

The costs associated with measuring, evaluating, or auditing products or services to ensure conformance to quality standards and performance requirements. These include the costs of incoming and source inspection/test of purchased material, in process and final inspection/test, product, process, or service audits, calibration of measuring and test equipment, and the costs of associated supplies and materials.

**FAILURE COSTS**

The costs resulting from products or services not conforming to requirements or customer/user needs. Failure costs are divided into internal and external failure cost categories.

**INTERNAL FAILURE COSTS**

Failure costs occurring prior to delivery or shipment of the product, or the furnishing of a service, to the customer. Examples are the costs of scrap, rework, reinspection, retesting, material review, and downgrading.

**EXTERNAL FAILURE COSTS**

Failure costs occurring after delivery or shipment of the product, and during or after furnishing of a service, to the customer. Examples are the costs of processing customer complaints, customer returns, warranty claims, and product recalls.

**TOTAL QUALITY COSTS**

The sum of the above costs. It represents the difference between the actual cost of a product or service, and what the reduced cost would be if there was no possibility of substandard service, failure of products, or defects in their manufacture.

**Figure 2.2** Quality costs general—description.
(From Campanella, 1999. Copyright © 1999 by ASQ Quality Press.)

**Figure 2.3** Hidden cost of quality and the multiplier effect.
(From Campanella, 1990. Copyright © 1990 by ASQ Quality Press.)

Another advantage of early detection is that it provides more meaningful feedback to help identify root causes. The time lag between production and field failure makes it very difficult to trace the occurrence back to the process state that produced it. While field failure tracking is useful in *prospectively* evaluating a "fix," it is usually of little value in retrospectively evaluating a problem.

Quality cost measurement need not be accurate to the penny to be effective. The purpose of measuring such costs is to provide broad guidelines for management decision-making and action. The very nature of cost of quality makes such accuracy impossible. In some instances it will only be possible to obtain periodic rough estimates of such costs as lost customer goodwill, cost of damage to the company's reputation, etc. These estimates can be obtained using special audits, statistical sampling, and other market studies. These activities can be jointly conducted by teams of marketing, accounting, and Six Sigma personnel. Since these costs are often huge, these estimates must be obtained. However, they need not be obtained every month. Annual studies are usually sufficient to indicate trends in these measures.

## *Cost of Quality Examples*

I. **Prevention costs**—Costs incurred to prevent the occurrence of nonconformances in the future, such as the following:[1]
   A. Marketing/customer/user
      - Marketing research
      - Customer/user perception surveys/clinics
      - Contract/document review
   B. Product/service/design development
      - Design quality progress reviews
      - Design support activities
      - Product design qualification test
      - Service design qualification
      - Field tests
   C. Purchasing
      - Supplier reviews
      - Supplier rating
      - Purchase order tech data reviews
      - Supplier quality planning
   D. Operations (manufacturing or service)
      - Operations process validation
      - Operations quality planning

- Design and development of quality measurement and control equipment
- Operations support quality planning
- Operator quality education
- Operator SPC/process control

E. Six Sigma, Lean, and quality administration
- Administrative salaries
- Administrative expenses
- Program planning
- Performance reporting
- Education and training
- Improvement projects
- Process and product audits
- Other prevention costs

II. **Appraisal costs**—Costs incurred in measuring and controlling current production to ensure conformance to requirements, such as the following:
   A. Purchasing appraisal costs
   - Receiving or incoming inspections and tests
   - Measurement equipment
   - Qualification of supplier product
   - Source inspection and control programs
   
   B. Operations (manufacturing or service) appraisal costs
   - Planned operations inspections, tests, audits
   - Checking labor
   - Product or service quality audits
   - Inspection and test materials
   - Setup inspections and tests
   - Special tests (manufacturing)
   - Process control measurements
   - Laboratory support
   - Measurement equipment
   - Depreciation allowances
   - Measurement equipment expenses
   - Maintenance and calibration labor
   - Outside endorsements and certifications
   
   C. External appraisal costs
   - Field performance evaluation
   - Special product evaluations
   - Evaluation of field stock and spare parts

D. Review of tests and inspection data
E. Miscellaneous quality evaluations

**III. Internal failure costs**—Costs generated before a product is shipped as a result of non-conformance to requirements, such as the following:
  A. Product/service design failure costs (internal)
    ▼ Design corrective action
    ▼ Rework due to design changes
    ▼ Scrap due to design changes
  B. Purchasing failure costs
    ▼ Purchased material reject disposition costs
    ▼ Purchased material replacement costs
    ▼ Supplier corrective action
    ▼ Rework of supplier rejects
    ▼ Uncontrolled material losses
  C. Operations (product or service) failure costs
    ▼ Material review and corrective action costs
    ▼ Disposition costs
    ▼ Troubleshooting or failure analysis costs (operations)
    ▼ Investigation support costs
    ▼ Operations corrective action
    ▼ Operations rework and repair costs
    ▼ Rework
    ▼ Repair
    ▼ Reinspection/retest costs
    ▼ Extra operations
    ▼ Scrap costs (operations)
    ▼ Downgraded end product or service
    ▼ Internal failure labor losses
  D. Other internal failure costs

**IV. External failure costs**—Costs generated after a product is shipped as a result of non-conformance to requirements, such as the following:
  A. Complaint investigation/customer or user service
  B. Returned goods
  C. Retrofit costs
  D. Recall costs
  E. Warranty claims
  F. Liability costs

G. Penalties
H. Customer/user goodwill
I. Lost sales
J. Other external failure costs

## *Quality Cost Bases*

The guidelines for selecting a base for analyzing quality costs are as follows:

- The base should be related to quality costs in a meaningful way.
- The base should be well known to the managers who will receive the quality cost reports.
- The base should be a measure of business volume in the area where quality cost measurements are to be applied.
- Several bases are often necessary to get a complete picture of the relative magnitude of quality costs.
- These are some commonly used bases (Campanella, 1990):
  - A labor base (such as total labor, direct labor, or applied labor)
  - A cost base (such as shop cost, operating cost, or total material and labor)
  - A sales base (such as net sales billed, or sales value of finished goods)
  - A unit base (such as the number of units produced, or the volume of output)

Although actual dollars spent are usually the best indicator for determining where quality improvement projects will have the greatest impact on profits and where corrective action should be taken, unless the production rate is relatively constant, it will not provide a clear indication of quality cost improvement trends. Since the goal of the cost of quality program is improvement over time, it is necessary to adjust the data for other time-related changes such as production rate, inflation, etc. Total quality cost compared to an applicable base results in an index, which may be plotted and analyzed using control charts as described in Chap. 8.

For long-range analyses and planning, net sales is the base most often used for presentations to top management (Campanella, 1990). If sales are relatively constant over time, the quality cost analysis can be performed for relatively short spans of time. In other industries this figure must be computed over a longer time interval to smooth out large swings in the sales base. For example, in industries such as shipbuilding or satellite manufacturing, some periods may have no deliveries, while others have large dollar amounts. It is important that the quality costs incurred be related to the sales for the same period. Consider the sales as the "opportunity" for the quality costs to happen.

Some examples of cost of quality bases are (Campanella, 1990):

- Internal failure costs as a percentage of total production costs
- External failure costs as an average percentage of net sales
- Procurement appraisal costs as a percentage of total purchased material cost
- Operations appraisal costs as a percentage of total production costs
- Total quality costs as a percentage of production costs

# Benchmarking

Benchmarking is a topic of general interest in Six Sigma. Thus, the discussion here goes beyond the use of benchmarking in defining customer expectations, developing process or product requirements, and setting goals. In general terms, benchmarking can be defined as measuring your performance against that of best-in-class companies, determining how the best-in-class achieve those performance levels, and using the information as the basis for your own company's targets, strategies, and implementation.

Benchmarking involves research into the best practices at the industry, firm, or process level. Benchmarking goes beyond a determination of the "industry standard." It breaks the firm's activities down to process operations and looks for the best-in-class for a particular operation. For example, to achieve improvement in their parts distribution process, Xerox Corporation studied the retailer L.L. Bean.

Benchmarking goes beyond the mere setting of goals. It focuses on practices that produce superior performance. Benchmarking involves setting up partnerships that allow both parties to learn from each other. Competitors can also engage in benchmarking, providing they avoid proprietary issues.

Benchmarking projects are like any other major project. Benchmarking must have a structured methodology to ensure successful completion of thorough and accurate investigations. However, it must be flexible to incorporate new and innovative ways of assembling difficult-to-obtain information. It is a discovery process and a learning experience. It forces the organization to take an external view, to look beyond itself.

## *The Benchmarking Process*

Camp (1989) lists the following steps for the benchmarking process:

1. Planning
    1.1. Identify what is to be benchmarked.
    1.2. Identify comparative companies.
    1.3. Determine data collection method and collect data.

2. Analysis
    2.1. Determine current performance "gap."
    2.2. Project future performance levels.
3. Integration
    3.1. Communicate benchmark findings and gain acceptance.
    3.2. Establish functional goals.
4. Action
    4.1. Develop action plans.
    4.2. Implement specific actions and monitor progress.
    4.3. Recalibrate benchmarks.
5. Maturity
    5.1. Leadership position is attained.
    5.2. Practices are fully integrated into process.

The first step in benchmarking is determining what to benchmark. To focus the benchmarking initiative on critical issues, begin by identifying the process outputs most important to the customers of that process (i.e., the key quality characteristics). This step applies to every organizational function, since each one has outputs and customers. The QFD/customer needs assessment is a natural precursor to benchmarking activities.

## *Getting Started with Benchmarking*

The essence of benchmarking is the acquisition of information. It begins with the identification of the process that is to be benchmarked. The process chosen should be one that will have a major impact on the success of the business.

Once the process has been identified, contact a business library and request a search for information relating to your area of interest. The library will identify material from a variety of external sources, such as magazines, journals, and special reports. You should also conduct research, using the Internet and other electronic networking resources. However, be prepared to pare down what will probably be an extremely large list of candidates (e.g., an Internet search on the word "benchmarking" produced 20,000 hits). Don't forget your organization's internal resources. If your company has an intranet, use it to conduct an internal search. Set up a meeting with people in key departments, such as R&D. Tap the expertise of those in your company who routinely work with customers, competitors, suppliers, and other "outside" organizations. Often your company's board of directors will have an extensive network of contacts.

The search is, of course, not random. Look for the best of the best, not the average firm. There are many possible sources for identifying the elites. One approach is to build a compendium of business awards and citations of merit that organizations have received in business process improvement. Sources to consider are *IndustryWeek*'s Best Plants Award; National Institute of Standards and Technology's Malcolm Baldrige Award; *USA Today* and Rochester Institute of Technology's Quality Cup Award; European Foundation for Quality Management Award; Occupational Safety and Health Administration (OSHA); Federal Quality Institute; Deming Prize; Competitiveness Forum; *Fortune* magazine; and United States Navy's Best Manufacturing Practices, to name just a few. You may wish to subscribe to an "exchange service" that collects benchmarking information and makes it available for a fee. Once enrolled, you will have access to the names of other subscribers—a great source for contacts.

Don't overlook your own suppliers as a source for information. If your company has a program for recognizing top suppliers, contact these suppliers and see if they are willing to share their "secrets" with you. Suppliers are predisposed to cooperate with their customers; it's an automatic door-opener. Also contact your customers. Customers have a vested interest in helping you do a better job. If your quality, cost, and delivery performance improve, your customers will benefit. Customers may be willing to share some of their insights as to how their other suppliers compare with you. Again, it isn't necessary that you get information about direct competitors. Which of your customer's suppliers are best at billing? Order fulfillment? Customer service? Keep your focus at the process level, and there will seldom be any issues of confidentiality. An advantage to identifying potential benchmarking partners through your customers is that you will have a referral that will make it easier for you to start the partnership.

Another source for detailed information on companies is academic research. Companies often allow universities access to detailed information for research purposes. While the published research usually omits reference to the specific companies involved, it often provides comparisons and detailed analysis of what separates the best from the others. Such information, provided by experts whose work is subject to rigorous peer review, will often save you thousands of hours of work.

After a list of potential candidates is compiled, the next step is to choose the best three to five targets. A candidate that looked promising early in the process might be eliminated later based on the following criteria (Vaziri, 1992):

- Not the best performer
- Unwilling to share information and practices (i.e., doesn't view the benchmarking process as a mutually beneficial learning opportunity)
- Low availability and questionable reliability of information on the candidate

As the benchmarking process evolves, the characteristics of the most desirable candidates will be continually refined. This occurs as a result of a clearer understanding of your organization's key quality characteristics and critical success factors and an improved knowledge of the marketplace and other players.

This knowledge and the resulting actions tremendously strengthen an organization.

## Why Benchmarking Efforts Fail

The causes of failed benchmarking projects are the same as those for other failed projects (DeToro, 1995):

- **Lack of sponsorship**—A team should submit to management a one- to four-page benchmarking project proposal that describes the project, its objectives, and potential costs. If the team can't gain approval for the project or get a sponsor, it makes little sense to proceed with a project that's not understood or appreciated or that is unlikely to lead to corrective action when completed.
- **Wrong people on team**—Who are the right people for a benchmarking team? Individuals involved in benchmarking should be the same ones who own or work in the process. It's useless for a team to address problems in business areas that are unfamiliar or where the team has no control or influence.
- **Teams don't understand their work completely**—If the benchmarking team didn't map, flowchart, or document its work process, and if it didn't benchmark with organizations that also documented their processes, there couldn't be an effective transfer of techniques. The intent in every benchmarking project is for a team to understand how its process works and compare it to another company's process at a detailed level. The exchange of process steps is essential for improved performance.
- **Teams take on too much**—The task a team undertakes is often so broad that it becomes unmanageable. This broad area must be broken into smaller, more manageable projects that can be approached logically. A suggested approach is to create a functional flowchart of an entire area, such as production or marketing, and identify its processes. Criteria can then be used to select a process to be benchmarked that would best contribute to the organization's objectives.
- **Lack of long-term management commitment**—Since managers aren't as familiar with specific work issues as their employees, they tend to underestimate the time, cost, and effort required to successfully complete a benchmarking project. Managers should be informed that while it's impossible to know the exact time it will take for a typical benchmarking project, there is a rule of thumb that a team of four or five individuals requires one-third of their time for five months to complete a project.
- **Focus on metrics rather than processes**—Some firms focus their benchmarking efforts on performance targets (metrics) rather than processes. Knowing that a com-

petitor has a higher return on assets doesn't mean that its performance alone should become the new target (unless an understanding exists about how the competitor differs in the use of its assets and an evaluation of its process reveals that it can be emulated or surpassed).

- **Not positioning benchmarking within a larger strategy**—Benchmarking is one of many Six Sigma tools—such as problem solving, process improvement, and process reengineering—used to shorten cycle time, reduce costs, and minimize variation. Benchmarking is compatible with and complementary to these tools, and they should be used together for maximum value.
- **Misunderstanding the organization's mission, goals, and objectives**—All benchmarking activity should be launched by management as part of an overall strategy to fulfill the organization's mission and vision by attaining first the short-term objectives and then the long-term goals.
- **Assuming every project requires a site visit**—Sufficient information is often available from the public domain, making a site visit unnecessary. This speeds the benchmarking process and lowers the cost considerably.
- **Failure to monitor progress**—Once benchmarking has been completed for a specific area or process, benchmarks have been established and process changes implemented, managers should review progress in implementation and results.

The issues described here are discussed throughout this book. The best way of dealing with them is to prevent their occurrence by carefully planning and managing the project from the outset.

This list can be used as a checklist to evaluate project plans; if the plans don't clearly preclude these problems, then the plans are not complete.

## *The Benefits of Benchmarking*

The benefits of competitive benchmarking include the following:

- Creating a culture that values continuous improvement to achieve excellence
- Enhancing creativity by devaluing the not-invented-here syndrome
- Increasing sensitivity to changes in the external environment
- Shifting the corporate mindset from relative complacency to a strong sense of urgency for ongoing improvement
- Focusing resources through performance targets set with employee input
- Prioritizing the areas that need improvement
- Sharing the best practices between benchmarking partners

## Some Dangers of Benchmarking

Benchmarking is based on learning from others, rather than developing new and improved approaches. Since the process being studied is there for all to see, benchmarking cannot give a firm a sustained competitive advantage. Although helpful, benchmarking should never be the primary strategy for improvement.

Competitive analysis is an approach to goal setting used by many firms. This approach is essentially benchmarking confined to one's own industry. Although common, competitive analysis virtually guarantees second-rate quality because the firm will always be following their competition. If the entire industry employs the approach, it will lead to stagnation for the entire industry, setting them up for eventual replacement by outside innovators.

# Innovation

A recent *Harvard Business Review* blog laments the overuse of the word "innovation" in current times, noting with a hint of sarcasm how Kellogg's had cited the "innovation" behind their new peanut-butter flavored Pop Tart, while Hewlett-Packard had used the phrase over 70 times in a single conference call to describe their cost-cutting plans to streamline operations (Taylor). Clearly, these are not exemplary examples of innovation.

Instead, innovation is more closely associated with groundbreaking inventions, what Harvard Business School professor Clayton Christensen termed *disruptive-innovation* in his classic 1997 text *The Innovator's Dilemma*. These are ideas that fundamentally change the market, such as through price, accessibility, or convenience; they are game changers. Recent examples include personal computers, digital cameras, and smartphones. Henry Ford's ability to mass-produce automobiles, driving their cost to the point of affordability for the average worker, is a classic example.

Lack of innovation is cited as a key cause for the demise of industry behemoths Kodak and GM, as well as the source of a predicted doom for Microsoft (Silberzahn and Jones, 2012). The authors liken lack of innovation in an organization to Toynbee's theories on the demise of once powerful civilizations: Civilizations prosper when they effectively motivate their population to creatively contribute to society, which then attracts creative people from outside to further nourish the society. The civilization begins its descent when the creative capacity dissipates and creativity gives way to control. Kotter (2012) cites a complacency for innovation within Kodak's leadership ranks as the reason for their demise; they were apparently overly focused on the high (yet short-term) profitability margins associated with their film business (as one insider suggests).

Customers seldom spark true innovation, since they are usually unaware of state-of-the-art developments within their suppliers' field or the processes used within the suppliers' facilities. Nonetheless, their expectations often drive their suppliers' processes and products, as suggested by the Kano model.

## Kano Model

Noritaki Kano developed the following model of the relationship between customer satisfaction and quality (Fig. 2.4).

The Kano model shows that there is a basic level of quality that customers assume the product will have. For example, all automobiles have windows and tires. If asked, customers don't even mention the basic quality items; they take them for granted. However, if this quality level *isn't* met, the customer will be dissatisfied; note that the entire "Basic quality" curve lies in the lower half of the chart, representing dissatisfaction. However, providing basic quality isn't enough to create a satisfied customer.

The "Expected quality" line represents those expectations that customers explicitly consider, such as the length of time spent waiting in line at a checkout counter. The model shows that customers will be dissatisfied if their quality expectations are not met; satisfaction increases as more expectations are met.

The "Exciting quality" curve lies entirely in the satisfaction region. This is the effect of innovation. Exciting quality represents *unexpected* quality items. The customers receive more than they expected. For example, Cadillac pioneered a system where the headlights

**Figure 2.4** Kano model.

stay on long enough for the owner to walk safely to the door. When first introduced, the feature excited people.

Competitive pressure will constantly raise customer expectations. Today's exciting quality is tomorrow's basic quality. Firms that seek to lead the market must innovate constantly. Conversely, firms that seek to offer standard quality must constantly research customer expectations to determine the currently accepted quality levels. It is not enough to track competitors since expectations are influenced by outside factors as well. For example, the quality revolution in manufacturing has raised expectations for service quality as well.

Defining quality as merely the lack of nonconforming product reflects a limited view of quality. Motorola, of course, never intended to define quality this way in developing their Sigma Level metric. However, some have misinterpreted the Six Sigma program in this way.

One problem with "garden variety" Six Sigma is that it addresses only half of the Kano model. By focusing on customer expectations and prevention of nonconformances and defects, basic Six Sigma programs address only the portion of the Kano model on and below the line labeled "Expected Quality." While there is nothing wrong with improving these aspects of business performance, it will not ensure that the organization remains viable in the long term. Long-term success requires that the organization innovate. Innovation is the result of creative activity.

See online video describing the Kano Model at www.mhprofessional.com/ssh5.

## *Quality Function Deployment*

Quality Function Deployment (QFD) is a customer-driven process for planning innovation in products and services. It starts with the voice of the customer, which becomes the basis for setting requirements. QFD matrices, sometimes called "the house of quality," are graphical displays of the result of the planning process. QFD matrices vary a great deal and may show such things as competitive targets and process priorities. The matrices are created by interdepartmental teams, thus overcoming some of the barriers that exist in functionally organized systems.

QFD is also a system for design of a product or service based on customer demands, a system that moves methodically from customer requirements to specifications for the product or service. QFD involves the entire company in the design and control activity. Finally, QFD provides documentation for the decision-making process. The QFD approach involves four distinct phases (King 1987):

1. **Organization phase**—Management selects the product or service to be improved, the appropriate interdepartmental team, and defines the focus of the QFD study.

2. **Descriptive phase**—The team defines the product or service from several different directions such as customer demands, functions, parts, reliability, cost, and so on.
3. **Breakthrough phase**—The team selects areas for improvement and finds ways to make them better through new technology, new concepts, better reliability, cost reduction, etc., and monitors the bottleneck process.
4. **Implementation phase**—The team defines the new product and how it will be manufactured.

QFD is implemented through the development of a series of matrices. In its simplest form QFD involves a matrix that presents customer requirements as rows and product or service features as columns. The cell, where the row and column intersect, shows the correlation between the individual customer requirement and the product or service requirement. This matrix is sometimes called the "requirement matrix." When the requirement matrix is enhanced by showing the correlation of the columns with one another, the result is called the "house of quality." Figure 2.5 shows one commonly used house of quality layout.

The house of quality relates, in a simple graphical format, customer requirements, product characteristics, and competitive analysis. It is crucial that this matrix be developed

**Figure 2.5** The house of quality.

carefully since it becomes the basis of the entire QFD process. By using the QFD approach, the customer's demands are "deployed" to the final process and product requirements.

One rendition of QFD, called the Macabe approach, proceeds by developing a series of four related matrices (King, 1987): product planning matrix, part deployment matrix, process planning matrix, and production planning matrix. Each matrix is related to the previous matrix as shown in Fig. 2.6.

Figure 2.7 shows an example of an actual QFD matrix.

## Data Collection and Review of Customer Expectations, Needs, Requirements, and Specifications

Another approach to QFD is based on work done by Yoji Akao. Akao (1990, pp. 7–8) presents the following 11-step plan for developing the quality plan and quality design, using QFD.

1. First, survey both the expressed and latent quality demands of consumers in your target marketplace. Then decide what kinds of "things" to make.

**Figure 2.6** QFD methodology: Macabe approach.

**Figure 2.7** QFD matrix for an aerospace firm.
(From Wahl and Bersbach, 1991. Reprinted with permission.)

2. Study the other important characteristics of your target market and make a demanded quality function deployment chart that reflects both the demands and characteristics of that market.
3. Conduct an analysis of competing products on the market, which we call a competitive analysis. Develop a quality plan and determine the selling features (sales points).
4. Determine the degree of importance of each demanded quality.

5. List the quality elements and make a quality elements deployment chart.
6. Make a quality chart by combining the demanded quality deployment chart and the quality elements deployment chart.
7. Conduct an analysis of competing products to see how other companies perform in relation to each of these quality elements.
8. Analyze customer complaints.
9. Determine the most important quality elements as indicated by customer quality demands and complaints.
10. Determine the specific design quality by studying the quality characteristics and converting them into quality elements.
11. Determine the quality assurance method and the test methods.

## *Translating Customer Demands*

Once customers have made their demands known, it is important that these be translated into internal requirements and specifications. The term "translation" is used to describe this process because the activity literally involves interpreting the words in one language (the customer's) into those of another (the employee). For example, regarding the door of her automobile the customer might say "I want the door to close completely when I push it, but I don't want it swinging closed from just the wind." The engineer working with this requirement must convert it into engineering terminology such as pounds of force required to move the door from an open to a closed position, the angle of the door when it's opened, and so on. Care must be taken to maintain the customer's intent throughout the development of internal requirements. The purpose of specifications is to transmit the voice of the customer throughout the organization.

In addition to the issue of maintaining the voice of the customer, there is the related issue of the importance assigned to each demand by the customer. Design of products and services always involves tradeoffs: gasoline economy suffers as vehicle weight increases, but safety improves as weight increases. The importance of each criterion must be determined by the customer. When different customers assign different importance to criteria, design decisions are further complicated.

It becomes difficult to choose from competing designs in the face of ambiguity and customer-to-customer variation. Add to this the differences between internal personnel and objectives—department versus department, designer versus designer, cost versus quality, etc.—and the problem of choosing a design alternative quickly becomes complex. A rigorous process for deciding which alternative to settle on is helpful in dealing with the complexity.

## Structured Decision-Making

The first step in deciding upon a course of action is to identify the goal. For example, let's say you're the owner of the Product Development process for a company that sells software to help individuals manage their personal finances. The product, let's call it DollarWise, is dominant in its market and your company is well respected by its customers and competitors, in large part because of this product's reputation. The business is profitable and the leadership naturally wants to maintain this pleasant set of circumstances and to build on it for the future. The organization has committed itself to a strategy of keeping DollarWise the leader in its market segment so it can capitalize on its reputation by launching additional new products directed toward other financially oriented customer groups, such as small businesses. They have determined that Product Development is a core process for deploying this strategy.

As the process owner, or Business Process Executive, you have control of the budget for product development, including the resources to upgrade the existing product. Although it is still considered the best personal financial software available, DollarWise is getting a little long in the tooth and the competition has steadily closed the technical gap. You believe that a major product upgrade is necessary and want to focus your resources on those things that matter most to customers. Thus, your goal is

GOAL: Determine where to focus product upgrade resources

Through a variety of "listening posts" (focus groups, user laboratories, Internet forums, trade show interviews, conference hospitality suites, surveys, letters, and technical support feedback), you have determined that customers ask questions and make comments like the following:

- Can I link a DollarWise total to a report in my word processor?
- I have a high speed connection and I'd like to be able to download big databases of stock information to analyze with DollarWise.
- I like shortcut keys so I don't have to always click around in menus.
- I only have a 56K connection and DollarWise is slow on it.
- I use the Internet to pay bills through my bank. I'd like to do this using DollarWise instead of going to my bank's Web site.
- I want an interactive tutorial to help me get started.
- I want printed documentation.
- I want the installation to be simple.
- I want the user interface to be intuitive.
- I want to be able to download and reconcile my bank statements.

- I want to be able to upgrade over the Internet.
- I want to manage my stock portfolio and track my ROI.
- I'd like to have the reports I run every month saved and easy to update.
- It's a pain to set up the different drill downs every time I want to analyze my spending.
- It's clunky to transfer information between DollarWise and Excel.
- When I have a minor problem, I'd like to have easy-to-use self-help available on the Internet or in the help file.
- When it's a problem I can't solve myself, I want reasonably priced, easy to reach technical support.
- You should be making patches and bug-fixes available free on the Internet.

The first step in using this laundry list of comments is to see if there's an underlying structure embedded in them. If these many comments address only a few issues, it will simplify the understanding of what the customer actually wants from the product. While there are statistical tools to help accomplish this task (e.g., structural equation modeling, principal components analysis, factor analysis), they are quite advanced and require that substantial data be collected using well-designed survey instruments. An alternative is to create an "affinity diagram," which is a simple procedure described elsewhere in this text. After creating the affinity diagram, the following structure was identified:

1. Easy to learn.
    1.1. I want the installation to be simple.
    1.2. I want an interactive tutorial to help me get started.
    1.3. I want printed documentation.
    1.4. I want the user interface to be intuitive.
2. Easy to use quickly after I've learned it well.
    2.1. I like shortcut keys so I don't have to always click around in menus.
    2.2. I'd like to have the reports I run every month saved and easy to update.
    2.3. It's a pain to set up the different drill downs every time I want to analyze my spending.
3. Internet connectivity.
    3.1. I use the Internet to pay bills through my bank. I'd like to do this using DollarWise instead of going to my bank's Web site.
    3.2. I only have a 56K connection and DollarWise is slow on it.
    3.3. I have a high speed connection and I'd like to be able to download big databases of stock information to analyze with DollarWise.
    3.4. I want to be able to download and reconcile my bank statements.
    3.5. I want to manage my stock portfolio and track my ROI.
4. Works well with other software I own.

4.1. It's clunky to transfer information between DollarWise and Excel.
4.2. Can I link a DollarWise total to a report in my word processor?
5. Easy to maintain.
    5.1. I want to be able to upgrade over the Internet.
    5.2. You should be making patches and bug-fixes available free on the Internet.
    5.3. When I have a minor problem, I'd like to have easy-to-use self-help available on the Internet or in the help file.
    5.4. When it's a problem I can't solve myself, I want reasonably priced, easy to reach technical support.

The reduced model shows that five key factors are operationalized by the many different customer comments (Fig. 2.8).

Next, we must determine importance placed on each item by customers. There are a number of ways to do this.

- Have customers assign importance weights using a numerical scale (e.g., "How important is 'Easy self-help' on a scale between 1 and 10?").
- Have customers assign importance using a subjective scale (e.g., unimportant, important, very important, etc.).
- Have customers "spend" $100 by allocating it among the various items. In these cases it is generally easier for the customer to first allocate the $100 to the major categories, and then allocate another $100 to items within each category.
- Have customers evaluate a set of hypothetical product offerings and indicate their preference for each product by ranking the offerings, assigning a "likely to buy" rating, etc. The product offerings include a carefully selected mix of items chosen from the list of customer demands. The list is selected in such a way that the relative value the customer places *on each item in the offering* can be determined from the preference values. This is known as *conjoint analysis*, an advanced technique that is covered in most texts on marketing statistics.
- Have customers evaluate the items in pairs, assigning a preference rating to one of the items in each pair, or deciding that both items in a pair are equally important. This is less tedious if the major categories are evaluated first, then the items within each category. The evaluation can use either numeric values or descriptive labels that are converted to numeric values. The pairwise comparisons can be analyzed to derive item weights using a method known as the Analytic Hierarchical Process (AHP) to determine the relative importance assigned to all of the items. The series of matrices constructed in this AHP process are often referred to as a *prioritization matrix*, given their application in defining priorities.

Chapter 2  Recognizing and Capitalizing on Opportunity | 97

**Demands**

1.1. Simple install
1.2. Interactive tutorial
1.3. Printed documentation
1.4. Intuitive interface

2.1. Shortcut keys
2.2. Saved reports
2.3. Saved drill downs

3.1. Bank
3.2. 56K
3.3. High-speed features
3.4. Download reconcile
3.5. Manage portfolio

4.1. Excel interface
4.2. Word link

5.1. Internet upgrades
5.2. Free Internet fixes
5.3. Easy self-help
5.4. Good tech support value

**Categories**

1. Easy to learn
2. Easy to use after learned
3. Internet connectivity
4. Works with other software
5. Easy to maintain

Satisfied customer → Voice of customer

**Figure 2.8** Customer demand model.

All of the above methods have their advantages and disadvantages. The simple methods are easy to use but less powerful (i.e., the assigned weights are less likely to reflect actual weights). The more advanced conjoint and AHP methods require special skills to analyze and interpret properly. We will illustrate the use of AHP for our hypothetical software product. AHP is a powerful technique that has been proven in a wide variety of applications. In addition to its use in determining customer importance values, it is useful for decision-making in general.

## Category Importance Weights

We begin our analysis by making pairwise comparisons at the top level. The affinity diagram analysis identified five categories: easy to learn, easy to use quickly after I've learned it, Internet connectivity, works well with other software I own, and easy to maintain. Arrange these items in a matrix as shown in Fig. 2.9.

For our analysis we will assign verbal labels to our pairwise comparisons; the verbal responses will be converted into numerical values for analysis. Customers usually find it easier to assign verbal labels than numeric labels. All comparisons are made relative to the customer's goal of determining which product he will buy, which we assume is synonymous with our goal of determining where to focus product upgrade efforts. The highlighted cell in the matrix compares the "easy to learn" attribute and the "easy to use quickly after I've learned it" attribute. The customer must determine which is more important to him, or if the two attributes are of equal importance. In Fig. 2.9 this customer indicates that "easy to learn" is moderately to strongly preferred over "easy to use quickly after I've

**Figure 2.9** Matrix of categories for pairwise comparisons.
(Created using Expert Choice 2000 Software, www.expertchoice.com.)

*Although the analysis is easier with special software, you can obtain a good approximation using a spreadsheet. See Appendix 17 for details.*

|  | Easy to lea | Easy to us( | Internet co | Works well | Easy to ma |
|---|---|---|---|---|---|
| Easy to learn |  | 4.0 | 1.0 | 3.0 | 1.0 |
| Easy to use quickly after I've learned it |  |  | 0.20 | 0.33 | 0.25 |
| Internet connectivity |  |  |  | 3.0 | 3.0 |
| Works well with other software I own |  |  |  |  | 0.33 |
| Easy to maintain | Incon: 0.05 |  |  |  |  |

**Figure 2.10** Completed top-level comparison matrix.

learned it" and the software has placed a +4 in the cell comparing these two attributes. (The scale goes from 1/9 to +9, with "equal" being identified as a +1.) The remaining attributes are compared one by one, resulting in the matrix shown in Fig. 2.10. The shaded bars over the attribute labels provide a visual display of the relative importance of each major item to the customer. Numerically, the importance weights are:[2]

- Easy to learn: 0.264 (26.4%)
- Easy to use quickly after I've learned it: 0.054 (5.4%)
- Internet connectivity: 0.358 (35.8%)
- Works well with other software I own: 0.105 (10.5%)
- Easy to maintain: 0.218 (21.8%)

These relative importance weights can be used in QFD as well as in the AHP process that we are illustrating here. In our allocation of effort, we will want to emphasize those attributes with high-importance weights over those with lower weights.

## Subcategory Importance Weights

The process used for obtaining category importance weights is repeated for the items within each category. For example, the items interactive tutorial, good printed documentation, and intuitive interface are compared pairwise within the category "easy to learn." This provides weights that indicate the importance of each item on the category. For example, within the "easy to learn" category, the customer weights might be:

- Interactive tutorial: 11.7%
- Good documentation: 20.0%
- Intuitive interface: 68.3%

If there were additional levels below these subcategories, the process would be repeated for them. For example, the intuitive interface subcategory might be subdivided into "number of menus," "number of submenus," "menu items easily understood," etc. The greater the level of detail, the easier the translation of the customers' demands into internal specifications. The tradeoff is that the process quickly becomes tedious and may end up with the customer being asked for input he isn't qualified to provide. In the case of this example, we'd probably stop at the second level.

## Global Importance Weights

The subcategory weights just obtained tell us how much importance the item has with respect to the *category*, not with respect to the ultimate goal. Thus, they are often called *local importance weights*. However, the subcategory weights don't tell us the impact of the item on the overall goal, which is called its *global impact*. This is determined by multiplying the subcategory item weight by the weight of the category in which the item resides. The global weights for our example are shown in Table 2.2 in descending order.

The global importance weights are most useful for the purpose of allocating resources to the overall goal: *Determine where to focus product upgrade efforts*. For our example,

**Table 2.2**  Local and Global Importance Weights

| Category | Subcategory | Local Weight | Global Weight |
|---|---|---|---|
| Easy to learn | Intuitive interface | 68.3% | 18.0% |
| Internet connectivity | Online billpay | 43.4% | 15.5% |
| Internet connectivity | Download statements | 23.9% | 8.6% |
| Internet connectivity | Download investment information | 23.9% | 8.6% |
| Works well with other software | Hotlinks to spreadsheet | 75.0% | 7.9% |
| Easy to maintain | Free Internet patches | 35.7% | 7.8% |
| Easy to maintain | Great, free self-help technical assistance on the Internet | 30.8% | 6.7% |
| Easy to learn | Good documentation | 20.0% | 5.3% |
| Easy to maintain | Reasonably priced advanced technical support | 20.0% | 4.4% |
| Internet connectivity | Works well at 56K | 8.9% | 3.2% |
| Easy to learn | Interactive tutorial | 11.7% | 3.1% |
| Easy to maintain | Automatic Internet upgrades | 13.5% | 2.9% |
| Works well with other software | Edit reports in word processor | 25.0% | 2.6% |
| Easy to use quickly after I've learned it | Savable frequently used reports | 43.4% | 2.3% |
| Easy to use quickly after I've learned it | Shortcut keys | 23.9% | 1.3% |
| Easy to use quickly after I've learned it | Short menus showing only frequently used commands | 23.9% | 1.3% |
| Easy to use quickly after I've learned it | Macro capability | 8.9% | 0.5% |

Internet connectivity obviously has a huge customer impact. "Easy to use quickly after I've learned it" has relatively low impact. "Easy to learn" is dominated by one item: the user interface. These weights will be used to assess different proposed upgrade plans. Each plan will be evaluated on each subcategory item and assigned a value depending on how well it addresses the item. The values will be multiplied by the global weights to arrive at an overall score for the plan. The scores can be rank-ordered to provide a list that the process owner can use when making resource allocation decisions. Or, more proactively, the information can be used to develop a plan that emphasizes the most important customer demands. Table 2.3 shows part of a table that assesses project plans using the global weights. The numerical rating used in the table is 0 = No Impact, 1 = Some Impact, 3 = Moderate Impact, 5 = High Impact. Since the global weights sum to 1 (100%), the highest possible score is 5. Of the five plans evaluated, plan C has the highest score. It can be seen that plan C has a high impact on the six most important customer demands. It has at least a moderate impact on 10 of the top 11 items, with the exception of "Reasonably priced advanced technical support." These items account for almost 90% of the customer demands.

The plan's customer impact score is, of course, only one input into the decision-making process. The rigor involved usually makes the score a very valuable piece of information. It is also possible to use the same approach to incorporate other information, such as cost, timetable, feasibility, etc., into the final decision. The process owner would make pairwise comparisons of the different inputs (customer impact score, cost, feasibility, etc.) to assign weights to them, and use the weights to determine an overall plan score. Note that this process is a mixture of AHP and QFD.

**Download prioritization matrix template in Excel at www.mhprofessional.com/ssh5.**

**Download example in Excel at www.mhprofessional.com/ssh5.**

## Creative Destruction

A study of nearly a hundred years of market data by Innosight, a leading global strategy and innovation consulting firm cofounded by innovation pioneer Clayton Christensen, notes the recent fragility in the length of time companies stayed on the list of Standard and Poor's top 500 stock market capitalizations (i.e., the S&P 500 Index): While the average firm stayed on the list for 61 years in 1958, their term dropped to 25 years by 1980 and 18 years by 2010. Nearly half the S&P 500 companies have been replaced over the last decade alone (Innosight, 2012). A loss in the company's market value (including but not limited to bankruptcy) is a leading cause of the delisting, along with acquisition by another firm, or in some cases privatization.

Table 2.3  Example of Using Global Weights in Assessing Alternatives

| Item | Plan Customer Impact Score | Intuitive Interface | Online Billpay | Download Statements | Download Investment Information | Hotlinks to Spreadsheet | Free Internet Patches | Great, Free Self-help Technical Assistance on the Internet | Good Documentation | Reasonably Priced Advanced Technical Support | Works Well at 56K | Interactive Tutorial |
|---|---|---|---|---|---|---|---|---|---|---|---|---|
| Global weight | | 18.0% | 15.5% | 8.6% | 8.6% | 7.9% | 7.8% | 6.7% | 5.30% | 4.4% | 3.2% | 3.1% |
| Plan A | 3.57 | 3 | 5 | 1 | 1 | 3 | 3 | 4 | 5 | 5 | 5 | 5 |
| Plan B | 2.99 | 1 | 1 | 1 | 3 | 3 | 5 | 5 | 5 | 5 | 5 | 5 |
| Plan C | 4.15 | 5 | 5 | 5 | 5 | 5 | 5 | 3 | 3 | 1 | 3 | 3 |
| Plan D | 3.36 | 3 | 3 | 3 | 3 | 3 | 3 | 3 | 5 | 5 | 5 | 5 |
| Plan E | 2.30 | 5 | 0 | 0 | 0 | 5 | 5 | 1 | 1 | 0 | 1 | 1 |

Foster (of Innosight) asserts that a corporation's life span is influenced by their ability to balance three somewhat competing management objectives: (1) creation of new business opportunities; (2) effective and efficient operation of the core business; and (3) selling or closing of business units that are not sustainable based on their expectations for growth or return. This "create, operate, and trade" model, also known by the title of Foster's 2001 text *Creative Destruction*, can create tensions within the organization, as various units compete for resources or units are traded to divert resources to longer-term strategies. Foster cites several examples of companies that have adopted the "create, operate, and trade" model, including GE, IBM, Johnson and Johnson, and Procter and Gamble, which have each well outpaced the S&P 500 Index over the ten-year period from 2002 to 2012. Foster cites their ability to innovate as a clear differentiator, suggesting that the leadership team of an organization should consider three key issues (Innosight, 2012):

- "Are our operations world-class?" Foster asserts that operational excellence is a prerequisite to effective creation of new businesses, or trading of lackluster units, so this should be tackled first.
- "How fast do we have to change to maintain our position in the industry?" This rate of change is industry-specific, although Foster points out that markets that change more slowly than the economy are ripe for replacement by low-cost competitors, typically from abroad.
- "Do our control systems work effectively?" Here, he refers to operational controls, which maintain the standards in the core businesses.

The Innosight study suggests that the rate of creative destruction within an economy is linked to the overall economic growth. When growth is stagnant, demand for innovation through creative destruction builds. Their recommendation is that organizational leaders embrace innovation through creative destruction, since the alternative is to become a victim of the inevitable creative destruction demand of the market.

A recent Accenture survey of over 500 large organizations (in excess of $100M revenue) in Europe and the United States cited two obstacles to successful innovation: a focus on simple product extensions, rather than groundbreaking new products; and an inability to transform new ideas into a credible, working business model (Koetzier and Alon, 2013). Citing evidence that those with formal innovation systems fare much better than others, the authors recommend managing innovation as a separate value stream, with emphasis on speed and flexibility. Innovation should also concentrate on business solutions (i.e., the entire process of meeting customer needs) rather than focusing on products alone.

The innovation efforts at Procter and Gamble (P&G) are well documented, partly because of their association with the Innosight group, but more likely because of their overwhelming market success, which in turn is largely attributed to their innovation.

P&G spends nearly $2 billion annually on research and development (approximately 50% more than their closest competitor), and an additional $400 million on consumer research involving nearly 5 million consumers and hundreds of innovation teams. While only 15% of their innovation efforts met profit and revenue targets in 2000, 50% met them in 2010 (Brown and Anthony, 2011), a resounding success given the inherent risks assumed in at least a portion of innovation projects. A recent synopsis of their efforts by the P&G CTO Bruce Brown and Innosight's Scott Anthony cite the following issues for management to address to successfully spur growth through innovation:

- **Closely coordinate the innovation process with the core business**—The core business generates the cash needed for investment, so any new growth largely depends on the success of the core business. The core business is also rich with capabilities, resources, tools, and market knowledge essential to the innovation efforts.
- **Develop a portfolio of innovation projects, each of which is managed separately in accordance with the relative investment and its anticipated return**—P&G defines four fundamental types of innovation projects, with an emphasis on sustaining and transformational-sustaining projects that support and grow the core business:
    a. **Sustaining**—These are incremental improvements to existing products and services that help maintain the product or service as relevant in the market.
    b. **Commercial**—These are essentially marketing or promotional campaigns to increase market share, which may not even be tied to specific products.
    c. **Transformational-sustaining**—These represent drastic changes to an existing business that fundamentally redefine the market. An example cited was the introduction of the anti-aging cream Olay Pro-X, which (while expensive relative to other P&G brands) directly competed with more expensive prescription products.
    d. **Disruptive**—These are groundbreaking products or services that radically upset the market, such as P&G's Swiffer brand.
- **Start small and grow carefully**—Innovation projects are not quick fixes, but require long-term investment with uneven cycles of progress and much experimentation. Depending on the project type (noted above) and its relative integration with the core business, there may be projects that should be run like start-ups, entirely separate from the existing business units. Flexibility in organizational structure and resource allocation is necessary to invest in the project relative to its current needs.
- **Create new tools for gauging new businesses**—P&G developed new approaches to evaluating market potential for new products, using small-scale experimentation to gather consumer input and gauge market interest.
- **Ensure the right people are doing the right work**—Training in the tools of innovation is an important consideration, as is the allocation of experienced staff to the

project teams on a full-time basis. As noted for Six Sigma project teams elsewhere in this book, a small and nimble team with full-time membership is preferred.

- **Encourage intersections**—P&G actively works with noncompeting organizations to improve their external perspective and benchmark best practices. They also actively hire from outside the organization where needed skill sets can be quickly acquired to complement their in-house expertise. Elon Musk, CEO of innovative Tesla Motors and SpaceX, cites the transfer of ideas across his two companies, enabled by breaking down functional silos, as a key source of the innovation needed to mass-produce the longest-range all-electric automobile available (Endler, 2013). "You don't see a lot of airplanes made out of steel . . . it was an obvious thing to make a car out of aluminum, but the Model S is the only car made out of aluminum in North America," says Musk.

A key aspect of the P&G experience noted above is the portfolio approach to innovation. While breakthrough products or services successfully developed through disruptive or transformational-sustaining innovation typically yield enormous return, they often require much higher investment, over longer term, and with higher risk (i.e., a smaller probability of attaining any return).

Kanter (2006) identifies four key areas for innovation success:

- **Strategy**—Use a pyramid approach for project investment: a relatively few disruptive ideas, supplemented with many more promising midlevel projects (sized in terms of investment and return) and a wide array of sustaining projects (that require much smaller investment). This provides ongoing return from the high volume of quick-hit projects, even though each project had relatively small returns (and investment). Focus not just on new product, but also on services and processes.
- **Process**—Add flexibility to budgets and planning, so that leaders can respond rapidly to the current environment. Don't set too stringent requirements, expecting returns to match those of existing (mature) product lines. She cites the UK hit TV series *The Office* as an example of an innovation resulting from the flexibility afforded by the budget process: Money that had been put aside for special projects was diverted when the opportunity arose.
- **Structure**—Innovation should not be conducted in the dark, nor should there be undue prestige afforded those working in innovation projects. Rather, it's important to foster communication and collaboration between the core business and the innovation projects, in both directions, since the projects will benefit from capabilities that cut across the business and the resulting innovation will provide revenue to the organization as a whole.
- **Skills**—Communication and collaboration skills are fundamental to innovation, so training in these methods is clearly helpful, if not necessary. Team members should

be committed to working together for the life of the project, and midproject reassignments should be avoided.

These recommendations largely agree with the P&G findings above in all four areas. Christensen and Overdorf further develop the skills and resource requirement for project teams based on how well the innovation fits with current processes and organizational values. In this context, value is limited to the organization's propensity for risk, and appetite for growth. By definition, larger organizations require larger opportunities to maintain a specified growth rate. For example, a $10 billion organization requires $1 billion in new business for a 10% growth rate, while a $1 million company requires only $100,000 in additional revenue. In most cases, larger companies also have a higher tolerance for risk.

When the innovation fits well with both existing processes and values, the successful innovation team can be assembled from various functional areas (i.e., a cross-functional team) and managed in-house, within existing organizational structures. In any other case, the team should be dedicated full-time to the innovation project, with complete responsibility for its own management and success. If the innovation fits poorly with both existing processes and values, the project should operate in a separate spin-off or skunk-works type of operation, outside the organizational structure, to foster the adaptation of new processes within a new set of values. If value alone is the conflict and the processes mesh well, then the innovation can be developed within the organization to leverage the existing processes but then spun off for commercialization (Christensen and Overdorf, 2000).

As discussed above, innovation can take many forms, from relatively simple extensions of existing products or services to quite dramatic disruptions of markets. Yet, innovation of any magnitude is valuable simply because it provides a perceived competitive advantage for the organization. Porter (1996) suggests that *strategic positioning* emerges as a benefit to the organization when organizations make decisions to serve either the narrow interests of a broad customer base or the broad interests of a narrow customer base (or a narrow market), often due to innovation. These decisions lead to trade-offs, where the organization must choose to pursue one market at the expense of others, due to considerations such as resource restrictions or brand image. This then becomes a strategy, which can only be attained when tightly integrated with the organization's activities. This is the objective of strategic planning.

## Strategic Planning[3]

Despite the inevitability of the future, it cannot be predicted. Nonetheless, long-range planning has valuable benefits, providing opportunity for managers to critically question (1) whether the effects of present trends can be extended into the future; (2) assumptions that today's products, services, markets, and technologies will be the products, services, markets, and technologies of tomorrow; and (3) perhaps most important, the usefulness of devoting their energies and resources to the defense of yesterday (Drucker, 1974).

Traditional strategic planning starts by answering two simple questions: "What *is* our business?" and "What *should* it be?"

Strategic planning is not forecasting, which, as Drucker (1974) pointedly noted, "is not a respectable human activity and not worthwhile beyond the shortest of periods." Strategic planning is necessary precisely because we cannot forecast the future. It deliberately seeks to upset the probabilities by innovations and organizational change.

Strategic planning is the continuous process of making present entrepreneurial decisions systematically and with the greatest knowledge of their futurity, organizing systematically the efforts needed to carry out these decisions, and measuring the results of these decisions against the expectations through organized, systematic feedback (Drucker, 1974).

## *Organizational Vision*

The answer to these questions leads an organization to develop *value* and *mission* statements to explain the organization's broad (or sometimes quite specific) goals. Lean organizations may refer to this as true north; the vision defines the ideal state. The successful organization will outlive the people who are currently its members. Thus, the mission of the successful organization must provide vision for the long term, describing why the organization exists. No organization exists merely to "make a profit." Profits accrue to organizations that produce value in excess of their costs; that is, profits are an effect of productive existence, not a cause. Consider these examples:

- Matsushita electric industrial company is one of the world's largest firms. Its stated mission is to eliminate poverty in the world by making their products available to the people of the world at the lowest possible cost (Mintzberg and Quinn, 1991).
- Henry Ford's mission was to provide low-cost transportation to the common person.

One might go a step further and ask why the organization was created to fulfill its mission. The answer, at least in the beginning, might lie in the values of the organization's founder. Henry Ford, for whatever reason, felt that it was important (i.e., valued) to provide the farmer with affordable and reliable motorized transportation. Furthermore, to elicit the cooperation of the members of the organization, the values of the organization must be compatible with the values of its members.

Organizational leaders are responsible for defining the organization's vision. Defining the vision requires developing a mental image of the organization at a future time. The future organization will more closely approximate the ideal organization, where "ideal" is defined as that organization that completely achieves the organization's values. How will such an organization "look"? What will its employees do? Who will be its customers? How will it behave toward its customers, employees, and suppliers? Developing a lucid image of this organization will help the leader see how she should proceed with her primary duty of transforming the present organization. Without such an image in her mind, the executive will lead the organization through a maze with a thousand dead ends. Conversely, with her vision to guide her, the transformation process will proceed on course. This is not to say that the transformation is ever "easy." But when there is a leader with a vision, it's as if the organization is following an expert scout through hostile territory. The path is clear, but the journey is still difficult.

When an individual has a vision of where he wants to go himself, he can pursue this vision directly. However, when one is dealing with an organization, simply having a clear vision is not enough. The leader must *communicate* the vision to the other members of the organization. Communicating a vision is a much different task from communicating instructions or concrete ideas.

Organizational visions that embody abstract values are necessarily abstract in nature. To effectively convey the vision to others, the leader must convert the abstractions to concretes. One way to do this is by living the vision. The leader demonstrates her values in every action she takes, every decision she makes, which meetings she attends or ignores, when she pays rapt attention and when she doodles absentmindedly on her notepad. Employees who are trying to understand the leader's vision will pay close attention to the behavior of the leader.

Another way to communicate abstract ideas is to tell stories. In organizations there is a constant flow of events. Customers encounter the organization through employees and systems, suppliers meet with engineers, literally thousands of events take place every day. From time to time an event occurs that captures the essence of the leader's vision. A clerk provides exceptional customer service, an engineer takes a risk and makes a mistake, a supplier keeps the line running through a mighty effort. These are concrete examples of what the leader wants the future organization to become. She should repeat these stories to others and publicly recognize the people who made the stories. She should also create stories of her own, even if it requires staging an event. There is nothing dishonest about creating a situation with powerful symbolic meaning and using it to communicate a vision. For example, Nordstrom has a story about a sales clerk who accepted a customer return of a defective tire. This story has tremendous symbolic meaning because Nordstrom doesn't sell tires! The story illustrates Nordstrom's policy of allowing employees to use their own best judgment in all situations, even if they make "mistakes," and of going the extra mile to satisfy customers. However, it is doubtful that the event ever occurred. This is irrelevant. When employees hear this story during their orientation training, the message is clear. The story serves its purpose of clearly communicating an otherwise confusing abstraction.

## *Strategy Development*

After the objectives of the business are specified via the vision and mission statements, the next activity in strategic planning is to develop the strategies (i.e., the plan) to achieve these objectives. Yet planning must go beyond simply coming up with new things the business can do in the future. We must also ask of each present activity, product, process, or market, "If we weren't already doing this, would we start?" If the answer is no, then the organization should develop plans to stop doing it, ASAP.

The planning aims to make organizational changes: changes in the way people work, changes in the systems to meet customers' future needs. The plan shows how to allocate scarce resources and builds accountability. Deadlines are necessary, as are feedback on progress and measurement of the final result.

A traditional basis of strategy formation is the comparison of internal Strengths and Weaknesses to external Opportunities and Threats (SWOT). As shown in Fig. 2.11, strategy is created at the intersection of an external appraisal of the threats and opportunities facing an organization in its environment, considered in terms of key factors for success, and an internal appraisal of the strengths and weaknesses of the organization itself, distilled into a set of distinctive competencies. Outside opportunities are exploited by inside strengths, while threats are avoided (or addressed) and weaknesses circumvented (or addressed). Opportunities and threats are identified by understanding customers and their markets. Internal strengths and weaknesses are evaluated through rigorous organizational assessments. Taken into consideration, both in the creation of the strategies and in their subsequent evaluation, are the values of the leadership as well as its social responsibility.

Once a strategy has been chosen, it is implemented as a cross-functional improvement project using the improvement methods described in Part II. The structured DMAIC improvement model, an improvement to Shewhart's plan-do-study-act, effectively builds buy-in among the stakeholder groups as well as the key middle management.

Projects are defined to achieve the vision established by the leadership team, with understanding of the customer needs and expectations. Organizational assessments, as well as customer data, provide the basis for the Measure stage of DMAIC. Key drivers of the outcomes are established in the Analyze stage. An improvement strategy is implemented in the Improve stage and continually verified in the Control stage.

Larger strategies are executed primarily at the executive level. The implementation of organization-wide initiatives in the Improve stage may be best broken into smaller projects aligned with a specific substrategy and executed at lower levels of the organization.

Figure 2.11　SWOT model of strategy formulation (Mintzberg, 1994).

## *Strategic Styles*

Strategic plans are necessarily specific to a particular organization, at a particular time in its life cycle. What works for a Fortune 100 firm, in a developed market, will not likely apply directly to a start-up in a niche market. Reeves et al. (2012) define four broad categories of strategic style:

- **Classical**—This traditional approach to defining longer-term plans should be limited to organizations in fairly predictable, mature environments. Examples include oil exploration and production; air freight/logistics; beverages; utilities; paper products; and tobacco.
- **Adaptive**—This flexible approach fosters experimentation to develop strategy within unpredictable environments in which you have little ability to change. Many of today's markets are continually changing due to competition, innovation, and/or economic

uncertainty, making some aspects of a strategic plan obsolete or irrelevant after perhaps only a few months. In these environments, firms must become learning organizations, essentially integrating their strategic planning with operations to quickly develop and then track the results of each strategic iteration. Examples include biotechnology, communications equipment, specialty retail, and computer hardware.

- **Shaping**—When the environment is unpredictable, but you have some ability to change the environment, a shaping strategy is recommended. In this case, the organization develops strategy that seeks to influence the market, such as through innovation. This may disrupt a stagnant market or a fragmented market, but in either case the objective is to redefine the market to the company's advantage. The authors cite Facebook as an example of successful deployment of the shaping strategy. Their ability to overtake an initially dominant competitor in MySpace was complemented by redefining the social media space with applications, such as games. Other examples include software development, airlines, catalog retail, consumer services, and wireless services.
- **Visionary**—When an organization can both shape the environment and reliably predict the future, they can develop bold and decisive plans to create new markets or redefine an organization. Since the market is predictable, the organization can take the time and commit resources to develop and execute a complete plan. The authors cite examples such as XM satellite radio; UPS as "the enablers of global e-commerce"; aerospace and defense; and food products.

Unfortunately, the authors noted that most organizations surveyed were using the visionary (40%) or the classical (35%) strategic styles, which rely on predictable markets, when assumptions of predictability were clearly unwarranted (Reeves et al., 2012).

## *Possibilities-Based Strategic Decisions*

Clearly, defining and implementing organizational vision involves elements of creativity. Yet, a rigorous, scientific approach is necessary to ensure that a full breadth of options is explored and evaluated. Lafley et al. (2012) define the following seven steps to strategy making, which differ from traditional methods in clearly articulating possibilities:

1. **Frame a choice**—Define the issue using two or more mutually exclusive options. This moves the discussion from investigating issues to evaluating solutions and making decisions. It further ingrains the team with the notion that they have choices. It's often useful to include the status quo as an option, to explore the assumptions necessary in maintaining current practices, which often highlights the need for action.
2. **Generate possibilities**—Creatively brainstorm to develop additional options. At this point, options are not evaluated beyond general plausibility, but sufficient detail is

necessary so that the team can understand the nature of the option. Practically, the authors recommend three to five options.

3. **Specify conditions**—Define limitations of each option to describe the conditions under which the option would be strategically desirable. Note that this is not the time to argue the merits of any option, or whether these necessary conditions exist now or could exist in the future. Instead, it is an opportunity to define the issues that would have to be evaluated to make the option attractive to the team. When conditions have been completely defined for an option, the team should be in agreement that if all conditions were met, the option would be acceptable. If a given condition is desirable, but not necessary, it should be removed. The ultimate goal at this step is to understand the limitations of every option before analysis begins.

4. **Identify barriers**—Determine which conditions are least probable. The focus in this step is to identify the conditions that are most troublesome to the team members: Which conditions would you be most likely to be concerned about attaining?

5. **Design tests**—For each key barrier condition, construct tests that, when implemented, would persuade the team that the conditions can be sufficiently met. What is the standard of proof required for the team to assert, with an acceptable degree of confidence, that the condition can be met? This may involve surveys of customers or suppliers, or honest discussions with key customers or suppliers, depending on the team's confidence in the condition. The authors suggest that the team members with greatest skepticism for a given option be tasked with developing the tests for that option.

6. **Conduct the tests**—Implement the tests defined in the preceding step. The authors recommend starting with the barrier condition that the team has the least confidence in, on the premise that the condition can be quickly dismissed without additional testing. They refer to this as the "lazy man's approach" and cite it as a key driver for reducing the costs and effort of the analysis.

7. **Make the choice**—The option with the fewest key barriers will naturally surface as the preferred option by the team.

Lafley, a former chairman and CEO of Procter & Gamble, cites the use of the possibilities-based approach to strategic decision-making in P&G's transformation of the Olay product line to a premium "prestige" brand, eventually accounting for $2.5 billion in annual sales. When the option was considered of marketing a prestige product to younger clientele through mass-market channels, the following necessary conditions were defined (Lafley et al., 2012):

▲ **Industry segmentation**—A sufficiently large number of women want to "fight the seven signs of aging."

- **Industry structure**—The emerging masstige (i.e., a prestige product in mass market) segment will be at least as structurally attractive as the current mass-market segment.
- **Channel**—Mass retailers will embrace the idea of creating a masstige experience to attract prestige customers.
- **Consumers**—A pricing sweet spot exists that will induce mass consumers to pay a premium and prestige shoppers to purchase in the mass channel.
- **Business model capabilities**—P&G can create prestigelike brand positioning, packaging, and in-store promotions in the mass channel; and P&G can build strong partnerships with mass retailers to create and exploit a masstige segment.
- **Costs**—P&G can create a prestigelike product with a cost structure that enables it to hit the pricing sweet spot.
- **Competitors**—Because of channel conflict, prestige competitors will not try to follow Olay into the masstige segment; and mass competitors will find it hard to follow because the lower price point is covered by the basic Olay Complete line.

In this example, Lafley cites the pricing sweet spot as the most challenging issue. Analysis focused on evaluating the price points at which consumers perceived prestige, but were willing to purchase in the mass market. This particular challenge, and the analysis resulting from the possibility-based conditions, proved the most influential to ensuring success of the strategy.

The authors note the key differences in the possibilities-based approach compared to more traditional strategic planning:

- Rather than ask, "What should we do?" ask, "What might we do?" Whereas the former leads to hasty decisions, the latter fosters introspective thought, which can then be scrutinized.
- The specification of conditions leads the team to consider the assumptions necessary for the option to be desirable. This forces the team to imagine possibilities, rather than deal directly with the perceived limitations.
- The team is essentially tasked with focusing on defining the right questions that lead them to the best decision, rather than trying to jump to the best solution quickly. The focus on inquiry is a fundamental aspect of the scientific approach to problem solving.

## Strategic Development Using Constraint Theory[4]

More than ever before, operational leaders are finding themselves in need of system-level tools to sustain the business success they've fought so hard to achieve. One such system-level tool is constraint management. Constraint management acknowledges that quality is but one important element in the business equation. Constraint management seeks to help managers at all levels of an organization maintain proper focus on the factors that are most critical to overall success: *system constraints*. In some systems, these might be quality related. In other systems, they may extend well beyond the traditional territory of quality.

There are many types of constraints. Some are not *physical* (e.g., lack of space, not enough resources, etc.). In many cases they derive from *policies*: the laws, regulations, rules, or procedures that determine what we can or can't do. Who hasn't heard it said, "That's the way we do things around here"? Or, alternatively, "That's *not* the way we do things around here." What you're hearing is the verbalization of a policy, possibly unwritten, but accepted as traditional practice nonetheless. When a policy of any kind inhibits what we need (or want) to accomplish, it, too, constitutes a constraint.

The practice of constraint management tacitly recognizes that constraints limit what we can do in any circumstance, and it provides the vehicle to understand why this happens and what can be done about the constraints we face.

Constraint management is an outgrowth of the Theory of Constraints (TOC), a set of principles and concepts introduced by Eliyahu M. Goldratt, an Israeli physicist, in the 1980s in a book entitled *The Goal* (Goldratt, 1986). These principles and concepts are a blend of both existing and new ideas. The new ideas build upon older ones to produce a robust, holistic approach to understanding and managing complex systems. To extend the theoretical principles and concepts into application, Goldratt developed three classes of tools, which will be described in greater detail later. For now, the important point to remember is that TOC, and constraint management as a whole, constitutes a *systems management* methodology.

## The Systems Approach

What do we mean by *systems management*? Throughout the twentieth century, management thought was largely activity-oriented. In the middle 1900s, Frederick Taylor's scientific management (Taylor, 1947) focused on dividing and subdividing work into discrete tasks or activities that could be closely monitored, measured, and "tweaked" to produce the most efficient performance from each activity. By the second half of the century, the focus had enlarged somewhat to encompass managing processes composed of several activities. At some level, these processes could become quite large and complex, such as a production process, a purchasing process, or a marketing and sales process. One way of dealing with complexity is to compartmentalize it—to cut it up into "manageable bites." Organizations typically do that by creating functional departments. Each department is responsible for some function that constitutes a part of the whole system. One could even say that these "parts" are actually individual processes. This is an orderly way to come to grips with the issue of complexity.

Throughout the 1980s and early 1990s, the meteoric rise of the quality movement reinforced the idea that success lay in continuous refinement of processes. The ultimate objective became "Six Sigma," a level of defect-free performance unheard of twenty years before. Unquestionably, both commercial and noncommercial organizations needed this focus. Poor product quality (which is usually a result of faulty process quality) can bring down an organization faster than just about anything else. But many companies, despite herculean efforts and the expenditure of significant amounts of money, were disappointed to find that their payback wasn't what they expected. The idea that if you "build a better mousetrap, the world will beat a path to your door" worked exceptionally well for the companies whose overriding constraint had been product quality. Yet for other companies, the strategy seemed to be somewhat underwhelming.

Despite the admonition to "consider your internal customer," many departments still behave as if they're in a "silo" by themselves. They pay lip service to the idea, but for a variety of reasons they don't practice it very well. Their focus remains inward, on individual measures of performance and efficiency. Most efforts are spent improving the links of the supply chain, with little effort devoted to the linkages, or interfaces between links, and the operation of the chain as a whole.

## Systems Thinking

What these companies failed to appreciate was that a higher level of thinking was needed: *systems thinking*. Once the quality of individual processes is put reasonably well into line, other factors emerge to warrant attention. Consider the analogy of a football team.

Major professional sports spend a lot of time and money on process improvement, even though they probably don't look at it that way. A team owner can spend millions on a contract for a star quarterback. By applying natural talent, they expect to "improve the passing process." But in many cases, the touchdowns don't appear, despite the huge sums spent on star quarterbacks. At some point in the "process failure mode effects analysis," the coaches discover it's impossible for this highly valued quarterback to complete passes from flat on his back. They find that the offensive line needs shoring up. Or a good blocking back is needed, or a better game plan, or any number of other factors.

The point is that any organization, like a football team, succeeds or fails as a complete system, not as a collection of isolated, independent parts or processes. In the same way that a motion picture clip tells us much more about a situation than an instantaneous snapshot, systems thinking gives us a clearer picture of the whole organizational dynamic. In *The Fifth Discipline* (Senge, 1990), Peter Senge proposes that the only sustainable competitive advantage comes from transforming a company into a "learning organization." The keys to doing this, Senge maintains, are five basic disciplines that every organization striving for success must master: systems thinking, personal mastery, mental models, building a shared vision, and team learning. Guess which one he considers the most important? Though he numbers it fifth, he lists it first, and he titled his book after it.

**System Optimization Versus Process Improvement**

If one "thinks system," the question inevitably arises: What do we do with process improvement? Do we ignore it now that we're thinking at a higher level? No, process improvement is *still* important. It constitutes the building blocks upon which system performance is based. But like the football team alluded to above, once you have a "star performer" at every position, you have a challenge of a different sort: coordinating and synchronizing the efforts of every component in the system to produce the best *system* result. In other words, once the ducks are in line, the task is to make them march in step together. We refer to this as *system optimization*.

How important is it to optimize the system, rather than its component parts? Deming himself answered that question in *The New Economics for Industry, Government, Education* (Deming, 1993, pp. 53, 100). He observed

> Optimization is the process of orchestrating the efforts of all components toward achievement of the stated aim. Optimization is management's job. Everybody wins with optimization.
>
> Anything less than optimization of the system will bring eventual loss to every component in the system. Any group should have as its aim optimization of the larger system that the group operates in.
>
> The obligation of any component is to contribute its best to the system, not to maximize its own production, profit, or sales, nor any other competitive measure. Some components may operate at a loss themselves in order to optimize the whole system, including the components that take a loss.

This is a powerful indictment of the way most companies have been doing business since Frederick Taylor's time, not excluding the "quality enlightenment" era of the 1980s and 1990s. In essence, Deming said that maximizing local efficiencies everywhere in a system is not necessarily a good thing to do.

**Systems as Chains**

To express the concept of system constraints more simply, Goldratt has equated systems to chains (Goldratt, 1990, p. 53):

> We are dealing here with "chains" of actions. What determines the performance of a chain? The strength of the chain is determined by the strength of its weakest link. How many weakest links exist in a chain? As long as statistical fluctuations prevent the links from being totally identical, there is only one weakest link in a chain.

Goldratt goes on to suggest that there are as many constraints in a system as there are truly independent chains. Realistically, in most systems there aren't very many truly independent chains. The dictionary (Barnes & Noble, 1989) defines *system* as "an assemblage or combination of things or parts forming a complex or unitary whole; the structure or organization, society, business."

Thomas H. Athey defines a system as any set of components that could be seen as working together for the overall objective of the whole (Athey, 1982, p. 12). The underlying theme in these definitions is an interrelatedness or interdependency. By definition, then, a system can't have too many truly independent chains. So if there aren't too many independent chains in a particular system—whether a manufacturing, service, or government system—at any given time, only a very few variables truly determine the performance of the system.

This idea has profound implications for managers. If only a very few variables determine system performance, the complexity of managers' jobs can be dramatically reduced. Look at it in terms of the Pareto rule, which suggests that only 20 percent of a system accounts for 80 percent of the problems within it. If this is a valid conclusion, managers should be able to concentrate most of their attention on that critical 20 percent. Goldratt's concept of chains and "weakest links" takes the Pareto concept a step further: the weakest link accounts for 99 percent of the success or failure of a system to progress toward its goal (Goldratt, 1990, p. 53).

## *Basic Constraint Management Principles and Concepts*

Constraint management exhibits the theoretical foundation that Deming considered so important to effective management action (Deming, 1986). Theories can be either *descriptive* or *prescriptive*. A descriptive theory generally tells only why things are the way they are. It doesn't provide any guidance for what to *do* about the information it provides. An example of descriptive theory might be Newton's laws of gravitation, or Einstein's theories of relativity. Prescriptive theory describes, too, but it also guides through prescribed actions. Most management theories are prescriptive. The Deming philosophy prescribes through its fourteen points (Deming, 1986). Ken Blanchard's *One-Minute Manager* (Blanchard, 1988) and Peter Senge's *The Fifth Discipline* (Senge, 1990) explain prescriptive theories in detail. Constraint management is prescriptive as well. It provides a common definition of a constraint, four basic underlying assumptions, and five focusing steps to guide management action.

## Definition of a Constraint

Simply put, a constraint is *anything that limits a system* (company or agency) *in reaching its goal* (Goldratt, 1990, pp. 56–57). This is a very broad definition, because it encompasses a wide variety of possible constraining elements. Constraints could be physical (equipment, facilities, material, people), or they could be policies (laws, regulations, or the way we choose to do business—or choose *not* to do business). Frequently, policies cause physical constraints to appear.

## Types of Constraints

Identifying and breaking constraints becomes a little easier if there is an orderly way of classifying them. From the preceding discussion, we know that system constraints can be considered either physical or policy. Within those two broad categories, there are seven basic types (Schragenheim and Dettmer, 2000, Chap. 4):

- **Market**—Not enough demand for a product or service.
- **Resource**—Not enough people, equipment, or facilities to satisfy the demand for products or services.
- **Material**—Inability to obtain required materials in the quantity or quality needed to satisfy the demand for products or services.
- **Supplier/vendor**—Unreliability (inconsistency) of a supplier or vendor, or excessive lead time in responding to orders.
- **Financial**—Insufficient cash flow to sustain an operation. An example is a company that can't produce more until payment has been received for work previously completed, because they might need that revenue to purchase materials for a firm order that's waiting.
- **Knowledge/competence**—Knowledge: information or knowledge to improve business performance is not resident within the system or organization. Competence: people don't have the skills (or skill levels) necessary to perform at higher levels required to remain competitive.
- **Policy**—Any law, regulation, rule, or business practice that inhibits progress toward the system's goal.

*Note*: In most cases, a policy is most likely behind a constraint from any of the first six categories. For this reason, the theory of constraints assigns a very high importance to policy analysis, which will be discussed in greater detail under "The Logical Thinking Process," below.

Not all these types apply to all systems. Material and supplier/vendor constraints might not apply to service organizations. Market constraints are generally not relevant in not-for-profit systems, such as government agencies. But resource, financial, knowledge/competence, and policy constraints can potentially affect all types of organizations.

## Four Underlying Assumptions

Constraint management is based on four assumptions about how systems function (Schragenheim and Dettmer, 2000, Chap. 2). These assumptions are as follows:

- Every system has a goal and a finite set of necessary conditions that must be satisfied to achieve that goal. Effective effort to improve system performance is not possible without a clear understanding and consensus about what the goal and necessary conditions are.
- The sum of a system's local optima does not equal the global system optimum. In other words, the most effective system does not come from maximizing the efficiency of each system component individually, without regard to its interaction with other components.
- Very few variables—maybe only one—limit the performance of a system at any one time. This is equivalent to the "weakest link" concept discussed earlier.
- All systems are subject to logical cause and effect. There are natural and logical consequences to any action, decision, or event. For those events that have already occurred, these consequences can be visually mapped to aid in situation or problem analysis. For those decisions that have yet to occur, or which are contemplated, the outcomes of these actions, decisions, or events can be logically projected into the future and visually mapped as well.

All the descriptions and prescriptions contained in constraint management are predicated on these assumptions.

## Goal and Necessary Conditions

The first assumption above holds that every system has a goal and a set of necessary conditions that must be satisfied to achieve that goal (Schragenheim and Dettmer, 2000, Chap. 2). The philosopher Friedrich Nietzsche once observed that by losing your goal, you have lost your way. Or there is another way of putting it: *if you don't know what the destination is, then any path will do.*

While this assumption is undoubtedly valid in most cases, there are obviously some organizations that have not expended the time or effort to clearly and unequivocally define what their goal is. And even if they have defined a goal, most have not gone the extra step to define the minimum necessary conditions, or critical success factors, for achieving that goal.

For example, most for-profit companies have something financial as their goal. Goldratt contends that the goal of for-profit companies is to "make more money, now and in the future" (Goldratt, 1990, p. 12). Another way of saying this is *profitability*. This, of course, would not be an appropriate goal for a government agency, such as the Department of Defense or Department of Education. Nonfinancial goals would have to be developed for such agencies. But it works quite well for most companies engaged in commercial business.

However, having profitability as a goal isn't enough. For any organization to be profitable, and for those profits to consistently increase, there is a discrete set of necessary conditions it must satisfy. Some of these will be unique to the industry that the company is in; others will be generic to all for-profit companies. But one thing that all organizations will have in common: there will be very few of these necessary conditions, maybe fewer than five.

Necessary conditions are critical success factors. They are actually required to achieve the goal. For instance, customer satisfaction is unquestionably essential to continued progress toward a financial goal. Employee satisfaction might be considered necessary for achieving the goal. Including necessary conditions like these as part of the goal hierarchy gives it credibility, identifying it as something that is not just temporary but must be satisfied throughout the lifetime of the organization. Figure 2.12 illustrates a typical goal/necessary condition hierarchy.

Necessary conditions differ from the goal. While the goal itself usually has no limit (it's normally worded in such a way that it's not likely ever to be fully realized), necessary conditions are more finite. They might be characterized as a "zero-or-one" situation: It's either there or it isn't (a "yes-or-no" state). For example, a for-profit organization might want to make as much money as it can—no limits. But employee security and satisfaction, as a necessary condition, should be established at a well-defined minimum level. A for-profit company's goal can't expect to satisfy its employees without limit, but the organization should recognize the need to achieve a certain level of employee security and satisfaction as one minimum requirement for achieving the goal. This is not to say that all necessary conditions are zero-or-one in nature. Some, such as customer satisfaction, can be increased, and doing so can be expected to improve progress toward the goal. But even such variable necessary conditions have practical limits.

The importance of identifying a system's (organization's) goal and necessary conditions is that they become the standard by which all results are judged and all contemplated decisions are evaluated. Did yesterday's, or last month's, actions better satisfy a necessary condition or contribute to realizing the goal? If so, the organization knows that it's making progress in the right direction. Can we expect the decisions contemplated today, or next week, to advance the company toward its goal, or to satisfy a necessary condition? If so, the decision is a good one from the system perspective.

**Figure 2.12** Hierarchy of goal and necessary conditions.

## The Five Focusing Steps

Once the necessary conditions are established, constraint theory prescribes applying five "focusing steps" in order to continuously proceed inexorably toward satisfying those necessary conditions (Goldratt, 1986, p. 307).

Goldratt created the five focusing steps as a way of making sure management "keeps its eye on the ball"—what's really important to success: the system constraint. In one respect, these steps are similar to the Shewhart Cycle: plan-do-check/study-act (Deming, 1986). They constitute a continuous cycle. You don't stop after just one "rotation." The Five Focusing Steps (Schragenheim and Dettmer, 2000, Chap. 2) are as follows:

1. **Identify**—The first step is to *identify the system's constraint*. What limits system performance now? Is it inside the system (a resource or policy) or is it outside (the market, material supply, a vendor, . . . or another policy)? Once the system constraint is identified, if it can be broken without much investment, immediately do so, and revert to the first step again. If it can't be easily broken, proceed to the second step.
2. **Exploit**—*Decide how to exploit the system's constraint.* "*Exploit*" means to "get the most" out of the constraining element without additional investment. In other words, change the way you operate so that the maximum financial benefit is achieved from the constraining element. For example, if the system constraint is market demand (not enough sales), it means catering to the market so as to win more sales. On the other hand, if the constraint is an internal resource, it means using that resource in the best way to maximize its marginal contribution to profit. This might mean process quality improvement, reengineering the flow of work through the process, or changing the product mix. Exploitation of the constraint should be the kernel of tactical planning, ensuring the best performance the system can draw now. For this reason, the responsibility for exploitation lies with the line managers who must provide that plan and communicate it, so that everyone else understands the exploitation scheme for the immediate future.
3. **Subordinate**—Once the decision on how to exploit the constraint has been made, *subordinate everything else to that decision*. This is, at the same time, the most important and the most difficult of the five focusing steps to accomplish. Why is it so difficult? It requires everyone and every part of the system not directly involved with the constraint to subordinate, or "put in second place," their own cherished success measures, efficiencies, and egos. It requires everyone, from top management on down, to accept the idea that excess capacity in the system at most locations is not just acceptable—it's actually a good and necessary thing!

    Subordination formally relegates all parts of the system that are not constraints (referred to as "nonconstraints") to the role of supporters of the constraint. This can create behavioral problems at almost all levels of the company. It's very difficult for most people to accept that they and/or their part of the organization isn't just as critical to the success of the system as any other. Consequently, most people at nonconstraints will resist doing the things necessary to subordinate the rest of the system to the constraint. This is what makes the third step so difficult to accomplish.

    What makes the constraint more critical to the organization is its *relative weakness*. What distinguishes a nonconstraint is its *relative strength*, which enables it to be more flexible. So the current performance of the organization really hinges on the weak point. While the other parts of the system could do more, because of that weak point *there is no point* in doing more. Instead, the key to better performance is wisely subordinating the stronger points so that the weak point can be exploited in full.

Subordination actually redefines the objectives of every process in the system. Each process is supposed to accomplish a mission that's necessary for the ultimate achievement of the goal. But among processes there may be conflicting priorities, such as competition for the same resources. Subordinating nonconstraints actually focuses the efforts of every process on truly supporting the goal of organization. It allows the constraint to be exploited in the best way possible.

Consider a raw material warehouse. What is its objective? The storing and releasing of material is needed as a "bridge" between the time materials arrive from the vendors and the time the same materials are needed on the production floor. When a specific work center is the constraint, any materials needed by that particular work center should be released precisely at the required time. If market demand is the only constraint, any order coming in should trigger material release.

However, even if no new orders enter the system, shop foremen often like to continue working, so as to keep their efficiency high. But if the nonconstraints in a production system are properly subordinated, material should *not* be released. The material release process must be subordinated to the needs of the system constraint, not to arbitrary efficiency measurements. Maintaining the order in the warehouse is part of the subordination process. Release of materials not immediately needed for a firm order should be treated as a lower priority than the quick release of materials the constraint will soon need to fulfill a definite customer requirement.

Subordination serves to focus the efforts of the system on the things that help it to maximize its current performance. Actions that contradict the subordination rationale should be suppressed.

It's possible that, after completion of the third step, the system constraint might be broken. If so, it should be fairly obvious. Output at the system level will usually take a positive jump, and some other part of the system might start to look like a "bottleneck." If this is the case, go back to the first step and begin the Five Focusing Steps again. Identify which new factor has become the system constraint, determine how best to exploit that component, and subordinate everything else.

4. **Elevate**—However, if, after completing step 3, the original constraint is still the system constraint, at this point the best you can be assured of is that you're wringing as much productivity out of it as possible—it's not possible for the system to perform any better than it is without additional management action. In taking this action, it's necessary to proceed to the fourth step to obtain better performance from the system. That step is to *evaluate alternative ways to elevate the constraint* (or constraints, in the unlikely event that there is more than one). Elevate means to "increase capacity." If the constraint is an internal resource, this means obtaining more time for that resource to do productive work. Some typical alternatives for doing this might be to acquire more

machines or people, or to add overtime or shifts until all 24 hours of the day are used. If the constraint is market demand (lack of sales), elevation might mean investing in an advertising campaign, or a new-product introduction to boost sales. In any case, elevating invariably means "spend more money to make more money."

Notice the use of the word "evaluate" in this step. This word is emphasized for a good reason. From the preceding examples—buying more equipment or adding shifts, or overtime—it should be clear that there's more than one way to skin a cat. Some alternatives are less expensive than others. Some alternatives are more attractive for reasons that can't be measured directly in financial terms (easier to manage, for example). In any case, a choice on the means to elevate will usually be required, so jumping on the first option that you think of might not necessarily be a good idea.

One of the reasons to favor one elevation alternative over another is the identity of the next potential constraint. Constraints don't "go away" per se. When a constraint is broken, some other factor, either internal or external to the system, becomes the new system constraint—albeit at a higher level of overall system performance, but a constraint nonetheless. It's possible that the next potential constraint might be more difficult to manage than the one we currently have—it might reduce the margin of control we have over our system.

It's also possible these alternatives might drive the system constraint to different locations, one of which might be preferable to the other. Or dealing with the potential new constraint might require a much longer lead time than breaking the current constraint. In this case, if we decided to break the current constraint, we would want to get a head start on the tasks needed to exercise some control over the new constraint.

***Ineffective Elevation: An Example.*** For example, one company involved in the manufacture of solid-state circuit boards found its constraint to be the first step in its process: a surface-mount (gaseous diffusion) machine (Schragenheim and Dettmer, 2000, Chap. 2). Without considering which other resource might become the new constraint, they opted to purchase another surface-mount machine. This certainly relieved the original constraint. But the automated test equipment (ATE)—about eight steps down the production line—became the new constraint, and managing the constraint at this location was no easy task. It was more complex to schedule at that point, and it suffered more problems. Moreover, moving the constraint out of the ATE section was even more challenging. Buying more ATE was more expensive than buying additional surface-mount equipment. Finding qualified ATE operators was also more difficult.

In short, it took more time, effort, and money to manage or break the ATE constraint than it did to break the surface-mount constraint. Had the company been able to anticipate that ATE would become the system constraint, they could have chosen to either (1) leave the constraint where it was—at the surface-mount machine, or (2) begin long-lead-time acquisition of ATE and ATE operators to boost the ATE section's capacity *before* increasing the surface-mount capacity. Doing so would have increased system performance, yet preserved the system constraint at a location that was far easier to manage.

Another important factor to consider is return on investment. Once the company described above broke the surface-mount constraint, there was potential to generate more throughput, but how much? If the ATE's capacity was only slightly more than that of the original surface-mount machine, the company might have gained only a small increase in throughput as a payback for the cost of the new surface-mount unit. This could become a definite disappointment.

As long as the next constraint poses a substantially higher limit than the existing one, it's probably safe to say that the company did the right thing. Even if exploiting the ATE is more difficult, the increase in throughput might be worth the aggravation. The ATE could always be loaded a little less, and the company would still realize more money. What's the lesson here? Assessing the real return on investment from an *elevation* action requires an understanding of constraint theory, where the next constraint will be, and how much throughput will increase before hitting the next constraint. So the "evaluate" part of the elevation step can be extremely important. It's important to know where the new constraint will occur, because it could affect our decision on how to elevate.

*How to Determine Where the Next Constraint Will Be.* The easiest way to do this is to apply the first three of the five focusing steps "in our heads," before actually elevating for the first time. In other words, identify the next most-limiting factor, inside or outside the system, that will keep the whole system from achieving better performance after the current constraint is broken. Then determine what actions will be necessary to exploit that new constraint in the future and how the rest of the system will have to act to subordinate itself to the exploitation of the new constraint.

Once this is done, the ramifications of each alternative to elevate should be obvious, and a better-informed decision is possible about which alternative to choose—and it might not be the obvious choice or the cheapest one!

5. **Go back to step 1, but beware of "inertia"**—Even if the *exploit* and *subordinate* steps don't break the system constraint, the *elevate* step very likely will, unless a conscious decision is made to curtail elevation actions short of that point. In either case, after the subordinate or elevate steps it's important to go back to the first step (identify) to verify where the new system constraint is, or to determine that is has not migrated away from the original location. Sometimes a constraint moves, not as a result of intentional actions, but as a result of a change in the environment. For instance, a

change in preferences of the market might drive a company to change its product mix to such an extent that the constraint moves elsewhere. While such external changes don't happen very frequently, it's worth the effort to go back to the first step from time to time, just to verify that what we believe to be the constraint still is, in fact, the system's limiting factor.

The warning about inertia says: "Don't become complacent." There are two reasons for this. First, when the constraint moves, the actions or policies we put into place to exploit and subordinate the rest of the system to the "old" constraint may no longer be the best things to do for the benefit of the whole system. If we hadn't reevaluated where the new system constraint was, this deficiency would never have been noticed. Second, there is often a tendency to say, "Well, we've solved that problem. There's no need to revisit it again." But today's solution eventually becomes tomorrow's historical curiosity. An organization that's too lazy (or distracted by other demands for its attention) to revisit old solutions can be sure that eventually—probably sooner, rather than later—it won't be getting the best possible performance from its system.

## *Tools of Constraint Management*

Success or failure in any endeavor often relies on the selection and proper use of the right tools. Constraint management is no exception. While the Five Focusing Steps are effective guidelines for the tactical and strategic management of any kind of system, in specific situations the nature of constraints and the problems associated with them call for different tools and procedures. Exploiting a constraint would be done differently in a service environment than in a production process. Subordination would be different in a heavy manufacturing company that produces standardized products than it would be in a small job shop. Wouldn't it be useful to have an aid that could point us toward the right constraint management actions for each situation?

### The Logical Thinking Process

With so many different kinds of constraints, and with policy constraints underlying most of them, how can we identify what specific changes we should be working on? Many of these constraints aren't easy to identify. Often, they're not physical, or they're not easy to measure. They sometimes extend beyond the boundaries of production processes alone, although they still affect manufacturing, and sometimes—especially if they're policies—they pervade the whole organization.

To facilitate the analysis of complex systems, Goldratt created a logical thinking process. The thinking process is composed of six logic diagrams, or "trees" (Dettmer, 1997, 1998). It was specifically designed to analyze the policies of an organization and determine which one(s) might constitute a constraint to better performance.

This thinking process is unique from one perspective: it's one of the few (maybe the only) problem-solving methodologies that goes beyond problem identification and solution generation, and into solution verification and implementation planning. The components of the thinking process include the following:

1.  **The Current Reality Tree (CRT)**—Designed to help identify the system constraint, especially when that constraint is a policy of some kind. Figure 2.13 shows an example of a typical Current Reality Tree.

**Figure 2.13** Current Reality Tree—manufacturing example.
(Adapted from Schragenheim and Dettmer, 2000.)

2. **The "Evaporating Cloud" (EC)**—A kind of conflict resolution diagram. It helps create breakthrough solutions to resolve hidden, underlying conflicts that tend to perpetuate the constraint. Figure 2.14 illustrates a typical Evaporating Cloud.

**Assumptions:**
1. The *fast* supplier is more expensive than other suppliers.
2. Buying from the *fast* supplier is *always* more costly.
3. Purchase costs are significant.
4. All cost savings are important.

```
                    R1
              Keep material  ←——————  P1
              costs as low            Don't buy from
              as possible             the fast supplier
                  ↑
         Injection #1: Establish
         an "emergency" stock
    O    level; reorder from    Injection #2:
  High due-date  fast supplier when  Emergency stock
  performance   penetrated.     is set to a level at
  at low costs                  which it's not      (Conflict)
         Injection #3: Accept raw  likely to be
         material cost increases as exhausted within
         long as ΔT remains positive. the fast
                                  supplier's QLT.
                                          ↓
                    R2                    P2
              Ensure material  ←——————  Buy from the
              availability            fast supplier
```

**Assumptions:**
5. The regular supplier is unreliable.
6. The regular supplier takes too long to deliver.
7. We never know about peak demands in time to order from the regular supplier.
8. Purchase cost is less important than the cost of a missed delivery.

**Figure 2.14** "Evaporating Cloud"—unreliable supplier. (Adapted from Schragenheim and Dettmer, 2000.)

3. **The Future Reality Tree (FRT)**—Tests and validates potential solutions. It provides logical verification that a proposed solution will actually deliver the desired results. Figure 2.15 depicts a Future Reality Tree.

**Figure 2.15** Future Reality Tree—"red-line control" (manufacturing control). (Adapted from Schragenheim and Dettmer, 2000.)

**130** | The Six Sigma Handbook

4. **The Negative Branch (NB)**—Actually a subset of the FRT. It helps identify and avoid any new, devastating effects that might result from the solution. Figure 2.16 represents a notional example of a Negative Branch, and how it might have been used to anticipate the disastrous consequences of a very high-profile decision. Notice that this example underscores the fact that application of the thinking process tools is not confined to commercial business situations alone.

**Figure 2.16** Los Angeles District Attorney's Negative Branch— "If it does not fit, you must acquit." (Adapted from Dettmer, 1998.)

Chapter 2 Recognizing and Capitalizing on Opportunity | 131

5. **The Prerequisite Tree (PRT)**—It helps to surface and eliminate obstacles to implementation of a chosen solution. Also it time-sequences the actions required to achieve the objective. Figure 2.17 shows a typical Prerequisite Tree.

**Figure 2.17** Prerequisite Tree—example. (Adapted from Dettmer, 1997.)

**6. The Transition Tree (TT)**—It can facilitate the development of step-by-step implementation plans. Also it helps explain the rationale for the proposed actions to those responsible for implementing them. This can be especially important when those charged with executing a plan are not the same people who developed it. Figure 2.18 contains a typical Transition Tree. Either the Transition Tree or the Prerequisite Tree can form the basis of a project activity network for implementation of change.

**Figure 2.18** Transition Tree—example. (Adapted from Dettmer, 1997.)

Four of these trees—the Current Reality Tree, Future Reality Tree, Negative Branch, and Transition Tree are cause-and-effect trees. They're read using "If . . . then . . . ." The Evaporating Cloud and Prerequisite Tree are referred to as "necessary condition" trees. They're read a little differently, using "In order to have . . . we must . . . ."

These tools are specifically designed to help answer the three major questions inherent in the first three of the five focusing steps:

▲ *What* to change?
▲ What to change *to*?
▲ How to *cause* the change?

## "Drum-Buffer-Rope" Production Scheduling

Probably the best-known of the constraint management tools developed by Goldratt is called "Drum-Buffer-Rope" (DBR) (Figure 2.19). The origin of this name dates back to the analogy Goldratt and Cox used in *The Goal* (Goldratt, 1986) to describe a system with dependencies and statistical fluctuations. The analogy was a description of a boy scout hike. The drum was the pace of the slowest boy scout, which dictated the pace for the others. The buffer and rope were additional means to ensure all the boy scouts walked at approximately the pace of the slowest boy.

Goldratt and Fox, in *The Race* (Goldratt and Fox, 1986), describe in detail the manufacturing procedure that stems from the concepts of a drum, buffer, and rope originally

**Figure 2.19** Basic DBR concept. (Adapted from Schragenheim and Dettmer, 2000.)

introduced through the boy scout hike. The DBR method provides the means for synchronizing an entire manufacturing process with the "weakest link" in the production chain. Figure 2.19 illustrates the DBR concept.

***The "Drum."*** In a manufacturing or service company, the "drum" is the schedule for the resource or work center with the most limited capacity: the capacity-constrained resource (CCR). The reason the CCR is so important is that it determines the maximum possible output of the entire production system. It also represents the whole system's output, since the system can't produce any more than its least-capable resource.

The "buffer" and the "rope" ensure that this resource is neither starved for work nor overloaded (causing backlogs). In constraint management, buffers are composed of *time*, not *things*.

***The "Buffer."*** Starvation can result from upstream process variability, which might delay the transfer of work-in-process beyond its expected time. To ensure a CCR is not starved for work, a buffer time is established to protect against variability. This is a period of time in advance of the scheduled "start processing" time that a particular job arrives at the CCR. For example, if the CCR schedule calls for this valuable resource to begin processing a particular work order at 3:00 p.m. on Tuesday, the material for that job might be released early enough to allow all preceding processing steps to be completed at 3:00 p.m. on Monday (a full workday ahead of the time required). The buffer time serves to "protect" the most valuable resource from having no work to do—a serious failing, since the output of this resource is equivalent to the output of the entire system.

Note that only critical points in a service or production process are protected by buffers (refer to Fig. 2.19). These critical points are the capacity-constrained resource, any subsequent process step where assembly with other parts occurs, and the shipping schedule. Because the protection against variability is concentrated only at the most critical places (and eliminated everywhere else in the process), actual lead time can be shortened considerably, sometimes by 50% or more, without compromising due-date reliability. Shorter lead times and higher delivery reliability are important service characteristics that customers often look for.

***The "Rope."*** The rope is constraint management's safeguard against overloading the CCR. In essence, it's a material release schedule that prevents work from being introduced into the system at a rate faster than the CCR can process it. The rope concept is designed to prevent the backlog of work at most points in the system (other than the planned buffers at the critical protected points). This is important, because work-in-process queues are one of the chief causes of long delivery lead times.

When the entire drum-buffer-rope concept is applied, delivery reliability of 100% is not an unreasonable target, and actual lead time reductions of 70% are common. Bethlehem Steel's Sparrows Point plant increased delivery reliability from 49% to nearly

100% while they reduced actual lead times from 16 weeks to about 4 weeks (Dettmer, 1998, Chap. 1).

## Critical Chain

Another valuable asset in the constraint management toolbox is called "critical chain." Also the title of a book by Goldratt (Goldratt, 1997), the critical chain concept provides an effective way to schedule project activities by accommodating uncertainty and resolving simultaneous needs (contentions) for the same resource. Critical chain constitutes the application to one-time projects of the same principles that DBR applies to repetitive production. The result of applying critical chain scheduling and resource allocation is a higher probability of completing projects on time and, in some cases, actually shortening total project duration. Originally applied to the management of a single project, the critical chain method has been expanded to multiproject environments, based on the concept of the "drum," described in "Drum-Buffer-Rope," above.

Since projects aren't quite the same as repetitive production, some differences in employing critical chain project planning are inevitable. But the concepts are much the same as those of DBR. What distinguishes critical chain from PERT/CPM and other traditional project management approaches?

First, critical chain recognizes and accounts for some human behavioral phenomena that traditional project management methods don't (Leach, 2000; Newbold, 1998). These phenomena include

- The tendency of technical professionals to "pad" their time estimates for individual tasks, in an effort to protect themselves from late completion.
- The so-called student syndrome—waiting until the last minute to begin work on a task with a deadline.
- Parkinson's law (ensuring that an activity consumes every bit of the estimated time, no matter how quickly the associated tasks can actually be completed).
- Multitasking—the tendency of management to assign people more than one deadline activity simultaneously. Multitasking can create a devastating effect. Project personnel switch back and forth between several tasks, causing "drag" in all of them. The result is that other resources that depend on these task completions for their inputs are delayed. Delays cascade when several simultaneous projects are involved.

The delays that result from this kind of behavior are caused by a flawed management assumption: *The only way to ensure on-time completion of a project is to ensure that every activity will finish on time.* This common management belief prompts management to attribute unwarranted importance to meeting the scheduled completion time for each separate activity in a project. People then pad their estimates of task times to ensure they can get everything done in time. But at the same time, they try to not finish much earlier than their own inflated estimation, so as not to be held to shorter task times on subsequent projects. All these "human machinations" cause a vicious circle, dragging projects out longer and longer, but the reliability of meeting the original schedules isn't improved.

Second, to solve this problem, critical chain takes most of the protective time out of each individual activity and positions some of it at key points in the project activity network: at convergence points and just ahead of project delivery. Since accumulating protection on an entire chain is much more effective than protecting every activity, only half of the aggregated "protective pad" extracted from individual activities is put back in at the key locations. The rest can contribute to earlier project completion. In traditional project execution, if protective time in a specific activity isn't used, it's lost forever— unusable by later activities that might need more protection than they were originally assigned. This formerly "lost time" is, in many cases, usable in critical chain.

Third, critical chain devotes more attention to the availability of critical resources when they're needed for specific activities. Leveling the resources on any single project is mandatory. The Critical Chain is really the longest sequence in the project that considers *both* dependent, sequential activity links *and* resource links. The critical path reflects only the sequential linking of dependent tasks.

The key elements of critical chain project management include the following:

- ▲ **The critical chain**—The set of tasks that determines project duration, considering both task precedence *and* resource dependencies (Newbold, 1998). The longest sequence of dependent activities is the "critical path" in a PERT/CPM approach. But when the duration of this sequence is adjusted for optimum resource availability (resource leveling), the whole definition of the critical path becomes irrelevant. What results is a different time duration, based on resource usage. This is the critical chain. (Refer to Fig. 2.20.) Because the critical chain now constitutes the constraint that determines the earliest date that a project can finish, it's crucial to monitor progress along the critical chain, because it reflects the progress of the entire project.
- ▲ **The project buffer**—A project buffer is established at the end of the final activity on the critical chain, and before the required delivery date. It's designed to protect against extreme variability and uncertainty that may impact the critical chain.
- ▲ **Feeding buffers**—Each activity (or sequence of activities) that feeds the critical chain is buffered with some reasonable amount of time to protect against variability in that particular task or sequence of tasks. Feeding buffers protect the critical chain from delays occurring in activities that are not on the critical chain. Variability is *all* that these buffers protect against. They don't protect against *multitasking*—the tendency of organizations to assign technical personnel tasks for different projects or obligations

Chapter 2 Recognizing and Capitalizing on Opportunity | 137

**Figure 2.20** Critical chain versus critical path.

simultaneously. In fact, one of the determinants of success or failure in applying critical chain is the willingness of management to move away from multitasking.
- **Buffer management**—A means of control that, at any time in the project life, affords project managers the opportunity to determine how much of various buffers has been used and to take action as soon as the project appears to be in danger of exceeding its scheduled time. Task completion delays will cause "penetration" into buffers, but warning of this happening comes so much earlier that it's often possible to prevent schedule overruns sooner, with considerably less (or less extreme) corrective action. Monitoring the buffers, especially the project buffer, results in a higher probability that the project will complete on time.
- **The drum**—This concept, similar to that of DBR, applies only to multiproject situations. The drum concept requires choosing one of the most heavily loaded resources as the "drum" and staggering multiple projects according to the availability of that resource.

The ultimate effects of critical chain project management are higher delivery due-date reliability, more frequent earlier delivery, less "crashing," and conservation of project costs. Buffer management, in particular, provides much better focused information on current project status.

## Constraint Management Measurements

One of the unique contributions of the theory of constraints to the management body of knowledge is the measurements used to assess progress toward the system's goal. Goldratt recognized some deficiencies inherent in traditional measurement systems and conceived of a different—more reliable—way to measure results and evaluate decisions.

### Dilemma: System or Process?

The measurement issue harks back to the earlier discussion about systems versus processes and the fallacy of assuming that the sum of local efficiencies is the system optimum. Traditional rationale maintains that achieving the highest possible productivity in every discrete function of the system equates to good management. Productivity is typically represented as the ratio of outputs to inputs. These inputs and outputs are sometimes expressed in financial terms. Managers often spend inordinate time chasing higher productivity for their own departments, without much concern for whether the whole system benefits. This underscores the heart of the problem: How can we be sure that the decisions we make day to day truly benefit the system as a whole? In other words, how can local decisions be related to the global performance of the company?

This is not necessarily an easy question to answer. Consider yourself a production manager for a moment. A sales manager comes to you and asks you to interrupt your current production run (i.e., break a setup) to process a small but urgent order for a customer. How will doing what the sales manager wants affect the company's bottom line? What will it cost to break the setup (both financially and to the production manager's productivity figures)? How much will it benefit the company? Or the production department? These are not easy questions to answer, yet throughout many companies people are called upon to make such decisions daily.

**New Financial Measures**

Assuming that a company's goal is to make more money, Goldratt conceived of three simple financial measures to ensure that local decisions line up effectively with this goal. These measures are easy to apply by anyone at virtually any level of a company: *throughput*, *inventory* or *investment*, and *operating expense* (Goldratt, 1990, pp. 19–51).

*Throughput T* is defined as the rate at which a system generates money through sales (Goldratt, 1990, p. 19). Another way to think about it is as the marginal contribution of sales to profit. Throughput can be assessed for the entire company over some period of time, or it can be broken out by product line, or even by individual unit of product sold. Mathematically, throughput equates to sales revenue (SR) minus variable cost (VC).

$$T = SR - VC$$

*Inventory* (I) is defined as all the money the system invests in purchasing things it intends to sell (presumably after adding some value to them) (Goldratt, 1990, p. 23). Because Goldratt's concept of inventory includes fixed assets, such as equipment, facilities, and real estate, the term *I* has come to represent "investment" as well, rather than just "inventory" alone. Inventory/investment certainly includes the materials that the company will turn into finished products or services. But it also includes the assets of a company, which are eventually sold off at depreciated or scrap value and replaced with new assets. This is even true of factory buildings themselves. However, for day-to-day decisions, most managers consider *I* to represent the consumable inventory of materials that will be used to produce finished products or services.

*Operating expense* (OE) is defined as all the money the system spends turning inventory into throughput (Goldratt, 1990, p. 29). Notice that overhead is not included in the throughput formula (or the definition). Overhead, and most other kinds of fixed costs, is included in operating expense. Constraint management measurements deliberately segregate fixed costs from the throughput calculation for a valid reason: allocating fixed costs to units of product sold produces a distorted concept of actual product costs in most day-to-day situations.

For example, let's say you're a small manufacturer of precision-machined parts. You're working on an order for 100 units of a particular part for an original equipment manufacturer. In the middle of this run, the customer calls and asks you to add ten more units to the order. This will increase your production time by 53 minutes. How much more have these additional ten units cost you? As long as you're not backlogged with work, the cost of the extra units is the value of the raw materials alone! You didn't pay any more in salary to the machine operator (he or she works by the hour, not by the piece). In most cases you wouldn't pay any more in electricity costs. You have to turn the lights on for business for the whole day anyway. And the cost of the general manager's company car didn't change just because you produced ten additional units of the product. The real cost of the increase in production volume is limited primarily to the cost of materials alone.

Labor costs are also considered an operating expense, because in almost all cases they are paid by some fixed unit of time, not by the individual unit of product produced. We pay people by the hour, week, month, or year, whether they are actively producing a product for sale or not. Moreover, the capacity to produce a product or service (the resources: people, facilities, equipment, etc.) is obtained in "chunks." It's really difficult to hire six-tenths of a person, or to buy three-quarters of a machine.

So the expenditure of cost for capacity usually comes in sizable increments—a step function. Products, on the other hand, are normally priced and sold by the unit—smaller steps, perhaps, but closer to a continuous function. All this makes it difficult to attach an accurate allocation of fixed costs to a unit of product. This means that those who do so obtain a distorted impression of product costs—not a good basis from which to make operating decisions.

For daily management decisions, which we'd like to be able to relate to the system's goal, $T$, $I$, and OE are much more useful than the traditional organizational success measures of net profit (NP), return on investment (ROI), and cash flow (CF). Yet there has to be a connection between the two types of financial measures. And here it is:

$$NP = T - OE$$
$$ROI = (T - OE)/I$$
$$CF = T - OE \pm \Delta I$$

Net profit is the difference between throughput and operating expense. Return on investment is net profit (throughput minus operating expense) divided by inventory/investment. And cash flow is net profit (throughput minus operating expense) plus or minus the change in inventory.

If throughput is increased, net profit increases, even if operating expense remains the same. If operating expense is reduced, net profit also increases (as long as throughput at least remains constant). If inventory is reduced, ROI increases, even if there are no changes to throughput or operating expense. These measures relate to operational management decisions much better than net profit and return on investment, keeping daily decisions more in line with the system's goal than abstract efficiency measures such as machine utilization, units produced per day/week, etc.

For example, here's how a manager might use $T$, $I$, and OE to evaluate a decision he or she is contemplating (Dettmer, 1998, p. 33):

- Will the decision result in a better use of the worst-constrained resource (i.e., more units of product available to sell in the same or less time)?
- Will it make full use of the worst-constrained resource?
- Will total sales revenue increase because of the decision?
- Will it speed up delivery to customers?
- Will it provide a characteristic of product or service that our competitors don't have (e.g., speed of delivery)?
- Will it win repeat or new business for us?
- Will it reduce scrap or rework?
- Will it reduce warranty or replacement costs?
- Will we be able to divert some people to do other work (work we couldn't do before) that we can charge customers for? If it does any of these things, the decision will improve throughput.
- Will we need less raw material or purchased parts?
- Will we be able to keep less material on hand?
- Will it reduce work-in-process?
- Will we need less capital facilities or equipment to do the same work? If the answer is yes, the decision will reduce inventory or investment.
- Will overhead go down?
- Will payments to vendors decrease?

If so, the decision will decrease operating expense.

Let's return to the example mentioned earlier about the production manager faced with a decision to break a setup in response to a sales manager's request. Using conventional reasoning, an efficiency-oriented production manager would see her productivity figures suffering because of the time lost to doing the new setup. Inserting this urgent order would also disrupt a formal schedule, slipping every subsequent order's scheduled starting and finishing times.

But a constraint-oriented manager would look at it a little differently. His first question would be, "Am I internally constrained by a shortage of resources?" The answer should be fairly obvious, because the manager would already be aware of the size of the backlog, if any. If the answer is no, the production process has excess capacity to be able to accommodate the urgent order without delaying scheduled work. The cost of the additional order is only the raw materials used and possibly the loss of materials from a unit of the job currently on the machine when the setup is broken. The machine operators' time doesn't cost any more. They're paid by the hour, whether they're doing setups or producing products. The direct increase in net profit to the company by agreeing to the sales manager's request would be the throughput (sales revenue minus variable costs) for the new order. And it's possible that providing this expedited service to a customer might win more repeat business in the future. So as long as the manufacturing process is not internally constrained, the production manager cognizant of constraint theory would probably accept this new job, while a traditional manager might tell the salesperson, "Get in line! We operate on a first-in, first-out basis." And this manager's "numbers" would look good at the end of the month—but what would that decision have done for the company?

The use of $T$, $I$, and OE for making management decisions is not a replacement for generally accepted accounting procedures (GAAP). Those are required, and will probably always be, for external reporting purposes—annual reports to stockholders, securities and exchange filings, tax reporting, etc. But throughput, inventory, and operating expense are considerably easier for most line managers to understand and use in gauging the financial effects of their daily decisions.

### The Strategic Implications of $T$, $I$, and OE

What makes throughput, inventory/investment, and operating expense even more beneficial is the strategic implications of their application by senior managers and executives. Of course, every company executive wants to improve net profit, return on investment, and cash flow. However, in standard accounting $T$, $I$, and OE are "embedded" in these terms and often difficult to single out, even though they are better managed separately.

Consider the bar graph in Fig. 2.21. It shows three bars representing $T$, $I$, and OE. Most companies devote extraordinary effort to reducing costs (both fixed and variable). The cost-of-quality concept emphasizes this as a justification for pursuing quality in the first place. Lean manufacturing does, too. The same is true of reducing inventory. In fact, in many corporate strategies, cost saving features high on the priority list.

The graph tells us a different story, however. While there might well be savings to be had in these areas (OE and $I$ reductions), these savings have a point of diminishing returns. There's a practical level below which neither inventory nor operating expense can be reduced without hurting a company's ability to generate throughput. Beyond that point, you're not cutting "fat" anymore, you're cutting "muscle." And the truth is that most managers can't really say where that point lies. Moreover, many companies have been engaged in cost reduction efforts for so long that all the OE (and even the variable costs of throughput) have largely been wrung out of the system. To improve financial performance, there's nowhere else to go except to throughput. And look at the throughput bar in

**Figure 2.21** Limits to T, I, and OE and why chasing cost reduction is nonproductive. (Schragenheim and Dettmer, 2000.)

Fig. 2.21. The improvement potential may not be infinite, but from a practical perspective there's a lot more potential to improve profitability by increasing T than there ever will be by reducing I and OE. So what does it make sense to prioritize: rearranging the deck chairs on the Titanic, or steering away from the iceberg? In summary, not many companies can "save their way" to prosperity.

Using T, I, and OE is variously referred to as throughput accounting, constraint accounting, or throughput-based decision support. They aren't rigorous enough to replace GAAP for standard accounting needs. But they're simpler and usually more effective for decision-making than traditional management accounting. More details on how to use T, I, and OE may be found in Goldratt (1990), Noreen et al. (1995), Corbett (1998), and Smith (2000).

## *Summary and Conclusion*

To summarize, constraint management:

- Is a systems management methodology
- Separates the "critical few" from the "trivial many"
- Emphasizes attention to the critical few factors that determine system success. It's based on four underlying assumptions:

- Every system has a goal and a finite set of necessary conditions that must be satisfied to achieve that goal.
- The sum of a system's local optima does not equal the global system optimum.
- Very few variables—maybe only one—limit the performance of a system at any one time.
- All systems are subject to logical cause and effect.

The theory of constraints, as conceived by Goldratt, is embodied in the Five Focusing Steps (Identify, Exploit, Subordinate, Elevate, Repeat/Inertia), a set of financial progress measurements (throughput, inventory, and operating expense), and three generic tools (the logical thinking process, drum-buffer-rope, and critical chain) that can be applied in a variety of organizational situations.

# CHAPTER 3

# Data-Driven Management

Management decisions based on objective data is a key component of a Six Sigma initiative. On a project-by-project basis, Six Sigma projects provide a means of analyzing process data to achieve process improvements. In the larger organizational view, these process improvements are initiated so that the organization can achieve its organizational priorities. The priorities are developed based on analysis of key stakeholder needs and wants, including the customer, shareholder, and employee groups. In this way, data-driven management provides a means of achieving organizational objectives by quantifying needs or wants of stakeholder groups relative to current baselines, and acting upon the data to reduce those critical gaps in performance.

## Attributes of Good Metrics

The choice of what to measure is crucial to the success of the organization. Improperly chosen metrics lead to suboptimal behavior and can lead people away from the organization's goals instead of toward them. Joiner (1994) suggests three system-wide measures of performance—overall customer satisfaction, total cycle time, and first-pass quality. An effective metric for quantifying first-pass quality is total cost of poor quality (later described in this chapter). Once chosen, the metrics must be communicated to the members of the organization. To be useful, the employee must be able to influence the metric through his performance, and it must be clear precisely how the employee's performance influences the metric.

Rose (1995) lists the following attributes of good metrics:

- ▲ They are customer centered and focused on indicators that provide value to customers, such as product quality, service dependability, and timeliness of delivery, or are associated with internal work processes that address system cost reduction, waste reduction, coordination and teamwork, innovation, and customer satisfaction.
- ▲ They measure performance across time, which shows trends rather than snapshots.

- They provide direct information at the level at which they are applied. No further processing or analysis is required to determine meaning.
- They are linked with the organization's mission, strategies, and actions. They contribute to organizational direction and control.
- They are collaboratively developed by teams of people who provide, collect, process, and use the data.

Rose also presents a performance measurement model consisting of eight steps:

- **Step 1: Performance category**—This category is the fundamental division of organizational performance that answers the question: What do we do? Sources for determining performance categories include an organization's strategic vision, core competencies, or mission statement. An organization will probably identify several performance categories. These categories define the organization at the level at which it is being measured.
- **Step 2: Performance goal**—The goal statement is an operational definition of the desired state of the performance category. It provides the target for the performance category and, therefore, should be expressed in explicit, action-oriented terms. An initial goal statement might be right on the mark, so complex that it needs further division of the performance category, or so narrowly drawn that it needs some combination of performance categories. It might be necessary to go back and forth between the performance goals in this step and the performance categories in step 1 before a satisfactory result is found for both.
- **Step 3: Performance indicator**—This is the most important step in the model because this is where progress toward the performance goal is disclosed. Here irrelevant measures are swept aside if they do not respond to an organizational goal. This is where the critical measures—those that communicate what is important and set the course toward organizational success—are established. Each goal will have one or more indicators, and each indicator must include an operational definition that prescribes the indicator's intent and makes its role in achieving the performance goal clear. The scope of the indicator might be viewed differently at various levels in the organization.
- **Step 4: Elements of measure**—These elements are the basic components that determine how well the organization meets the performance indicator. They are the measurement data sources—what is actually measured—and are controlled by the organization. Attempting to measure things that are beyond organizational control is a futile diversion of resources and energy because the organization is not in a position to respond to the information collected. This would be best handled in the next step.
- **Step 5: Parameters**—These are the external considerations that influence the elements of measure in some way, such as context, constraint, and boundary. They are

not controlled by the organization but are powerful factors in determining how the elements of measure will be used. If measurement data analysis indicates that these external considerations present serious roadblocks for organizational progress, a policy change action could be generated.

▲ **Step 6: Means of measurement**—This step makes sense out of the preceding pieces. A general, how-to action statement is written that describes how the elements of measure and their associated parameters will be applied to determine the achievement level in the performance indicator. This statement can be brief, but clarifying intent is more important than the length.

▲ **Step 7: Notional metrics**—In this step, conceptual descriptions of possible metrics resulting from the previous steps are put in writing. This step allows everyone to agree on the concept of how the information compiled in the previous steps will be applied to measuring organizational performance. It provides a basis for validating the process and for subsequently developing specific metrics.

▲ **Step 8: Specific metrics**—In this final step, an operational definition and a functional description of the metrics to be applied are written. The definition and description describe the data, how they are collected, how they are used, and, most importantly, what the data mean or how they affect organizational performance. A prototype display of real or imaginary data and a descriptive scenario that shows what actions might be taken as a result of the measurement are also made. This last step is the real test of any metric. It must identify what things need to be done and disclose conditions in sufficient detail to enable subsequent improvement actions.

Rose presents an application of his model used by the U.S. Army Materiel Command, which is shown in Fig. 3.1.

## *Measuring Causes and Effects*

Dashboard metrics are measurements of the results delivered by complex processes and systems. These results are, in a sense, "effects" caused by things taking place within the processes. For example, "cost per unit" might be a metric on a top-level dashboard. This is, in turn, composed of the cost of materials, overhead costs, labor, and so on. Cost of materials is a "cause" of the cost per unit. Cost of materials can be further decomposed into, say, cost of raw materials, cost of purchased subassemblies, and so on. At some level we reach a "root cause," or most basic reason behind an effect. Black Belts and Green Belts learn numerous tools and techniques to help them identify these root causes. However, the dashboard is the starting point for the quest.

In Six Sigma work, results are known as "$Y$'s" and root causes are known as "$X$'s." Six Sigma's historical roots are technical, and its originators generally came from engineering

### Application of the Performance Measurement Model

**Organizational performance metrics**

**Performance category:** Material acquisition

**Performance goal:** Provide technologically superior equipment in the shortest possible time, while ensuring best value in all procurements.

| Performance indicator | Elements of measure | Parameters | Means of measurement | Notional metrics | Specific metrics |
|---|---|---|---|---|---|
| Industrial base management (assessing and assuring adequate industrial facilities to meet military needs) (Other performance indicators) | Source of supply for material and services Quality of suppliers | Material and services requirements Procurement policy | Determine the level of industrial base support for required material and services Certify the quality of existing and potential suppliers | Number and capacity of existing and potential sector suppliers vs. material requirements Percentage of quality-certified sector suppliers | (Separate package containing operational definitions; functional descriptions of data collection, use, and meaning; and prototype display) |

(Other performance categories and goals)

(Optional: subordinate organizational performance metrics)

**Figure 3.1** Organizational performance metrics. (From Rose, 1995. Reprinted by permission.)

and scientific backgrounds. In the mathematics taught to engineers and scientists, equations are used that often express a relationship in the form:

$$Y = f(X) \tag{3.1}$$

This equation simply means that the value identified by the letter $Y$ is determined as a function of some other value $X$. The equation $Y = 2X$ means that if we know what $X$ is, we can find $Y$ if we multiply $X$ by 2. If $X$ is the temperature of a solution, then $Y$ might be the time it takes the solution to evaporate. Equations can become more complicated. For example, $Y = f(X_1, X_2)$ indicates that the value $Y$ depends on the value of two different $X$ variables. You should think of the $X$ in Eq. (3.1) as including any number of $X$ variables. There can be many levels of dashboards encountered between the top level $Y$, called the "big $Y$," and the root cause $X$'s. In Six Sigma work some special notation has evolved to identify whether a root cause is being encountered, or an intermediate result. Intermediate results are sometimes called "little $Y$'s."

In these equations think of $Y$ as the output of a process and the $X$ as inputs. The process itself is symbolized by $f()$. The process can be thought of as a *transfer function* that converts inputs into outputs in some way. An analogy is a recipe. Here's an example:

**Corn Crisp Recipe 12 servings**
¾ cup yellow stone-ground cornmeal
1 cup boiling water
½ teaspoon salt
3 tablespoons melted butter

Preheat the oven to 400°F. Stir the cornmeal and boiling water together in a large glass measuring cup. Add the salt and melted butter. Mix well and pour onto a cookie sheet. Using a spatula, spread the batter out as thin as you possibly can—the thinner, the crisper. Bake the cornmeal for 30 minutes or until crisp and golden brown. Break into 12 roughly equal pieces.

Here the big $Y$ is the customer's overall satisfaction with the finished corn crisp. Little $Y$ would include flavor ratings, "crunchiness" rating, smell, freshness, and other customer-derived metrics that drive the big $Y$. The $X$'s that drive the little $Y$'s might include thinness of the chips, the evenness of the salt, the size of each chip, the color of the chip, and other measurements of the finished product. The inputs could be determined at each major step, for example, actual measurement of the ingredients, the oven temperature, the thoroughness of stirring, how much the water cools before it is stirred with the cornmeal, actual bake time, and so on, as well as the oven used, the cookware, utensils, etc.

Finally, the way different cooks follow the recipe is the *transfer function* or actual process that converts the ingredients into corn crisps. Numerous sources of variation (more $X$'s) can probably be identified by observing the cooks in action. Clearly, even such a simple process can generate some very interesting discussions. If you haven't developed dashboards, it might be worthwhile to do so for the corn crisps as a practice exercise.

Figure 3.2 illustrates how dashboard metrics flow down until eventually linking with Six Sigma projects.

## The Balanced Scorecard

Given the magnitude of the difference between Six Sigma and the traditional three sigma performance levels, the decision to pursue Six Sigma performance obviously requires a radical change in the way things are done. The organization that makes this commitment will never be the same. Since the expenditure of time and resources will be huge, it is crucial that Six Sigma projects and activities are linked to the organization's top-level goals. It is even more important that these be the right goals. An organization that uses Six Sigma to pursue the wrong goals will just get to the wrong place more quickly. The organization's goals must ultimately come from the constituencies it serves: customers, shareholders or

$Y_1 = \boxed{X_{11}} + X_{12} + \cdots + X_{1n}$    CEO, Senior Leadership Team

$Y_2 = X_{21} + \boxed{X_{22}} + \cdots + X_{2n}$    General Managers

$Y_3 = \boxed{X_{31}} + X_{32} + \cdots + X_{3n}$    Directors

$Y_4 = \boxed{X_{41} + X_{42}} + X_{43} + \boxed{X_{44} + X_{45}}$    Managers/sponsors

Six Sigma Project    Six Sigma Project    Black Belts

**Figure 3.2** Flowdown of strategies to drivers and projects.

owners, and employees. Focusing too much on the needs of any one of these groups can be detrimental to all of them in the long run. For example, companies that look at shareholder performance as their only significant goal may lose employees and customers. To use the balanced scorecard senior management must translate these stakeholder-based goals into metrics. These goals and metrics are then mapped to a strategy for achieving them. *Dashboards* are developed to display the metrics for each constituency or *stakeholder*. Finally, Six Sigma is used to either close gaps in critical metrics, or to help develop new processes, products, and services consistent with top management's strategy.

    Tactical quality planning involves developing an approach to implementing the strategic quality plan. Sheridan (1993) describes policy deployment as the development of a measurement-based system as a means of planning for continuous quality improvement throughout all levels of an organization. Originally developed by the Japanese (and known by the name *Hoshin Kanri*, meaning "direction control," companies worldwide have adopted policy deployment as a means to clearly define the long-range direction of company development, as opposed to short-term.

    Balanced scorecards help the organization maintain perspective by providing a concise display of performance metrics with regard to four perspectives that correspond roughly to the major stakeholders—customer, financial, internal processes, and learning and growth (Kaplan and Norton, 1992). The simultaneous measurement from these four perspectives prevents local suboptimization, the common phenomenon where performance in one part of the organization is improved at the expense of performance in another part of the organization. This leads to the well-known loop where this year we focus on quality, driving up costs. Next year we focus on costs, hurting cycle time. When we look at cycle time people take shortcuts, hurting quality, and so on. This also happens on a larger scale, where we alternately focus on employees, customers, or shareholders at the expense of the stakeholders who are not the current focus. Clearly, such "firefighting" doesn't make anyone happy. We truly need the "balance" in balanced scorecards.

In *Using the Balanced Scorecard as a Strategic Management System*, Kaplan and Norton define four key processes that enable organizations to link everyday decisions to their organizational strategy, allowing for a constant check of how well strategy is deployed across the organization: translating the vision; communicating and linking; business planning; feedback and learning.

## *Translating the Vision*

The strategic vision will only be realized if the organization is focused on achieving its goals. The vision must be translated into objectives, which can guide the organization's activities and whose success can be measured. Defining the metrics that will indicate how successful the organization has been in addressing each strategy is not a simple process, but nonetheless necessary. This process forces serious deliberation and buy-in of the vision and strategy by upper and mid-level management. Selection of suitable metrics for each strategy forces reconsideration of lofty unattainable goals into practical objectives that can be measured. The result of this process is consensus by the top managers on the strategic vision, and a suitable set of metrics to monitor the organization's progress in attaining each strategic objective and (by extension) the overall vision.

### **Customer Perspective**

Let's take a closer look at each of the major perspectives on the balanced scorecard, starting with the customer. The balanced scorecard requires that management translate their vague corporate mission ("Acme will be #1 in providing customer value") into specific measures of factors that matter to customers. The customer scorecard answers the question: "How do our customers view us?"

To answer this, you must ask yourself two related questions: What things do customers consider when evaluating us? How do we know? While the only true way to answer these questions is to communicate with real customers, it is well established that customers in general tend to consider four broad categories of factors when evaluating an organization:

- **Quality**—How well do you keep your promises by delivering error-free service or defect-free product? Did I receive what I ordered? Was it undamaged? Are your promised delivery times accurate? Do you honor your warranty or pay your claims without a hassle?

▲ **Timeliness**—How fast is your service? How long does it take to have my order delivered? Do improvements appear in a timely manner?
▲ **Performance and service**—How do your products and services help me? Are they dependable?
▲ **Value**—What is the cost of buying and owning your product or service? Is it worth it?

The first step in the translation is to determine precisely *what* customers consider when evaluating your organization. This can be done by communicating with customers via one-on-one contacts, focus groups, questionnaires, chat rooms, forums, and so on. Management should see the actual, unvarnished words used by customers to describe what they think about the company, its products, and its services. Once management is thoroughly familiar with their target customer, they need to articulate their customer goals in words meaningful to them. For example, management might say:

▲ We will cut the time required to introduce a new product from nine months to three months.
▲ We will be the best in the industry for on-time delivery.
▲ We will intimately involve our customers in the design of our next major product.

These goals must be operationalized by designating metrics to act as surrogates for the goals. Think of the goals themselves as *latent* or hidden constructs. The objective is to identify observable things directly related to the goals that can be measured. These are *indicators* that guide you toward your goals. Table 3.1 shows examples of how the goals mentioned above might be operationalized.

Plans, budgets, goals, and targets are *key requirements* set by the leadership for the organization. If not done properly, the behavior driven by these key requirements may not be anywhere close to what the leadership desires, or expects. Key requirements are used to

**Table 3.1** Operationalizing Customer-centric Goals

| Goal | Candidate Metrics |
| --- | --- |
| We will cut the time required to introduce a new product from nine months to three months. | • Average time to introduce a new product for most recent month or quarter<br>• Number of new products introduced in most recent quarter |
| We will be the best in the industry for on-time delivery. | • Percentage of on-time deliveries<br>• Best in industry on-time delivery percentage divided by our on-time delivery percentage<br>• Percentage of late deliveries |
| We will intimately involve our customers in the design of our next major product. | • Number of customers on design team(s)<br>• Number of customer suggestions incorporated in new design |

assess employee performance, which is linked to promotions, pay increases, bonuses, and many other things that people care about a great deal.

The most common flaw in goal setting is the tendency to set goals that are mere wishes and hopes. The leadership looks at a metric and pontificates on what a "good" level of performance would be for it. If enough heads nod around the conference table, this becomes that metric's target.

A better way to arrive at goals for key requirements is to examine the actual history of the metric over time. This information should be plotted on a process behavior chart (see Chaps. 7 and 8). If the metric falls within the calculated limits the bulk of the time, then the process is considered predictable. Typically, unless the metric is operationalizing a differentiation strategy, the goal for predictable processes will be to maintain the historical levels. These metrics will not appear on the dashboards that the leadership reviews on a routine basis. However, their performance is monitored by process owners, and the leadership is informed on an exception basis if the process behavior deviates from expectation.

**Internal Process Perspective**

In the Internal Process section of the balanced scorecard we develop metrics that help answer the question: Which internal processes are critical to meet customer and shareholder goals? Internal process excellence is linked to customer perceived value, but the linkage is indirect and imperfect. It is often possible to hide internal problems from customers by throwing resources at problems; for example, increased inspection and testing. Also, customer perceived value is affected by factors other than internal processes such as price, competitive offerings, etc. Similarly, internal operations consume resources so they impact the shareholders. Here again, the linkage is indirect and imperfect. For example, sometimes it is in the organization's strategic interest to drive up costs in order to meet critical short-term customer demands or to head off competitive moves in the market. Thus, simply watching the shareholder or customer dashboards won't always give leadership a good idea of how well internal processes are performing. A separate dashboard is needed for this purpose.

This section of the scorecard gives operational managers the internal direction they need to focus on customer needs. Internal metrics should be chosen to support the leadership's customer strategy, with knowledge of what customers need from internal operations. Process maps should be created that show the linkage between suppliers, inputs, process activities, outputs, and customers (SIPOC). SIPOC is a flowcharting technique that helps identify those processes that have the greatest impact on customer satisfaction, as discussed in greater detail in Chap. 7.

Companies need to identify and measure their core competencies. These are areas where the company must excel. It is the source of their competitive advantage. Goals in these areas must be ambitious and challenging. This is where you "wow" your customer.

Other key areas will pursue goals designed to satisfy customers, perhaps by maintaining competitive performance levels. Table 3.2 shows how core competencies might drive customer value propositions. The metrics may be similar for the different companies, but the goals will differ significantly. For example, Company A would place greater emphasis on the time required to develop and introduce new services. Companies B and C would not ignore this aspect of their internal operations, but their goals would be less ambitious in this area than Company A's. Company A is the industry *benchmark* for innovation.

**Table 3.2** Customer Value Proposition Versus Core Competency

| Internal Process | Company A | Company B | Company C |
|---|---|---|---|
| Innovation | X | | |
| Customer relationship management | | X | |
| Operations and logistics | | | X |
| Customer value proposition | Product or service attributes | Flexibility, customization | Cost, dependability |

"X" indicates the company's core competency.

Of course, it is possible that your competitor will try to leapfrog you in your core competency, becoming the new benchmark and stealing your customers. Or you may find that your customer base is dwindling and the market for your particular competency is decreasing. Leadership must stay on the alert for such developments and be prepared to react quickly. Most companies will fight to maintain their position of industry leadership as long as there is an adequate market. Six Sigma can help in this battle because Six Sigma projects are usually of short duration strategically speaking, and Black Belts offer a resource that can be redeployed quickly to where they are most needed.

## Innovation and Learning Perspective

In the Innovation and Learning Perspective section of the balanced scorecard we develop metrics that help answer the question: Can we continue to improve and create value? Where is innovation required?

Success is a moving target. What worked yesterday may fail miserably tomorrow. Previous sections of the balanced scorecard have identified the metrics the leadership considers to be most important for success in the near future. But the organization must be prepared to meet the new and changing demands that the more distant future will surely bring. Building shareholder value is especially dependent on the company's ability to innovate, improve, and learn. The intrinsic value of a business is the discounted value of the cash that can be taken out of the business during its remaining life (Buffett, 1996). Intrinsic value is directly related to a company's ability to create new products and processes, to improve operating efficiency, to discover and develop new markets, and to increase revenues and margins. Companies able to do this well will throw off more cash over the long term than companies that do it poorly. The cash generated can be withdrawn by the owners or reinvested in the business.

Innovation and learning were the areas addressed by the *continuous improvement* (CI) initiatives of the past. Devotees of CI will be happy to learn that it's alive and well in the Six Sigma world. However, CI projects were often local in scope, while most Black Belt Six Sigma projects are cross-functional. Many so-called Green Belt projects are reminiscent of the CI projects in the past. Also, CI tended to focus narrowly on work processes, while Green Belt projects cover a broader range of business processes, products, and services. A well-designed Six Sigma program will have a mix of Green Belt and Black Belt projects addressing a range of enterprise and local process improvement issues.

Dashboards designed to measure performance in the area of Innovation and Learning often address three major areas: employee competencies, technology, and corporate culture. These are operationalized in a wide variety of ways. One metric is the average rate of improvement in the sigma level of an organizational unit. As described in Chap. 1, an organizational Six Sigma strategy to reduce mistakes, errors, and defects may target a 10 time reduction in the error rate every two years. This breakthrough rate of improvement is usually not attained instantly, and a metric of the actual rate is a good candidate for including on the Innovation and Learning dashboard. The rate of improvement is a measure of the overall maturity of the Six Sigma initiative. Other Innovation and Learning metric candidates might include such things as:

- R&D cycle time
- New product features released
- Percentage of revenue from new products or services
- Closure of gaps identified in the training needs audit

### Financial Perspective

Obsession with financial metrics has been the undoing of many improvement initiatives. When senior leaders look only at results they miss the fact that these results come from a complex chain of interacting processes that effectively and efficiently produce value for customers. Only by providing value that customers are willing to pay for can an enterprise generate sales, and only by creating these values at a cost less than their price can it produce profits for owners. For many companies the consequence of looking only at short-term financial results has been a long-term decline in business performance. Many companies have gone out of business altogether.

The result of this unfortunate history is that many critics have advocated the complete abandonment of the practice of using financial metrics to guide leadership action. The argument goes something like this: since financial results are determined by a combination of customer satisfaction and the way the organization runs its internal operations, if we focus on these factors the financial performance will follow in due course. This is throwing the baby out with the bathwater. The flaw in the logic is that it assumes that leaders and managers *know* precisely *how* customer satisfaction and internal operational excellence lead to financial results. This arrogance is unjustified. Too often we learn in retrospect that we are focusing on the wrong things and the financial results fail to materialize. For example, we may busily set about improving the throughput of a process that already has plenty of excess capacity. All we get from this effort is more excess capacity. Even well-managed improvement efforts won't result in bottom-line impact because management fails to take the necessary steps such as reducing excess inventory, downsizing extra personnel, selling off unneeded equipment, etc. As Toyota's Taiichi Ohno says:

> If, as a result of labor saving, 0.9 of a worker is saved, it means nothing. At least one person must be saved before a cost reduction results. Therefore, we must attain worker saving.
>
> TAIICHI OHNO
> *Toyota Production System: Beyond Large-Scale Production*

    The truth is, it's very difficult to lay people off and a poor reward for people who may have participated in creating the improvement. Most managers agree that this is the worst part of their job. However, simply ignoring the issue isn't the best way to deal with it. Plans must be made before starting a project for adjusting to the consequences of success. If there will be no bottom-line impact because there are to be no plans to convert the savings into actual reductions in resource requirements, the project shouldn't be undertaken in the first place. On the other hand, plans can often be made at the enterprise level for dealing with the positive results of Six Sigma by such means as hiring moratoriums, early retirement packages, etc. Better still are plans to increase sales or to grow the business to absorb the new capacity. This can often be accomplished by modifying the customer value proposition through more reliable products, lower prices, faster delivery time, lower cycle times, etc. These enhancements are made possible as a result of the Six Sigma improvements.

    There are many dangers in failing to monitor financial results. We may blindly pour resources into improving customer satisfaction inaccurately measured by a faulty or incomplete survey. In other cases, the competition may discover a new technology that makes ours obsolete. The list of issues that potentially disrupt the link between internal strategies and financial performance is endless. Financial performance metrics provide the feedback needed to test assumptions.

    Actual metrics for monitoring financial performance are numerous. The top-level dashboard will often include metrics in the areas of improved efficiency (e.g., cost per unit, asset utilization) or improved effectiveness (e.g., revenue growth, market share increase, profit per customer).

## Strategy Deployment Plan

Unlike traditional measurement systems, which tend to have a control bias, balanced scorecards are based on strategy. The idea is to realize the leadership vision using a set of linked strategies. Metrics operationalize these strategies and create a bond between the activities of the organization and the vision of the leadership.

Figure 3.3 illustrates these principles for a hypothetical organization, where the metrics are shown in rectangles on the left side of the figure. The strategy deployment plan clearly displays the linkage between the metrics and the larger issues of interest. These unobserved or latent constructs are shown in ellipses and are inferred from the metrics. This perspective helps leadership understand the limitations of metrics, as well as their value. If, for example, all of the metrics leading to shareholder perceived value are strongly positive, but surveys of the shareholders (Voice of Shareholder) indicate shareholder dissatisfaction, then the dashboard metrics are obviously inadequate and need to be revised.

The organization is pursuing a particular strategy, and the leadership team has selected five *differentiator metrics*, which are shown in boldface type in Figure 3.3:

- Cost per unit
- Revenues from new sources
- (Customer) service relationship
- Product introductions, (new product) revenues
- Research deployment time

Goals for these metrics will be set very high in an attempt to differentiate this organization from its competition. Goals for other metrics (key requirements) will be set to achieve competitiveness, usually by maintaining the historical levels for these metrics.

The organization's leaders believe their core competencies are in the areas of technology and customer service. They want their customers to think of them as *the company to go to for the very best products completely customized to meet extremely demanding needs.*

A comparison of the differentiator metrics with the leadership vision statement shows an inconsistency: achieving benchmark status for items 2 to 5 would conflict with a similar goal for item 1. This conflict indicates that the productivity strategy for this organization should be reevaluated. Unless the company is losing its market share due to uncompetitive prices, or losing its investors due to low profits, item 1 should probably be a key requirement maintained at historical levels. If costs are extremely out of line, cost per unit might be the focus of a greater than normal amount of attention to bring it down to reasonable levels. However, it should not be shown as a *differentiator* on the strategic dashboard. The

**158** | The Six Sigma Handbook

**Top Level Dashboard**

*Bold type indicates item is a differentiator, which we must excel at. Regular type indicates item is a standard requirement.*

*Excel in one customer value proposition. Maintain threshold levels for the other two.*

**STRATEGY DEPLOYMENT PLAN**

**Figure 3.3** Strategy deployment plan for a hypothetical organization.

company has no desire to become a cost leader in the eyes of customers or shareholders, as evidenced by the vision statement.

Six Sigma plays a vital role in achieving the leadership vision by providing the resources needed to facilitate change where it is needed. Six Sigma projects are linked to dashboard metrics through the project selection process discussed in Chap. 4. The process involves calculating the expected impact of the project on a dashboard metric. The metrics used for Six Sigma projects are typically on a lower-level dashboard, but since the lower-level dashboard metrics flow down from the top level, the linkage is explicit. The process begins by identifying the gap between the current state and the goal for each top-level dashboard metric; Master Black Belts responsible for assisting with the organizational deployment commonly assist with this activity. Six Sigma projects impacting differentiator dashboard metrics that show large gaps are prime project candidates. This information may also be useful in selecting Black Belt candidates. Candidates with backgrounds in areas where high-impact projects will be pursued may be given preference over equally qualified candidates from elsewhere in the organization.

## *Communicating and Linking*

Communicating the objectives throughout the organization is a critical part of attaining the organization's objectives. While this may seem obvious, it is by no means widespread, particularly at the level of communication needed for employees to understand the strategic objectives in terms of their daily operations. In this process, we seek to drill down on the overall strategic objectives to understand the relative contribution needed by each business unit, functional area, department, or even employee, where applicable. Yes, some organizations will even define balanced scorecards to assist in employee development with regard to training or customer service goals, for example. The fundamental idea is to allow each employee to realize how his or her actions fit into the overall strategic plan for the organization. These lower-level scorecards can conceivably feed into evaluation and performance incentives, for management and employee alike, although the authors note risks associated with this practice at the employee level, including potential for unintended consequences, gaming the system, or data validation at that level of use. Yet, it should be obvious that at each successive level, from business unit to functional area to department to employee, communication of the overall strategy to the party and the party's input in defining suitable objectives and metrics will greatly increase their buy-in and the subsequent probability of attaining the goals. Linking of rewards to these metrics is a fundamental improvement to simple revenue- or profit-based formulas, as the balance scorecard provides a longer-term view of the organization's performance than the short-term financial metrics (as previously discussed).

Defining the linkage between strategic objectives and dashboard metrics, operations, activities, and projects can be provided via a QFD matrix approach.

## The Strategy Deployment Matrix

The process for developing the strategy deployment matrix is as follows:

1. Create a matrix of the strategies and metrics.
2. Determine the strength of the relationship between each strategy and metric.
3. Calculate a weight indicating the relative importance of the metric.

To begin, we create a matrix where the rows are the strategies (what we want to accomplish) and the columns are the dashboard metrics (how we will operationalize the strategies and monitor progress). Note that this is the typical what-how QFD matrix layout, with a slightly different spin. In each cell (intersection of a row and a column) we will place a symbol assigning a weight to the relationship between the row and the column. A completed phase I strategy deployment matrix is shown in Fig. 3.4. The weights and symbols used are shown in Fig. 3.5.

The weight scale is somewhat arbitrary and is chosen to suit particular needs. In this example, the scale values increase more or less exponentially, which places a high emphasis on strong relationships, to more clearly establish priorities. Weights of 1-2-3 would treat the different relationship strengths as increasing linearly. Choose the weighting scheme you feel best fits your strategy.

After the relationships have been determined for each cell, we are ready to calculate scores for each row. Remember, the rows represent strategies. For example, the first row represents our productivity strategy. The strategy deployment plan indicated that the productivity strategy was operationalized by the metrics cost-per-unit and asset utilization, and a strong relationship is shown between these metrics and the productivity strategy. However, the matrix shows a strong relationship between the productivity strategy and inventory turns, which affects asset utilization. Critical to quality (CTQ) and profit per customer are somewhat related to this strategy. To get an overall score for the productivity strategy, just sum the weights across the first row; the answer is 29. These row (strategy) weights provide information on how well the dashboards measure the strategies. A zero would indicate that the strategy isn't measured at all. However, a relatively low score doesn't necessarily indicate a problem. For example, the regulatory compliance strategy has a score of 9, but that comes from a strong relationship between the regulatory compliance audit and the strategy. Since the audit covers all major compliance issues, it's entirely possible that this single metric is sufficient.

The columns represent the metrics on the top-level dashboard, although only the differentiator metrics will be monitored on an ongoing basis. The metric's target is shown at the

**Figure 3.4** Strategy deployment matrix.

| Relationship description | Weight | Symbol |
|---|---|---|
| Strong relationship | 9 | ⊙ |
| Moderate relationship | 3 | ○ |
| Some relationship | 1 | △ |
| Differentiator metric | 5 | ● |
| Key requirement metric | 1 | ✓ |

**Figure 3.5** Weights and symbols in strategy deployment matrix.

bottom of each column in the "how" portion of the matrix. This strategy deployment matrix will provide a reality check on the targets. We will use a similar QFD matrix to link the targets to specific Six Sigma activities designed to achieve them. At the project phase it is far easier to estimate the impact the projects will have on the metric. If the sum of the organization's activities, including projects and any other nonproject activities that impact the target, isn't enough to reach the target, either more effort is needed or the target must be revised.

In this example, leadership's vision is to be the supplier of choice for customers who desire state-of-the-art products customized to their demanding requirements. To achieve this vision, they will focus their strategy on four key differentiator metrics: new product introductions, revenue from new sources, intimate customer relationship, and R&D deployment time. With our chosen weighting scheme, differentiator columns have a strategic importance score of 5, indicated with a • symbol in the row labeled Strategic Importance Score. These are the metrics that leadership will focus on throughout the year, and the goals for them are set very high. Other metrics must meet less demanding standards and will be brought to the attention of leadership only on an exception basis. The row labeled Relative Metric Weight is the product of the criteria score and the strategic importance score as a percentage for each column. The four differentiator metrics have the highest relative scores, while product selection (i.e., having a wide variety of standard products for the customer to choose from) is the lowest.

It is vital when using QFD to focus on only the most important columns!

Columns identified with a checkmark in the row labeled Strategic Importance Score are not part of the organization's differentiation strategy. This isn't to say that they are unimportant. What it does mean is that targets for these metrics will probably be set at or near their historical levels as indicated by process behavior charts. The goals will be to maintain these metrics, rather than to drive them to new heights. An organization has only limited resources to devote to change, and these resources must be focused if they are to make a difference that will be noticed by customers and shareholders. This organization's complete dashboard has 20 metrics, which can hardly be considered a "focus." By limiting attention to the four differentiators, the organization can pursue the strategy that their leadership believes will make them stand out in the marketplace for customer and shareholder dollars.[1]

## Business Planning

The business planning process defines the key drivers of strategic objectives at each functional level and suitable time frames to measure their progress in achieving a suitable target. For example, an organization's strategic objective of increasing sales revenue may depend on increased sales in several (specified) product lines. This in turn may require increased capacity in production and delivery, as well as a change in marketing, for each affected product line, among a host of other activities needed to achieve the organizational goal. Only by drilling the organization's goals down to the functional levels will each function appreciate the local changes needed to drive the organization's strategy.

By focusing on the critical few drivers of the strategic plan at each functional level of the organization, each level will optimize its contribution to the organization's achievement of the strategic plan by effectively aligning its resources with the strategy. Metrics defined along the four scorecard perspectives (customer, financial, internal processes, and learning and growth), with targets to be achieved by specified dates (i.e., milestones), provide a means to consistently and continuously track performance.

Most importantly, the business planning process itself will force managers at all levels to ask critical, yet often tough, questions regarding the strategy and their part in achieving it. This exposure of the strategy across the organization removes much of the guesswork that might be required if only executive-level staff participated. It also builds consensus across the organization on how the strategy will be achieved, and local commitment to making it happen.

In organizations where the budgeting process has assumed the primary means of accountability of various levels of management (on a weekly, monthly, or quarterly basis), the business planning process forces an integration of the strategic plan with the operational budget, ensuring that operational resources are budgeted and assigned to obtaining the strategic objectives.

## Deploying Differentiators to Operations

To maintain the proper focus on the differentiator metrics, a second-level matrix linked only to the differentiators is created. This matrix relates the differentiator dashboard metrics to departmental support strategies; see Fig. 3.6.

To keep things simple, we only show the strategy linkage for three departments: engineering, manufacturing, and marketing; each department can prepare its own QFD matrix. Notice that the four differentiator metric columns now appear as rows in the matrix shown in Fig. 3.6. These are the QFD "whats." The software automatically brings over the criteria performance target, criteria scores, and relative criteria scores for each row. This information is used to evaluate the strategy support plans for the three departments.

The support plans for the three departments are shown as columns, the QFD "hows," or how these three departments plan to implement the strategies. The relationship between the whats and hows is determined as described above. For each column the sum of the relationship times the row criteria score is calculated and shown in the score row near the bottom of the chart. This information will be used to select and prioritize Six Sigma projects in the next phase of the QFD.

**164** | The Six Sigma Handbook

**Figure 3.6** Phase II matrix: differentiators.
(Chart produced using QFD Designer software. Qualsoft, LLC. www.qualisoft.com.)

Figure 3.6 also has a roof, which shows correlations between the whats. This is useful in identifying related Six Sigma projects, either within the same department or in different departments. For example, there is a strong relationship between two engineering activities: faster prototype development and improve concept-to-design cycle time. Perhaps faster prototype development should be a subproject under the broad heading of improve concept-to-design cycle time. This also suggests that "improve concept-to-design cycle time" may have too large a scope. The marketing strategy of "improve ability to respond to changing customer needs" is correlated with three projects in engineering and manufacturing. When a strategy support plan involves many cross-functional projects, it may indicate the existence of a core process. This suggests a need for high-level sponsorship, or the designation of a process owner to coordinate projects.

## Deploying Operations Plans to Projects

Figure 3.7 is a QFD matrix that links the department plans to Six Sigma projects. (In reality this may require additional flowdown steps, but the number of steps should be kept as small as possible.) The rows are the department plans. The software also carried over the numeric relative score from the bottom row of the previous matrix, which is a measure of the relative impact of the department plan on the overall differentiator strategy. The far right column, labeled Goal Score, is the sum of the relationships for the row. For this example only the top five department plans are deployed to Six Sigma projects. By summing the numeric relative scores we can determine that these five plans account for 86% of the impact. In reality you will also only capture the biggest hitters, although there's nothing magic about the number 5.

There are three Black Belts shown and eight projects. Each project is shown as a column in the matrix. The relationship between the project and each departmental plan is

**Figure 3.7** Phase III matrix: Six Sigma projects.
(Chart produced using QFD Designer software. Qualsoft, LLC. www.qualisoft.com.)

shown in the matrix. The bottom row shows the "project impact score," which is the sum of the relationships for the project's column times the row's numeric relative score.

*Interpretation.* Since the numeric relative scores are linked to department plans, which are linked to differentiator metrics, which are linked to strategies, the project impact score measures the project's impact on the strategy. The validity of these "carryover scores" has been questioned (Burke et al., 2002). Through the strategy deployment plan we can trace the need for the project all the way back to stakeholders. This logical thread provides those engaged in Six Sigma projects with an anchor to reality and the meaning behind their activities.

The Goal Score column can also be used to determine the support Six Sigma provides for each department plan. Note that the marketing plan to "identify target markets for new products" isn't receiving any support at all from Six Sigma projects (assuming that these eight projects are all of the Six Sigma projects). This may be okay, or it may not be. It all depends on how important the plan is to the strategic objectives, and what other activities are being pursued to implement the plan. The Executive Six Sigma Council may wish to examine project QFD matrices to determine if action is necessary to reallocate Six Sigma resources.

The Project Impact Score row is useful in much the same way. This row can be rank-ordered to see which projects have the greatest impact on the strategy. It is also useful in identifying irrelevant projects. The project Mike L is pursuing to improve "pin manufacturing capability" has no impact on any of the departmental plans. Unless it impacts some other strategy support plan that isn't shown in the QFD matrix, it should be abandoned as a Six Sigma project. The project may still be something the manufacturing department wants to pursue, perhaps to meet a goal for a key requirement. However, as a general rule, Six Sigma projects requiring a Black Belt should focus on plans that have a direct linkage to differentiator strategies.

## Feedback and Learning

The business planning process essentially established a set of hypotheses regarding the business strategy: If each functional area achieves its targets over the specified time frame, the organization will achieve its strategic objective. For example, the differentiators' matrix in Fig. 3.6 suggests that the revenue from new sources is tightly linked to the key drivers of *faster prototype development* and *improve concept-to-design cycle time* in engineering; and *improve ability to respond to changing customer needs* and *identify target markets for new products* in marketing. As progress is made in achieving these local targets in engineering and marketing, we expect to see progress toward our target of revenue from new sources. If not, then the hypothesis itself must be questioned. Perhaps a key driver is missing, or the strengths of the relationships are not as strong as suspected, or a fundamental change in the market (including competitor innovation) has upset the previously well-established cause-and-effect relationships at the core of our business model.

This ability to challenge root assumptions of the business model's assumed cause-and-effect relationships using real-time data is known as *double-loop learning* (Argyris, 1991). This enables an organization to be a learning organization, one that constantly tests the underlying assumptions it is using, and takes action to correct the assumptions, as necessary, to achieve its overall objectives. In some cases, this learning may question the strategy itself, perhaps as misguided given the current market and customer values.

## Dashboard Design

Strategies are operationalized by metrics displayed on dashboards. Dashboard displays are designed to standardize information throughout the organization. A process owner at any level of the organization should be able to quickly recognize the meaning of the dashboard data and accelerate the learning cycle. The strategy deployment plan is merely a hypothesis. Science-based management requires testing of the hypothesis to determine if it is in reasonable agreement with the facts, and take action; or revise the strategy deployment plan accordingly.

Dashboard metrics should embody the principles of good metrics discussed earlier in the chapter:

1. Display performance over time.
2. Include statistical guidelines to help separate signal (variation from an identifiable cause) from noise (variation similar to random fluctuations).
3. Show causes of variation when known.
4. Identify acceptable and unacceptable performance (defects).
5. Be linked to higher-level dashboards (goals and strategies) or lower-level dashboards (drivers) to guide strategic activity within the organization.

The statistical process control charts described in Chaps. 7 and 8 address the first two concerns: they are the proper tool to analyze the metric over time and include guidelines (i.e., statistical control limits) to separate real process changes from the random variation expected within a stable process. The statistical control limits operationally define when intervention is needed. Generally, when metrics fall within the limits, the process should be left alone. However, when a metric falls outside the limits, it indicates that something important has changed that requires attention. An exception to these general rules occurs when a deliberate intervention is made to achieve a goal. In this case the metric is *supposed to* respond to the intervention by moving in a positive direction. The limits will tell leadership if the intervention produced the desired result. If so, the metric will go beyond the proper control limit indicating improvement. Once the metric stabilizes at the new and improved level, the limits should be recalculated so they can detect slippage. SPC software can include more advanced features that allow for drill-down between upper and lower level metrics (principle 5), as well as drill-down to display and quantify causes of excessive variation (principles 3 and 4). An example of a business-level customer dashboard based on the example strategy deployment plan described earlier is shown in Fig. 3.8. The tabular dashboard (top left in the right-hand pane of the software application window) provides a quick view of top-level metrics, with status (either predictably in control or subject to unpredictable out of control signals) and sigma level over a given period of time. For example, the *Service Time* metric has a relatively low Sigma level of 1.115, corresponding to a defect rate of approximately 13%. Its *In Control* status further indicates the error rate is inherent to the process and not due to unpredictable fluctuations: the process must be fundamentally changed to reduce its error rate from 13%. (The interpretation of control charts are further discussed in Chaps. 7 and 8).

Each of the key dashboard metrics may be drilled down to display its behavior over time, as shown in Fig. 3.8 for the *Service Time* metric. The control chart in this example

**Figure 3.8** Example of a business level dashboard with drill down. (Courtesy of www.qualityamerica.com by permission.)

displays the observed process behavior for a 30-day window. The absence of data beyond the statistical Upper Control Limit (UCL) of 18.375 and Lower Control Limit (LCL) of 0.234 indicate the process is stable during this time period, as initially indicated in the dashboard. An interesting period of reduced variation is noted toward the middle of the chart. The large circles around the data points indicate this period of relatively low variability is statistically significant; further investigation of this period may translate to understanding of process design that could lead to reduced overall variation.

The process goal (i.e., the Upper Specification Limit) is conveniently shown on the chart as eight minutes. The visual display makes it easy to see there are 19 observations exceeding this requirement: 13 are plotted as triangles, 3 as boxes, and 3 as filled circles. Each of these symbol types has been linked to a potential cause identified during data entry by the customer support personnel: boxes associated with "System Slow;" triangles with "Required Approval"; and filled circles with "Customer Information Lacking." The time-ordered nature of the control chart shows these nonconformances to be somewhat equally distributed over time, rather than clumped in a specific time period.

Further drill-down provides one of several possible Pareto diagrams associated with the data. The Pareto diagram shown in Fig. 3.8 provides the product associated with each of the excessive service times (i.e., service time greater than eight minutes), further partitioned by stacked boxes within each product type based on the customer service error code. This particular drill-down indicates Product C has the highest number of errors (58%), followed by Product A (an additional 26%), with the remaining errors associated with Product D. Another Pareto easily created through drill-down (and not shown) indicates these percentages roughly correspond to the percent of total products processed. In other words, all things being equal, we might expect Product C to have about 60% of the errors, since it represents that percentage of the total product processed.

Interestingly, all of the Product D errors are associated with the "Customer Information Lacking" error; a majority of the Product A errors are associated with the "Required Approval" error; and neither of these errors occurred for Product C. While not statistically significant among themselves, these are potential issues useful for project directed improvements to the process. For example, a project may be designed to reduce errors due to "Customer Information Lacking," and further analysis of similar data over a longer time period may be used to isolate the products for which this error occurs.

In this way, the dashboard provides at first glance the means to monitor the status of the organizational transformation toward the goals of the strategic plan. The dashboard's drill-down capabilities provide a much richer source of information to further understand the process complexities and potentials. Clearly, it is beneficial to design an information management system capable of collecting and disseminating information reflective of the ever-changing organizational needs.

## Information Systems Requirements

Balanced scorecards begin with the highest level metrics. At any given level, dashboards will display a relatively small number of metrics. While this allows the user of the dashboard to focus on key items, it also potentially hampers their ability to understand the larger issues associated with a given process change. Information systems (IS) help address this issue through the "drill down" capability shown in the earlier section. Drill down involves disaggregating dashboard metrics into their component parts. For example, a cost-per-unit metric can be decomposed by division, plant, department, shift, worker, week, etc. These components of the higher-level metric are sometimes already on dashboards at lower levels of the organization, in which case the answer is provided in advance. However, if the established lower-level dashboard metrics can't explain the situation, other exploratory drill downs may be required. Online analytic processing (OLAP) cubes often ease the demands on the IS caused by drill down requests.

This raises an important point: in Six Sigma organizations the IS must be accessed by many more people. The attitude of many IS departments is "The data systems belong to us. If you want some data, submit a formal request." In a Six Sigma organization, this attitude is hopelessly outmoded. The demands on IS increase dramatically when Six Sigma is deployed. In addition to the creation of numerous dashboards, and the associated drill downs and problem investigations, the Black Belts and Green Belts make frequent use of IS in their projects. Six Sigma "show me the data" emphasis places more demands on the IS. In planning for Six Sigma success, companies need to assign a high-level champion to oversee the adaptation of IS to the new realities of Six Sigma. A goal is to make access as easy as possible while maintaining data security and integrity.

Although it's important to be timely, most Six Sigma data analyses don't require real-time data access. Data that are a day or a few days old will often suffice. The IS department may want to provide facilities for offline data analysis by Six Sigma team members and Belts. A few high-end workstations capable of handling large data sets or intensive calculations are also very useful at times, especially for data mining analyses such as clustering, neural networks, or classification and decision trees.

Six Sigma technical leaders work to extract actionable knowledge from an organization's information warehouse. To assure access to the needed information, Six Sigma activities should be closely integrated with the information systems (IS) of the organization. Obviously, the skills and training of Six Sigma technical leaders must be supplemented by an investment in software and hardware.

## Integrating Six Sigma with Other Information Systems Technologies

There are three information systems topics that are closely related to Six Sigma activities:

- Data warehousing
- Online analytic processing (OLAP)
- Data mining

Data warehousing relates to the data retained by the organization, and therefore available for use in Six Sigma activities. It impacts the data storage, which impacts ease of access for Six Sigma analyses. OLAP enables the analysis of large databases by persons who may not have the technical background of a Six Sigma technical leader. Data mining involves retrospective analysis of data using advanced tools and techniques. Each of these subjects will be discussed in turn.

While larger organizations clearly require an advanced approach to data warehousing, OLAP, and data mining, there is much to be gained by smaller organizations in implementing appropriately sized systems. Stated simply, an Excel spreadsheet, or more likely a proliferation of these spreadsheets, is ill-suited for the longer-term data management and analysis required by Six Sigma teams. Relatively simple database designs using MS Access or SQL Server back-ends provide immense benefits over spreadsheet data storage, particularly when leveraged with user interfaces to easily drill-down the database, or query associated databases, such as Laboratory Information Systems (LIMS) or Manufacturing Resource Planning (MRP) systems.

*Data Warehousing.* Data warehousing has progressed rapidly. Virtually nonexistent in 1990, now every large corporation has at least one data warehouse; some have several. Hundreds of vendors offer data warehousing solutions, from software to hardware to complete systems. Few standards exist, and there are as many data warehousing implementations as there are data warehouses. However, the multitiered approach to data warehousing is a model that appears to be gaining favor and recent advances in technology and decreases in prices have made this option more appealing to corporate users.

Multitiered data warehousing architecture focuses on how the data are used in the organization. While access and storage considerations may require summarization of data into multiple departmental warehouses, it is better for Six Sigma analysis if the warehouse keeps all of the detail in the data for historical analysis. The major components of this architecture are (Berry and Linoff, 1997):

- *Source systems* are where the data come from.
- *Data transport and cleansing* move data between different data stores.
- The *central repository* is the main store for the data warehouse.
- The *metadata* describes what is available and where.

▲ *Data marts* provide fast, specialized access for end users and applications.
▲ *Operational feedback* integrates decision support back into the operational systems.
▲ *End users* are the reason for developing the warehouse in the first place.

Figure 3.9 illustrates the multi-tiered approach.

Every data warehouse includes at least one of these building blocks. The data originate in the source systems and flows to the end users through the various components. The components can be characterized as hardware, software, and networks. The purpose is to deliver information, which is in turn used to create new knowledge, which is then acted on to improve business performance. In other words, the data warehouse is ultimately a component in a decision-support system.

**Figure 3.9** The multi-tiered approach to data warehousing. (Berry and Linoff, 1997. Used by permission of the publisher.)

*OLAP.* Online analytic processing, or OLAP, is a collection of tools designed to provide ordinary users with a means of extracting useful information from large databases. These databases may or may not reside in a data warehouse. If they do, then the user obtains the benefit of knowing the data has already been cleansed, and access is likely to be more efficient. OLAP consists of client-server tools that have an advanced graphical interface that accesses data arranged in "cubes." The cube is ideally suited for queries that allow users to slice-and-dice the data in any way they see fit. OLAP tools have very fast response times compared to SQL queries on standard relational databases, so they are particularly useful when large data warehouses are accessed.

The basic unit of OLAP is the *cube*. An OLAP cube consists of sub-cubes that summarize data from one or more databases. Each cube is composed of multiple dimensions that represent different fields in a database. For example, an OLAP cube might consist of warranty claims arranged by months, products, and region, as shown in Fig. 3.10.

*Data Mining.* Data mining is the exploration and analysis by automatic or semiautomatic means of large quantities of data in order to uncover useful patterns. These patterns are studied in order to develop performance rules, i.e., new and better ways of doing things. Data mining, as used in Six Sigma, is directed toward improving customer satisfaction, lowering costs, reducing cycle times, and increasing quality.

Data mining is a grab-bag of techniques borrowed from various disciplines. Like Six Sigma, data mining alternates between generating questions via knowledge discovery, and testing hypotheses via designed experiments. Six Sigma and data mining both look for the same things in evaluating data, namely *classification, estimation, prediction, affinity grouping, clustering,* and *description*. However, data mining tends to use a different set of tools than traditional Six Sigma tools and therefore it offers another way to look for improvement opportunities. Also, where Six Sigma tends to focus on internal business processes, data mining looks primarily at marketing, sales, and customer support. Since the object of Six Sigma is, ultimately, to improve customer satisfaction, the external focus

**Figure 3.10** An OLAP cube.

of data mining provides both feed forward data to the Six Sigma program and feedback data on its success.

Data mining is a process for retrospectively exploring business data. There is growing agreement on the steps involved in such a process, and any differences relate only to the detailed tasks within each stage.[2]

- **Goal definition**—This involves defining the goal or objective for the data mining project. This should be a business goal or objective that normally relates to a business event such as arrears in mortgage repayment, customer attrition (churn), energy consumption in a process, etc. This stage also involves the design of how the discovered patterns will result in action that leads to business improvement.
- **Data selection**—This is the process of identifying the data needed for the data mining project and the sources of these data.
- **Data preparation**—This involves cleansing the data, joining/merging data sources, and the derivation of new columns (fields) in the data through aggregation, calculations, or text manipulation of existing data fields. The end result is normally a flat table ready for the application of the data mining itself (i.e., the discovery algorithms to generate patterns). Such a table is normally split into two data sets: one set for pattern discovery and other set for pattern verification.
- **Data exploration**—This involves the exploration of the prepared data to get a better feel prior to pattern discovery and also to validate the results of the data preparation. Typically, this involves examining descriptive statistics (minimum, maximum, average, etc.) and the frequency distribution of individual data fields. It also involves field versus field scatter plots to understand the dependency between fields.
- **Pattern discovery**—This is the stage of applying the pattern discovery algorithm to generate patterns. The process of pattern discovery is most effective when applied as an exploration process assisted by the discovery algorithm. This allows business users to interact with and to impart their business knowledge to the discovery process. For example, if creating a classification tree, users can at any point in the tree construction examine/explore the data filtering to that path, examine the recommendation of the algorithm regarding the next data field to use for the next branch, then use their business judgment to decide on the data field for branching. The pattern discovery stage also involves analyzing the ability to predict occurrences of the event in data other than those used to build the model.
- **Pattern deployment**—This stage involves the application of the discovered patterns to solve the business goal of the data mining project. This can take many forms:
    - **Pattern presentation**—The description of the patterns (or the graphical tree display) and their associated data statistics are included in a document or presentation.

- **Business intelligence**—The discovered patterns are used as queries against a database to derive business intelligence reports.
- **Data scoring and labeling**—The discovered patterns are used to score and/or label each data record in the database with the propensity and the label of the pattern it belongs to.
- **Decision support systems**—The discovered patterns are used to make components of a decision support system.
- **Alarm monitoring**—The discovered patterns are used as norms for a business process. Monitoring these patterns will enable deviations from normal conditions to be detected at the earliest possible time. This can be achieved by embedding the data mining tool as a monitoring component, or through the use of a classical approach, such as control charts.
- **Pattern validity monitoring**—As a business process changes over time, the validity of patterns discovered from historic data will deteriorate. It is therefore important to detect these changes at the earliest possible time by monitoring patterns with new data. Significant changes to the patterns will point to the need to discover new patterns from more recent data.

*OLAP, Data Mining, and Six Sigma.*   OLAP is not a substitute for data mining. OLAP tools are a powerful means for reporting on data, while data mining focuses on finding hidden patterns in data. OLAP helps users explore existing theories by quickly presenting data to confirm or deny ad hoc hypotheses. It is, essentially, a semiautomated means of analysis. OLAP and data mining are complementary: both approaches complement the standard arsenal of tools and techniques used in Six Sigma. Both OLAP and data mining are used for *retrospective studies*: they are used to generate hypotheses by examining past data. Designed experiments, on the other hand, help users design *prospective studies*: they test the hypotheses generated by OLAP and data mining. Used together, Six Sigma, data mining and OLAP comprise a powerful collection of business improvement tools.

# CHAPTER 4

# Maximizing Resources

> The best Six Sigma projects begin not inside the business but outside it, focused on answering the question: How can we make the customer more competitive? What is critical to the customer's success? Learning the answer to that question and learning how to provide the solution is the only focus we need.
>
> JACK WELCH, CEO, General Electric

This chapter addresses Six Sigma project selection and the management support activities related to project success. Projects are the core activity driving change in the Six Sigma organization. Although change also takes place due to other efforts, such as *Kaizen*, project-based change is the force that drives breakthrough and cultural transformation. In a mature Six Sigma organization about 1% of the workforce is engaged full time in change activities, and each of these change agents will complete between three and seven projects in a year. In addition there are another 5% or so part-time change agents, each of whom will complete about two smaller projects per year. In a large organization with 10,000 employees, this translates to about 500 major projects and 1,000 smaller projects in any given year. Clearly, learning how to effectively select, manage, and execute projects is critical to Six Sigma success.

## Choosing the Right Projects

Projects must be focused on the *right* goals. This is the responsibility of the senior leadership, for example, the project sponsor, Executive Six Sigma Council, or equivalent group. Senior leadership is the only group with the necessary authority to designate cross-functional responsibilities and allow access to interdepartmental resources. Six Sigma projects will impact one of the major stakeholder groups: customers, shareholders, or employees. Although it is possible to calculate the impact of any given project on all three groups, I recommend that initially projects be evaluated separately for each group. This keeps the analysis relatively simple and ensures that a good stakeholder mix is represented in the project portfolio.

## Types of Projects

### Customer Value Projects

Many, if not most Six Sigma projects are selected because they have a positive impact on customers. To evaluate such projects one must be able to determine the linkage between business processes and customer-perceived value. Chapter 2 discussed how to create organizations that are customer-driven, which is essential. Customer-driven organizations, especially process enterprises, focus on customer value as a matter of routine. This focus will generate many Six Sigma customer value projects in the course of strategy deployment. However, in addition to the strategy-based linkage of Six Sigma projects described in Chap. 3, there is also a need to use customer demands directly to generate focused Six Sigma projects. The techniques for obtaining this linkage are the same as those used in Chap. 3. The difference is that the focus here is not on strategy deployment or budgeting, but on Six Sigma improvement projects focused on specific customer demands.

Learning what customers value is primarily determined by firsthand contact with customers through customer focus groups, interviews, surveys, etc. The connection between customer-perceived value and business processes, or "customer value streams," is established through business process mapping (see Chaps. 6 and 7) and quality function deployment (QFD). The Executive Six Sigma Council and project sponsors should carefully review the results of these efforts to locate the "lever points" where Six Sigma projects will have the greatest impact on customer value.

### Shareholder Value Projects

Six Sigma provides a "double whammy" by addressing both efficiency and revenues. Revenue is impacted by improving the customer value proposition, which allows organizations to charge premium prices for superior quality, or to keep prices competitive and increase sales volume and market share due to superior quality. Improved efficiency is achieved by reducing the cost of poor quality, reducing cycle time, or eliminating waste in business processes. To determine which Six Sigma projects address the issue of business process efficiency, evaluate the high-level business process maps (including SIPOC) and flowcharts.

### Other Six Sigma Projects

Some Six Sigma projects address intangibles, such as employee morale, regulatory concerns, or environmental issues. These projects can be just as important as those which address customer or shareholder value.

## *Analyzing Project Candidates*

You now have a list of candidate Six Sigma projects. Assuming that the organization has limited resources, the next task is to select a subset of these projects to fund and staff.

Projects cost money, take time, and disrupt normal operations and standard routines. For these reasons projects designed to improve processes should be limited to processes that are important to the enterprise. Furthermore, projects should be undertaken only when success is highly likely. *Feasibility* is determined by considering the scope and cost of a project and the support it receives from the process owner. In this section a number of techniques and approaches are presented to help identify those projects that will be chosen for Six Sigma.

### Benefit-Cost Analysis

Benefit-cost analysis can be as elaborate or as simple as the magnitude of the project expenditures demands. The Six Sigma manager is advised that most such analyses are easier to "sell" to senior management if done by (or reviewed and approved by) experts in the finance and accounting department. The plain fact is that the finance department has credibility in estimating cost and benefit that the Six Sigma department, and any other department, lacks. The best approach is to get the finance department to conduct the benefit-cost analysis with support from the other departments involved in the project. We will provide an overview of some principles and techniques that are useful in benefit-cost analysis.

A fundamental problem with performing benefit-cost analysis is that, in general, it is easier to accurately estimate costs than benefits. Costs can usually be quantified in fairly precise terms in a budget. Costs are claims on resources the firm already has. In contrast, benefits are merely predictions of future events, which may or may not actually occur. Also, benefits are often stated in units other than dollars, making the comparison of cost and benefit problematic. The problem is especially acute where quality improvement projects are concerned. For example, a proposed project may involve placing additional staff on a customer "hot line." The cost is easy to compute: X employees at a salary of $Y each, equipment, office space, supervision, etc. The benefit is much more difficult to determine. Perhaps data indicate that average time on hold will be improved, but the amount of the improvement and the probability that it will occur are speculations. Even if the time-on-hold improvement were precise, the impact on customer satisfaction would be an estimate. The association between customer satisfaction and revenues is yet another estimate. Despite these difficulties, reasonable cause-and-effect linkages can be established to form the basis for benefit-cost analysis. To compensate for the uncertainties of estimating benefits, it makes sense to demand a relatively high ratio of benefit to cost. For example, it is not unusual to have senior leadership demand a ROI of 100% in the first year on a Six Sigma project. Rather than becoming distressed at this "injustice," the Black Belt should realize that such demands are a response to the inherent difficulties in quantifying benefits.

## Types of Savings

The accounting or finance department should formally define the different categories of savings. Savings are typically placed in categories such as:

- *Hard savings* are actual reductions in dollars now being spent, such as reduced budgets, fewer employees, reduction of prices paid on purchasing contracts, and so on. Hard savings can be used to lower prices, change bid models, increase profits, or for other purposes where a high degree of confidence in the benefit is required.
- *Soft savings* are projected reductions that should result from the project. For example, savings from less inventory, reduced testing, lower cycle times, improved yields, lower rework rates, and reduced scrap.

It is important that savings be integrated into the business systems of the organization. If the institutional framework doesn't change, the savings could eventually be lost. For example, if a Six Sigma project improves a process yield, be sure the MRP system's calculations reflect the new yields.

## Calculating the Value of Customer Retention

Customers have value. This simple fact is obvious when one looks at a customer making a single purchase. The transaction provides revenue and profit to the firm. However, when the customer places a demand on the firm, such as a return of a previous purchase or a call for technical support, there is a natural tendency to see this as a loss. At these times it is important to understand that customer value must not be viewed on a short-term transaction-by-transaction basis. Customer value must be measured *over the lifetime of the relationship*. One method of calculating the lifetime value of a loyal customer, based on work by Frederick Reichheld of Bain and Co. and the University of Michigan's Claes Fornell, is as follows (Stewart, 1995):

1. Decide on a meaningful period of time over which to do the calculations. This will vary depending on your planning cycles and your business: A life insurer should track customers for decades, a disposable diaper maker for just a few years, for example.
2. Calculate the profit (net cash flow) customers generate each year. Track several samples—some newcomers, some old-timers—to find out how much business they gave you each year, and how much it cost to serve them. If possible, segment them by age, income, sales channel, and so on. For the first year, be sure to subtract the cost of acquiring the pool of customers, such as advertising, commissions, and back-office costs of

setting up a new account. Get specific numbers—profit per customer in year 1, year 2, and so on—not averages for all customers or all years. Long-term customers tend to buy more, pay more (newcomers are often lured by discounts), and create less bad debt.

3. Chart the customer "life expectancy," using the samples to find out how much your customer base erodes each year. Again, specific figures are better than an average such as "10% a year"; old customers are much less likely to leave than new ones. In retail banking, 26% of account holders defect in the first year; in the ninth year, the rate drops to 9%.

4. Once you know the profit per customer per year and the customer-retention figures, it's simple to calculate net present value (NPV). Pick a discount rate—if you want a 15% annual return on assets, use that. In year 1, the NPV will be profit ÷ 1.15. Next year, NPV = (year 2 profit × retention rate) ÷ $(1.15)^2$. In year $n$, the last year in your figures, the NPV is the year $n$ adjusted profit ÷ $(1.15)^n$. The sum of the years 1 through $n$ is how much your customer is worth—the net present value of all the profits you can expect from his or her tenure. (These calculations are discussed in greater detail in the "Deliverables" section of Chap. 6.)

This is very valuable information. It can be used to find out how much to spend to attract new customers, and which ones. Better still, you can exploit the leverage customer satisfaction offers. Take your figures and calculate how much more customers would be worth if you increased retention by 5%. Figure 4.1 shows the increase in customer NPV for a 5% increase in retention for three industries.

Once the lifetime value of the customer is known, it forms the basis of loyalty-based management[SM] of the customer relationship. According to Reichheld (1996), loyalty-based management is the practice of carefully selecting customers, employees, and investors and

**Figure 4.1** Increase in customer NPV for a 5% increase in customer retention.

then working hard to retain them. There is a tight, cause-and-effect connection between investor, employee, and customer loyalty. These are the human assets of the firm.

## A System for Assessing Six Sigma Projects

Assessing Six Sigma projects is an art as well as a science. It is also critical to the success of Six Sigma, and to the individual Black Belt. Far too many Black Belts fail because they are not discriminating enough in their selection of projects. If project selection is systematically sloppy, the entire Six Sigma effort can fail.

The approach offered here is quantitative in the sense that numbers are determined and an overall project score calculated. It is subjective to the degree it requires interpretation of the situation, estimated probabilities, costs, and commitments. However, the rigor of completing this assessment process allows for better judgments regarding projects. The numbers (weights, scores, acceptable length of projects, dollar cutoffs, etc.) are recommendations which can and should be replaced by the organization's leadership to reflect organizational priorities. The scale ranges from 0 to 9 for each criterion, and the weights sum to 1.00, so the highest possible weighted score for a project is 9.

The Six Sigma department or Process Excellence function can compile summary listings of project candidates from these assessments. Sorting the list in descending order provides a guide to the final decision as to which projects to pursue. Each Black Belt or Green Belt will probably have their own list, which can also be sorted and used to guide their choices.

**Worksheet 1**  Six Sigma Project Evaluation

| Project Name: | Date of Assessment: |
|---|---|
| Black Belt: | Master Black Belt: |
| Weighted Overall Project Score: | Project Number: |

| Criteria | Score | Weight | Weighted Score* |
|---|---|---|---|
| 1. Sponsorship | | 0.23 | |
| 2. Benefits (specify main beneficiary)<br>• 2.1 External customer:<br>• 2.2 Shareholder:<br>• 2.3 Employee or internal customer:<br>• 2.4 Other (e.g., supplier, environment): | Overall Benefit Score | 0.19 | |
| 3. Availability of resources other than team | | 0.16 | |
| 4. Scope in terms of Black Belt effort | | 0.12 | |
| 5. Deliverable | | 0.09 | |

| | | | |
|---|---|---|---|
| 6. Time to complete | | 0.09 | |
| 7. Team membership | | 0.07 | |
| 8. Project charter | | 0.03 | |
| 9. Value of Six Sigma approach | | 0.02 | |
| TOTAL (sum of weighted score column) | | 1.00 | |

*Note: Any criterion scores of zero must be addressed before project is approved.*

*\*Weighted score = project's score for each criterion times the weight.*

**Worksheet 2** Six Sigma Project Evaluation Guidelines

### 1.0 Sponsorship

| Score | Interpretation |
|---|---|
| 9 | Director-level sponsor identified, duties specified, and sufficient time committed and scheduled |
| 3 | Director-level sponsor identified, duties specified, and sufficient time committed but not scheduled |
| 1 | Willing director-level sponsor who has accepted charter statement |
| 0 | Director-level sponsor not identified, or sponsor has not accepted the charter |

### 2.0 Stakeholder Benefits*

"Tangible and verifiable benefits for a major stakeholder"

### 2.1 Stakeholder: External Customer

#### 2.1.1 Customer Satisfaction

| Score | Interpretation |
|---|---|
| 9 | Substantial and statistically significant increase in *overall* customer satisfaction or loyalty |
| 3 | Substantial and statistically significant increase in a *major subcategory* of customer satisfaction |
| 1 | Substantial and statistically significant increase in a *focused area* of customer satisfaction |
| 0 | Unclear or nonexistent customer satisfaction impact |

#### 2.1.2 Quality Improvement (CTQ)

| Score | Interpretation |
|---|---|
| 9 | 10 × or greater improvement in critical to quality (CTQ) metric |
| 5 | 5 × to 10 × improvement in CTQ metric |
| 3 | 2 × to 5 × improvement in CTQ metric |
| 1 | Statistically significant improvement in CTQ metric, but less than 2 × magnitude |
| 0 | Project's impact on CTQ metrics undefined or unclear |

*(continued on next page)*

**Worksheet 2**  Six Sigma Project Evaluation Guidelines (*continued*)

### 2.2  Stakeholder: Shareholder

#### 2.2.1  Financial Benefits

| Score | Interpretation |
|---|---|
| 9 | Hard net savings (budget or bid model change) greater than $500K. Excellent ROI |
| 5 | Hard net savings between $150K and $500K. Excellent ROI |
| 3 | Hard net savings between $50K and $150K, or cost avoidance greater than $500K. Good ROI |
| 1 | Hard savings of at least $50K, or cost avoidance of between $150K and $500K. Acceptable ROI |
| 0 | Project claims a financial benefit but has hard savings less than $50K, cost avoidance less than $150K, or unclear financial benefit |

#### 2.2.2  Cycle Time Reduction

| Score | Interpretation |
|---|---|
| 9 | Cycle time reduction that improves revenue, bid model, or budget by more than $500K. Excellent ROI |
| 5 | Cycle time reduction that improves revenue, bid model, or budget by $150K to $500K. Excellent ROI |
| 3 | Cycle time reduction that improves revenue, bid model, or budget by $50K to $150K, or creates a cost avoidance of more than $500K. Good ROI |
| 1 | Cycle time reduction that results in cost avoidance between $150K and $500K. Acceptable ROI |
| 0 | Project claims a cycle time improvement but has hard savings less than $50K, cost avoidance less than $150K, or unclear financial benefit from the improvement in cycle time |

#### 2.2.3  Revenue Enhancement

| Score | Interpretation |
|---|---|
| 9 | Significant increase in revenues, excellent ROI |
| 3 | Moderate increase in revenues, good ROI |
| 1 | Increase in revenues with acceptable ROI |
| 0 | Unclear or nonexistent revenue impact |

*(continued on next page)*

### 2.3 Stakeholder: Employee or Internal Customer
#### 2.3.1 Employee Satisfaction

| Score | Interpretation |
|---|---|
| 9 | Substantial and statistically significant increase in *overall* employee satisfaction |
| 3 | Substantial and statistically significant increase in *a major element* of employee satisfaction |
| 1 | Substantial and statistically significant increase in *a focused area* of employee satisfaction |
| 0 | Unclear or nonexistent employee satisfaction impact |

### 2.4 Stakeholder: Other
#### 2.4.1 Specify Stakeholder: _____

**Benefits**

| Score | Interpretation |
|---|---|
| 9 | |
| 5 | |
| 3 | |
| 1 | |
| 0 | Unclear or nonexistent benefit |

### 3.0 Availability of Resources Other Than Team

| Score | Interpretation |
|---|---|
| 9 | Needed resources available when needed |
| 3 | Limited or low priority access to needed to resources |
| 0 | Resources not available, or excessive restrictions on access to resources |

### 4.0 Scope in Terms of Black Belt Effort

| Score | Interpretation |
|---|---|
| 9 | Projected return substantially exceeds required return |
| 3 | Projected return exceeds required return |
| 1 | Projected return approximately equals required return |
| 0 | Projected return not commensurate with required return |

Required return can be calculated as follows:*
   (1) Length of project (months) = _____
   (2) Proportion of Black Belt's time required (between 0 and 1) = _____
   (3) Probability of success (between 0 and 1) = _____
Required return** = $83,333 × (1) × (2) ÷ (3) = $ _____
Projected return: $_____

*(continued on next page)*

**Worksheet 2**  Six Sigma Project Evaluation Guidelines (*continued*)

### 5.0 Deliverable (Scope)

| Score | Interpretation |
|---|---|
| 9 | New or improved process, product, or service to be created is clearly and completely defined |
| 3 | New or improved process, product or service to be created is defined |
| 0 | Deliverable is poorly or incorrectly defined. For example, a "deliverable" that is really a tool such as a process map |

### 6.0 Time to Complete

| Score | Interpretation |
|---|---|
| 9 | Results realized in less than 3 months |
| 3 | Results realized in between 3 and 6 months |
| 1 | Results realized in 7 to 12 months |
| 0 | Results will take more than 12 months to be realized |

### 7.0 Team Membership

| Score | Interpretation |
|---|---|
| 9 | Correct team members recruited and time commitments scheduled |
| 3 | Correct team members recruited, time committed but not scheduled |
| 1 | Correct team members recruited |
| 0 | Team members not recruited or not available |

### 8.0 Project Charter

| Score | Interpretation |
|---|---|
| 9 | All elements of the project charter are complete and acceptable. Linkage between project activities and deliverable is clear |
| 3 | Project charter acceptable with minor modifications |
| 0 | Project charter requires major revisions |

### 9.0 Value of Six Sigma Approach (DMAIC or equivalent)

| Score | Interpretation |
|---|---|
| 9 | Six Sigma approach essential to the success of the project. Black Belt/Green Belt skill set required for success |
| 3 | Six Sigma approach helpful but not essential. Black Belt/Green Belt skill set can be applied |
| 0 | Usefulness of Six Sigma approach not apparent. Specific Black Belt or Green Belt skills are not necessary |

See online video describing the Consensus Criteria Method at www.mhprofessional.com/ssh5.

See online video describing Project Selection Using a Prioritization Matrix at www.mhprofessional.com/ssh5.

Download example templates for a basic Matrix Diagram, the Prioritization Consensus Method, and/or the Prioritization Full Analytic Method at www.mhprofessional.com/ssh5.

### Other Methods of Identifying Promising Projects

Projects should be selected to support the organization's overall strategy and mission. Because of this global perspective most projects involve the efforts of several different functional areas. Not only do individual projects tend to cut across organizational boundaries, different projects are often related to one another. To effectively manage this complexity it is necessary to integrate the planning and execution of projects across the entire enterprise. One way to accomplish this is QFD, which is discussed in detail in Chap. 2. In addition to QFD and the scoring method described above, a number of other procedures are presented here to help identify a project's potential worth.

### *Using Pareto Analysis to Identify Six Sigma Project Candidates*

*Pareto principle* refers to the fact that a small percentage of processes cause a large percentage of the problems. The Pareto principle is useful in narrowing a list of choices to those few projects that offer the greatest potential. When using Pareto analysis keep in mind that there may be hidden "pain signals." Initially problems create pain signals such as schedule disruptions and customer complaints. Often these *symptoms* are treated rather than their underlying "diseases"; for example, if quality problems cause schedule slippages which lead to customer complaints, the "solution" might be to keep a large inventory and sort the good from the bad. The result is that the schedule is met and customers stop complaining, but at huge cost. These opportunities are often greater than those currently causing "pain," but they are now built into business systems and therefore very difficult to see. One solution to the hidden problem phenomenon is to focus on processes rather than symptoms. Some guidelines for identifying dysfunctional processes for potential improvement are shown in Table 4.1.

The "symptom" column is useful in identifying problems and setting priorities. The "disease" column focuses attention on the underlying causes of the problem, and the "cure" column is helpful in chartering quality improvement project teams and preparing mission statements.

**Table 4.1** Dysfunctional Process Symptoms and Underlying Diseases

| Symptom | Disease | Cure |
|---|---|---|
| Extensive information exchange, data redundancy, rekeying | Arbitrary fragmentation of a natural process | Discover why people need to communicate with each other so often; integrate the process |
| Inventory, buffers, and other assets stockpiled | System slack to cope with uncertainty | Remove the uncertainty |
| High ratio of checking and control to value-added work (excessive test and inspection, internal controls, audits, etc.) | Fragmentation | Eliminate the fragmentation, integrate processes |
| Rework and iteration | Inadequate feedback in a long work process | Process control |
| Complexity, exceptions and special causes | Accretion onto a simple base | Uncover original "clean" process and create new process(es) for special situations; eliminate excessive standardization of processes |

## Prioritizing Projects with the Pareto Priority Index

After a serious search for improvement opportunities the organization's leaders will probably find themselves with more projects to pursue than they have resources. The Pareto Priority Index (PPI) is a simple way of prioritizing these opportunities. The PPI is calculated as follows (Juran and Gryna , 1993):

$$\text{PPI} = \frac{\text{Saving} \times \text{probability of success}}{\text{Cost} \times \text{time to completion (years)}} \tag{4.1}$$

A close examination of the PPI equation shows that it is related to return on investment adjusted for probability of success. The inputs are, of course, estimates, and the result is totally dependent on the accuracy of the inputs. The resulting number is an index value for a given project. The PPI values allow comparison of various projects. If there are clear standouts the PPI can make it easier to select a project. Table 4.2 shows the PPIs for several hypothetical projects.

The PPI indicates that resources be allocated first to reducing wave solder defects, then to reducing customer delivery complaints, and so on. The PPI may not always give such a clear ordering of priorities. When two or more projects have similar PPIs a judgment must be made on other criteria.

**Table 4.2** Illustration of the Pareto Priority Index (PPI)

| Project | Savings, $ Thousands | Probability | Cost, $ Thousands | Time, Weeks | PPI |
|---|---|---|---|---|---|
| Reduce wave solder defects 50% | $700 | 0.7 | $15 | 15 | 2.18 |
| NC machine capability improvement | $500 | 0.9 | $20 | 20 | 1.13 |
| ISO 9001 certification | $1,500 | 0.9 | $75 | 78 | 0.23 |
| Eliminate customer delivery complaints | $2,500 | 0.5 | $35 | 26 | 1.37 |
| Reduce assembly defects 50% | $900 | 0.7 | $30 | 20 | 1.05 |

## *Throughput-Based Project Selection*

While careful planning and management of projects is undeniably important, they matter little if the projects being pursued have no impact on the bottom line (throughput). As you will see in the following section, if you choose the wrong projects it is possible to make big "improvements" in quality and productivity that have no impact whatever on the organization's net profit. Selecting which projects to pursue is of critical importance. In this section we will use the theory of constraints (TOC) to determine which project(s) to pursue.

### Theory of Constraints

Every organization has constraints. Constraints come in many forms. When a production or service process has a resource constraint (i.e., it lacks a sufficient quantity of some resource to meet the market demand), then the sequence of improvement projects should be identified using very specific rules. According to Goldratt (1990), the rules are

1. **Identify the system's constraint(s)**—Consider a fictitious company that produces only two products, P and Q (Fig. 4.2). The market demand for P is 100 units per week and P sells for $90 per unit. The market demand for Q is 50 units per week and Q sells for $100 per unit. Assume that A, B, C, and D are workers who have different non-interchangeable skills and that each worker is available for only 2,400 minutes per week (8 hours per day, 5 days per week). For simplicity, assume that there is no variation, waste, etc. in the process. Assume this process has a constraint, worker B. This fact has profound implications for selecting Six Sigma projects.
2. **Decide how to exploit the system's constraint(s)**—Look for Six Sigma projects that minimize waste of the constraint. For example, if the constraint is (feeding) the market demand (i.e., a capacity constraint), then we look for Six Sigma projects that provide 100% on-time delivery. Let's not waste anything! If the constraint is a machine,

**Figure 4.2** A simple process with a constraint.

or process step, as in this example, focus on reducing setup time, eliminating errors or scrap, and keeping the process step running as much as possible.

3. **Subordinate everything else to the above decision**—Choose Six Sigma projects to maximize throughput of the constraint. After completing step 2, choose projects to eliminate waste from downstream processes; once the constraint has been utilized to create something we don't want to lose it due to some downstream blunder. Then choose projects to ensure that the constraint is always supplied with adequate nondefective resources from upstream processes. We pursue upstream processes last because by definition they have slack resources, so small amounts of waste upstream that are detected before reaching the constraint are less damaging to throughput.

4. **Elevate the system's constraint(s)**—Elevate means "lift the restriction." This is step 4, not step 2. Often the projects pursued in steps 2 and 3 will eliminate the constraint. If the constraint continues to exist after performing steps 2 and 3, look for Six Sigma projects that provide additional resources to the constraint. These might involve, for example, purchasing additional equipment or hiring additional workers with a particular skill.

5. If, in the previous steps, a constraint has been broken, go back to step 1. There is a tendency for thinking to become conditioned to the existence of the constraint. A kind of mental inertia sets in. If the constraint has been lifted, then you must rethink the entire process from scratch. Returning to step 1 takes you back to the beginning of the cycle.

## Comparison of TOC with Traditional Approaches

It can be shown that the TOC approach is superior to the traditional TQM approaches to project selection. For example, consider the data in the following table. If you were to apply Pareto analysis to scrap rates you would begin with Six Sigma projects that reduced the scrap produced by worker A. In fact, assuming the optimum product mix, worker A has about 25% slack time, so the scrap loss can be made up without shutting down worker B, who is the constraint. The TOC would suggest that the scrap loss of worker B and the downstream processes C and D be addressed first, the precise opposite of what Pareto analysis recommends.

| Process    | A  | B  | C  | D  |
|------------|----|----|----|----|
| Scrap rate | 8% | 3% | 5% | 7% |

Process Scrap Rates

Of course, before making a decision as to which projects to finance cost/benefit analyses are still necessary, and the probability of the project succeeding must be estimated. But by using the TOC you will at least know where to look first for opportunities.

## Using Constraint Information to Focus Six Sigma Projects

Applying the TOC strategy described earlier tells us *where* in the process to focus. Adding CTX information (see Table 4.3) can help tell us which type of project to focus on, that is, should we focus on quality, cost, or schedule projects? Assume that you have three Six Sigma candidate projects, all focusing on process step B, the constraint. The area addressed is correct, but which project should you pursue first? Let's assume that we learn that one project will primarily improve quality, another cost, and another schedule. Does this new information help? Definitely! Take a look at Table 4.3 to see how this information can be used. Projects in the same priority group are ranked according to their impact on throughput.

**Table 4.3** Throughput Priority of CTX Projects That Affect the Constraint

| Project Type | Discussion |
|---|---|
| CTQ | Any unit produced by the constraint is especially valuable because if it is lost as scrap additional constraint time must be used to replace it or rework it. Since constraint time determines throughput (net profit of the entire system), the loss far exceeds what appears on scrap and rework reports. CTQ projects at the constraint are the highest priority. |
| CTS | CTS projects can reduce the time it takes the constraint to produce a unit, which means that the constraint can produce more units. This directly impacts throughput. CTS projects at the constraint are the highest priority. |
| CTC | Since the constraint determines throughput, the unavailability of the constraint causes lost throughput *of the entire system*. This makes the cost of constraint down time extremely high. The cost of *operating* the constraint is usually miniscule by comparison. Also, CTC projects can have an adverse impact on quality or schedule. Thus, CTC projects at the constraint are low priority. |

The same thought process can be applied to process steps before and after the constraint. The results are shown in Table 4.4.

Table 4.4  Project Throughput Priority Versus Project Focus

| Focus of Six Sigma Project | | Before the constraint | At the constraint | After the constraint |
|---|---|---|---|---|
| CTX: Characteristic addressed is critical to ... | Quality (CTQ) | △ | ◎ | ◎ |
| | Cost (CTC) | ○ | ☐ | ○ |
| | Schedule (CTS) | ☐ | ◎ | ○ |

☐ Low throughput priority.
○ Moderate throughput priority.
◎ High throughput priority.

Note that Table 4.4 assumes that projects *before* the constraint do not result in problems *at* the constraint. Remember, impact should always be measured in terms of throughput. If a process upstream from the constraint has an adverse impact on throughput, then it can be considered a constraint. If an upstream process *average* yield is enough to feed the constraint on the average, it may still present a problem. For example, an upstream process producing 20 units per day with an average yield of 90% will produce, on average, 18 good units. If the constraint requires 18 units, things will be okay about 50% of the time, but the other 50% of the time things won't be okay. One solution to this problem is to place a work-in-process (WIP) inventory between the process and the constraint as a safety buffer. Then on those days when the process yield is below 18 units, the inventory can be used to keep the constraint running. However, there is a cost associated with carrying a WIP inventory. A Six Sigma project that can improve the yield will reduce or eliminate the need for the inventory and should be considered even if it doesn't impact the constraint directly, assuming the benefit-cost analysis justifies the project. On the other hand, if an upstream process can easily make up any deficit before the constraint needs it, then a project for the process will have a low priority.

Knowing the project's throughput priority will help you make better project selection decisions. Of course, the throughput priority is just one input into the project selection process, other factors may lead to a different decision. For example, impact on other projects, a regulatory requirement, a better payoff in the long term, etc.

## Multitasking and Project Scheduling

A Six Sigma enterprise will always have more projects to pursue than it has resources to do them. The fact that resources (usually Black Belts or Green Belts) are scarce means that projects must be scheduled, that is, some projects must be undertaken earlier than others. In such situations it is tempting to use *multitasking* of the scarce resource. Multitasking is defined as the assignment of a resource to several priorities during the same period of time. The logic is that by working on several projects or assignments simultaneously, the entire portfolio of work will be done more quickly. However, while this is true for independent resources working independent projects or subprojects in parallel, it is *not* true when applied to a single resource assigned to multiple projects or interdependent tasks within a project.

Consider the following situation. You have three Six Sigma projects, A, B, and C. A single-tasking solution is to first do A, then B, and then C. Here's the single-activity project schedule.

| A | B | C |
|---|---|---|
| (Complete in Wk 10) | (Complete in Wk 20) | (Complete in Wk 30) |

If each project takes 10 weeks to complete, then A will be completed in 10 weeks, B in 20 weeks, and C in 30 weeks. The average time to complete the three projects is (10 + 20 + 30)/3 = 60/3 = 20 weeks. The average doesn't tell the whole story, either. The benefits will begin as soon as the project is completed, and by the end of the 30-week period project A will have been completed for 20 weeks, and project B for 10 weeks.

Now let's consider a multitasking strategy. Here we split our time equally between the three projects in a given 10-week period. That way the sponsors of projects B and C will see activity on their projects much sooner than if we used a single-task approach to scheduling. The new schedule looks like this:

| A | B | C | A | B | C | A | B | C |
|---|---|---|---|---|---|---|---|---|

With this multitasking schedule project A will be completed in 23.3 weeks, project B in 26.7 weeks, and project C will still take 30 weeks. The completion time for project A went from 10 weeks to 23.3 weeks, for project B it went from 20 weeks to 26.7 weeks, and for project C it remained the same, 30 weeks. The overall average completion time went from 20 weeks to 26.67 weeks, a 33% deterioration in average time to complete. And this is a best-case scenario. In real life there is always some lost time when making the transition from one project to another. The Black Belt has to clear her head of what she was doing, review the next project, get the proper files ready, reawaken sponsors and team members, and so on. This can often take a considerable amount of time, which is added to the time needed to complete the projects.

## Critical Chain Project Portfolio Management

Critical chain project management avoids the multitasking problem by changing the way the organization manages groups of projects, and the way the individual projects are managed.

## Managing the Organization's Projects

First, at the organizational level, multitasking of key resources is stopped. People and other resources are allowed to focus on projects one at a time. This means that management must accept responsibility for prioritizing projects, and policies must be developed which mandate single-project focus and discourage multitasking. To be successful the organization must determine its capacity to complete projects. Every organization finds itself with more opportunities than it can successfully pursue with finite resources. This means that only a select portfolio of projects should be undertaken in any time interval. The constraining resource is usually a key position in the organization, say the time available by project sponsors, engineers, programmers, etc. This information can be used to determine organizational capacity and to schedule project start dates according to the availability of the key resource. This is called *project launch synchronization*, and the scarce resource that paces the project system is called a *synchronizer resource*.

## Synchronizer Resource Usage

Critical chain project management does not permit multitasking of scarce resources. People and equipment that are fully utilized on projects, *synchronizer resources*, are assigned to a sequence of single projects. The sequence of projects is based on enterprise priorities. If a project requires one or more synchronizer resources, it is vital that your project start dates integrate the schedules of these resources. In particular, this will require that those activities that require time from a synchronizer resource (and the project as a whole) stipulate "Start no earlier than" dates. Although synchronizer resources are protected by capacity buffers and might hypothetically start at a date earlier than specified, the usual practice is to utilize any unplanned excess capacity to allow the organization to pursue additional opportunities, thereby increasing the organization's capacity to complete projects. Note that human resources are defined in terms of the skills required for the activity, not in terms of individual people. In fact, the resource manager should refrain from assigning an activity to an individual until all predecessors have been completed and the activity is ready to begin. This precludes the temptation to multitask as the individual looks ahead and sees the activity start date drawing near.

Project start dates are determined by beginning with the highest priority project and calculating the end date for the synchronizing resource based on the estimated duration of all activities that require the synchronizing resource. The second highest priority project's start date will be calculated by adding a capacity buffer to the expected end date of the first project. The third highest priority project's start date is based on the completion date of the second, and so on. If, by chance, the synchronizing resource is available before the scheduled start date, the time can be used to increase the organization's capacity to complete more projects. Figure 4.3 illustrates this strategy.

**Figure 4.3** Critical chain scheduling illustration.

### Summary and Preliminary Project Selection

At this point you have evaluated project candidates using a number of different criteria. You must now rank the projects, and make your preliminary selections. You may use Worksheet 3 to assist you with this. The reason your selections are preliminary is that you lack complete data. As they work on the project, Six Sigma project teams will continuously reevaluate it, and they may uncover data which will lower or raise the project's priority. The project sponsor is responsible for coordinating changes in priority with the process owners.

## Ongoing Management Support

Once projects have been selected, management support continues through the life of the project in a number of ways. Chapter 5 discusses the project reporting requirements expected of the project team and its leader. Management sponsors provide the management interface necessary to ensure the project remains on course relative to its objectives, or to change objectives if necessary given new information discovered by the project team. At times, it will be necessary for management to reiterate its project support to clear roadblocks, as discussed below. Management must also evaluate the project results, as well as the team performance, to provide feedback to the management systems for identifying improvement opportunities. This is further discussed later in this chapter.

## Internal Roadblocks

Most organizations still have a hierarchical, command-and-control organizational structure, sometimes called "smokestacks" or "silos." The functional specialists in charge of each smokestack tend to focus on optimizing their own functional area, often to the detriment of the organization as a whole. In addition, the hierarchy gives these managers a monopoly on the authority to act on matters related to their functional specialty. The combined effect is both a desire to resist change and the authority to resist change, which often creates insurmountable roadblocks to quality improvement projects.

It is important to realize that organizational rules are, by their nature, a barrier to change. The formal rules take the form of written standard operating procedures (SOPs). The very purpose of SOPs is to standardize behavior. Unfortunately, the quality profession has historically overemphasized formal documentation. Approaches such as ISO 9000 and ISO 14000 risk indoctrinating formal rules that are merely responses to problems that no longer exist after the reason for their existence has passed. In an organization that is serious about its written rules even senior leaders find themselves helpless to act without submitting to a sometimes burdensome rule-changing process. In those cases, the true power in such an organization is the bureaucracy that controls the procedures. If the organization falls into the trap of creating written rules for too many things, it can find itself moribund in a fast-changing external environment. This is a recipe for disaster. While electronic document control systems can remove some of these issues, it's critical to manage them as a means toward control only when it meets the overall needs of the system, rather than as an inefficient solution to localized problems.

Restrictive rules need not take the form of management limitations on itself; procedures that define hourly work in great detail also produce barriers, for example, union work rules. Projects almost always require that work be done differently, and such procedures prohibit such change. Organizations that tend to be excessive in SOPs also tend to be heavy on work rules. The combination is often deadly to quality improvement efforts.

Organization structures preserve the status quo in other ways besides formal, written restrictions in the form of procedures and rules. Another effective method of limiting change is to require permission from various departments, committees, councils, boards, experts, and so on. Even though the organization may not have a formal requirement that "permission" be obtained, the effect may be the same, for example, "You should run that past accounting" or "Ms. Reimer and Mr. Evans should be informed about this project." When permission for vehicles for change (e.g., project budgets, plan approvals) is required from a group that meets infrequently it creates problems for project planners. Plans may be rushed so they can be presented at the next meeting, lest the project be delayed for months. Plans that need modifications may be put on hold until the next meeting, months away. Or, projects may miss the deadline and be put off indefinitely.

## *External Roadblocks*

Modern organizations do not exist as islands. Powerful external forces take an active interest in what happens within the organization. Government bodies have created a labyrinth of rules and regulations that the organization must negotiate to utilize its human resources without incurring penalties or sanctions. The restrictions placed on modern businesses by outside regulators are challenging to say the least. When research involves people, ethical and legal concerns sometimes require that external approvals be obtained. The approvals are contingent on such issues as informed consent, safety, cost and so on.

Many industries have "dedicated" agencies to deal with. For example, the pharmaceutical industry must deal with the Food and Drug Administration (FDA). These agencies must often be consulted before undertaking projects. For example, a new treatment protocol involving a new process for treatment of pregnant women prior to labor may involve using a drug in a new way (e.g., administered on an outpatient basis instead of on an inpatient basis).

Many professionals face liability risks that are part of every decision. Often these fears create a "play it safe" mentality that acts as a barrier to change. The fear is even greater when the project involves new and untried practices and technology.

## *Individual Barriers to Change*

Perhaps the most significant change, and therefore the most difficult, is to change ourselves. It seems to be a part of human nature to resist changing oneself. By and large, we worked hard to get where we are, and our first impulse is to resist anything that threatens our current position. Forsha (1992) provides the process for personal change shown in Fig. 4.4.

The adjustment path results in preservation of the status quo. The action path results in change. The well-known PDCA cycle can be used once a commitment to action has been made by the individual. The goal of such change is continuous *self*-improvement.

Within an organizational context, the individual's reference group plays a part in personal resistance to change. A reference group is the aggregation of people a person thinks of when they use the word "we." If "we" refers to the company, then the company is the individual's reference group and he or she feels connected to the company's success or failure. However, "we" might refer to the individual's profession or trade group, for example, "We doctors," "We engineers," "We union members." In this case the leaders shown on the formal organization chart will have little influence on the individual's attitude toward the success or failure of the project. When a project involves external reference groups with competing agendas, the task of building buy-in and consensus is daunting indeed.

**Figure 4.4** The process of personal change. (From Forsha, 1992. Copyright © 1992 by ASQ Quality Press, Milwaukee, WI. Used by permission.)

## *Ineffective Management Support Strategies*

**Strategy 1: Command people to act as you wish**—With this approach the senior leadership simply commands people to act as the leaders wish. The implication is that those who do not comply will be subjected to disciplinary action. People in less senior levels of an organization often have an inflated view of the value of raw power. The truth is that even senior leaders have limited power to rule by decree. Human beings by their nature tend to act according to their own best judgment. Thankfully, commanding that they do otherwise usually has little effect. The result of invoking authority is that the decision-maker must constantly try to divine what the leader wants them to do in a particular situation. This leads to stagnation and confusion as everyone waits on the leader. Another problem with commanding as a form of "leadership" is the simple communication problem. Under the best of circumstances people will often simply misinterpret the leadership's commands.

**Strategy 2: Change the rules by decree**—When rules are changed by decree the result is again confusion. What are the rules today? What will they be tomorrow? This leads again to stagnation because people don't have the ability to plan for the future. Although rules make it difficult to change, they also provide stability and structure that may serve some useful purpose. Arbitrarily changing the rules based on force (which is what "authority" comes down to) instead of a set of guiding principles does more harm than good.

**Strategy 3: Authorize circumventing of the rules**—Here the rules are allowed to stand, but exceptions are made for the leader's "pet projects." The result is general disrespect for and disregard of the rules, and resentment of the people who are allowed to violate rules that bind everyone else. An improvement is to develop a formal method for circumventing the rules, for example, deviation request procedures. While this is less arbitrary, it adds another layer of complexity and still doesn't change the rules that are making change difficult in the first place.

**Strategy 4: Redirect resources to the project**—Leaders may also use their command authority to redirect resources to the project. A better way is to develop a fair and easily understood system to ensure that projects of strategic importance are adequately funded as a matter of policy. In our earlier discussion of project scheduling we discussed "crash scheduling" as a means of completing projects in a shorter time frame. However, the assumption was that the basis for the allocation was cost or some other objective measure of the organization's best interest. Here we are talking about political clout as the basis of the allocation.

## *Effective Management Support Strategies*

**Strategy 1: Transform the formal organization and the organization's culture**—By far the best solution to the problems posed by organizational roadblock is to transform the organization to one where these roadblocks no longer exist. As discussed earlier, this process can't be implemented by decree. As the leader helps project teams succeed, he will learn about the need for transformation. Using his persuasive powers the leader-champion can undertake the exciting challenge of creating a culture that embraces change instead of fighting it.

**Strategy 2: Mentor**—In Greek mythology, Mentor was an elderly man, the trusted counselor of Odysseus, and the guardian and teacher of his son Telemachus. Today the term, "mentor" is still used to describe a wise and trusted counselor or teacher. When this person occupies an important position in the organization's hierarchy, he or she can be a powerful force for eliminating roadblocks. Modern organizations are complex and confusing. It is often difficult to determine just where one must go to solve a problem or obtain a needed resource. The mentor can help guide the project manager through this maze by clarifying lines of authority. At the same time, the mentor's senior position enables him to see the implications of complexity and to work to eliminate unnecessary rules and procedures.

**Strategy 3: Identify informal leaders and enlist their support**—Because of their experience, mentors often know that the person whose support the project really needs is not the one occupying the relevant box on the organization chart. The mentor can direct the project leader to the person whose opinion really has influence. For example, a project may need the approval of, say, the vice president of engineering. The engineering VP may be balking because his senior metallurgist hasn't endorsed the project.

**Strategy 4: Find legitimate ways around people, procedures, resource constraints, and other roadblocks**—It may be possible to get approvals or resources through means not known to the project manager. Perhaps a minor change in the project plan can bypass a cumbersome procedure entirely. For example, adding an engineer to the team might automatically place the authority to approve process experiments within the team rather than in the hands of the engineering department.

## *Cross-Functional Collaboration*

This section will address the impact of organizational structures on management of Six Sigma projects.

Six Sigma projects are process-oriented, and most processes that have significant impact on quality cut across several different departments. Modern organizations, however, are hierarchical, that is, they are defined by superior/subordinate relationships. These organizations tend to focus on specialized functions (e.g., accounting, engineering). But adding value for the customer requires that several different functions work together. The ideal solution is the transformation of the organization into a structure designed to produce value without the need for a hierarchical structure. However, until that is accomplished, Six Sigma project managers will need to deal with the conflicts inherent in doing cross-functional projects in a hierarchical environment.

Project managers "borrow" people from many departments for their projects, which create matrix organizational structures. The essential feature of a matrix organization is that some people have two or more bosses or project customers. These people effectively report to multiple bosses, for example, the project manager and their own boss. Ruskin and Estes refer to people with more than one boss as *multi-bossed individuals*, and their bosses and customers *as multiple bosses*. Somewhere in the organization is a *common boss*, who resolves conflicts between multiple bosses when they are unable to do so on their own. Of course, multiple bosses can prevent conflicts by cooperation and collaboration before problems arise.

Often multi-bossed individuals are involved with several projects, further complicating the situation. Collaborative planning between the multiple bosses is necessary to determine how the time of multi-bossed individuals, and other resources, will be shared. For additional discussion of more complex matrix structures, see Ruskin and Estes, 1995.

Good communication is helpful in preventing problems. Perhaps the most important communication is frequent, informal updates of all interested parties by the project manager. More formal status reports should also be specified in the project plan and sent to people with an interest in the project. The project manager should determine who gets what information, which is often tricky due to the multi-boss status of the project manager. Some managers may not want "too much" information about their department's "business" shared with their peers from other areas. Other managers may be offended if they receive less information than everyone else. The project manager's best diplomatic skills may be required to find the right balance.

Status reports invariably indicate that the project plan is less than perfect. The process by which the plans will be adjusted should be understood in advance. The process should specify who will be permitted to make adjustments, when the adjustments will be allowed, and how much authority the bosses and project manager have in making adjustments.

Negotiated agreements should be documented, while generating the minimum possible amount of additional red tape and paperwork. The documentation will save the project manager a great deal of time in resolving disputes down the road regarding who agreed to what.

## Tracking Six Sigma Project Results

It is vital that information regarding results be accumulated and reported. This is useful for a variety of purposes:

- Evaluating the effectiveness of the Six Sigma project selection system
- Determining the overall return on investment
- Setting budgets

**Worksheet 3**   Project Assessment Summary

| Project Description or ID Number | Project Score | PPI Priority | ROI Priority | Throughput Priority | Comments |
|---|---|---|---|---|---|
|  |  |  |  |  |  |
|  |  |  |  |  |  |
|  |  |  |  |  |  |

- Appraising individual and group performance
- Setting goals and targets

- Identifying areas where more (or less) emphasis on Six Sigma is indicated
- Helping educate newcomers on the value of Six Sigma
- Answering skeptics
- Quieting cynics

A major difference between Six Sigma and failed programs of the past is the emphasis on tangible, measurable results. Six Sigma advocates a strong point of the fact that projects are selected to provide a mixture of short- and long-term paybacks that justify the investment and the effort. Unless proof is provided, any statements regarding paybacks are nothing more than empty assertions.

Data storage is becoming so inexpensive that the typical organization can afford to keep fairly massive amounts of data in databases. The limiting factor is the effort needed to enter the data into the system. This is especially important if highly trained change agents such as Master Black Belts, Black Belts, or Green Belts are needed to perform the data entry (Table 4.5).

Usually viewing access is restricted to the project data according to role played in the project, position in the organization, etc. Change access is usually restricted to the project sponsor, leader, or Black Belt. However, to the extent possible, it should be easy to "slice-and-dice" this information in a variety of ways. Periodic reports might be created summarizing results according to department, sponsor, Black Belt, etc. The system should also allow ad hoc views to be easily created, such as the simple list shown in Table 4.6.

**Table 4.5** Possible Information To Be Captured

| |
|---|
| Charter information (title, sponsor, membership, deadline etc.) |
| Description of project in ordinary language |
| Project status |
| Savings type (hard, soft, cost avoidance, CTQ, etc.) |
| Process or unit owner |
| Key accounting information (charge numbers, etc.) |
| Project originator |
| Top-level strategy addressed by project |
| Comments, issues |
| Lessons learned |
| Keywords (for future searches) |
| Related documents and links |
| Audit trail of changes |
| Project task and schedule information |

Table 4.6  A Typical View of Six Sigma Projects

| Project ID | Project Title | Status | Black Belt | Sponsor | Due | Savings Type | Total Savings | Costs |
|---|---|---|---|---|---|---|---|---|
| 76 | Cup Dipole Antenna | Pending approval | J Jones | Jane Doe | 3/1/04 | Hard | $508,000 | $5,900 |
| 33 | Tank assembly | Define | B Olson | Sam Smith | 9/30/03 | Hard | $250,000 | $25,000 |
| 35 | SSPA | Completed | N Hepburn | Sal Davis | 10/31/03 | Cost avoidance | $1.3 Million | $13,000 |
| 37 | FCC RFI compliance | Control | M Littleton | Henry Little | 9/30/03 | Other | NA | $1,500 |
| • | • | • | • | • | • | • | • | • |
| • | • | • | • | • | • | • | • | • |
| • | • | • | • | • | • | • | • | • |

203

## Financial Results Validation

Six Sigma financial benefits claimed for every project must be confirmed by experts in accounting or finance. Initial savings estimates may be calculated by Black Belts or sponsors, but final results require at least the concurrence of the finance department. This should be built in from the start. The finance person assigned to work with the team should be listed in the project charter. Without this involvement the claimed savings are simply not credible. Aside from the built-in bias involved in calculating the benefit created from one's own project, there is the issue of qualifications. The best qualified people to calculate financial benefits are generally those who do such calculations for a living.

This is not to imply that the finance expert's numbers should go unchallenged. If the results appear to be unreasonable, either high or low, then they should be clearly explained in terms the sponsor understands. The Six Sigma Leader also has an interest in ensuring that the numbers are valid. Invalid results pose a threat to the viability of the Six Sigma effort itself. For example, on one project the Black Belt claimed savings of several hundred thousand dollars for "unpaid overtime." A finance person concurred. However, the Six Sigma Leader would not accept the savings, arguing quite reasonably that the company hadn't saved anything if it had never paid the overtime. This isn't to say that the project didn't have a benefit. Perhaps morale improved or turnover declined due to the shorter working hours. Care must be taken to show the benefits properly.

## Team Performance Evaluation

Evaluating team performance involves the same principles as evaluating performance in general. Before one can determine how well the team's task has been done, a baseline must be established and goals must be identified. Setting goals using benchmarking and other means is discussed elsewhere in this book (see Chap. 3). Records of progress should be kept as the team pursues its goals.

Performance measures generally focus on group tasks, rather than on internal group issues. Typically, financial performance measures show a payback ratio of between 2:1 and 8:1 on team projects. Some examples of tangible performance measures are

- Productivity
- Quality
- Cycle time
- Grievances
- Medical usage (e.g., sick days)
- Absenteeism
- Service

- Turnover
- Dismissals
- Counseling usage

Many intangibles can also be measured. Some examples of intangibles affected by teams are

- Employee attitudes
- Customer attitudes
- Customer compliments
- Customer complaints

The performance of the team process should also be measured. Project failure rates should be carefully monitored. A $p$ chart can be used to evaluate the causes of variation in the proportion of team projects that succeed. Failure analysis should be rigorously conducted.

Aubrey and Felkins (1988) list the effectiveness measures shown below:

- Leaders trained
- Number of potential volunteers
- Number of actual volunteers
- Percent volunteering
- Projects started
- Projects dropped
- Projects completed/approved
- Projects completed/rejected
- Improved productivity
- Improved work environment
- Number of teams
- Inactive teams
- Improved work quality
- Improved service
- Net annual savings

## *Team Recognition and Reward*

Recognition is a form of employee motivation in which the company identifies and thanks employees who have made positive contributions to the company's success. In an ideal company, motivation flows from the employees' pride of workmanship. When employees are enabled by management to do their jobs and produce a product or service of excellent quality, they will be motivated.

The reason recognition systems are important is not that they improve work by providing incentives for achievement. Rather, they make a statement about what is important to the company. Analyzing a company's employee recognition system provides a powerful insight into the company's values in action. These are the values that are actually driving employee behavior. They are not necessarily the same as management's stated values. For example, a company that claims to value customer satisfaction but recognizes only sales achievements probably does not have customer satisfaction as one of its values in action.

Public recognition is often better for two reasons:

- Some (but not all) people enjoy being recognized in front of their colleagues.
- Public recognition communicates a message to all employees about the priorities and function of the organization.

The form of recognition can range from a pat on the back to a small gift to a substantial amount of cash. When substantial cash awards become an established pattern, however, it signals two potential problems:

- It suggests that several top priorities are competing for the employee's attention, so that a large cash award is required to control the employee's choice.
- Regular, large cash awards tend to be viewed by the recipients as part of the compensation structure, rather than as a mechanism for recognizing support of key corporate values.

Carder and Clark (1992) list the following guidelines and observations regarding recognition:

- **Recognition is not a method by which management can manipulate employees**—If workers are not performing certain kinds of tasks, establishing a recognition program to raise the priority of those tasks might be inappropriate. Recognition should not be used to get workers to do something they are not currently doing because of conflicting messages from management. A more effective approach is for management to first examine the current system of priorities. Only by working on the system can management help resolve the conflict.
- **Recognition is not compensation**—In this case, the award must represent a significant portion of the employee's regular compensation to have significant impact. Recognition and compensation differ in a variety of ways:
    - Compensation levels should be based on long-term considerations such as the employee's tenure of service, education, skills, and level of responsibility. Recognition is based on the specific accomplishments of individuals or groups.
    - Recognition is flexible. It is virtually impossible to reduce pay levels once they are set, and it is difficult and expensive to change compensation plans.

- ▼ Recognition is more immediate. It can be given in timely fashion and therefore relate to specific accomplishments.
- ▼ Recognition is personal. It represents a direct and personal contact between employee and manager. Recognition should not be carried out in such a manner that implies that people of more importance (managers) are giving something to people of less importance (workers).
- ▲ **Positive reinforcement is not always a good model for recognition**—Just because the manager is using a certain behavioral criterion for providing recognition, it doesn't mean that the recipient will perceive the same relationship between behavior and recognition.
- ▲ **Employees should not believe that recognition is based primarily on luck**—An early sign of this is cynicism. Employees will tell you that management says one thing but does another.
- ▲ **Recognition meets a basic human need**—Recognition, especially public recognition, meets the needs for belonging and self-esteem. In this way, recognition can play an important function in the workplace. According to Abraham Maslow's theory, until these needs for belonging and self-esteem are satisfied, self-actualizing needs such as pride in work, feelings of accomplishment, personal growth, and learning new skills will not come into play.
- ▲ **Recognition programs should not create winners and losers**—Recognition programs should not recognize one group of individuals time after time while never recognizing another group. This creates a static ranking system, with all of the problems discussed earlier.
- ▲ **Recognition should be given for efforts, not just for goal attainment**—According to Imai (1986), a manager who understands that a wide variety of behaviors are essential to the company will be interested in criteria of discipline, time management, skill development, participation, morale, and communication, as well as direct revenue production. To be able to effectively use recognition to achieve business goals, managers must develop the ability to measure and recognize such process accomplishments.
- ▲ **Employee involvement is essential in planning and executing a recognition program**—It is essential to engage in extensive planning before instituting a recognition program or before changing a bad one. The perceptions and expectations of employees must be surveyed.

## *Lessons-Learned Capture and Replication*

It is often possible to apply the lessons learned from a project to other processes, either internally or externally. Most companies have more than one person or organizational unit performing similar or identical tasks. Many also have suppliers and outsourcers who do work similar to that being done internally. By replicating the changes done during a project the benefits of Six Sigma can be multiplied manyfold, often at very minimal cost. Think of it as a form of benchmarking. Instead of looking for the best-in-class process for you to learn from, the Six Sigma team *created* a best-in-class process and you want to teach the new approach to others.

Unlike benchmarking, where the seeker of knowledge is predisposed to change what they are doing, the process owners who might benefit from the knowledge gained during a Six Sigma project may not even be aware that they can benefit from a change. This needs to be accounted for when planning the program for sharing lessons learned. The process is a combination of motivation, education, and selling the target audience on the new approach. Chances are that those who worked the project are *not* the best ones to sell others on the new approach. They can serve as technical advisers to those who will carry the message to other areas. The Six Sigma function (Process Excellence) usually takes the lead in developing a system for replication and sharing of lessons learned.

In addition to the lessons learned about business processes, a great deal will be learned about how to conduct successful projects. In a few years even a moderately sized Six Sigma effort will complete hundreds or thousands of projects. These project lessons learned should be captured and used to help other project teams. The information is usually best expressed in simple narratives by the project Black Belt. The narratives can be indexed by search engines and used by other Black Belts in the organization. The lessons learned database is an extremely valuable asset to the Six Sigma organization.

# PART II

# Six Sigma Tools and Techniques

## Chapter 5
Project Management Using DMAIC and DMADV

## Chapter 6
The Define Phase

## Chapter 7
The Measure Phase

## Chapter 8
Process Behavior Charts

## Chapter 9
Measurement Systems Evaluation

## Chapter 10
Analyze Phase

## Chapter 11
The Improve/Design Phase

## Chapter 12
Control/Verify Phase

# CHAPTER 5

# Project Management Using DMAIC and DMADV

Part II addresses the tools and techniques commonly used in Six Sigma. Many of these tools have been used by the quality professional and applied statistician for decades. Six Sigma formalizes the use of the tools within the DMAIC and DMADV project deployment methodologies, where they are applied to real-world projects designed to deliver tangible results for identified stakeholders.

## DMAIC and DMADV Deployment Models

When applied for performance improvement of an existing product, process, or service, the Define-Measure-Analyze-Improve-Control, or DMAIC model is used. DMAIC is summarized in Fig. 5.1.

The DMAIC structure provides a useful framework for creating a "gated process" for project control, as shown in Fig. 5.2. Criteria for completing a particular phase are defined and projects reviewed to determine if all of the criteria have been met before the next phase is begun. If all criteria have been satisfied, the gate (e.g., define) is "closed."

Table 5.1 shows a partial listing of tools often found to be useful in a given stage of a project. There is considerable overlap in practice.

When the project goal is the development of a new or radically redesigned product, process or service, the Define-Measure-Analyze-Design-Verify, or DMADV, model is used (Fig. 5.3). DMADV is a part of the design for Six Sigma (DFSS) toolkit. Note the similarities between the tools used, as well as the objectives.

Figure 5.4 illustrates the relationship between DMAIC and DMADV.

Projects are the means through which processes and products are systematically changed; the bridge between the planning and the doing.

| | |
|---|---|
| D | Define the goals of the improvement activity, and incorporate into a Project Charter. Obtain sponsorship and assemble team. |
| M | Measure the existing system. Establish valid and reliable metrics to help monitor progress toward the goal(s) defined at the previous step. Establish current process baseline performance using metric. |
| A | Analyze the system to identify ways to eliminate the gap between the current performance of the system or process and the desired goal. Use exploratory and descriptive data analysis to help you understand the data. Use statistical tools to guide the analysis. |
| I | Improve the system. Be creative in finding new ways to do things better, cheaper, or faster. Use project management and other planning and management tools to implement the new approach. Use statistical methods to validate the improvement. |
| C | Control the new system. Institutionalize the improved system by modifying compensation and incentive systems, policies, procedures, MRP, budgets, operating instructions and other management systems. You may wish to utilize standardization such as ISO 9000 to ensure that documentation is correct. Use statistical tools to monitor stability of the new systems. |

**Figure 5.1**  Overview of DMAIC.

**DMAIC**

**① DEFINE**
- Define project Scope, Objective & Schedule — BlackBelt Sponsor
- Define Process (top-level) and Stakeholders — BlackBelt Sponsor
- Select Team Members — BlackBelt Sponsor
- Obtain Authorization from Sponsors — BlackBelt Sponsor
- Assemble and Train team — BlackBelt Sponsor ProjectTeam

**② MEASURE**
- Define Process — BlackBelt ProjectTeam
- Define Metric — BlackBelt ProjectTeam
- Establish Process Baseline — BlackBelt ProjectTeam
- Evaluate Measurement System — BlackBelt ProjectTeam

**③ ANALYZE**
- Analyze Value Stream — BlackBelt ProjectTeam
- Analyze Sources of Variation — BlackBelt
- Determine process drivers — BlackBelt
- Benchmark against best in class — BlackBelt

**④ IMPROVE**
- Prioritize Improvement Opportunities — BlackBelt ProjectTeam Sponsor
- Define new process or product design — BlackBelt ProjectTeam
- Optimize Process/Product Settings — BlackBelt ProjectTeam
- Evaluate for risks and failure modes — BlackBelt ProjectTeam

**⑤ CONTROL**
- Validate new process/product — BlackBelt ProjectTeam
- Develop and Implement Control Plan — BlackBelt ProjectTeam
- Document Lessons Learned — BlackBelt
- Approve Deliverables — Sponsor Finance

**Figure 5.2**  Using DMAIC on a Six Sigma project.

**Table 5.1** Six Sigma Tools Commonly Used in Each Phase of a Project

| Project Phase | Candidate Six Sigma Tools |
|---|---|
| Define | • Project Charter<br>• VOC tools (surveys, focus groups, letters, comment cards)<br>• Process map<br>• QFD, prioritization matrix, matrix diagram<br>• SIPOC<br>• Benchmarking<br>• Project planning and management tools<br>• Work breakdown structure<br>• Pareto analysis |
| Measure | • Measurement systems analysis, including R&R study and linearity analysis<br>• Process behavior charts (SPC)<br>• Exploratory data analysis<br>• Descriptive statistics, including histograms and goodness of fit tests<br>• Data mining<br>• Pareto analysis |
| Analyze | • Cause-and-effect diagrams<br>• Box-whisker plots<br>• Pareto diagrams<br>• Work breakdown structures<br>• Tree diagrams<br>• Spaghetti diagrams<br>• Brainstorming, including affinity diagrams<br>• PERT and Gantt analysis of critical path (cycle time reduction)<br>• Process behavior charts (SPC)<br>• Process maps<br>• Design of experiments<br>• Enumerative statistics (hypothesis tests, confidence intervals, contingency tables, histograms)<br>• Inferential statistics (X's and Y's)<br>• Simulation |
| Improve | • Force field diagrams<br>• Lean tools of 5S and level loading<br>• FMEA<br>• 7M tools<br>• Project planning and management tools<br>• Prototype and pilot studies<br>• Simulations<br>• Optimization techniques including response surface analysis<br>• Process behavior charts (SPC) |
| Control | • Process behavior charts (SPC)<br>• Measurement systems analysis<br>• Process maps<br>• FMEA<br>• ISO 900$\times$<br>• Change budgets, bid models, cost estimating models<br>• Reporting system |

| | |
|---|---|
| **D**efine | **Define** the goals of the design activity. |
| **M**easure | **Measure** customer input to determine what is critical to quality from the customers' perspective. Use special methods when a completely new product or service is being designed (see the Kano Model discussions in Chap. 2). Translate customer requirements into project goals. |
| **A**nalyze | **Analyze** innovative concepts for products and services to create value for the customer. Determine performance of similar best-in-class designs. |
| **D**esign | **Design** new processes, products, and services to deliver customer value. Use predictive models, simulation, prototypes, pilot runs, etc. to validate the design concept's effectiveness in meeting goals. |
| **V**erify | **Verify** that new systems perform as expected. Create mechanisms to ensure continued optimal performance. |

**Figure 5.3** Overview of DMADV.

**Figure 5.4** DMAIC and DMADV.

Frank Gryna makes the following observations about projects (Juran and Gryna, 1988, pp. 22.18–22.19):

- An agreed-upon project is also a legitimate project. This legitimacy puts the project on the official priority list. It helps to secure the needed budgets, facilities, and personnel. It also helps those guiding the project to secure attendance at scheduled meetings, to acquire requested data, to secure permission to conduct experiments, etc.
- The project provides a forum of converting an atmosphere of defensiveness or blame into one of constructive actions.
- Participation in a project increases the likelihood that the participant will act on the findings.
- All breakthrough is achieved *project by project*, and in no other way.
- Effective project management will prevent a number of problems that result in its absence.
  - Projects have little or no impact on the organization's success, even if successful, no one will really care.
  - Missions overlap the missions of other teams. For example, Team A's mission is to reduce solder rejects, Team B's mission is to reduce wave solder rejects, Team C's mission is to reduce circuit board assembly problems.
  - Projects improve processes that are scheduled for extensive redesign, relocation, or discontinuation.
  - Studying a huge system ("patient admitting"), rather than a manageable process ("outpatient surgery preadmission").
  - Studying symptoms ("touch-up of defective solder joints") rather than root causes ("wave solder defects").
  - Project deliverables are undefined. For example, "study TQM" rather than "reduce waiting time in urgent care."

There are several reasons why one should plan carefully before starting a project (Ruskin and Estes, 1995, p. 44):

- The plan is a simulation of prospective project work, which allows flaws to be identified in time to be corrected.
- The plan is a vehicle for discussing each person's role and responsibilities, thereby helping direct and control the work of the project.
- The plan shows how the parts fit together, which is essential for coordinating related activities.
- The plan is a point of reference for any changes of scope, thereby helping project managers deal with their customers.
- The plan helps everyone know when the objectives have been reached and therefore when to stop.

The official plan and authorization for the project is summarized in the Six Sigma Project Charter, which is initially developed in the Define stage of the project, as outlined in Chap. 6.

## *Project Scheduling*

There are a wide variety of tools and techniques available to help the project manager develop a realistic project timetable, allocate resources based on the timetable, and track progress during the implementation of the project plan. We will review two of the most common here: Gantt charts and PERT-type systems.

### Gantt Charts

A Gantt chart shows the relationships among the project tasks, along with time constraints. The horizontal axis of a Gantt chart shows the units of time (days, weeks, months, etc.). The vertical axis shows the activities to be completed. Bars show the estimated start time and duration of the various activities. Figure 5.5 illustrates a Gantt chart, modified to reflect the DMAIC objectives' milestones, as developed using MS Project software.

The *milestone* symbol represents an event rather than an activity; it does not consume time or resources. When Gantt charts are modified in this way they are sometimes called "milestone charts."

**Figure 5.5** Enhanced Gantt chart (milestone chart).

Gantt charts and milestone charts can be modified to show additional information, such as who is responsible for a task, why a task is behind schedule, remedial action planned or already taken, etc.

Although every project is unique, Six Sigma projects that use the DMAIC or DMADV frameworks have many tasks in common, at least at a general level. Many people find it helpful if they have a generic "template" they can use to plan their project activities. This is especially true when the Black Belt or Green Belt is new and has limited project management experience. Table 5.2 can be used as a planning tool by Six Sigma teams. It shows the details used to construct the Gantt chart shown above, including the duration of each task, the task or tasks immediately preceding the task (i.e., its predecessor, identified by task ID), and the resource responsibilities for each task. The duration for each DMAIC stage (the summary task IDs 1, 6, 12, 17, and 23) are calculated based on the tasks within the stage and their predecessor relationships. The resource designated as "leader" would typically be the Black Belt assigned to the project; however, in cases where a suitably trained Green Belt serves as project leader, the tasks that typically require Black Belt expertise are noted as such. This template project has a total duration of 76 working days, or approximately 3½ calendar months, based on 22 working days per month.

**Download the MS Project DMAIC template at www.mhprofessional.com/ssh5.**

**Table 5.2** Typical DMAIC Project Tasks and Responsibilities

| Task ID | Task | Duration (days) | Predecessor (Task ID #'s) | Resource |
|---|---|---|---|---|
| 1 | DEFINE | 10 | | |
| 2 | Define project, develop charter | 6 | | Leader, sponsor |
| 3 | Top-level process definition | 2 | 2 | Leader |
| 4 | Team formation | 3 | 2 | Leader, sponsor |
| 5 | Charter approval | 1 | 4, 3 | Leader, sponsor |
| 6 | MEASURE | 15 | 1 | |
| 7 | Detailed process definition | 3 | 5 | Leader, team |
| 8 | Process metric definition | 2 | 7 | Leader, team |
| 9 | Measurement system analysis | 3 | 8 | Leader, team |
| 10 | Process baseline definition | 10 | 8 | Leader, team |
| 11 | Project status update | 0 | 10, 9 | Leader, sponsor |

*(continued on next page)*

**Table 5.2** Typical DMAIC Project Tasks and Responsibilities (*continued*)

| Task ID | Task | Duration (days) | Predecessor (Task ID #'s) | Resource |
|---|---|---|---|---|
| 12 | ANALYZE | 18 | 6 | |
| 13 | Analyze value stream | 8 | 11 | Leader, team |
| 14 | Analyze sources of variation | 3 | 13 | Leader, Black Belt |
| 15 | Determine process drivers | 7 | 14 | Leader, Black Belt |
| 16 | Project status update | 0 | 15 | Leader, sponsor |
| 17 | IMPROVE | 25 | 12 | |
| 18 | Determine new process operating conditions | 10 | 16 | Leader, team, Black Belt |
| 19 | Estimate benefits from new process | 3 | 18 | Leader, team |
| 20 | Define and address process failure modes | 5 | 18 | Leader, team |
| 21 | Implement and verify changes at process level | 10 | 20,19 | Leader, team |
| 22 | Project status update | 0 | 21 | Leader, sponsor |
| 23 | CONTROL | 8 | 17 | |
| 24 | Implement process control plan | 7 | 22 | Leader, team |
| 25 | Implement monthly savings tracking plan | 3 | 22 | Leader, team |
| 26 | Document lessons learned | 2 | 22 | Leader, team |
| 27 | Formal project completion and approval | 1 | 24, 25, 26 | Leader, sponsor, team |

## PERT-CPM

While useful, the traditional Gantt charts provided limited project schedule analysis capabilities. The successful management of large-scale projects requires more rigorous planning, scheduling, and coordinating of numerous interrelated activities. To aid in these tasks, formal procedures based on the use of networks and network techniques were developed beginning in the late 1950s. The most prominent of these procedures have been PERT (**P**rogram **E**valuation and **R**eview **T**echnique) and CPM (**C**ritical **P**ath **M**ethod), both of which are also incorporated into MS Project. The two approaches are usually referred to as PERT-type project management systems. The most important difference between PERT and CPM is that originally the time estimates for the activities were assumed deterministic in CPM and were probabilistic in PERT. In other words, while a time estimate for a given task in CPM is fixed (say, 5 days), the time estimate for the same task in PERT would recognize variability in the time estimate (e.g., from 3 to 7 days). In most applications, the

PERT time estimate for a task is specified (as an input) in terms of the *most likely time*, the *optimistic time*, and the *pessimistic time*. The tasks' *expected* times are calculated from these estimates, using statistical theory, as described below. Today, the terms *PERT* and *CPM* are often used interchangeably, even though there are obvious differences between the deterministic and probabilistic approaches.

Project scheduling by PERT-CPM consists of four basic phases: planning, scheduling, improvement, and controlling. The planning phase involves breaking the project into distinct activities. The time estimates for these activities are then determined and a network (or arrow) diagram is constructed with each activity being represented by an arrow.

PERT-type systems are used to:

- Aid in planning and control of projects
- Determine the feasibility of meeting specified deadlines
- Identify the most likely bottlenecks in a project
- Evaluate the effects of changes in the project requirements or schedule
- Evaluate the effects of deviating from schedule
- Evaluate the effect of diverting resources from the project, or redirecting additional resources to the project

The ultimate objective of the scheduling phase is to construct a time chart showing the start and finish times for each activity as well as its relationship to other activities in the project. The schedule must identify activities that are "critical" in the sense that they *must* be completed on time to keep the project on schedule.

It is vital not to merely accept the schedule as a given. The information obtained in preparing the schedule can be used to improve the project schedule. Activities that the analysis indicates to be critical are candidates for improvement. Pareto analysis can be used to identify those critical elements that are most likely to lead to significant improvement in overall project completion time. Cost data can be used to supplement the time data, and the combined time/cost information analyzed using Pareto analysis.

The final phase in PERT-CPM project management is project control. This includes the use of the network diagram and Gantt chart for making periodic progress assessments.

***CPM Example.*** The activities involved, and their estimated task times, were presented in Table 5.2. It is important that certain of these activities be done in a particular order: there is a *precedence relationship*, which also provides information on where tasks can take place simultaneously. For example, tasks 3 and 4 (*top-level process definition* and *team formation*) each require completion of task 2 (*define project, develop charter*) before either can begin. Yet, task 4 may be started before task 3 is completed, and vice versa. It should be clear that conducting tasks 3 and 4 simultaneously, rather than sequentially, would tend to reduce the total project cycle time. The network diagram graphically displays the precedence relationships involved, as shown in Fig. 5.6 (an arrow diagram).

There are two time values of interest calculated within the analysis for each event: its *earliest time of completion (also known as earliest finish time, EF)* and its *latest time of completion (also known as latest finish time, LF)*. The earliest completion time for a given event is the calculated time at which the event will be completed if the preceding activities are started as early as possible (i.e., each event's *earliest start time*, ES). The latest completion time for an event is the calculated time the event can be completed without delaying the completion of the project beyond its earliest (possible) completion time. Earliest start and completion times of events are calculated by starting at the initial event and working forward, successively calculating the time at which each event will occur if each immediately preceding event occurs at its earliest time and each intervening activity uses only its estimated time. Latest start times (LS) are calculated by starting at the final event's latest completion time (LF, which for the last task is set equal to its earliest finish time EF) and working backward, successively subtracting the event time from the event's latest completion time.

Table 5.3 shows the calculated earliest start (ES), earliest finish (EF), latest start (LS), and latest finish (LF) for each task presented in Table 5.2. (Note that the summary tasks are not shown in this table, to clarify the calculations). The calculations are performed as follows:

- **Earliest Start (ES)**—Initial task set to 0. Subsequent tasks are set equal to the maximum EF for tasks listed in the predecessor column. For example, task 5 has predecessor tasks 4 and 3. The EF for task 4 is 9; the EF for task 3 is 8; the maximum EF for tasks 3 and 4 is 9, so the ES for task 5 is 9.
- **Earliest Finish (EF)**—Equals the task's ES plus its duration. For example, task 3 (*top-level process definition*) has a duration of 2 days and an ES of 6, so its EF equals 6 + 2 = 8.
- **Latest Finish (LF)**—The last task's LF is set equal to its EF. Moving backward (upward in the table), the LF for each predecessor is set equal to the minimum LS for its subsequent tasks. For example, task 27 has predecessor tasks 24, 25, and 26, so the LF for these tasks is set equal to the LS for task 27 (a value of 75). Task 22 is a predecessor for tasks 24, 25, and 26. The minimum LS for these tasks is 68, so the LF for task 22 is 68.
- **Latest Start (LS)**—Calculated by subtracting the task duration from its latest finish. For example, the LS for task 25 is 75 − 3 = 72.

*Slack time* for an event is the difference between the latest and earliest times for a given event. Thus, assuming everything else remains on schedule, the slack for an event indicates how much delay in reaching the event can be tolerated without delaying the project completion.

Table 5.3  Activities, Durations, and Critical Path Calculations for the DMAIC Project

| Task ID | Task | Duration (days) | Predecessor (Task ID #'s) | ES | EF | LF | LS |
|---|---|---|---|---|---|---|---|
| 2 | Define project, develop charter * | 6 | | 0 | 6 | 6 | 0 |
| 3 | Top-level process definition | 2 | 2 | 6 | 8 | 9 | 7 |
| 4 | Team formation * | 3 | 2 | 6 | 9 | 9 | 6 |
| 5 | Charter approval * | 1 | 4, 3 | 9 | 10 | 10 | 9 |
| 7 | Detailed process definition * | 3 | 5 | 10 | 13 | 13 | 10 |
| 8 | Process metric definition * | 2 | 7 | 13 | 15 | 15 | 13 |
| 9 | Measurement system analysis | 3 | 8 | 15 | 18 | 25 | 22 |
| 10 | Process baseline definition * | 10 | 8 | 15 | 25 | 25 | 15 |
| 11 | Project status update * | 0 | 10, 9 | 25 | 25 | 25 | 25 |
| 13 | Analyze value stream * | 8 | 11 | 25 | 33 | 33 | 25 |
| 14 | Analyze sources of variation * | 3 | 13 | 33 | 36 | 36 | 33 |
| 15 | Determine process drivers * | 7 | 14 | 36 | 43 | 43 | 36 |
| 16 | Project status update * | 0 | 15 | 43 | 43 | 43 | 43 |
| 18 | Determine new process operating conditions * | 10 | 16 | 43 | 53 | 53 | 43 |
| 19 | Estimate benefits from new process | 3 | 18 | 53 | 56 | 58 | 55 |
| 20 | Define and address process failure modes * | 5 | 18 | 53 | 58 | 58 | 53 |
| 21 | Implement and verify changes at process level * | 10 | 20,19 | 58 | 68 | 68 | 58 |
| 22 | Project status update * | 0 | 21 | 68 | 68 | 68 | 68 |
| 24 | Implement process control plan * | 7 | 22 | 68 | 75 | 75 | 68 |
| 25 | Implement monthly savings tracking plan | 3 | 22 | 68 | 71 | 75 | 72 |
| 26 | Document lessons learned | 2 | 22 | 68 | 70 | 75 | 73 |
| 27 | Formal project completion and approval * | 1 | 24, 25, 26 | 75 | 76 | 76 | 75 |

*Critical path tasks

**Figure 5.6** Project network diagram for a DMAIC project.

Events and activities with slack times equal to zero lie on the *critical path* for the project. A critical path for a project is defined as a path through the network such that the activities on this path have *zero slack*. All activities and events having zero slack must lie on a critical path, but no others can. Figure 5.6 shows the activities on the critical path in hexagon-shaped boxes with thick borders. (Note that only a portion of the network diagram can be displayed; however readers may download the MS Project file to view the complete network diagram for the example.) The critical path tasks are also labeled with an asterisk in Table 5.3.

These calculations are rarely done by hand, as software such as MS Project automates the calculations and identifies the critical path, as indicated with the colored task boxes in Fig. 5.5. The identification of the critical path is an important aspect of project management (or the analysis of any time-based sequential activities), as further discussed in the schedule slippage section below.

*PERT Example.* The example can be further developed using a PERT analysis, where the estimated duration for each task is specified (i.e., input) as a most likely time estimate, a pessimistic time estimate, and an optimistic time estimate. An expected (or mean) duration for each task, as well as its standard deviation, can be calculated from these estimates, using the assumption of a beta distribution for the task time as follows:

$$\mu = [\text{Optimistic time} + (4 \times \text{Most likely time}) + \text{Pessimistic Time}] / 6$$
$$\sigma = (\text{Pessimistic time} - \text{Optimistic time}) / 6$$

The calculated values of the mean (expected) duration and the standard deviation of each task time are shown in Table 5.4 for all nonzero tasks. The critical path is determined based on the calculated expected duration for each task, which includes the influence of the optimistic and pessimistic estimates. For these data, the critical path is unchanged; however, that may not be the case for all data. The total expected project duration, calculated by summing the expected durations of the tasks on the critical path, is 86.6 days (approximately 14% larger than the deterministic value of 76 days presented earlier). While the task standard deviations cannot be added to obtain the overall standard deviation, the task variances (i.e., the task standard deviation squared) can be added. In this case, the sum of the variances equals 40.3. Taking the square root of the total variance results in a standard deviation of the total project duration of 6.3 days.

With these estimates of the project duration and standard deviation, further estimates of the project duration can be made assuming the total project duration is normally distributed. (This is a reasonable assumption due to the central limit theorem, since the total duration is the sum of the individual task durations.) The properties of the well-documented normal distribution can thus be used to answer questions of the form, What is the probability of completing the project in $x$ days? Or, How many days would be required to complete the project 90% of the time?

For example, to calculate the probability of completing the project in 76 days (or less), the $z$ value is calculated as:

$$z = (x - \mu) / \sigma$$

$$z = (76 - 86.6) / 6.3 = -1.67$$

From App. 2, the cumulative probability associated with this $z$ value is 4.8%. In other words, based on this probabilistic estimate, there is only a 5% chance of completing the project in the 76 days indicated by the deterministic estimate above. (The Excel function Norm.Dist, or Normdist in earlier releases, yields a similar value.) Similarly, the tables (or Excel's Norm.Inv function) can be used to estimate a 90% probability of completing the project within 94.7 days. (These calculations for the normal distribution are covered in detail in Chap. 8, and are provided here only to show the value of the methods in project planning.)

## Control and Prevention of Schedule Slippage

Project managers can use the project schedule analysis in a variety of ways to help them manage their projects. One way is, of course, to pay close attention to the activities that lie on the critical path. Any delay in these activities will result in a delay for the project. Furthermore, reduction in task time for activities not on the critical path will yield no overall reduction in project duration.

**Table 5.4** Activities, Durations, and Critical Path Calculations for the DMAIC Project

| Task ID | Task | Most Likely Duration (days) | Optimistic Duration (days) | Pessimistic Duration (days) | Expected Duration (days) | Duration Std. Dev. | Duration Variance |
|---|---|---|---|---|---|---|---|
| 2 | Define project, develop charter* | 6 | 2 | 10 | 6.0 | 1.3 | 1.8 |
| 3 | Top-level process definition | 2 | 1 | 3 | 2.0 | 0.3 | 0.1 |
| 4 | Team formation* | 3 | 1 | 5 | 3.0 | 0.7 | 0.4 |
| 5 | Charter approval* | 1 | 0.5 | 6 | 1.8 | 0.9 | 0.8 |
| 7 | Detailed process definition* | 3 | 1 | 6 | 3.2 | 0.8 | 0.7 |
| 8 | Process metric definition* | 2 | 1 | 5 | 2.3 | 0.7 | 0.4 |
| 9 | Measurement system analysis | 3 | 2 | 5 | 3.2 | 0.5 | 0.3 |
| 10 | Process baseline definition* | 10 | 7 | 25 | 12.0 | 3.0 | 9.0 |
| 13 | Analyze value stream* | 8 | 5 | 15 | 8.7 | 1.7 | 2.8 |
| 14 | Analyze sources of variation* | 3 | 2 | 10 | 4.0 | 1.3 | 1.8 |
| 15 | Determine process drivers* | 7 | 5 | 20 | 8.8 | 2.5 | 6.3 |
| 18 | Determine new process operating conditions* | 10 | 5 | 20 | 10.8 | 2.5 | 6.3 |
| 19 | Estimate benefits from new process | 3 | 2 | 6 | 3.3 | 0.7 | 0.4 |
| 20 | Define and address process failure modes* | 5 | 3 | 15 | 6.3 | 2.0 | 4.0 |
| 21 | Implement and verify changes at process levels* | 10 | 5 | 15 | 10.0 | 1.7 | 2.8 |
| 24 | Implement process control plan* | 7 | 5 | 15 | 8.0 | 1.7 | 2.8 |
| 25 | Implement monthly savings tracking plan | 3 | 2 | 5 | 3.2 | 0.5 | 0.3 |
| 26 | Document lessons learned | 2 | 2 | 4 | 2.3 | 0.3 | 0.1 |
| 27 | Formal project completion and approval* | 1 | 1 | 5 | 1.7 | 0.7 | 0.4 |

*Critical path tasks

The project manager should also review the diagram for opportunities to modify the project plan to result in a reduction in the total project completion time. Since the network times are based on *estimates*, it is likely that the completion times will vary. When this occurs it often happens that a new critical path appears. Thus, the network should be viewed as a dynamic entity which should be revised as conditions change.

Primary causes of slippage include poor planning and poor management of the project. Software such as MS Project can reduce these issues, with their resource planning and active reporting tools. Tight schedules in particular must be tightly managed to avoid slippage.

Outside forces beyond the control of the project manager will also play a role. However, it isn't enough to be able to simply identify "outside forces" as the cause and beg forgiveness. Astute project managers will anticipate as many such possibilities as possible and prepare contingency plans to deal with them. The process decision program chart (PDPC) is a useful tool for identifying possible events that might be encountered during the project. A portion of a PDPC, applied to one of the critical path tasks of a DMAIC project, is shown in Fig. 5.7. The emphasis of PDPC is the impact of the "failures" (problems) on project schedules. The PDPC seeks to describe specific actions to be taken to prevent the problems from occurring in the first place, and to mitigate the impact of the problems if they do occur. An enhancement to classical PDPC is the Fault Tree Analysis described in Chap. 11, which assigns subjective probabilities to the various problems and to use these to help assign priorities. The amount of detail that should go into contingency plans is a judgment call. The project manager should consider both the seriousness of the potential problem and the likelihood of its occurring.

**Figure 5.7** A portion of the PDPC for the DMAIC project example; constructed using Mind Genius software (www.mindgenius.com by permission).

## Cost Considerations in Project Scheduling

Most project schedules can be compressed, if one is willing to pay the additional costs. For the analysis here, costs are defined to include direct elements only. Indirect costs (administration, overhead, etc.) will be considered in the final analysis. Assume that a straight-line relationship exists between the cost of performing an activity on a *normal schedule*, and the cost of performing the activity on a *crash schedule*. Also assume that there is a crash time beyond which no further time saving is possible, regardless of cost. Figure 5.8 illustrates these concepts.

For a given activity the cost-per-unit-of-time saved is found as:

$$\frac{\text{crash cost} - \text{normal cost}}{\text{normal time} - \text{crash time}} \quad (5.1)$$

When deciding which activity on the critical path to improve, one should begin with the activity that has the smallest cost-per-unit-of-time saved. The project manager should be aware that once an activity time has been reduced there may be a new critical path. If so, the analysis should proceed using the updated information, that is, activities on the new critical path should be analyzed.

The data for the example are shown in Table 5.5, with additional data for costs and crash schedule times for each activity on the critical path; only critical path activities are being considered since only they can produce an improvement in overall project duration. Thus, the first activity to consider improving would be *Determine Process Drivers*, which

**Figure 5.8** Cost-time relationship for an activity.

Table 5.5  Schedule Costs for Activities Involved in the DMAIC Project Example

| Activity | Normal schedule Time (days) | Normal schedule Cost | Crash schedule Time (days) | Crash schedule Cost | Slope |
|---|---|---|---|---|---|
| Define project, develop charter | 6 | 1,000 | 5 | 3,000 | 2,000 |
| Team formation | 3 | 1,600 | 2 | 3,400 | 1,800 |
| Charter approval | 1 | 7,500 | 1 | 7,500 | 0 |
| Detailed process definition | 3 | 3,000 | 1 | 6,000 | 1,500 |
| Process metric definition | 2 | 4,400 | 1 | 6,000 | 1,600 |
| Process baseline definition | 10 | 13,750 | 8 | 17,500 | 1,875 |
| Analyze value stream | 8 | 3,500 | 6 | 7,000 | 1,750 |
| Analyze sources of variation | 3 | 3,000 | 1 | 6,000 | 1,500 |
| Determine process drivers | 7 | 9,200 | 5 | 11,600 | 1,200* |
| Define new process levels | 10 | 3,000 | 8 | 5,500 | 1,250 |
| Define and mitigate failure | 5 | 4,800 | 1 | 11,000 | 1,550 |
| Implement and verify | 10 | 4,900 | 6 | 12,000 | 1,775 |
| Define control plan | 7 | 5,600 | 3 | 12,000 | 1,600 |
| Project approval | 1 | 4,500 | 1 | 4,500 | 0 |

costs $1,200 per day saved on the schedule [identified with an asterisk (*) in Table 5.5]. If additional resources could be directed toward this activity it would produce the best "bang for the buck" in terms of reducing the total time of the project. Assuming the critical path doesn't change, the next activities for cost consideration would be *Define New Process Levels*, then *Detailed Process Definition*, etc.

As activities are addressed one by one, the time it takes to complete the project will decline, while the direct costs of completing the project will increase. Figure 5.9 illustrates the cost-duration relationship graphically.

Conversely, *indirect costs* such as overhead, etc., are expected to *increase* as projects take longer to complete. When the indirect costs are added to the direct costs, total costs will generally follow a pattern similar to that shown in Fig. 5.10.

To optimize resource utilization, the project manager will seek to develop a project plan that produces the minimum cost schedule. Of course, the organization will likely have multiple projects being conducted simultaneously, which places additional constraints on resource allocation.

**Figure 5.9** Direct costs as a function of project duration.

**Figure 5.10** Total costs as a function of project duration.

## *Project Reporting*

Project information should be collected on an ongoing basis as the project progresses. Information obtained should be communicated in a timely fashion to stakeholders and sponsors, who can often help the project manager to maintain or recover the schedule.

The Project Charter (further developed in Chap. 6) provides an important feedback tool in the form of a formal, written report. It provides the background for periodic reporting to various stakeholder groups:

▲ **Status reports**—Formal, periodic written reports, often with a standardized format, indicating the current project status, relative to the plan outlined in the latest distributed revision of the Charter. Where project performance does not match planned performance, the reports include information as to the cause of the problem and corrective actions to align with the plan. Remedial action may, at times, involve revising the plan, subject to the approval of the sponsor. When the project is not meeting the plan due to obstacles which the project team cannot overcome, the status report will request senior management intervention.

▲ **Management reviews**—These meetings, scheduled in advance, provide the project leader with the opportunity to interact with key members of the management team. The chief responsibility for these meetings is management's. The purpose is to brief management on the status of the project, review the Project Charter and project team mission, discuss those management activities that are likely to have an impact on the progress of the team, etc. This is the appropriate forum for addressing systems barriers encountered by the team: While the team must work within existing systems, management has the authority to change the systems, as necessary. At times a minor system change can dramatically enhance the ability of the team to progress.

▲ **Budget reviews**—While budget reports are included in each status report, a budget review is a formal evaluation of actual resource utilization with respect to budgeted utilization. Budget review may also involve revising budgets, either upward or downward, based on developments since the original budget approval. Among those unschooled in the science of statistics there is an unfortunate tendency to react to every random tick in budget variances as if they were due to a special cause of variation. Six Sigma managers should coach finance and management personnel on the principles of variation to preclude tampering with the budgeting process (see "Project Budgets" for further discussion of budgets).

▲ **Customer audits**—The "customer" in this context refers to the senior management of the principal stakeholder group for the project. The project deliverables are designed to meet the objectives of this customer, and the customer should play an active role in keeping the project on track to the stated goals.

▲ **Updating plans and timetables**—The purpose of feedback is to provide information to form a basis for modifying future behavior. Since that behavior is documented in the project plans and schedules, these documents must be modified to ensure that the appropriate action is taken.

▲ **Resource redirection**—The modifications made to the plans and timetables will result in increasing or decreasing resource allocation to the project, or accelerating or decelerating the timetable for resource utilization. The impact of these resource redirections on other projects should be evaluated by management in view of the organization's overall objectives.

## Project Budgets

The process of allocating resources to be expended in the future is called *budgeting*. A listing of project expenditures, broken out into specific categories, is called the project budget.

Ruskin and Estes (1995) list the following types of project-related budgets:

- **Direct labor budgets** are usually prepared for each work element in the project plan, then aggregated for the project as a whole. Control is usually maintained at the work element level to ensure the aggregate budget allowance is not exceeded. Budgets may be in terms of dollars or some other measure of value, such as direct labor hours expended.
- **Support services budgets** need to be prepared because, without budgets, support services tend to charge based on actual costs, without allowances for errors, rework, etc. The discipline imposed by making budget estimates and being held to them often leads to improved efficiency and higher quality.
- **Purchased items budgets** cover purchased materials, equipment, and services. The budgets can be based on negotiated or market prices. The issues mentioned for support services also apply here.

Budgets allocate resources to be used in the future. No one can predict the future with certainty. Thus, an important element in the budgeting process is tracking actual expenditures after the budgets have been prepared. Project managers, sometimes in conjunction with their contacts in finance, are responsible for evaluating expenditures on a periodic basis. Often, variance reports are compiled to compare the actual and budgeted expenditures. (The term "variance" is used here in the accounting sense, not the statistical sense. In accounting, a variance is simply the difference between the budgeted amount and the actual amount. An accounting variance may or may not indicate a special cause of variation; statistical techniques are required to make this determination.)

Variance reports can appear in a variety of formats. Most common are simple tables that show the actual/budgeted/variances by budget item, overall for the current period, and cumulatively for the project. Since it is unlikely that variances will be zero, an allowance is usually made; for example, 5% over or under is allowed without the need for explanations. A better approach is to plot historical data on control charts to set allowances and/or spot patterns or trends.

The project manager should review the variance data for patterns which contain useful information. Ideally, the pattern will be a mixture of positive and negative but minor variances. Assuming that this pattern is accompanied by an on-schedule project, this indicates a reasonably good budget, that is, an accurate forecasting of expenditures. Variances should be evaluated separately for each type of budget (direct labor, materials, etc.). However, the variance report for the entire project is the primary source of information concerning the status of the project in terms of resource utilization.

Since budgeted resources are generally scarce, overspending represents a serious threat to the project and, perhaps, to the organization itself. When a project overspends its budget, it depletes the resources available for other activities and projects. The project team and team leader and sponsors should design monitoring systems to detect and correct overspending before it threatens the project or the organization. Overspending is often a symptom of other problems with the project, for example, paying extra in an attempt to "catch up" after falling behind schedule, additional expenses for rework, etc.

Underspending is potentially as serious as overspending. If the project budget was prepared properly then the expenses reflect a given schedule and quality level. Underspending may reflect "cutting corners" or allowing suppliers an allowance for slower delivery. The reasons for any significant departure from the plan should be explained.

## *Project Records*

Project records provide information that is useful both while the project is underway and afterwards. Project records serve three basic purposes: cost accounting requirements, legal requirements, and learning.

Project records should be organized and maintained as if they were part of a single database, even if it isn't possible to keep all of the records in a single location. There should be one "official" copy of the documentation, and a person designated to act as the caretaker of this information while the project is active. Upon completion of the project, the documentation should be sent to the organization's archives. Large quality improvement projects are expensive and time-consuming undertakings. The process involved is complex and often confusing to those unschooled in the techniques. However, a great deal can be learned from studying the "project process" where data exists. Archives of a number of projects can be used to identify common problems and patterns between the various projects. For example, project schedules may be consistently too optimistic or too pessimistic.

The following records should be kept:

- Statement of work
- Plans and schedules for projects and subprojects
- Correspondence (written and electronic)
- Written agreements

- Meeting minutes
- Action items and responsibilities
- Budgets and financial reports
- Cost-benefit analyses
- Status reports
- Presentation materials
- Documentation of changes made to plans and budgets
- Procedures followed or developed
- Notes of significant lessons learned

It is good practice for the project team to have a final meeting to perform a "post mortem" of the project. The meeting should be conducted soon after the project's completion, while memories are still fresh. The meeting will cover the lessons learned from conducting the project, and recommendations for improving the process. The minutes from these meetings should be used to educate project managers.

Low-cost document control and project management software can automatically catalog the information and provide the ability to search the entire database quickly and easily. There seems to be little reason not to store complete project information indefinitely.

## Six Sigma Teams

Six Sigma teams working on projects are the primary means of deploying Six Sigma and accomplishing the goals of the enterprise. Six Sigma teams are sometimes led by the Black Belt, but the team leader may also be a properly trained Green Belt or a Six Sigma champion who has a passion for the project. In these latter cases, the team must include a Black Belt to oversee the analysis of data, which is not a part of the Green Belt and Champion training.

Six Sigma teams are composed of groups of individuals who bring authority, knowledge, skills, abilities, and personal attributes to the project. There is nothing particularly special about Six Sigma teams compared with other work teams. They are people with different backgrounds and talents pursuing a common short-term goal. Like all groups of people, there are dynamics involved that must be understood if the mission of the team is to be accomplished. This section addresses the practices that Black Belts, Green Belts, sponsors, champions, facilitators, and leaders can employ to ensure that Six Sigma teams are successful. It focuses on:

- Stages in learning to work as a team
- The difference between group maintenance roles and group task roles

- Identifying and encouraging productive roles essential to team success
- Identifying and discouraging counterproductive behavior on teams
- Facilitating team meetings
- Dealing constructively with conflicts
- Evaluating, recognizing, and rewarding teams

## *Team Membership*

The structure of modern organizations is based on the principle of division of labor. Most organizations today consist of a number of departments, each devoted to their own specialty. A fundamental problem is that the separate functional departments tend to optimize their own operations, often to the detriment of the organization as a whole.

Traditional organizations, in effect, create barriers between departments. Departmental managers are often forced to compete for shares of limited budgets; in other words, they are playing a "zero sum game" where another manager's gain is viewed as their department's loss. Behavioral research has shown that people engaged in zero sum games think in terms of win-lose. This leads to self-destructive and cutthroat behavior. Overcoming this tendency requires improved communication and cooperation between departments.

Interdepartmental teams are groups of people with the skills needed to deliver the value desired. Processes are designed by the team to create the value in an effective and efficient manner. Management must see to it that the needed skills exist in the organization. It is also management's job to see that they remove barriers to cooperation, as discussed in Chap. 2.

## *Team Dynamics Management, Including Conflict Resolution*

Conflict management is usually the responsibility of the Six Sigma project team leader. If the team also includes a facilitator, then the facilitator can assist the leader by assuring that creative conflict is not repressed, but encouraged. Explore the underlying reasons for the conflict. If "personality disputes" are involved that threaten to disrupt the team meeting, arrange one-on-one meetings between the parties and attend the meetings to help mediate.

The first step in establishing an effective group is to create a consensus decision rule for the group, namely:

> No judgment may be incorporated into the group decision until it meets at least tacit approval of every member of the group.

This minimum condition for group movement can be facilitated by adopting the following behaviors:

- **Avoid arguing for your own position**—Present it as lucidly and logically as possible, but be sensitive to and consider seriously the reactions of the group in any subsequent presentations of the same point.
- **Avoid "win-lose" stalemates in the discussion of opinions**—Discard the notion that someone must win and someone must lose in the discussion; when impasses occur, look for the next most acceptable alternative for all the parties involved.
- **Avoid changing your mind only to avoid conflict and to reach agreement and harmony**—Withstand pressures to yield that have no objective or logically sound foundation. Strive for enlightened flexibility; but avoid outright capitulation.
- **Avoid conflict-reducing techniques such as the majority vote, averaging, bargaining, coin-flipping, trading out, and the like**—Treat differences of opinion as indicative of an incomplete sharing of relevant information on someone's part, either about task issues, emotional data, or gut level intuitions.
- **View differences of opinion as both natural and helpful rather than as a hindrance in decision-making**—Generally, the more the ideas expressed, the greater the likelihood of conflict will be; but the richer the array of resources will be as well.
- **View initial agreement as suspect**—Explore the reasons underlying apparent agreements; make sure people have arrived at the same conclusions for either the same basic reasons or for complementary reasons before incorporating such opinions into the group decision.
- **Avoid subtle forms of influence and decision modification**—For example, when a dissenting member finally agrees, don't feel that he must be rewarded by having his own way on some subsequent point.
- **Be willing to entertain the possibility that your group can achieve all the foregoing and actually excel at its task**—Avoid doomsaying and negative predictions for group potential.

Collectively, the above steps are sometimes known as the "consensus technique." In tests it was found that 75% of the groups who were instructed in this approach significantly outperformed their best individual resources.

## Stages in Group Development

Groups of many different types tend to evolve in similar ways. It often helps to know that the process of building an effective group is proceeding normally. Tuckman (1965) identified four stages in the development of a group: forming, storming, norming, and performing.

During the *forming* stage a group tends to emphasize procedural matters. Group interaction is very tentative and polite. The leader dominates the decision-making process and plays a very important role in moving the group forward.

The *storming* stage follows forming. Conflict between members, and between members and the leader, are characteristic of this stage. Members question authority as it relates to the group objectives, structure, or procedures. It is common for the group to resist the attempts of its leader to move them toward independence. Members are trying to define their role in the group.

It is important that the leader deal with the conflict constructively. There are several ways in which this may be done:

- Do not tighten control or try to force members to conform to the procedures or rules established during the forming stage. If disputes over procedures arise, guide the group toward new procedures based on a group consensus.
- Probe for the true reasons behind the conflict and negotiate a more acceptable solution.
- Serve as a mediator between group members.
- Directly confront counterproductive behavior.
- Continue moving the group toward independence from its leader.

During the *norming* stage the group begins taking responsibility, or ownership, of its goals, procedures, and behavior. The focus is on working together efficiently. Group norms are enforced on the group by the group itself.

The final stage is *performing*. Members have developed a sense of pride in the group, its accomplishments, and their role in the group. Members are confident in their ability to contribute to the group and feel free to ask for or give assistance.

Table 5.6 lists some common problems with teams, along with recommended remedial action (Scholtes, 1988).

**Table 5.6** Common Team Problems and Remedial Action

| Problem | Action |
| --- | --- |
| Floundering | • Review the plan<br>• Develop a plan for movement |
| The expert | • Talk to offending party in private<br>• Let the data do the talking<br>• Insist on consensus decisions |
| Dominating participants | • Structure participation<br>• Balance participation<br>• Act as gatekeeper |
| Reluctant participants | • Structure participation<br>• Balance participation<br>• Act as gatekeeper |

*(continued on next page)*

**Table 5.6** Common Team Problems and Remedial Action (*continued*)

| Problem | Action |
|---|---|
| Using opinions instead of facts | • Insist on data<br>• Use scientific method |
| Rushing things | • Provide constructive feedback<br>• Insist on data<br>• Use scientific method |
| Attribution (i.e., attributing motives to people with whom we disagree) | • Don't guess at motives<br>• Use scientific method<br>• Provide constructive feedback |
| Ignoring some comments | • Listen actively<br>• Train team in listening techniques<br>• Speak to offending party in private |
| Wanderlust | • Follow a written agenda<br>• Restate the topic being discussed |
| Feuds | • Talk to offending parties in private<br>• Develop or restate ground rules |

## *Member Roles and Responsibilities*

### Productive Group Roles

There are two basic types of roles assumed by members of a group: task roles and group maintenance roles. Group task roles are those functions concerned with facilitating and coordinating the group's efforts to select, define, and solve a particular problem. The group task roles shown in Table 5.7 are generally recognized.

**Table 5.7** Group Task Roles

| Role I.D. | Description |
|---|---|
| Initiator | Proposes new ideas, tasks, or goals; suggests procedures or ideas for solving a problem or for organizing the group. |
| Information seeker | Asks for relevant facts related to the problem being discussed. |
| Opinion seeker | Seeks clarification of values related to problem or suggestion. |
| Information giver | Provides useful information about subject under discussion. |
| Opinion giver | Offers his/her opinion of suggestions made. Emphasis is on values rather than facts. |
| Elaborator | Gives examples. |
| Coordinator | Shows relationship among suggestions; points out issues and alternatives. |
| Orientor | Relates direction of group to agreed-upon goals. |

| | |
|---|---|
| Evaluator | Questions logic behind ideas, usefulness of ideas, or suggestions. |
| Energizer | Attempts to keep the group moving toward an action. |
| Procedure technician | Keeps group from becoming distracted by performing such tasks as distributing materials, checking seating, etc. |
| Recorder | Serves as the group memory. |

Another type of role played in small groups is the group maintenance roles. Group maintenance roles are aimed at building group cohesiveness and group-centered behavior. They include those behaviors shown in Table 5.8.

**Table 5.8**  Group Maintenance Roles

| Role I.D. | Description |
|---|---|
| Encourager | Offers praise to other members; accepts the contributions of others. |
| Harmonizer | Reduces tension by providing humor or by promoting reconciliation; gets people to explore their differences in a manner that benefits the entire group. |
| Compromiser | This role may be assumed when a group member's idea is challenged; admits errors, and offers to modify his/her position. |
| Gatekeeper | Encourages participation, suggests procedures for keeping communication channels open. |
| Standard setter | Expresses standards for group to achieve, evaluates group progress in terms of these standards. |
| Observer/commentator | Records aspects of group process; helps group evaluate its functioning. |
| Follower | Passively accepts ideas of others; serves as audience in group discussions. |

The development of task and maintenance roles is a vital part of the team-building process. Team building is defined as the process by which a group learns to function as a unit, rather than as a collection of individuals.

## Counterproductive Group Roles

In addition to developing productive group-oriented behavior, it is also important to recognize and deal with individual roles that may block the building of a cohesive and effective team. These roles are shown in Table 5.9.

The leader's role includes that of process observer. In this capacity, the leader monitors the atmosphere during group meetings and the behavior of individuals. The purpose is to identify counterproductive behavior. Of course, once identified, the leader must tactfully and diplomatically provide feedback to the group and its members. The success of Six Sigma is, to a great extent, dependent on the performance of groups.

**Table 5.9** Counterproductive Group Roles

| Role I.D. | Description |
|---|---|
| Aggressor | Expresses disapproval by attacking the values, ideas, or feelings of others. Shows jealousy or envy. |
| Blocker | Prevents progress by persisting on issues that have been resolved; resists attempts at consensus; opposes without reason. |
| Recognition seeker | Calls attention to himself/herself by boasting, relating personal achievements, etc. |
| Confessor | Uses group setting as a forum to air personal ideologies that have little to do with group values or goals. |
| Playboy | Displays lack of commitment to group's work by cynicism, horseplay, etc. |
| Dominator | Asserts authority by interrupting others, using flattery to manipulate, and claiming superior status. |
| Help seeker | Attempts to evoke sympathy and/or assistance from other members through "poor me" attitude. |
| Special interest pleader | Asserts the interests of a particular group. This group's interest matches his/her self-interest. |

## *Management's Role*

Perhaps the most important thing management can do for a group is to give it time to become effective. This requires, among other things, that management work to maintain consistent group membership. Group members must not be moved out of the group without very good reason. Nor should there be a constant stream of new people temporarily assigned to the group. If a group is to progress through the four stages described earlier in this chapter, to the crucial performing stage, it will require a great deal of discipline from both the group and management.

Another area where management must help is creating an atmosphere within the company where groups can be effective.

## *Facilitation Techniques*

### When to Use an Outside Facilitator

It is not always necessary to have an outside party facilitate a group or team. While facilitators can often be of benefit, they may also add cost and the use of facilitators should, therefore, be carefully considered. The following guidelines can be used to determine if outside facilitation is needed (Schuman, 1996):

▲ **Distrust or bias**—In situations where distrust or bias is apparent or suspected, groups should make use of an unbiased outsider to facilitate (and perhaps convene) the group.

- **Intimidation**—The presence of an outside facilitator can encourage the participation of individuals who might otherwise feel intimidated.
- **Rivalry**—Rivalries between individuals and organizations can be mitigated by the presence of an outside facilitator.
- **Problem definition**—If the problem is poorly defined, or is defined differently by multiple parties, an unbiased listener and analyst can help construct an integrated, shared understanding of the problem.
- **Human limits**—Bringing in a facilitator to lead the group process lets members focus on the problem at hand, which can lead to better results.
- **Complexity or novelty**—In a complex or novel situation, a process expert can help the group do a better job of working together intellectually to solve the problem.
- **Timelines**—If a timely decision is required, as in a crisis situation, the use of a facilitator can speed the group's work.
- **Cost**—A facilitator can help the group reduce the cost of meeting—a significant barrier to collaboration.

### Selecting a Facilitator

Facilitators should possess four basic capabilities (Schuman, 1996):

- He or she should be able to anticipate the complete problem-solving and decision-making processes.
- He or she should use procedures that support both the group's social and cognitive process.
- He or she should remain neutral regarding content issues and values.
- He or she should respect the group's need to understand and learn from the problem-solving process.

Facilitation works best when the facilitator:

- Takes a strategic and comprehensive view of the problem-solving and decision-making processes and selects, from a broad array, the specific methods that match the group's needs and the tasks at hand.
- Supports the group's social and cognitive processes, freeing the group members to focus their attention on substantive issues.
- Is trusted by all group members as a neutral party who has no biases or vested interest in the outcome.
- Helps the group understand the techniques being used and helps the group improve its own problem-solving processes.

## Principles of Team Leadership and Facilitation

Human beings are social by nature. People tend to seek out the company of other people. This is a great strength of our species, one that enabled us to rise above and dominate beasts much larger and stronger than ourselves. It is this ability that allowed men to control herds of livestock to hunt swift antelope, and to protect themselves against predators. However, as natural as it is to belong to a group, there are certain behaviors that can make the group function more (or less) effectively than their members acting as individuals.

We will define a group as a collection of individuals who share one or more common characteristics. The characteristic shared may be simple geography, that is, the individuals are gathered together in the same place at the same time. Perhaps the group shares a common ancestry, like a family. Modern society consists of many different types of groups. The first group we join is, of course, our family. We also belong to groups of friends, sporting teams, churches, PTAs, and so on. The groups differ in many ways. They have different purposes, different time frames, and involve varying numbers of people. However, all effective groups share certain common features. In their work, *Joining Together*, Johnson and Johnson (1999) list the following characteristics of an effective group:

- Group goals must be clearly understood, be relevant to the needs of group members, and evoke from every member a high level of commitment to their accomplishment.
- Group members must communicate their ideas and feelings accurately and clearly. Effective, two-way communication is the basis of all group functioning and interaction among group members.
- Participation and leadership must be distributed among members. All should participate, and all should be listened to. As leadership needs arise, members should all feel responsibility for meeting them. The equalization of participation and leadership makes certain that all members will be involved in the group's work, committed to implementing the group's decisions, and satisfied with their membership. It also ensures that the resources of every member will be fully utilized, and increases the cohesiveness of the group.
- Appropriate decision-making procedures must be used flexibly if they are to be matched with the needs of the situation. There must be a balance between the availability of time and resources (such as members' skills) and the method of decision-making used for making the decision. The most effective way of making a decision is usually by consensus (see below). Consensus promotes distributed participation, the equalization of power, productive controversy, cohesion, involvement, and commitment.
- Power and influence need to be approximately equal throughout the group. They should be based on expertise, ability, and access to information, not on authority. Coalitions that help fulfill personal goals should be formed among group members on the basis of mutual influence and interdependence.

- Conflicts arising from opposing ideas and opinions (controversy) are to be *encouraged*. Controversies promote involvement in the group's work, quality, creativity in decision-making, and commitment to implementing the group's decisions. Minority opinions should be accepted and used. Conflicts prompted by incompatible needs or goals, by the scarcity of a resource (money, power), and by competitiveness must be negotiated in a manner that is mutually satisfying and does not weaken the cooperative interdependence of group members.
- Group cohesion needs to be high. Cohesion is based on members liking each other, each member's desire to continue as part of the group, the satisfaction of members with their group membership, and the level of acceptance, support, and trust among the members. Group norms supporting psychological safety, individuality, creativeness, conflicts of ideas, growth, and change need to be encouraged.
- Problem-solving adequacy should be high. Problems must be resolved with minimal energy and in a way that eliminates them permanently. Procedures should exist for sensing the existence of problems, inventing and implementing solutions, and evaluating the effectiveness of the solutions. When problems are dealt with adequately, the problem-solving ability of the group is increased, innovation is encouraged, and group effectiveness is improved.
- The interpersonal effectiveness of members needs to be high. Interpersonal effectiveness is a measure of how well the consequences of your behavior match intentions.

These attributes of effective groups apply regardless of the activity in which the group is engaged. It really doesn't matter if the group is involved in a study of air defense, or planning a prom dance. The common element is that there is a group of human beings engaged in pursuit of group goals.

### Facilitating the Group Task Process

Team activities can be divided into two subjects: task-related and maintenance-related. Task activities involve the reason the team was formed, its charter, and its explicit goals.

The facilitator should be selected before the team is formed, and he or she should assist in identifying potential team members and leaders, and in developing the team's charter.

The facilitator also plays an important role in helping the team develop specific goals based on their charter. Goal-setting is an art, and it is not unusual to find that team goals bear little relationship to what management actually had in mind when the team was formed. Common problems are goals that are too ambitious, goals that are too limited, and goals that assume a cause and effect relationship without proof. An example of the latter would be a team chartered to reduce scrap assuming that Part X had the highest scrap loss (perhaps based on a week's worth of data) and setting as its goal the reduction of scrap for that part. The facilitator can provide a channel of communication between the team and management.

Facilitators can assist the team leader in creating a realistic schedule for the team to accomplish its goals. The issue of scheduling projects is covered in Chap. 6.

Facilitators should ensure that adequate records are kept on the team's projects. Records should provide information on the current status of the project. Records should be designed to make it easy to prepare periodic status reports for management. The facilitator should arrange for clerical support with such tasks as designing forms, scheduling meetings, obtaining meeting rooms, and securing audio visual equipment and office supplies.

Other activities where the facilitator's assistance is needed include

- **Meeting management**—Schedule the meeting well ahead of time. Be sure that key people are invited and that they plan to attend. Prepare an agenda and stick to it! Start on time. State the purpose of the meeting clearly at the outset. Take minutes. Summarize from time-to-time. Actively solicit input from those less talkative. Curtail the overly talkative members. Manage conflicts. Make assignments and responsibilities explicit and specific. End on time.
- **Communication**—The idea that "the quality department" can "ensure" or "control" quality is now recognized as an impossibility. To achieve quality the facilitator must enlist the support and cooperation of a large number of people outside of the team. The facilitator can relay written and verbal communication between the team and others in the organization. Verbal communication is valuable even in the era of instantaneous electronic communication. A five-minute phone call can provide an opportunity to ask questions and receive answers that would take a week exchanging email and faxes. Also, the team meeting is just one communication forum; the facilitator can assist team members in communicating with one another between meetings by arranging one-on-one meetings, acting as a go-between, etc.

## Facilitating the Group Maintenance Process

Study the group process. The facilitator is in a unique position to stand back and observe the group at work. Are some members dominating the group? Do facial expressions and body language suggest unspoken disagreement with the team's direction? Are quiet members being excluded from the discussion?

When these problems are observed, the facilitator should provide feedback and guidance to the team. Ask the quiet members for their ideas and input. Ask if anyone has a problem with the team's direction. Play devil's advocate to draw out those with unspoken concerns.

# CHAPTER 6

# The Define Phase

The key objectives within the Define phase are

- Develop the Project Charter
    - Define scope, objectives, and schedule
    - Define the process (top-level) and its stakeholders
    - Select team members
    - Obtain authorization from sponsor
- Assemble and train the team

## Project Charters

The official plan and authorization for the project is summarized in the Six Sigma Project Charter, as shown in Fig. 6.1. The Project Charter is a contract between the project team and its sponsor. As such, any changes in the critical elements of scope, objectives, or schedule require approval from the sponsor and consensus of the team.

The charter documents the *why, how, who,* and *when* of a project, include the following elements:

- Problem statement
- Project objective or purpose, including the business need addressed
- Scope
- Deliverables (i.e., objective measures of success that will be used to evaluate the effectiveness of the proposed changes, as discussed below)
- Sponsor and stakeholder groups
- Team members
- Project schedule (using Gantt or PERT as an attachment)
- Other resources required

These items are largely interrelated: as the scope increases, the timetable and the deliverables also expand. Whether initiated by management or proposed by operational per-

## Project Charter

| Project Name/Title: | Order Processing Efficiency | Start Date: 9/17/07 |
|---|---|---|

**Problem/Project Description:**
Current capacity in Sales/Customer Support area is constrained, while there are untapped opportunities for increased sales. We should limit, wherever possible, Sales involvement in order processing to free up resource for active lead follow-up and sales generation. In addition, errors and/or gaps in information acquired during Order Processing procedure have a negative impact on time required to generate, and/or receipt rate of, email marketing and software renewals to existing clients. This has an especially large potential impact, since it requires correction by senior sales staff, who might otherwise have more time to engage with clients, develop marketing efforts, or work with product development staff.

**Project Scope (Process, Product, functional areas):**
Limited to software products.

| Project Objectives & Goals: | Metric | Baseline | Goal |
|---|---|---|---|
| To decrease cycle time & costs of specific Sale Department activities:<br>➢ Order Processing by 50%+<br>➢ Marketing to existing clients by 80+%<br>➢ Software renewals by 80+% | Cost/Order<br><br>Time/campaign<br>Time/update | $32 download<br>$40 shipped<br>2-4 hours<br>2-4 hours | $16 download<br>$20 shipped<br>20 minutes<br>20 minutes |

**Business Need**

*Customer Impact:*
Improved notification rate for renewals & upgrades; reduction in total cycle time as procedure more streamlined.

*Shareholder Impact:*
Increased sales potential, immediately on upgrades, but also for future sales with availability of sales staff; Reduced cost for order processing. Reduced costs for marketing & renewal campaigns.

*Employee Impact:*
Clearer responsibilities; Less interruption in process flow.

| Project Sponsor:<br>Peter Keene, VP | Stakeholder Group:<br>Sales & Operations | Signature / Date |
|---|---|---|
| Team Black Belt:<br>Patrick Killihan | | |
| Team Members:<br>Don Debuski | Customer Support | |
| Helen Winkleham | Shipping & Packaging | |
| Anne Sheppard | Accounting | |

| DEFINE | MEASURE | ANALYZE | IMPROVE | CONTROL |
|---|---|---|---|---|
| Objective DateComplete | Objective DateComplete | Objective DateComplete | Objective DateComplete | Objective Date Complete |
| ➢ Project Def. 9/17/07<br>➢ Top level Process Def. 9/19/07<br>➢ Team Formation 9/19/07 | ➢ Process Definition ____<br>➢ Metric Def. ____<br>➢ Estimate Baseline ____ | ➢ Value Stream Analysis ____<br>➢ Analyze Variation ____<br>➢ Determine Drivers ____ | ➢ Implement Process ____<br>➢ Assess Benefits ____<br>➢ Evaluate Failure Mode ____ | ➢ Standardize Methods ____<br>➢ Control Plan ____<br>➢ Lessons Learned ____ |

**Figure 6.1** Example Project Charter.

sonnel, many projects initially have too broad a scope. As the project cycle time increases, the tangible cost of the project deployment, such as cost due to labor and material usage, will increase. The intangible costs of the project will also increase: frustration due to lack of progress, diversion of manpower away from other activities, and delay in realization of project benefits, to name just a few. When the project cycle time exceeds six months or so, these intangible costs may result in the loss of critical team members, causing additional delays in the project completion. These "world peace" projects, with laudable but unrealistic goals, generally serve to frustrate teams and undermine the credibility of the Six Sigma program.

Readers familiar with lean implementations will likely notice the similarity of the charter to the lean A3 problem-solving report. An example of the A3 report from Toyota is shown in Fig. 6.2. (The A3 gets its unique name from the size of the paper customarily used to print the report). The charter shares many elements with the A3, and both are meant as living documents to communicate a current summary of the project. Where the charter devotes more space to economic justification of the proposal, the A3 puts more emphasis on the analysis and improvement stage findings, which would generally be summarized on a second page of the charter.

**Download an example Project Charter at www.mhprofessional.com/ssh5.**

**Figure 6.2** Example A3 Problem-Solving Report (Liker and Ross, 2017).

## Project Decomposition

Large projects must be broken down into smaller projects and, in turn, into specific work elements and tasks. The process of going from project objectives to tasks is called decomposition. The result of decomposition is the *project scope*: the particular area of interest and focus for the project, in process terms.

### *Work Breakdown Structures*

Ruskin and Estes (1995) offer work breakdown structures (WBS) as a process for defining the final and intermediate products of a project and their relationships. Defining project tasks is typically complex and accomplished by a series of decompositions followed by a series of aggregations. For example, a software project to develop an SPC software application would disaggregate the customer requirements into very specific analytic requirements (e.g., the customer's requirement that the product create $X$-bar charts would be decomposed into analytic requirements such as subroutines for computing subgroup means and ranges, plotting data points, drawing lines, etc.). Aggregation would involve linking the various modules to produce an $X$-bar chart displayed on the screen.

The WBS can be represented in a tree diagram, as shown in Fig. 6.3. Tree diagrams are used to break down or stratify ideas in progressively greater detail. The objective is to partition a big idea or problem into its smaller components, to reach a level where projects are "tiny." By doing this, you will make the idea easier to understand, or the problem easier to solve. The basic idea is that, at some level, a problem's solution becomes relatively easy to find, or is isolated enough in scope as to be simpler than other solutions necessary to satisfy all possible conditions. This is the tiny level. Work takes place on the smallest elements in the tree diagram.

For example, an order processing project team for a software sales and training firm decomposed the order processing tasks based on Software or Professional Service; then by product family (e.g., Product Family A, Product Family B, Product Family C), then by the type of sale within that family (Download, Shipment, Support Renewal). The breakdown allows the team to then further consider which of these somewhat unique elements should be considered for improvement.

While the WBS provides a graphical view of the potential project elements, it provides no insight into the relative benefits to be obtained in improving a given element. Here, a Pareto analysis is often helpful. Pareto diagrams are useful in the Define, Measure, and Analyze stages to focus project resources on the areas, defects, or causes yielding the highest return.

**Figure 6.3** Example WBS for Order Processing project; constructed using Mind Genius software (www.qualityamerica.com by permission).

## *Pareto Analysis*

A Pareto diagram is a vertical bar graph showing problems (or perhaps more directly *opportunities*) in a prioritized order, so it can be determined which problems or opportunities should be tackled first. The categories for the vertical bars represent mutually exclusive categories of interest. The categories are sorted in decreasing order from left to right in the Pareto diagram by their count or cost, whichever is being displayed.

In the Define phase, the Pareto diagram is used in conjunction with the WBS to quantify the opportunities inherent in each of the *tiny* elements resulting from the WBS.

For example, the data in Table 6.1 have been recorded for orders received in the Order Processing function considered in the WBS.

**Table 6.1** Raw Data for Pareto Analysis

| Product Family | Download | Shipment | Support |
|---|---|---|---|
| A | 27 | 3 | 30 |
| C | 33 | 5 | 8 |
| B | 13 | 5 | 0 |

The data in Table 6.1 is shown in a massaged format in Table 6.2, to demonstrate how it is analyzed for the Pareto diagram of Fig. 6.4. In this case, a "stacked" Pareto diagram is used to display the type of order that is processed: the bottom section of each bar represents *Downloads*; the middle section *Shipments*; and the top section *Support Renewals*.

**Table 6.2** Data Organized for Pareto Analysis

| Rank | Product Family | Count | Percentage | Cum % |
|---|---|---|---|---|
| 1 | A | 60 | 48.39 | 48.39 |
| 2 | C | 46 | 37.10 | 85.49 |
| 3 | B | 18 | 14.52 | 100.01 |

Note that, as often happens, the final percentage is slightly different than 100%. This is due to round-off error and is nothing to worry about.

## Deliverables

Six Sigma projects focus on one or more of the following three critical deliverables: cost, quality, and schedule. Factors Critical to Cost (CTC) include parameters that impact work in progress, finished goods inventory, overhead, delivery, material, and labor, even when the costs can be passed on to the customer. Critical to Quality (CTQ) factors are perhaps most familiar to operational personnel since they directly impact the functional requirements specified by the internal and external customers. Critical to Schedule (CTS) factors impact the delivery time of the product or service.

While there are other CTx factors that can be considered, such as Critical to Safety, these factors are often best expressed in terms of Quality, Schedule, and/or Cost. These three prime factors are most meaningful to organizational objectives, so they remain the focus of the vast majority of projects.

Most estimates of quality, schedule, and often cost are based on statistical samples. That is, it is often cost prohibitive to sample each and every potential unit (i.e., 100% sam-

**Number Orders**
Downloaded; Shipped; Support Renewals

**Figure 6.4** Example Pareto diagram constructed using Green Belt XL software (www.qualityamerica.com by permission).

pling) of product or service for all critical characteristics, so a sub-sample of the process is taken to estimate its quality, cost, or schedule metric. In that regard, it is critical to properly evaluate the metrics as statistical estimates, as discussed further in Chaps. 7 through 10. In the Define stage, rough estimates based on historical data are often used for expediency, with the expectation they will be validated and refined in the Measure stage through a thorough statistical analysis. As explained in Chaps. 7 and 8, processes are best analyzed using the principles and techniques of Statistical Process Control (SPC). Valid predictions for process error rates can only be made when the process is stable (i.e., in a state of statistical control), as determined through a suitable SPC control chart. Samples from unstable processes provide little information as to the future state of the process; since the process is unstable, it is nearly impossible to predict its future course. Samples from unstable processes provide only limited information about the time period from which the sample was drawn. Even this estimate is highly suspect, with strong arguments that 100% sampling is the only way to properly estimate its properties.

## Critical to Quality Metrics

There are a number of ways to quantify Critical to Quality (CTQ) process performance, as discussed below. CTQ metrics are derived by comparing process observations against the process requirements. In the simplest of cases, the process data is reduced to the resulting comparison: did the process meet expectations (i.e., it was acceptable), or fail to meet expectations (i.e., unacceptable, or defective). In this case, the actual process measure is not recorded; only the count of failures, errors, or defects. This *attribute* or *count* data provides less information than the actual *variable* measure of the process. This concept is further discussed in Chaps. 7 and 8 and summarized as follows.

Consider a process to grind the diameter of a pin to a size between 0.995 and 1.005 inches. A pin is sampled from the process, and measured at 1.001 inches. Since its size is between the maximum and minimum diameter allowed, it meets the requirements and is not counted as a defect. Likewise, a pin measuring 1.00495 meets requirements, since the diameter is less than the 1.005 inch maximum. Yet, even if these pins are equally acceptable in their application (which is questionable), the information content of the measured value is much higher than that of the count of defects. If, for example, a vast majority of the sampled pieces have diameters approaching the maximum size, then it is possible, if not likely, that some pieces have diameters exceeding the maximum since all pieces are not sampled. That is, even though the number of pins exceeding the maximum in the sample is zero, we could use the measured variables data to reliably predict that additional samples would yield a certain percentage larger than the maximum size. Using variables (i.e., measurement) data, we can use the statistical properties of the samples to predict errors even when there is no direct evidence of errors, demonstrating the higher informational content of the variables measurements over the attribute counts.

The CTQ metrics below use an estimate of the proportion of defects in the sample to estimate the metric. The percent defective may be estimated using the actual data (e.g., the measured diameters), as well as the count of items that fail to meet the requirements. In light of the arguments provided in the preceding paragraph, it should be clear that the estimates using the counts are less reliable than those using the variables data. These concepts are discussed in further detail in Chap. 8.

Process defect rates are best estimated using SPC control charts (detailed in Chap. 8). For attribute (count) data, the error rate for a given process characteristic is best estimated as the centerline of the $p$ control chart ($p$-bar: the average error percentage). The centerline on the attribute chart is the long-term expected quality level of the process, for example, the average proportion defective for the case of the $p$ chart. For variables (measurement) data, the defect rate for a process characteristic is estimated using a variable control chart (e.g., the $X$-bar control chart) and assumptions regarding the distribution of the process observations.

The Define stage estimates of error rates will depend on the operational data available, so unless the organization is using control charts to monitor their operations on an ongoing basis, the control chart estimate of the process performance will wait until the Measure stage. At that time, the initial estimates will be verified with the scrutiny provided by the control chart.

Also note that the terms *defects* and *errors* apply universally to all processes with customers, regardless of industry. A service process, such as the filling of a prescription at the local pharmacy, may have a customer expectation of delivery within 30 minutes of drop-off. In this case, the delivery time might be considered a critical quality characteristic, since it relates to the quality of the service as perceived by the customer. Of course, the content of the prescription refill (i.e., whether the correct medicine of the correct potency with the correct dosage instructions) could also be considered a critical quality characteristic. It should be clear from Chaps. 2 and 3 that quality metrics may be established for a wide variety of process parameters, including those related to schedule and cost, to achieve a customer requirement, expectation, or excitement (as in the Kano model). These limits may be defined by external customers, internal management, contract specifications, or regulation.

## Defects-per-Million Opportunities and Sigma Level

When considering the defects per million opportunities (DPMO) for a single process characteristic, DPMO is directly calculated using the estimated defect rate (i.e., percent defective; defects per hundred) by multiplying by $10^6$. For example, a process with an estimated error rate of 1.2% (0.012) will have a DPMO of 12,000. DPMO is the same as the parts per million calculated by Minitab, when calculated for a single characteristic (i.e., defect type).

In some cases, analysts seek to estimate an overall DPMO for a process. Consider the pharmacy example cited above, where there are four critical characteristics (delivery time, medicine formula, formula potency, and dosage). If a sample of 1,000 prescriptions are checked for all of the four critical characteristics, and 12 prescriptions are identified that have an error with at least one of the critical characteristics, the overall number of defects is 12,000 per million, as in the case cited in the above paragraph. Yet, the DPMO is calculated as 3,000 defects-per-million *opportunities*, since there are four opportunities for failure in each of the 12,000 samples. Observant readers will question the validity of this metric, since the DPMO could be arbitrarily lessened by increasing the number of critical characteristics (e.g., six critical characteristics using same data results in 2,000 DPMO). For these reasons, DPMO estimates for anything beyond a single characteristic should be treated with caution at the least.

A calculated DPMO can be easily converted to a sigma level using Appendix 14. The following example illustrates the calculations used to populate the appendix table.

In the example data presented in Table 6.3, some 600 items (e.g., purchase orders, manufactured pieces, software changes, etc.) are processed through the five process steps in a serial fashion. Items identified with one or more errors are reworked to correct the error(s).

The DPMO for each process step is calculated as:

$$\text{DPMO} = [(\text{\# errors}) / (\text{\#critical characteristics}) \times (\text{\#items})] \times 10^6 \quad (6.1)$$

The DPMO for process step 1 is calculated as:

$$\text{DPMO} = [74 / (4 \times 600)] \times 10^6 = 30833$$

The DPMO for each subsequent step is similarly calculated for this example. In some processes, as is common in manufacturing, items may be discarded (or scrapped) if they cannot be reworked. In this case, the number of items in the DPMO denominator for each subsequent step is decreased by the number of items scrapped in the prior steps. The number of opportunities at each process step varies to reflect the actual number of items processed at that step.

The sigma level for each process step is calculated using the properties of the normal distribution. (The normal distribution and the methods shown here are formally presented in Chap. 8.) The percentile associated with the DPMO is calculated for the right tail of the normal distribution. This percentile is converted to the standard normal distribution's $z$ value, which will be a positive number for the right-hand tail of the distribution. (In other words, the calculated alpha is subtracted from 1 in Eq. (6.3), resulting in a positive $z$ value.) The $z$ value is then corrected by the 1.5 sigma shift (discussed in Chap. 1; see also Dr. Mikel Harry's comment later in this section).

$$\alpha = \frac{\text{DPMO}}{1,000,000} \quad (6.2)$$

$$\text{Sigma Level} = z_{1-\alpha} + 1.5 \quad (6.3)$$

The sigma level for process step 1 is calculated as follows, using Appendix 2 to determine the $z$ value of 1.87 at the 96.92 percentile:

$$\alpha = \frac{30{,}833}{1{,}000{,}000} = 0.0308 \qquad \text{Sigma Level} = z_{.9692} + 1.5 = 1.87 + 1.5 = 3.37$$

The sigma level is also conveniently calculated in MS Excel as:

$$=(\text{NORM.S.INV}(1 - (\text{DPMO}/1000000))) + 1.5 \quad (6.4)$$

*Note:* **Calculation shown uses formula function beginning with Excel 2013. Prior versions used the function Normsinv.**

**Table 6.3** Example Data and CTQ Calculation Results for 600 Units Processed

| Process Step | Critical Characteristics | Errors Detected | Items with Error(s) | DPMO | Sigma Level | First-Pass Yield (%) | DPU | Throughput Yield (%) |
|---|---|---|---|---|---|---|---|---|
| 1 | 4 | 74 | 51 | 30,833 | 3.37 | 92 | 0.12 | 88 |
| 2 | 3 | 103 | 98 | 57,222 | 3.08 | 84 | 0.17 | 83 |
| 3 | 1 | 67 | 66 | 111,667 | 2.72 | 89 | 0.11 | 89 |
| 4 | 3 | 157 | 86 | 87,222 | 2.86 | 86 | 0.26 | 74 |
| 5 | 1 | 44 | 43 | 73,333 | 2.95 | 93 | 0.07 | 93 |

The sigma level calculations shown above reflect the impact of the 1.5 sigma shift, which, although customary and necessary to achieve the quoted 3.4 DPMO at a Six Sigma level of performance, has been the subject of much reflection and debate within the Six Sigma community. A variety of reasons for applying the shift have been provided over the years, the most credible being that even when processes are well controlled, they can still be subject to unidentified process shifts over the long term, which could impact the customer.

Dr. Mikel Harry, a cofounder of the practices known as Six Sigma while at Motorola, waded into the debate in a series of LinkedIn posts beginning late December 2013. Building from an article he had written earlier in the year, he explained:

> In short, the Six Sigma Model (which includes the shift factor) is a design engineering goal, not a process measure, nor should it be used as a surrogate for actual process capability data. As a goal, it stands without need of proof—it's simply a goal....
>
> Throughout this discussion thread I have repeatedly stated that the 1.5 shift factor is a design engineering tool, not to be used as a surrogate for actual data, especially when evaluating the capability or control of a production process. Design engineers should treat the 1.5 sigma shift as a constant so as to provide a consistent bias to their comparative Monte Carlo simulations (with the aim of evaluating design robustness). As a general rule, process engineers should not treat the 1.5 sigma shift as a constant—owing to the unique behaviors and patterns of live processes. In such cases, the process engineer must not treat the 1.5 sigma shift as a surrogate for actual process capability data.

These recent revelations would seem to conflict with the practices presented in his widely circulated book *Six Sigma: The Breakthrough Management Strategy Revolutionizing the World's Top Corporations*. In the first chapter, numerous examples are provided (pages 14 to 16: carpet cleaning; airline baggage handling; billing and charging) where the performance of a process is evaluated, including a detailed example template ("How to Approximate the Sigma Capability for One of Your Processes"). In each of these examples, process data are used to estimate the DPMO, which is then converted to a sigma level including the 1.5 sigma shift (as shown in Eqs. (6.3) and (6.4) above).

Obviously, there is potential for confusion with the sigma level metric. Many practitioners have opted to exclude the shift; others include it. This is not a major issue, so long as its definition is standardized within an organization, or otherwise presented clearly to the audience. The best overall recommendation is to avoid its use. The DPMO metric cited above is a clear alternative. A better alternative is the process capability index discussed in Chap. 8. This metric serves as a predictor for stable processes. When a process is not stable, it cannot be predicted by any metric. The best alternative in this case is the process performance index, which provides an estimate for a given sample of process.

See online video describing the DPMO and Sigma Level Calculations at www.mhprofessional.com/ssh5.

## Throughput Yield

For a process, throughput is a measure of the productivity of the process in producing error-free product or service. For a product, throughput is a measure of the quality of the entire product as a function of the quality of its various features.

The throughput yield is calculated based on the defects per unit (DPU) for each process step:

$$\text{DPU} = \frac{\#\text{errors}}{\#\text{items}} \tag{6.5}$$

$$Y_T = 1 - \text{DPU} \tag{6.6}$$

The DPU and throughput yield for process step 1 are calculated as:

$$\text{DPU} = \frac{74}{600} = .12 \qquad Y_T = 1 - 0.12 = .88 = 88\%$$

Note that throughout yield is not the same as first-pass yield, which may be calculated as follows:

$$Y_{FP} = 1 - \frac{\#\text{items with errors}}{\#\text{items}} \tag{6.7}$$

Here, the number of items with errors is considered before any rework takes place (i.e., after the first pass of processing and before any additional processing). Whereas throughput yield focuses on the total number of errors as an indication of severity of process errors, first-pass yield considers only the number of items impacted. The difference can be significant when each item has the potential for more than one error or deficiency, even when there is only a single critical characteristic (e.g., surface imperfections on a disk drive).

### Rolled Throughput Yield

The rolled throughput yield (RTY) summarizes stepwise throughput yields for a multiple-step process or multiple-feature product. RTY is a measure of the overall quality level: the likelihood of passing a unit of product or service through all steps error-free (Harry and Shroeder, 2000).

Rolled throughput yield $Y_{RT}$ for an $N$-step process (or $N$-characteristic product) is calculated using the throughput yield $Y_T$ for steps 1 through $N$ as:

$$Y_{RT} = Y_{T1} \times Y_{T2} \times Y_{T3} \times \ldots \times Y_{TN} \tag{6.8}$$

The rolled throughput yield for the five-step process shown in Table 6.3 is calculated as:

$$Y_{RT} = 0.88 \times 0.83 \times 0.89 \times 0.74 \times 0.93 = 0.44$$

Note that the rolled throughput yield is worse than the worst-case throughput yield of any process step. It is also worse than the average throughput yield of 85%.

### Normalized Yield

To compute the normalized yield, which is a kind of average, for a $k$-step process or $k$-product department or organization, use Eq. (6.9):

$$Y_N = \sqrt[k]{Y_{RT}} \tag{6.9}$$

For the five-step process noted in Table 6.3, the throughput yield was calculated as 0.44 (or 44%). The normalized yield for this process is calculated as:

$$Y_N = \sqrt[5]{0.44} = 0.85 \tag{6.10}$$

Normalized yield should be considered a handy accounting device for measuring overall system quality. Because it is a type of average it is not necessarily indicative of any particular product or process yield or of how the organization's products will perform in the field. To calculate these, refer to "Rolled Throughput Yield" and "Defects-per-Million Opportunities and Sigma Level" earlier in this chapter.

Unfortunately, finding the RTY isn't always as straightforward as described above. In the real world you seldom find a series of process steps all neatly feeding into one another in a nice, linear fashion. Instead, you have different supplier streams, each with different volumes and different yields. There are steps that are sometimes taken and sometimes not. There are test and inspection stations, with imperfect results. There is rework and repair. The list goes on and on. In such cases it is sometimes possible to trace a particular batch of inputs through the process, monitoring the results after each step. However, this is often exceedingly difficult to control. The production and information systems are not designed to provide the kind of tracking needed to get accurate results. The usual outcome of such attempts is questionable data and disappointment.

High-end simulation software offers an alternative. With simulation you can model the individual steps, then combine the steps into a process using the software. The software will monitor the results as it "runs" the process as often as necessary to obtain the accuracy needed. Figure 6.5 shows an example. Note that the Properties dialog box is for step 12 in

**Figure 6.5** Finding RTY using simulation software (iGrafx Process for Six Sigma, Corel Corporation).

the process ("Right Med?"). The model is programmed to keep track of the errors encountered as a Med Order works its way through the process. Statistics are defined to calculate dpu and RTY for the process as a whole (see the Custom Statistics box in the lower-right section of Fig. 6.5). Since the process is nonlinear (i.e., it includes feedback loops) it isn't a simple matter to determine which steps would have the greatest impact on RTY. However, the software lets the Black Belt test multiple what-if scenarios to determine this. It can also link to Minitab or Excel to allow detailed data capture and analysis.

See online video describing the Rolled Throughput Yield Calculations at www.mhprofessional.com/ssh5.

## *Critical to Schedule Metrics*

Critical to Schedule (CTS) metrics are related to Cycle Time and scheduling efficiencies, including: Cycle Time, Process Cycle Efficiency, Process Velocity, and Overall Equipment Effectiveness.

### Cycle Time

The Cycle Time for a process usually refers to the total elapsed time for the process (from process start to process completion). Note that this may not equal the total task time for the process, since there may be additional time where no task is being performed (e.g., queue time between tasks). This total elapsed time more accurately reflects the process from a customer's perspective, even though it may not reflect the internal costs of providing the service. In practice, it is suggested to clearly define the Cycle Time so that your audience is clear on its usage in your discussions.

### Process Cycle Efficiency

Process Cycle Efficiency is a metric useful for prioritizing improvement opportunities. Process Cycle Efficiency is calculated by dividing the value-added time associated with a process by the total lead time of the process (George, 2002). Note that this value-added time should be expressed in the same units as lead time. In this calculation, value-added time is *not* expressed as a percentage of the total cycle time, as is sometimes the case.

If the process consists of only value-added activities, then the Process Cycle Efficiency would reach a theoretical maximum of 100%. In practice, Process Cycle Efficiencies will exceed 25% for processes that have been improved through the use of Lean methods. Typical Process Cycle Efficiencies for various processes are shown in Fig. 6.6.

| Process Type | Typical Efficiency | World-Class Efficiency |
|---|---|---|
| Machining | 1% | 20% |
| Fabrication | 10% | 25% |
| Assembly | 15% | 35% |
| Continuous | 30% | 80% |
| Transactional | 10% | 50% |
| Creative | 5% | 25% |

**Figure 6.6** Typical and world-class process efficiencies (George, 2002).

The key to improving Process Cycle Efficiency is to reduce the Lead Time, the denominator of the equation.

## Process Lead Time

The Process Lead Time is the delay that occurs from the time the order is received until the unit of product or service can be processed. In some cases, it is known by inference. For example, the process cannot begin until the material arrives from the supplier and is processed through receiving. This typical six-day process is the lead time. For internal processes, it may be estimated by dividing the number of items in process by the completions per hour (George, 2002). For example, if it takes 2 hours on average to complete each Purchase Order, then there is 0.5 completion per hour. This is the denominator of the equation. If there are 10 Purchase Orders waiting in queue (the numerator), then the Process Lead Time is 20 hours (10 divided by 0.5 equals 20). In other words, new orders can't be processed until the 20-hour lead time has allowed the existing Work in Process to be completed.

## Velocity

Once Lead Time is known, Velocity can be calculated by dividing it into the number of value-added steps (George, 2002). In the preceding example, if there are five value-added process steps in the PO process, then the Velocity may be calculated as 5 divided by 20 equals 0.25 step per hour.

The Velocity of the process represents the responsiveness or flexibility of the process to customer demand. A long lead times result in slow velocity. Lead time is reduced, and velocity increased, when Work in Progress is reduced.

The rationale is simple: new orders from customers cannot be started until work (or items) in process is completed. Thus, the activity on new items is stalled. Consider your doctor's waiting room. The patients are Work in Progress. New patients aren't serviced by the doctor until those that arrived earlier are completed.

## Overall Equipment Effectiveness

The Overall Equipment Efficiency (OEE) is a Lean metric that incorporates process availability, performance, and quality in a single metric. The resulting OEE value can be used to compare and prioritize processes for improvement. It is also recommended to track and prioritize based on the individual Availability ($A$), Performance ($P$), and Quality ($Q$) components of the score, since their relative values may provide varying cost benefits.

$$\text{OEE} = A \times P \times Q \tag{6.11}$$

$A$ = actual operating time / planned time
$P$ = ideal cycle time / actual cycle time
$Q$ = acceptable output / total output

The OEE Availability is calculated as the actual time the process is producing product (or service) divided by the amount of time that is planned for production. The planned time, by definition, excludes all scheduled shutdowns when equipment is not operational, which may include lunches, breaks, plant shutdowns for holidays, and so on. The remaining portion of the planned time includes the available time for production and the downtime. Downtime is the loss in production time due to shift changeovers, part changeovers, waiting for material, equipment failures, and so on. It is any measurable time where the process is not available for production, when production had been planned. One hundred percent availability implies that the process is operational 100% of the time.

OEE Performance is the efficiency of the process in minimizing its Operating Time. Performance is calculated by dividing the ideal process cycle time by the actual cycle time. It accounts for process inefficiencies, such as due to poor quality materials, operator inefficiencies, equipment slowdown, etc. One hundred percent performance implies the process is running at maximum velocity.

OEE Quality is the percent of total output that meets the requirement without need for any repair or rework. It might otherwise be called the first-pass quality. One hundred percent quality implies the process is producing no errors.

## *Critical to Cost Metrics*

Metrics that are Critical to Quality and Critical to Schedule are (by definition) often Critical to Cost as well. Costs associated with the process issues must include the effect of losses due to Hidden Factory and Customer impact, such as delays in shipments or in information exchange.

Metrics used in CTC evaluations, often to quantify and compare opportunities, include the Net Present Value (NPV) and Internal Rate of Return (IRR). NPV and IRR calculations reflect the time value of money: over time, interest is earned on our initial investments, as well as on the interest that has already been accrued from the initial investment. This is known as *compounding*.

It's important to consider the time value of money, since improvements efforts will typically require an initial investment, whose benefits will be experienced at some future time. The initial investment represents a diversion of funds, either from a lender or from some other investment opportunity. The value of a particular investment, then, depends on the cost of that diversion, which varies based on market conditions and the length of time before the investment "pays off."

## Financial Analysis of Benefit and Cost

In performing benefit-cost analysis it is helpful to understand some of the basic principles of financial analysis, in particular, break-even analysis and the time value of money (TVM).

Let's assume that there are two kinds of costs:

- **Variable costs** are those costs that are expected to change at the same rate as the firm's level of sales. As more units are sold, total variable costs will rise. Examples include sales commissions, shipping costs, hourly wages, and raw materials.
- **Fixed costs** are those costs that are constant, regardless of the quantity produced, over some meaningful range of production. Total fixed cost *per unit* will decline as the number of units increases. Examples of fixed costs include rent, salaries, depreciation of equipment, etc.

These concepts are illustrated in Fig. 6.7.

**Figure 6.7**   Fixed and variable costs.

## Break-Even Points

We can define the *break-even point*, or *operating break-even point*, as the level of unit sales required to make earnings before interest and taxes (EBIT) equal to zero, that is, the level of sales where profits cover both fixed and variable costs.

Let $Q$ be the quantity sold, $P$ the price per unit, $V$ the variable cost per unit, and $F$ the total fixed costs. Then the quantity $P - V$ represents the variable profit per unit and:

$$Q(P - V) - F = \text{EBIT} \tag{6.12}$$

If we set EBIT equal to zero in Eq. (6.12) and solve for the break-even quantity $Q^*$, we get

$$Q^* = \frac{F}{P - V} \tag{6.13}$$

## Example Break-Even Analysis

A publishing firm is selling books for $30 per unit. The variable costs are $10 per unit and fixed costs total $100,000. The break-even point is

$$Q^* = \frac{F}{P - V} = \frac{\$100,000}{\$30 - \$10} = 5,000 \text{ units} \tag{6.14}$$

Of course, management usually wishes to earn a profit rather than to merely break even. In this case, simply set EBIT to the desired profit rather than zero in Eq. (6.12) and we get the production quantity necessary to meet management's target

$$Q^*_{\text{TARGET}} = \frac{F + \text{EBIT}_{\text{TARGET}}}{P - V} \tag{6.15}$$

For example, if the publisher mentioned above wishes to earn a $5,000 profit, then the break-even level of sales becomes

$$Q^*_{\text{TARGET}} = \frac{F + \text{EBIT}_{\text{TARGET}}}{P - V} = \frac{\$100,000 + \$5,000}{\$30 - \$10} = 5,250 \text{ units} \tag{6.16}$$

In project benefit-cost analysis these break-even quantities are compared to the sales forecasts to determine the probability that the expected return will actually be earned.

## The Time Value of Money

Because money can be invested to grow to a larger amount, we say that money has a "time value." The concept of time value of money underlies much of the theory of financial decision-making. We will discuss two TVM concepts: future value and present value.

*Future Value.* Assume that you have $1,000 today and that you can invest this sum and earn interest at the rate of 10% per year. Then, one year from today, your $1,000 will have grown by $100 and it will be worth $1,100. The $1,100 figure is the *future value* of your $1,000. The $1,000 is the *present value*. Let's call the future value FV, the present value PV and the interest rate $i$, where $i$ is expressed as a proportion rather than as a percentage. Then we can write this example algebraically as follows:

$$FV = PV + PV \times i = PV \times (1 + i) \qquad (6.17)$$

Now, let's say that you could invest at the 10% per year rate for two years. Then your investment would grow as follows:

| Year | Starting Amount | Interest | Ending Amount |
| --- | --- | --- | --- |
| 1 | $1,000 | $100 | $1,100 |
| 2 | $1,100 | $110 | $1,210 |

Observe that in year 2 you earned interest on your original $1,000 *and* on the $100 interest you earned in year 1. The result is that you earned more interest in year 2 than in year 1. This is known as *compounding*. The year time interval is known as the *compounding period*. Thus, the FV after two years is $1,210. Algebraically, here's what happened:

$$FV = [\$1,000 \times (1.10)] \times (1.10) = \$1,000 \times (1.10)^2 \qquad (6.18)$$

Where the value between the [ ] characters represents the value at the end of the first year. This approach can be used for any number of $N$ compounding periods. The equation is

$$FV = PV \times (1 + i)^N \qquad (6.19)$$

Of course, Eq. (6.19) can be solved for PV as well, which gives us the present value of some future amount of money at a given rate of interest.

$$PV = \frac{FV}{(1 + i)^N} \qquad (6.20)$$

*Nonannual Compounding Periods.* Note that $N$ can be stated in any time interval, it need not be in years. For example, if the compounding period was quarterly, then $N$ would be the number of quarters. Of course, the interest rate would also need to be stated in quarters. For example, if the $1,000 were invested for 2 years at 10% per year, compounded quarterly, then:

$$FV = PV \times (1 + i)^N = \$1,000 \times \left(1 + \frac{0.1}{4}\right)^{2 \times 4} = \$1,000 \times (1 + 0.025)^8 = \$1,218.40 \qquad (6.21)$$

***Continuous Compounding.*** Note that the FV is greater when a greater number of compounding periods are used. The limit is an infinite number of compounding periods, known as continuous compounding. For continuous compounding the PV and FV equations are

$$FV = PV \times e^{i \times t} \qquad (6.22)$$

$$PV = \frac{FV}{e^{i \times t}} \qquad (6.23)$$

Where *t* is the length of time (in years) the sum is compounded, *e* is a constant 2.71828, and all other terms are as previously defined. For our example, we have a two-year period, which gives

$$FV = PV \times e^{i \times t} = \$1,000 \times 2.7182818^{0.1 \times 2} = \$1,221.40 \qquad (6.24)$$

***Net Present Value.*** When evaluating project costs and benefits, it often happens that both costs and benefits come in *cash flow streams*, rather than in lump sums. Furthermore, the cash flow streams are uneven, that is, the amounts vary from one period to the next. The approach described above can be used for uneven cash flow streams as well. Simply compute the PV (or FV) of each cash flow separately and add the various results together. The result of applying this procedure is called *the net present value*, or NPV. The procedure, while conceptually easy to grasp, becomes tedious quite quickly. Fortunately, most spreadsheets have a built-in capability to perform this analysis.

Assume that a proposed project has the projected costs and benefits shown in the following table.

| Year | Cost | Benefit |
| --- | --- | --- |
| 1 | $10,000 | $0 |
| 2 | $2,000 | $500 |
| 3 | $0 | $5,000 |
| 4 | $0 | $10,000 |
| 5 | $0 | $15,000 |

Also assume that management wants a 12% return on their investment. What is the NPV of this project?

There are two ways to approach this question, both of which produce the same result (Fig. 6.8). One method would be to compute the net difference between the cost and benefit

| | A | B | C | D |
|---|---|---|---|---|
| 1 | Rate | 12% | | |
| 2 | | | | |
| 3 | Year | Cost | Benefit | Net |
| 4 | 1 | $10,000 | $0 | ($10,000) |
| 5 | 2 | $2,000 | $500 | ($1,500) |
| 6 | 3 | $0 | $5,000 | $5,000 |
| 7 | 4 | $0 | $10,000 | $10,000 |
| 8 | 5 | $0 | $15,000 | $15,000 |
| 9 | NPV | $10,523 | $18,824 | $8,301 |

**Figure 6.8** Using Excel to find the net present value of a project.

for each year of the project, then find the NPV of this cash flow stream. The other method is to find the NPV of the cost cash flow stream and benefit cash flow stream, then subtract.

The NPV of the cost column is $10,523; the NPV of the benefits is $18,824. The project NPV can be found by subtracting the cost NPV from the benefit NPV, or by finding the NPV of the yearly benefit minus the yearly cost. Either way, the NPV analysis indicates that this project's net present value is $8,301.

*Internal Rate of Return.* Often in financial analysis of projects, it is necessary to determine the yield of an investment in a project given its price and cash flows. For example, this may be the way by which projects are prioritized. When faced with uneven cash flows, the solution to this type of problem is usually done by computer. For example, with Microsoft Excel, we need to make use of the internal rate of return (IRR) function. The IRR is defined as the rate of return, which equates the present value of future cash flows with the cost of the investment. To find the IRR the computer uses an iterative process. In other words, the computer starts by taking an initial "guess" for the IRR, determines how close the computed PV is to the cost of the investment, and then adjusts its estimate of the IRR either upward or downward. The process is continued until the desired degree of precision has been achieved.

*IRR Example.* A quality improvement team in a hospital has been investigating the problem of lost surgical instruments. They have determined that in the rush to get the operating room cleaned up between surgeries many instruments are accidentally thrown away with the surgical waste. A test has shown that a $1,500 metal detector can save the following amounts:

| Year | Savings |
|------|---------|
| 1 | $750 |
| 2 | $1,000 |
| 3 | $1,250 |
| 4 | $1,500 |
| 5 | $1,750 |

After five years of use the metal detector will have a scrap value of $250. To find the IRR for this cash flow stream, we set up the Excel spreadsheet and solve the problem as illustrated in Fig. 6.9.

The Excel formula, shown in the window at the top of the figure, was built using the Insert Formula "wizard," with the cash flows in cells B2:B7 and an initial guess of 0.1 (10%). Note that in year 5 the $250 salvage value is added to the expected $1,750 in savings on surgical instruments. The cost is shown as a negative cash flow in year 0. Excel found the IRR to be 63%. The IRR can be one of the criteria for prioritizing projects, as an alternative to, or in addition to, using the PPI.

|   | A | B | C |
|---|---|---|---|
| 1 | Year | Cash Flow | |
| 2 | 0 | ($1,500) | |
| 3 | 1 | $750 | |
| 4 | 2 | $1,000 | |
| 5 | 3 | $1,250 | |
| 6 | 4 | $1,500 | |
| 7 | 5 | $2,000 | |
| 8 | IRR | 63% | |

B8  =IRR(B2:B7,0.1)
Fig. 5.13 IRR exar

**Figure 6.9** Using Excel to find the internal rate of return for a project.

## Top-Level Process Definition

In the Define stage, a top-level view of the process is created to identify the broad scope of the process to be evaluated. This 30,000 foot view of the process provides critical reference for discussions on specific project objectives, calculations of project deliverables, and definition of key stakeholder groups.

## Process Maps

A Process Map is used in the Define stage to document top-level process activities and their stakeholders. A *stakeholder* (also known as a *stakeholder group*) is a department, customer, and/or vendor influenced by the activity or its outcome.

In the top-level Process Map shown in Fig. 6.10, each of the broad process activities are placed in the appropriate "swim lane," where each swim lane indicates a unique stakeholder group. Note that the process activities as documented in the top-level map provide little of the detail necessary to completely define the existing process. The detailed map of process tasks and decisions is developed in the Measure stage, and referenced in the Analyze stage to uncover process complexities. The new process is fully documented on a revised Process Map in the Improve and Control stages to aid in communication with process stakeholders.

Large projects impact a large number of stakeholder groups within the organization. Of greatest importance in the Define stage is to identify the key stakeholders, which are the groups that can make or break any change effort associated with the process. As discussed in Chap. 1, the key stakeholder groups must have buy-in to the process change for successful implementation of the change effort.

**Figure 6.10** Top-level process map for order processing example.

## Assembling the Team

As soon as possible the project manager and project sponsor should arrange a short, informal meeting with managers of all stakeholder groups to inform them that a project is proposed in their functional area, with the permission and direction of the senior executives. This meeting represents an invitation for the middle managers to participate in the project by contributing a member of their team, or to challenge the scope, objective, or business need addressed by the project proposal. It is important to allow the managers time to consider and act on the leadership's decision to pursue the project. If concrete information suggests that tampering or sabotage is occurring, the project manager or process owner should immediately bring it to the attention of the senior executives who approved the Project Charter. The senior leadership should resolve the issue promptly.

If a week or so passes without clear opposition to the project, the project manager should proceed with the implementation of the project plan. Of course, the lines of communication should remain open throughout the implementation of the project plan.

The project team will ideally have one team member from each of the key stakeholder groups. It is important to select a stakeholder representative that is credible with their group and other stakeholder groups, has local management support, is enthusiastic for the change, and is capable and willing to serve on the team. In some cases, we may select a team member that is skeptical of the change, if he or she has all the other characteristics, particularly those of credibility and capability. In this case, we select them so as to build buy-in within the stakeholder group.

Effective teams are generally limited to five to seven participants. Larger teams are more difficult to manage, and members may lose a sense of responsibility to the team. Additional team members may be ad hoc members from non–key stakeholder groups, who only participate as needed, such as for process expertise.

The objectives of the project should be aligned with the interests of stakeholders. If this is not the case, when stakeholders act according to their own interests they will be acting to sabotage the project, intentionally or unintentionally.

Sell all stakeholders on the merits of the project. People resist changes unless they see the value in them and the urgency to take action. Stakeholders must be identified and their needs analyzed so that an action plan can be created to meet the needs and gain commitment. To avoid problems, the project team must constantly communicate with the stakeholders.

Stakeholder focus groups are a method that allows group members to evaluate the potential impact of a plan by identifying the stakeholders affected by or having influence over the project plan. The focus group approach is a highly structured method in which the project team first identifies the stakeholders and their assumptions, then brings those identified together to elicit their responses to the proposed project (see Chap. 2 for a discussion of the focus group technique). The team then rates these assumptions for importance to the stakeholders and importance to the plan. A stakeholder satisfaction plan may be developed to ensure the support of key individuals and groups.

# CHAPTER 7

# The Measure Phase

The objectives of the Measure stage include

1. Process definition: to ensure the specific process under investigation is clearly defined.
2. Metric definition: to define a reliable means of measuring the process, relative to the project deliverables.
3. Establish the process baseline: to quantify the current operating results as a means of verifying previously defined business needs, and to properly substantiate improvement results.
4. Evaluate measurement system: to validate the reliability of data for drawing meaningful conclusions.

For DFSS applications, where DMADV is used, the objectives of the Measure stage will be limited to defining the key metrics and development of a measurement system and plan for obtaining measurements once the new design becomes operational.

An argument can be made for asserting that quality begins with measurement. Only when quality is quantified, can meaningful discussion about improvement begin. Conceptually, measurement is quite simple: measurement is the assignment of numbers to observed phenomena according to certain rules. Measurement is a requirement of any science, including management science.

## Process Definition

A process consists of repeatable tasks, carried out in a specific order. If processes cannot be defined as a series of repeatable tasks, then there may be multiple processes in effect, even an infinite number of processes, or simply the lack of a well-defined process.

It's not uncommon to discover that situation when interviewing process personnel. Operational workers may customize a process to address situations seen in practice, which may not get communicated to all the relevant parties. In this way, customers will experience significant variation depending on the shift or even the specific personnel processing their order. Sometimes this results in improved product or service, and sometimes not. In any event, since we seek to understand the actual process in the Measure stage, the input of the process personnel is necessary. Later, in the Improve stage, we will document a desired process after receiving input from all stakeholders.

There are several useful tools available for defining the process.

- ▲ Flowcharts are particularly useful for highlighting process complexities.
- ▲ Process maps provide an additional level of detail to indicate functional responsibilities for each process step. Process maps were previously discussed in Chap. 6.
- ▲ SIPOC is a tool for identifying the process inputs, outputs and stakeholders.

Generally, these tools will be used in conjunction with one another.

## *Flowcharts*

A flowchart is a simple graphical tool for documenting the flow of a process. In the flowchart, each task is represented by a symbol. There is an ANSI standard that lists symbol types, primarily for computing processes, but most practitioners find rectangles appropriate for most tasks, and diamonds for decision tasks. Decisions should have only two outcomes (yes or no), so decision points must be phrased in this manner. For example, rather than having three options A, B, and C at a decision point, the decision would be constructed as a series of decisions, each of which has a simple yes or no resolution. The first decision might be Option A?, whose No path leads to a second decision of the form Option B? Since Option C is the only remaining choice, the No path from the second decision leads to Option C.

A rather simple flowchart is shown in Fig. 7.1. Notice the diamond decision points have two outcomes: one outcome continues down the main process flow, while the other diverts to a secondary path. Also note that these secondary paths may result in a jump to a later point in the process (as shown in the first decision's "yes" path) or even to a prior point in the process. Decision paths, as well as endpoints for processes, may also branch to other process flowcharts, as indicated by the double-border in the box containing the final "To Shipping" step.

Flowcharts are used in the Measure stage to document the current (as-is) process. In the Analyze phase, the flowchart will be reviewed to uncover complexities in the form of an excessive number of decision points that may contribute to delays or even defects. We can use symbol color or shape to indicate process delays, functional responsibility for each step (for example, oval is customer service), or points in the process where measurements are taken.

## Order Processing: Measure Stage Process Definition

**Figure 7.1** Example flowchart.

## SIPOC

Virtually all Six Sigma projects address business processes that have an impact on a top-level enterprise strategy. In previous chapters, a great deal of attention was devoted to developing a list of project candidates by meticulously linking projects and strategies using dashboards, QFD, structured decision making, business process mapping, and many other tools and techniques. However, Six Sigma teams usually find that although this approach succeeds in identifying important projects, these projects tend to have too large a scope to be completed within the time and budget constraints. More work is needed to clearly define that portion of the overall business process to be improved by the project. One way to do this is to apply process flowcharting or mapping to subprocesses until reaching the part of the process that has been assigned to the team for improvement. A series of questions are asked, such as:

- For which stakeholder does this process primarily exist?
- What value does it create? What output is produced?
- Who is the owner of this process?

- ▲ Who provides inputs to this process?
- ▲ What are the inputs?
- ▲ What resources does this process use?
- ▲ What steps create the value?
- ▲ Are there subprocesses with natural start and end points?

These questions, which are common to nearly all processes addressed by Six Sigma projects, have been arranged into a standard format known as SIPOC. SIPOC stands for Suppliers-Inputs-Process-Outputs-Customers.

SIPOCs begin with people who know something about the process. This may involve people who are not full-time members of the Six Sigma team. Bring the people together in a room and conduct a "focused brainstorming" session. To begin, briefly describe the process and obtain consensus on the definition. For example:

- ▲ "Make it easy for the customer to reach technical support by phone"
- ▲ "Reduce the space needed to store tooling"
- ▲ "Reduce the downtime on the Niad CNC machine"
- ▲ "Get roofing crew to the work site on time"
- ▲ "Reduce extra trips taken by copier maintenance person"

Post flip charts labeled suppliers, inputs, process, outputs, customers. Once the process has been described, create the SIPOC diagram as follows:

1. Create a simple, high-level process map of the process. Display this conspicuously while the remaining steps are taken to provide a reminder to the team.
2. Using brainstorming rules, identify the outputs of this process. Record all ideas on the outputs flip chart without critiquing them.
3. Using brainstorming rules, identify the customers who will receive the outputs. Record all ideas on the customers' flip chart without critiquing them.
4. Using brainstorming rules, identify the inputs needed for the process to create the outputs. Record all ideas on the tnputs flip chart without critiquing them.
5. Using brainstorming rules, identify the suppliers of the inputs. Record all ideas on the suppliers flip chart without critiquing them.
6. Clean up the lists by analyzing, rephrasing, combining, moving, etc.
7. Create a SIPOC diagram.
8. Review the SIPOC with the project sponsor and process owner. Modify as necessary.

### SIPOC Example

A software company wants to improve overall customer satisfaction (Big $Y$). Research has indicated that a key component of overall satisfaction is satisfaction with technical support (Little Y). Additional drill down of customer comments indicates that one important driver of technical support satisfaction is the customer's perception that it is easy to contact technical support. There are several different types of technical support available, such as self-help built into the product, the web, or the phone. The process owner commissioned Six Sigma projects for each type of contact. This team's charter is telephone support.

To begin, the team created the process map shown in Fig. 7.2.

Next the team determined that there were different billing options and created a work breakdown structure with each billing option being treated as a subproject. For this example we will follow the subproject relating to the billing-by-the-minute (BBTM) option. After completing the process described above, the team produced the SIPOC shown in Fig. 7.3.

**Figure 7.2** Process map for contacting technical support by telephone.

**Easy to Contact BBTM**

| SUPPLIERS | INPUTS | PROCESS | OUTPUTS | CUSTOMERS |
|---|---|---|---|---|
| • Phone company<br>• Product<br>• Internal computer systems<br>• Internal phone systems | • Question, or issue<br>• Manual<br>• Phone directory<br>• Web | Customer tries to find number | • Customer finds number | • User<br>• Technician |
| | • Dial phone correctly | Customer calls call center | • System gets caller ID<br>• VRU gets call | |
| | • Customer makes VRU selections | Enter VRU | • VRU routes call to CSA | |
| | • CSA sees customer info on screen<br>• Customer presents question or issue | CSA answers call | • Listen to customer<br>• Collect routing info<br>• Decide where to send call | |
| | • Customer gives CC info<br>• Customer purchases premium support<br>• Customer opts for BBTM | CSA obtains credit card information. Upsell. | • CSA gives info on support options<br>• CSA takes order | |
| | • CSA selects transfer option | CSA transfers call to technician | • Call goes to technician<br>• Call goes to queue | |

**Figure 7.3** SIPOC for easy to contact BBTM.

Note that the process is mapped at a very low level. At this level the process map is usually linear, with no decision boxes shown. The typical SIPOC shows the process as it is supposed to behave. Optionally, the SIPOC can show the unintended or undesirable outcomes, as shown in Fig. 7.4.

This "bad process" SIPOC is used only for team troubleshooting. It helps the team formulate hypotheses to be tested during the Analyze phase.

SIPOC analyses focus on the $X$'s that drive the $Y$s. It helps the team understand which "dials to turn" to make the top-level dashboard's Big $Y$ move. In the example, let's assume that the team collects information and determines that a significant percentage of the customers can't find the phone number for technical support. A root cause of the problem then is the obscure location of the support center phone number. Improving overall customer satisfaction is linked to making it easier for the customer to locate the correct number, perhaps by placing a big, conspicuous sticker with the phone number on the cover of the manual. The Big $Y$ and the root cause $X$ are separated by several levels, but the process mapping and SIPOC analysis chain provides a methodology for making the connection.

## NOT Easy to Contact BBTM

| SUPPLIERS | INPUTS | PROCESS | OUTPUTS | CUSTOMERS |
|---|---|---|---|---|
| • Phone company<br>• Product<br>• Internal computer systems<br>• Internal phone systems | • Customer doesn't know where to look for number<br>• Number in manual is incorrect<br><br>• Dial phone incorrectly<br><br>• Customer makes wrong VRU selections<br><br>• CSA sees no customer info<br>• Customer presents incorrect question or issue<br><br>• Customer won't give CC info<br>• Customer won't listen to upsell<br><br>• CSA selects wrong transfer option<br>• Phone system error | Customer tries to find number<br>↓<br>Customer calls call center<br>↓<br>Enter VRU<br>↓<br>CSA answers call<br>↓<br>CSA obtains credit card information. Upsell.<br>↓<br>CSA transfers call to technician | • Customer can't find number<br>• Customer gets 'wrong number' message<br><br>• No caller ID<br>• VRU drops call<br><br>• VRU routes call incorrectly<br>• VRU loses call<br><br>• Wrong person answers call<br>• Call queued<br><br>• Customer annoyed by upsell<br><br>• Call goes to wrong person<br>• Call is lost<br>• Call goes to queue | • User<br>• Technician |

**Figure 7.4** SIPOC for undesirable outcomes.

The SIPOC will be further reviewed in the Analyze phase to remove non-value-added outputs, inputs, or tasks.

## Metric Definition

A key objective of the Measure stage is to establish a process baseline, as discussed later in this chapter. The process baseline provides a quantifiable measure of the process performance before any improvement efforts have been initiated. In establishing a process baseline, one or more suitable process *metrics*, related to the previously defined project deliverable, is defined and measured. These metrics are sometimes easily defined with existing process expertise. For example, if key customers are concerned with the *time to delivery* for an order, then that may be a suitable metric, assuming it meets the conditions discussed below. In some case, however, the customers' key metrics do not provide a meaningful measure of process performance, as related to the deliverables previously defined. For example, in chemical and some manufacturing processes, it's not uncommon for the customers' key characteristics to be process endpoint measurements, whose properties are defined much earlier by internal process settings unknown to the end users. In those cases, the internal metrics, or the project deliverables themselves, may provide suitable process performance estimates. In all cases, the metric must be evaluated for its validity and accuracy, as discussed later in this chapter.

A *measurement* is simply a numerical assignment to an observation. Measurements convey certain information about the relationship between the element observed and other elements. Measurement involves a theoretical domain, an area of substantive concern represented as an empirical relational system, and a domain represented by a particular selected numerical relational system. There is a mapping function that carries us from the empirical system into the numerical system. The numerical system is manipulated and the results of the manipulation are studied to help the analyst better understand the empirical system.

In reality, measurement is problematic: the analyst can never know the "true" value of the element being measured. The numbers provide information on a certain scale and they represent measurements of some unobservable variable of interest. Some measurements are richer than others, that is, some measurements provide more information than other measurements. The information content of a number is dependent on the scale of measurement used. This scale determines the types of statistical analyses that can be properly employed in studying the numbers. Until one has determined the scale of measurement, one cannot know if a given method of analysis is valid.

## *Measurement Scales*

The four measurement scales are: nominal, ordinal, interval, and ratio. Harrington (1992) summarizes the properties of each scale in Table 7.1.

Numbers on a *nominal scale* aren't measurements at all; they are merely *category labels* in numerical form. Nominal measurements might indicate membership in a group (1 = male, 2 = female) or simply represent a designation (John Doe is 43 on the team). Nominal scales represent the simplest and weakest form of measurement. Nominal variables are perhaps best viewed as a form of classification rather than as a measurement scale. Ideally, categories on the nominal scale are constructed in such a way that all objects in the universe are members of one and only one class. Data collected on a nominal scale are called *attribute data*. The only mathematical operations permitted on nominal scales are = (equality, which shows that an object possesses the attribute of concern) or ≠ (inequality).

An ordinal variable is one that has a natural ordering of its possible values, but for which the distances between the values are undefined. An example is product preference rankings such as good, better, or best. Ordinal data can be analyzed with the mathematical operators, = (equality), ≠ (inequality), > (greater than), and < (less than). There are a wide

**Table 7.1** Types of Measurement Scales and Permissible Statistics

| Scale | Definition | Example | Statistics |
|---|---|---|---|
| Nominal | Only the presence/absence of an attribute; can only count items | Go/no go; success/fail; accept/reject | Percent; proportion; chi-square tests |
| Ordinal | Can say that one item has more or less of an attribute than another item; can order a set of items | Taste; attractiveness | Rank-order correlation |
| Interval | Difference between any two successive points is equal; often treated as a ratio scale even if assumption of equal intervals is incorrect; can add, subtract, order objects | Calendar time; temperature | Correlations; *t*-tests; *F*-tests; multiple regression |
| Ratio | True zero point indicates absence of an attribute; can add, subtract, multiply and divide | Elapsed time; distance; weight | *t*-test; *F*-test; correlations; multiple regression |

*From Harrington (1992), p. 516. Copyright © 1992. Used by permission of the publisher, ASQ Quality Press, Milwaukee, Wisconsin.*

variety of statistical techniques that can be applied to ordinal data including the Pearson correlation, discussed in Chap. 10. Other ordinal models include odds-ratio measures, log-linear models and logit models, both of which are used to analyze cross-classifications of ordinal data presented in contingency tables. In quality management, ordinal data are commonly converted into nominal data and analyzed using binomial or Poisson models. For example, if parts were classified using a poor-good-excellent ordering, the quality analyst might plot a *p chart* of the proportion of items in the poor category.

Interval scales consist of measurements where the ratios of differences are invariant. For example, 90°C = 194°F, 180°C = 356°F, 270°C = 518°F, 360°C = 680°F. Now 194°F/90°C ≠ 356°F/180°C, but:

$$\frac{356°F - 194°F}{680°F - 518°F} = \frac{180°C - 90°C}{360°C - 270°C}$$

Conversion between two interval scales is accomplished by the transformation:

$$y = ax + b \qquad a > 0$$

For example,

$$°F = 32 + \left(\frac{9}{5} \times °C\right)$$

where $a = 9/5$, $b = 32$. As with ratio scales, when permissible transformations are made statistical, results are unaffected by the interval scale used. Also, 0° (on either scale) is arbitrary. In this example, zero does not indicate an absence of heat.

*Ratio scale* measurements are so called because measurements of an object in two different metrics are related to one another by an invariant ratio. For example, if an object's mass was measured in pounds ($x$) and kilograms ($y$), then $x/y = 2.2$ for all values of $x$ and $y$. This implies that a change from one ratio measurement scale to another is performed by a transformation of the form $y = ax, a > 0$; for example, pounds = 2.2 × kilograms. When permissible transformations are used, statistical results based on the data are identical regardless of the ratio scale used. Zero has an inherent meaning: in this example it signifies an absence of mass.

### Discrete and Continuous Data

A more general classification of measurements may also be made, which is also useful in defining suitable probability distributions and analysis tools discussed later in this chapter. Data are said to be *discrete* when they take on only a finite number of points that can be represented by the nonnegative integers. An example of discrete data is the number of defects in a sample. Data are said to be *continuous* when they exist on an interval, or on several intervals. An example of continuous data is the measurement of pH.

For most purposes, nominal and ordinal data are considered discrete, while intervals and ratios possess qualities of continuous data. While discrete data may take on integer form only, a continuous data value may be defined theoretically to an infinite number of decimal places, assuming it could be measured as such. In the real world, even though length or time, for example, could theoretically be measured to an infinite number of decimal places, we are be limited by our measurement system. If the variation in length or time is smaller than the smallest unit of measure, the resulting data is essentially discrete, since the same data value is recorded for the majority of the data. Similar information content could be obtained by counting the number of items with that dimension.

Fundamentally, any item measure should meet two tests:

▲ The item measures what it is intended to measure (i.e., it is *valid*).
▲ A remeasurement would order individual responses in the same way (i.e., it is *reliable*).

The remainder of this chapter describes techniques and procedures designed to ensure that measurement systems produce numbers with these properties.

## Process Baseline Estimates

The process baseline is best described as "what were things like before the project?" There are several reasons for obtaining this information:

▲ **To determine if the project should be pursued**—Although the project charter provides a business case for the project, it sometimes happens that additional, detailed information fails to support it. It may be that the situation isn't as bad as people think, or the project may be addressing an unimportant aspect of the problem.

- **To orient the project team**—The process baseline helps the team identify CTQs and other hard metrics. The information on the historic performance of these metrics may point the team to strategies. For example, if the process is erratic and unstable the team would pursue a different strategy than if it was operating at a consistently poor level.

- **To provide data that will be used to estimate savings**—Baseline information will be invaluable when the project is over and the team is trying to determine the magnitude of the savings or improvement. Many a Black Belt has discovered after the fact that the information they need is no longer available after the completion of the project, making it impossible to determine what benefit was obtained. For example, a project that streamlined a production control system was aimed at improving morale by reducing unpaid overtime worked by exempt employees. However, no measure of employee morale was obtained ahead of time; nor was the unpaid overtime documented anywhere. Consequently, the Black Belt wasn't able to substantiate his claims of improvement and his certification (and pay increase) was postponed.

Process baselines are a critical part of any process improvement effort, as they provide the reference point for assertions of benefits attained. In the absence of a proper baseline estimate, there can be no credible evidence of sustainable improvements, except perhaps in the most obvious of cases. In the vast majority of cases, when the proper measurement system is applied, process observations will vary to the extent that real changes in the process cannot be evidenced without the use of meaningful statistics.

Consider the following scenario: An improvement team uses Lean techniques to reduce the time to process an order. A random sample of 25 orders before the change had an average time to process of 3.5 hours. A random sample of 25 orders after the change has an average time to process of 2 hours. The team asserts they have decreased the order processing time by more than 40%. Is that credible? Would it be more credible if a confidence interval for the order processing time was calculated for the original 25 orders, and the improvement asserted only if the new average of 2 hours fell outside of that original interval?

The correct answer is *NO*, in both cases. An improvement cannot be asserted without showing that the new process is significantly different, from a statistical point of view, and the use of confidence intervals is the wrong statistical tool for analyzing the process. It makes the mistake of applying enumerative statistical concepts to an analytic statistical situation. There is no evidence that the estimate of 3.5 hours for the first 25 samples is a valid estimate of the process, since we have not shown that the process is stable (consistent over time). It could be that a portion of the first 25 orders had processing times quite a bit larger than 3.5 hours, and another portion quite a bit less that 3.5 hours, perhaps because of the difference between clerical staff or the type or number of line items included in the order. The confidence interval calculation assumes the data is from a single population, yet fails to prove the process is in fact stable, suitable of being represented by a single distribution.

Some appropriate analytic statistics questions might be:

- Is the process central tendency stable over time?
- Is the process dispersion stable over time?
- Is the process distribution consistent over time?

If any of the above are answered "no," then what is the cause of the instability? To help answer this question, ask "what is the nature of the variation as revealed by the patterns?" when plotted in time-sequence and stratified in various ways.

If none of the above are answered "no," then, and only then, we can ask such questions as:

- Is the process meeting the requirements?
- Can the process meet the requirements?
- Can the process be improved by recentering it?
- How can we reduce variation in the process?

## *Enumerative and Analytic Studies*

Deming (1975) defines enumerative and analytic studies as follows:

- **Enumerative study**—A study in which action will be taken on the universe.
- **Analytic study**—A study in which action will be taken on a process to improve performance in the future.

The term "universe" is defined in the usual way: the entire group of interest, for example, people, material, units of product, which possess certain properties of interest. An example of an enumerative study would be sampling an isolated lot to determine the quality of the lot.

In an analytic study the focus is on a *process* and how to improve it. The focus is the *future*. Thus, unlike enumerative studies that make inferences about the universe actually studied, analytic studies are interested in a universe that has yet to be produced. Table 7.2 compares analytic studies with enumerative studies (Provost, 1988).

Deming (1986) points out that "Analysis of variance, *t*-tests, confidence intervals, and other statistical techniques taught in the books, however interesting, are inappropriate because they provide no basis for prediction and because they bury the information con-

**Table 7.2** Important Aspects of Analytic Studies

| Item | Enumerative Study | Analytic Study |
| --- | --- | --- |
| Aim | Parameter estimation | Prediction |
| Focus | Universe | Process |
| Method of access | Counts, statistics | Models of the process (e.g., flowcharts, cause and effects, mathematical models) |
| Major source of uncertainty | Sampling variation | Extrapolation into the future |
| Uncertainty quantifiable | Yes | No |
| Environment for the study | Static | Dynamic |

tained in the order of production." These traditional statistical methods have their place, but they are widely abused in the real world. When this is the case, statistics do more to cloud the issue than to enlighten.

Analytic study methods provide information for *inductive thinking*, rather than the largely *deductive* approach of enumerative statistics. Analytic methods are primarily graphical devices such as run charts in the simplest case or statistical control charts in the more general case. Analytic statistics provide operational guidelines, rather than precise calculations of probability. Thus, such statements as "There is a 0.13% probability of a Type I error when acting on a point outside a three-sigma control limit" are false (the author admits to having made this error in the past). The future cannot be predicted with a known level of confidence. Instead, based on knowledge obtained from every source, including analytic studies, one can state that one has a certain degree of belief (e.g., high, low) that such and such will result from such and such action on a process.

Another difference between the two types of studies is that enumerative statistics proceed from predetermined hypotheses while analytic studies try to help the analyst generate new hypotheses. In the past, this extremely worthwhile approach has been criticized by some statisticians as "fishing" or "rationalizing." However, this author believes that using data to develop plausible explanations retrospectively is a perfectly legitimate way of creating new theories to be tested. To refuse to explore possibilities suggested by data is to take a very limited view of the scope of statistics in quality improvement and control.

Although most applications of Six Sigma are analytic, there are times when enumerative statistics prove useful. These enumerative methods will be discussed in more detail in the Analyze phase, where they are useful in quantifying sources of variation. The analyst should keep in mind that analytic methods will be needed to validate the conclusions developed with the use of the enumerative methods, to ensure their relevance to the process under study.

In the Measure stage of DMAIC, statistical process control (SPC) charts (i.e., analytical statistics) are used to define the process baseline. If the process is statistically stable, as evidenced by the SPC charts, then process capability and sigma level estimates can be used to quantify the performance of the process relative to requirements.

If the process is not statistically stable, as evidenced by the SPC analysis, then clarity is sought regarding the causes of variation, as discussed in the following sections. If the special causes of variation can be easily identified and removed, and a stable baseline process established, then the requirements for the baseline objective have been met. If not, as is often the case, then our baseline estimate has provided critical information useful in our Analysis phase, where the causes of the process instability can be investigated under the controlled conditions of a designed experiment.

Of course, the baseline estimate will rely upon the validity of the measurement system, which is discussed in Chap. 9. While it may seem obvious that the measurement system should be evaluated before effort is spent in taking baseline measurements, the measurement system verification is presented after the baseline section for two reasons:

- A key analytical technique used for validating the measurement system is the control chart. Readers should become familiar with their use and application in process baselines before their special application toward measurement system validation is discussed.
- When a Six Sigma project's objectives include improvement to CTQ or CTS metrics, the Measure stage baseline provides validation of the initial conditions estimated in the Define stage. Significant improvements or changes to the measurement system before the baseline will bias the baseline relative to the initial conditions. Instead, the baseline should be estimated first using the existing methods, then repeated (after improvements to the measurement system) if the measurement system error is significant. Adherence to this recommendation will lend further credibility to the improvements, since the project team can demonstrate replication of the error as well as the resulting improvement. In some cases, the measurement error is found to be the most significant cause of process variation. In those cases, the focus of the Six Sigma project changes at this stage to concentrate on improvement to the measurement system, rather than the process itself.

The statistical control chart will provide estimates of the process location (i.e., its mean or median) and its variation (i.e., process standard deviation). Of key importance, however, is how these current conditions compare with existing requirements or project objectives. These considerations are answered with a proper understanding of the statistical distribution, as will be discussed.

## Principles of Statistical Process Control

A central concept in statistical process control (SPC) is that every measurable phenomenon is a statistical distribution. In other words, an observed set of data constitutes a sample of the effects of unknown common causes. It follows that, after we have done everything to eliminate special causes of variations, there will still remain a certain amount of variability exhibiting the state of control. Figure 7.5 illustrates the relationships between common causes, special causes, and distributions.

There are three basic properties of a distribution: location, spread, and shape. The *location* refers to the typical value of the distribution, such as the mean. The *spread* of the distribution is the amount by which smaller values differ from larger ones. The standard deviation and variance are measures of distribution spread. The *shape* of a distribution is its pattern—peakedness, symmetry, etc. A given phenomenon may have any one of a number of distribution shapes, for example, the distribution may be bell-shaped, rectangular-shaped, etc.

**Figure 7.5** Process control concepts illustrated. (From *Continuing Process Control and Process Capability Improvement*, p. 4a. Copyright 1983. Used by permission of the publisher, Ford Motor Company, Dearborn, Michigan.)

## Central Limit Theorem

The central limit theorem can be stated as follows:

Irrespective of the shape of the distribution of the population or universe, the distribution of average values of samples drawn from that universe will tend toward a normal distribution as the sample size grows without bound.

It can also be shown that the average of sample averages will equal the average of the universe and that the standard deviation of the averages equals the standard deviation of the universe divided by the square root of the sample size. Shewhart performed experiments that showed that small sample sizes were needed to get approximately normal distributions from even wildly non-normal universes. Figure 7.6 was created by Shewhart using samples of four measurements.

The practical implications of the central limit theorem are immense. Consider that without the central limit theorem effects, we would have to develop a separate statistical model for every non-normal distribution encountered in practice. This would be the only way to determine if the system were exhibiting chance variation. Because of the central limit theorem we can use *averages* of small samples to evaluate *any* process using the normal distribution. The central limit theorem is the basis for the most powerful of statistical process control tools, Shewhart control charts.

**Figure 7.6** Illustration of the central limit theorem. (From Shewhart, 1931, 1980. Figure 59. Copyright © 1931, 1980 by ASQC Quality Press. Used by permission of the publisher.)

## Common and Special Causes of Variation

Shewhart (1931, 1980) defined *control* as follows:

A phenomenon will be said to be controlled when, through the use of past experience, we can predict, at least within limits, how the phenomenon may be expected to vary in the future. Here it is understood that prediction within limits means that we can state, at least approximately, the probability that the observed phenomenon will fall within the given limits.

The critical point in this definition is that control is not defined as the complete absence of variation. Control is simply a state where all variation is *predictable*. A controlled process isn't necessarily a sign of good management, nor is an out-of-control process necessarily producing nonconforming product.

In all forms of prediction there is an element of risk. For our purposes, we will call any unknown random cause of variation a *chance cause* or a *common cause*; the terms are synonymous and will be used as such. If the influence of any particular chance cause is very small, and if the number of chance causes of variation is very large and relatively constant, we have a situation where the variation is predictable within limits. You can see from the definition above, that a system such as this qualifies as a controlled system. Where Dr. Shewhart used the term *chance cause*, Dr. W. Edwards Deming coined the term *common cause* to describe the same phenomenon. Both terms are encountered in practice.

Needless to say, not all phenomena arise from constant systems of common causes. At times, the variation is caused by a source of variation that is not part of the constant system. These sources of variation were called *assignable causes* by Shewhart, *special causes* of variation by Deming. Experience indicates that special causes of variation can usually be found without undue difficulty, leading to a process that is less variable.

Statistical tools are needed to help us effectively separate the effects of special causes of variation from chance cause variation. This leads us to another definition: statistical process control—the use of valid analytical statistical methods to identify the existence of special causes of variation in a process.

The basic rule of statistical process control is variation from common-cause systems should be left to chance, but special causes of variation should be identified and eliminated.

This is Shewhart's original rule. However, the rule should not be misinterpreted as meaning that variation from common causes should be ignored. Rather, common-cause variation is explored "off-line." That is, we look for long-term process improvements to address common-cause variation.

Figure 7.7 illustrates the need for statistical methods to determine the category of variation.

The answer to the question "should these variations be left to chance?" can only be obtained through the use of statistical methods. Figure 7.8 illustrates the basic concept.

**Figure 7.7** Types of variation.

**Figure 7.8** Should these variations be left to chance? (From Shewhart, 1931, 1980, p. 13. Copyright © 1931, 1980 by ASQC Quality Press. Used by permission of the publisher.)

In short, variation between the two "control limits" designated by the dashed lines will be deemed as variation from the common-cause system. Any variability beyond these fixed limits will be assumed to have come from special causes of variation. We will call any system exhibiting only common-cause variation, "statistically controlled." It must be noted that the control limits are not simply pulled out of the air, they are calculated from actual process data using valid statistical methods. Figure 7.7 is shown as Fig. 7.9, only with the control limits drawn on it; notice that process (a) is exhibiting variations from special causes, while process (b) is not. This implies that the type of action needed to reduce the variability in each case is of a different nature. Without statistical guidance there could be endless debate over whether special or common causes were to blame for variability.

**Figure 7.9** Charts from Fig. 7.7 with control limits shown. (From Shewhart, 1931, 1980, p. 13. Copyright © 1931, 1980 by ASQC Quality Press. Used by permission of the publisher.)

## Tampering Effects and Diagnosis

Tampering occurs when adjustments are made to a process that is in statistical control. Adjusting a controlled process will always increase process variability, an obviously undesirable result. The best means of diagnosing tampering is to conduct a process capability study and to use a control chart to provide guidelines for adjusting the process.

Perhaps the best analysis of the effects of tampering is from Deming (1986). Deming describes four common types of tampering by drawing the analogy of aiming a funnel to hit a desired target. These "funnel rules" are described by Deming (1986, p. 328):

Leave the funnel fixed, aimed at the target, no adjustment.

At drop $k$ ($k$ = 1, 2, 3, ...) the marble will come to rest at point $z_k$, measured from the target. (In other words, $z_k$ is the error at drop $k$.) Move the funnel the distance $-z_k$ from the last position. Memory 1.

Set the funnel at each drop right over the spot $z_k$, measured from the target. No memory.

Set the funnel at each drop right over the spot ($z_k$) where it last came to rest. No memory.

Rule 1 is the best rule for stable processes. By following this rule, the process average will remain stable and the variance will be minimized. Rule 2 produces a stable output but one with twice the variance of rule 1. Rule 3 results in a system that "explodes," that is, a symmetrical pattern will appear with a variance that increases without bound. Rule 4 creates a pattern that steadily moves away from the target, without limit (see Fig. 7.10).

At first glance, one might wonder about the relevance of such apparently abstract rules. However, upon more careful consideration, one finds many practical situations where these rules apply.

**Figure 7.10** Funnel rule simulation results.

Rule 1 is the ideal situation, and it can be approximated by using control charts to guide decision-making. If process adjustments are made only when special causes are indicated and identified, a pattern similar to that produced by rule 1 will result.

Rule 2 has intuitive appeal for many people. It is commonly encountered in such activities as gage calibration (check the standard once and adjust the gage accordingly) or in some automated equipment (using an automatic gage, check the size of the last feature produced and make a compensating adjustment). Since the system produces a stable result, this situation can go unnoticed indefinitely. However, as shown by Taguchi (1986), increased variance translates to poorer quality and higher cost.

The rationale that leads to rule 3 goes something like this: "A measurement was taken and it was found to be 10 units above the desired target. This happened because the process was set 10 units too high. I want the average to equal the target. To accomplish this I must try to get the next unit to be 10 units too low." This might be used, for example, in preparing a chemical solution. While reasonable on its face, the result of this approach is a wildly oscillating system.

A common example of rule 4 is the "train-the-trainer" method. A master spends a short time training a group of "experts," who then train others, who train others, etc. An example is on-the-job training. Another is creating a setup by using a piece from the last job. Yet another is a gage calibration system where standards are used to create other standards, which are used to create still others, and so on. Just how far the final result will be from the ideal depends on how many levels deep the scheme has progressed.

## *Estimating Process Baselines Using Process Capability Analysis*

This section presents the general method of analyzing the data using a statistical control chart and determining the extent to which the process meets requirements.

1. Collect samples from 25 or more subgroups of consecutively produced units.
2. Plot the results on the appropriate control chart. If all groups are in statistical control, go to the step 3. Otherwise, attempt to identify the special cause of variation by observing the conditions or time periods under which they occur. If possible, take action to eliminate the special cause and document it as part of the process improvement effort. Note that a special cause might be beneficial. Beneficial activities can be "eliminated" as special causes by doing them all of the time. A special cause is "special" only because it comes and goes, not because its impact is either good or bad.
3. Using the control limits from the previous step (called *operational control limits*), put the control chart to use for a period of time. Once you are satisfied that sufficient time has passed for most special causes to have been identified and eliminated, as verified by the control charts, go to step 4.

4. The process capability is estimated using the calculations provided in Chap. 6. This estimate should validate the error rates originally estimated in the Define stage that justified the project deployment. Note that when process variation is limited to common causes (i.e., the process is in statistical control), "problem solving" (e.g., studying each defective) won't help, and it may result in tampering. Tampering with a stable process is proven to increase process variation, which is of course the exact opposite effect it attempts to correct.

The application and selection of the control chart used for baseline analysis is covered in detail in Chap. 8.

# CHAPTER 8

# Process Behavior Charts

## Distributions

Distributions are a set of numbers (or data) collected from a well-defined universe of possible measurements arising from a property or relationship under study. Defining the distribution of the data allows us to associate probabilities with the numbers being studied. Assuming an appropriate distribution for a stable process or population allows one to answer such questions as:

- What is the probability that $x$ will occur?
- What is the probability that a value less than $x$ will occur?
- What is the probability that a value greater than $x$ will occur?
- What is the probability that a value will occur that is between $x$ and $y$?
- How rapidly or slowly do probabilities change over a given range of values?

In short, distributions provide a great deal of information. We will discuss the application and interpretation of some common distributions after a brief discussion of sampling.

### Methods of Enumeration

Enumeration involves counting techniques for very large numbers of possible outcomes. This occurs for even surprisingly small sample sizes. In Six Sigma, these methods are commonly used in a wide variety of statistical procedures.

The basis for all of the enumerative methods described here is the multiplication principle. The multiplication principle states that the number of possible outcomes of a series of experiments is equal to the product of the number of outcomes of each experiment. For example, consider flipping a coin twice. On the first flip there are two possible outcomes (heads/tails), and on the second flip there are also two possible outcomes. Thus, the series of two flips can result in $2 \times 2 = 4$ outcomes. Figure 8.1 illustrates this example.

**Figure 8.1** Multiplication principle applied to coin flips.

An ordered arrangement of elements is called a *permutation*. Suppose that you have four objects and four empty boxes, one for each object. Consider how many different ways the objects can be placed into the boxes. The first object can be placed in any of the four boxes. Once this is done there are three boxes to choose from for the second object, then two boxes for the third object, and finally one box left for the last object. Using the multiplication principle, you find that the total number of arrangements of the four objects into the four boxes is 4 × 3 × 2 × 1 = 24. In general, if there are $n$ positions to be filled with $n$ objects there are

$$n(n-1)\ldots(2)(1) = n! \tag{8.1}$$

possible arrangements. The symbol $n!$ is read $n$ factorial. By definition, $0! = 1$.

In applying probability theory to discrete variables in quality control we frequently encounter the need for efficient methods of counting. One counting technique that is especially useful is combinations. The combination formula is shown in Eq. (8.2).

$$C_r^n = \frac{n!}{r!(n-r)!} \tag{8.2}$$

Combinations tell how many unique ways you can arrange $n$ objects taking them in groups of $r$ objects at a time, where $r$ is a positive integer less than or equal to $n$. For example, to determine the number of combinations we can make with the letters X, Y, and Z in groups of 2 letters at a time, we note that $n = 3$ letters, $r = 2$ letters at a time and use the above formula to find

$$C_2^3 = \frac{3!}{2!(3-2)!} = \frac{3 \times 2 \times 1}{(2 \times 1)(1)} = \frac{6}{2} = 3$$

The three combinations are XY, XZ, and YZ. Notice that this method does not count reversing the letters as separate combinations, that is, XY and YX are considered to be the same.

## *Frequency and Cumulative Distributions*

A frequency distribution is an empirical presentation of a set of observations. If the frequency distribution is *ungrouped*, it simply shows the observations and the frequency of each number. If the frequency distribution is *grouped*, then the data are assembled into cells, each cell representing a subset of the total range of the data. The frequency in each cell completes the grouped frequency distribution. Frequency distributions are often graphically displayed in histograms or stem-and-leaf plots.

While histograms and stem-and-leaf plots show the frequency of specific values or groups of values, analysts often wish to examine the *cumulative frequency* of the data. The cumulative frequency refers to the total up to and including a particular value. In the case of grouped data, the cumulative frequency is computed as the total number of observations up to and including a cell boundary. Cumulative frequency distributions are often displayed on an *ogive*, as depicted in Fig. 8.2.

**Figure 8.2** Ogive of rod diameter data.

## Sampling Distributions

In most Six Sigma projects involving enumerative statistics, we deal with samples, not populations. We now consider the estimation of certain characteristics or parameters of the distribution from the data.

The empirical distribution assigns the probability $1/n$ to each $X_i$ in the sample, thus the mean of this distribution is:

$$\bar{X} = \frac{1}{n}\sum_{i=1}^{n} X_i \tag{8.3}$$

The symbol $\bar{X}$ is read as "X bar." Since the empirical distribution is determined by a sample, $\bar{X}$ is simply called the *sample mean*.

The sample variance is given by:

$$S^2 = \frac{1}{n-1}\sum_{i=1}^{n}(X_i - \bar{X})^2 \tag{8.4}$$

This equation for $S^2$ is commonly referred to as the *unbiased sample variance*. The *sample standard deviation* is given by:

$$S = \sqrt{S^2} = \sqrt{\frac{\sum_{i=1}^{n}(X_i - \bar{X})^2}{n-1}} \tag{8.5}$$

Another sampling statistic of special interest in Six Sigma is the standard deviation of the sample average, also referred to as the *standard error of the mean* or simply the standard error. This statistic is given by:

$$S_{\bar{X}} = \frac{S}{\sqrt{n}} \tag{8.6}$$

As can be seen, the standard error of the mean is inversely proportional to the square root of the sample size. That is, the larger the sample size, the smaller the standard deviation of the sample average. This relationship is shown in Fig. 8.3. It can be seen that averages of $n = 4$ have a distribution half as variable as the population from which the samples are drawn. This simple fact will be used later in this chapter to show how the performance of the X-bar control charts is improved relative to the plotting of individual values in an Individual-X control chart: As the plotted subgroups move from the centerline due to a process shift, the shift is more readily detected by the X-bar chart's narrower control limits.

**Figure 8.3** Effect of sample size on the standard error.

## *Binomial Distribution*

Assume that a process is producing some proportion of nonconforming units, which we will call $p$. If we are basing $p$ on a sample we find $p$ by dividing the number of nonconforming units in the sample by the number of items sampled. The probability of getting exactly $x$ defectives in a sample of $n$ units is shown by Eq. (8.7).

$$P(x) = C_x^n p^x (1-p)^{n-x} \tag{8.7}$$

This equation provides the probability mass function for the *binomial probability distribution*. In addition to being useful as the exact distribution of nonconforming units for processes in continuous production, it is also an excellent approximation to the cumbersome hypergeometric probability distribution when the sample size is less than 10% of the lot size.

If you were interested in knowing the probability of three or less defects, you could add the probability of zero defects, one defect, two defects, and three defects. This provides the cumulative distribution function.

### Example of Applying the Binomial Probability Distribution

A process is producing glass bottles on a continuous basis. Past history shows that 1% of the bottles has one or more flaws. If we draw a sample of 10 units from the process, what is the probability that there will be 0 nonconforming bottles?

Using the above information, $n = 10$, $p = .01$, and $x = 0$. Substituting these values into Eq. (8.7) gives us:

$$p(0) = C_0^{10} 0.01^0 (1 - 0.01)^{10-0} = 1 \times 1 \times 0.99^{10} = 0.904 = 90.4\%$$

Another way of interpreting the above example is that a sampling plan to "inspect 10 units, accept the process if no nonconformances are found" has a 90.4% probability of accepting a process that is averaging 1% nonconforming units.

### Example of Binomial Probability Calculations Using Microsoft Excel

Microsoft Excel has a built-in capability to analyze binomial probabilities. To solve the above problem using Excel, enter the number of successes in trials (0), the number of trials (10), the probability of success in each trial (0.01), and the Cumulative parameter (False) as shown in Fig. 8.4, where the formula result is shown near the bottom of the figure. The Cumulative property is used to indicate whether the result reflects the probability mass function (the probability of exactly $x$ successes) or the cumulative distribution function (the probability of less than $x$ successes). Figure 8.4 shows the cumulative property set to false to obtain the probability mass function, so the result provides the probability of exactly zero successes. The statistical term *success* here refers to a sample that meets the condition of interest, in this case a nonconformity. (A statistical success does not imply good or bad.) If the question asked the probability of less than two nonconformances in the sample, we enter 1 as the number of successes in trials, and set the Cumulative parameter to True, resulting in value 0.996 (or 99.6%). (Prior to Excel 2013, the Binomdist function was used, with identical parameters.)

**Figure 8.4** Example of finding binomial probability using Microsoft Excel.

## Poisson Distribution

Another situation encountered frequently concerns the number of occurrences of a condition of interest. For example, the number of accidents per month. The correct probability distribution for evaluating occurrences is the *Poisson distribution*. The probability mass function is given in Eq. (8.8).

$$p(x) = \frac{\mu^x e^{-\mu}}{x!} \tag{8.8}$$

In Eq. (8.8), $\mu$ is the average number of occurrences per unit, $x$ is the number of occurrences in the sample, and $e$ is the constant approximately equal to 2.7182818. The function $P(x)$ gives the probability of exactly $x$ occurrences in the sample.

### Example of Applying the Poisson Distribution

Past history shows that on the average a facility has three safety incidents a month. What is the probability that there will be zero safety incidents next month?

We have $\mu = 3$, $x = 0$. Substituting these values into Eq. (8.8) gives us:

$$P(0) = \frac{3^0 e^{-3}}{0!} = \frac{1 \times 0.05}{1} = 0.05 = 5\%$$

Thus, there is only a 5% chance of zero safety incidents, or a 100% − 5% = 95% probability there will be at least one safety incident in any month.

The Poisson distribution, in addition to being the exact distribution for the number of occurrences, is also a good approximation to the binomial distribution in certain cases. To use the Poisson approximation, you simply let $\mu = np$ in Eq. (8.8). Juran (1988) recommends considering the Poisson approximation if the sample size is at least 16, the population size is at least 10 times the sample size, and the probability of occurrence $p$ on each trial is less than 0.1. The major advantage of this approach is that it allows you to use the tables of the Poisson distribution, such as in Appendix 7. The approach is also useful for designing sampling plans.

### Example of Poisson Probability Calculations Using Microsoft Excel

Microsoft Excel has a built-in capability to analyze Poisson probabilities. To solve the above problem using Excel, enter the $x$ and average rate of occurrence values as shown in Fig. 8.5. Note the formula result near the bottom of the figure. (Prior to Excel 2013, the Excel Poisson function was used with identical parameters.)

**Figure 8.5** Example of finding Poisson probability using Microsoft Excel.

## *Hypergeometric Distribution*

Assume we have received a lot of 12 parts from a distributor. We need the parts badly and are willing to accept the lot if it has fewer than 3 nonconforming parts. We decide to inspect only 4 parts since we can't spare the time to check every part. Checking the sample, we find 1 part that doesn't conform to the requirements. Should we reject the remainder of the lot?

This situation involves sampling without replacement. We draw a unit from the lot, inspect it, and draw another unit from the lot. Furthermore, the lot is quite small, the sample of 4 units is 33% of the entire lot of 12 units. The probability mass function used to compute exact probabilities for the hypergeometric probability distribution is shown in Eq. (8.9).

$$p(x) = \frac{C_{n-x}^{N-m} C_x^m}{C_n^N} \quad (8.9)$$

In Eq. (8.9), $N$ is the size of the population, $m$ is the number of items in the population meeting the condition of interest, $n$ is the sample size, $x$ is the number of items in the sample meeting the condition of interest, and $P(x)$ is the probability of getting exactly $x$ items in the sample. Note that the numerator term $C_{n-x}^{N-m}$ gives the number of combinations of items not meeting the condition while $C_x^m$ is the number of items meeting the condition. Thus, the numerator gives the total number of arrangements of samples from populations of size $N$ with $m$ items meeting a condition where the sample of size $n$ contains exactly $x$

items meeting the condition. The term $C_n^N$ in the denominator is the total number of combinations of samples of size $n$ from populations of size $N$. Thus, the probability is a ratio of the likelihood of getting the result under the assumed conditions. (The interpretation of the combination function was described in Eq. (8.2).)

For our example, we must solve the above equation for $x = 0$ as well as $x = 1$, since we would also accept the lot if we had no defectives. The solution is shown as follows.

$$P(0) = \frac{C_{4-0}^{12-3} C_0^3}{C_4^{12}} = \frac{126 \times 1}{495} = 0.255$$

$$P(1) = \frac{C_{4-1}^{12-3} C_1^3}{C_4^{12}} = \frac{84 \times 3}{495} = \frac{252}{495} = 0.509$$

$$P(1 \text{ or less}) = P(0) + P(1)$$

Adding the two probabilities tells us the probability that our sampling plan will accept lots of 12 with 3 nonconforming units. The plan of inspecting 4 parts and accepting the lot if we have 0 or 1 nonconforming has a probability of 0.255 + 0.509 = 0.764, or 76.4%, of accepting this "bad" quality lot. This is the "consumer's risk" for this sampling plan. Such a high sampling risk would be unacceptable to most organizations.

### Example of Hypergeometric Probability Calculations Using Microsoft Excel

Microsoft Excel has a built-in capability to analyze hypergeometric probabilities. To solve the above problem, enter the number of successes in the sample (1), the size of the sample (4), the number of success in the population (3), the size of the population (12) and the cumulative property (True) as shown in Fig. 8.6. Note the formula result near the bottom of the figure (0.7636) provides the cumulative probability for $x = 1$ or less successes.

Prior to Excel 2013, the Excel Hypgeomdist function was used, but it lacked the cumulative parameter: To find the cumulative probability in the above example, sum the probability calculated for $x = 0$ with the probability calculated for $x = 1$.

## *Normal Distribution*

The most common continuous distribution encountered in Six Sigma work is the normal distribution. Sometimes the process itself produces an approximately normal distribution, other times a normal distribution can be obtained by performing a mathematical transformation on the data, sometimes as simple as using averages. The probability density function for the normal distribution is given by Eq. (8.10).

**Figure 8.6** Example of finding hypergeometric probability using Microsoft Excel.

$$f(x) = \frac{1}{\sigma\sqrt{2\pi}} e^{-(x-\mu)^2/2\sigma^2} \qquad (8.10)$$

If $f(x)$ is plotted versus $x$, the well-known "bell-shaped curve" of the cumulative normal distribution, also known as the Gaussian distribution, results as shown in Fig. 8.7.

In Eq. (8.10), $\mu$ is the population average or mean and $\sigma$ is the population standard deviation. These parameters have been discussed earlier in this chapter. In typical situations, sample data is used, so the population parameters $\mu$ and $\sigma$ in Eq. (8.10) are replaced with their sample counterparts, $\bar{x}$ and $s$, respectively.

**Figure 8.7** The normal/Gaussian curve.

## Example of Calculating Sample Mean and Standard Deviation

Find the sample mean and standard deviation for the sample data provided in Table 8.1.

**Table 8.1**  Example Sample Data

| |
|---|
| 110 |
| 103 |
| 97 |
| 96 |
| 105 |
| 110 |
| 100 |
| 93 |
| 90 |
| 103 |
| 97 |
| 103 |
| 90 |
| 97 |
| 99 |
| 106 |
| 90 |
| 96 |
| 109 |
| 90 |
| 91 |
| 108 |
| 96 |
| 101 |
| 96 |
| SUM: 2476 |

The equation for the sample mean is

$$\bar{x} = \frac{1}{N} \sum_{i=1}^{N} x_i \qquad (8.11)$$

To find the mean for our data, we compute

$$\bar{x} = \frac{1}{25}(2476) = 99.0$$

The variance and standard deviation are both measures of dispersion or spread. The equations for the sample variance $s^2$ and standard deviation $s$ are as follows:

$$s^2 = \sum_{i=i}^{N} \frac{(x_i - \bar{x})^2}{N}$$

$$s = \sqrt{s^2}$$

(8.12)

Referring to the data above with a mean $\bar{x}$ of 99, we compute the sample variance $s^2$ and sample standard deviation s as given in Table 8.2.

**Table 8.2** Example Sample Sigma Calculation

| i | $x_i$ | $x_i - \bar{X}$ | $(x_i - \bar{X})^2$ |
|---|---|---|---|
| 1 | 110 | 11.0 | 120.1 |
| 2 | 103 | 4.0 | 15.7 |
| 3 | 97 | −2.0 | 4.2 |
| 4 | 96 | −3.0 | 9.2 |
| 5 | 105 | 6.0 | 35.5 |
| 6 | 110 | 11.0 | 120.1 |
| 7 | 100 | 1.0 | 0.9 |
| 8 | 93 | −6.0 | 36.5 |
| 9 | 90 | −9.0 | 81.7 |
| 10 | 103 | 4.0 | 15.7 |
| 11 | 97 | −2.0 | 4.2 |
| 12 | 103 | 4.0 | 15.7 |
| 13 | 90 | −9.0 | 81.7 |
| 14 | 97 | −2.0 | 4.2 |
| 15 | 99 | 0 | 0 |
| 16 | 106 | 7.0 | 48.4 |
| 17 | 90 | −9.0 | 81.7 |
| 18 | 96 | −3.0 | 9.2 |
| 19 | 109 | 10.0 | 99.2 |
| 20 | 90 | −9.0 | 81.7 |

**Table 8.2** Example Sample Sigma Calculation (*continued*)

| i | $x_i$ | $x_i - \bar{X}$ | $(x_i - \bar{X})^2$ |
|---|---|---|---|
| 21 | 91 | −8.0 | 64.6 |
| 22 | 108 | 9.0 | 80.3 |
| 23 | 96 | −3.0 | 9.2 |
| 24 | 101 | 2.0 | 3.8 |
| 25 | 96 | −3.0 | 9.2 |
| SUM | 2476 | | 1033.0 |

$$s^2 = 1033 / (25 - 1) = 43.04$$

$$s = \sqrt{s^2} = \sqrt{43.04} \approx 6.6$$

The areas under the normal curve can be found by integrating Eq. (8.10) using numerical methods, but, more commonly, tables are used. Appendix 2 gives areas under the normal curve. The table is indexed by using the $Z$ transformation, which is

$$Z = \frac{x_i - \mu}{\sigma} \qquad (8.13)$$

for population data, or:

$$Z = \frac{x_i - \bar{x}}{s} \qquad (8.14)$$

for sample data.

By using the $Z$ transformation, we can convert any normal distribution into a normal distribution with a mean of 0 and a standard deviation of 1. Thus, we can use a single normal table to find probabilities.

### Example Using Normal Distribution Tables

Assume we have checked the breaking strength of a bonding process using the data in Table 8.1, whose average strength was determined in the section above to be 99 lbs. with a standard deviation of 6.6 lbs. Assuming the process distribution is well approximated by the normal distribution, if the engineering specification is 90 lbs. minimum, what percentage of the process will be below the low specification?

Since our data are a sample, we must compute $Z$ using Eq. (8.14).

$$Z = \frac{90 - 99.0}{6.6} = -1.36$$

Figure 8.8 illustrates this situation.

**Figure 8.8** Illustration of using Z tables for normal areas.

Entering in Appendix 2 for $Z = -1.36$, we find that 8.69% of the area is below this $Z$ value. Thus 8.69% of our breaking strengths will be below our low specification limit of 90 lbs. To improve the level of performance, we would have to improve the process by either raising the average breaking strength or reducing the process standard deviation, or both.

## Example of Normal Probability Calculations Using Microsoft Excel

Microsoft Excel has a built-in capability to analyze normal probabilities. To solve the above problem using Excel, enter the $X$ value (90), average (99), sigma (6.6), and the cumulative parameter (True) as shown in Fig. 8.9a. The formula result near the bottom of the figure provides the probability of being less than the minimum specification of 90 lbs. (Prior to Excel 2013, the Normdist function was used with identical parameters.)

MS Excel also provides the means of easily determining the x value corresponding to a specified percentile of the normal distribution. For example, to determine the lower specification that provides a 0.27% nonconformance rate use the Norm.Inv function as shown in Fig. 8.9b to determine that the lower specification would need to be at approximately 80 lbs.

## Evaluating Data for Normality

If sample data are approximately normally distributed, then it is easy to answer many questions about the sample, such as shown above. It is therefore beneficial to assume normality of the data, if the sample data can be shown to approximate the normal distribution.

**Figure 8.9a** Example of finding normal probability using Microsoft Excel.

**Figure 8.9b** Example of finding value corresponding to percentile of Normal distribution using MS Excel.

A convenient tool for evaluating normality is the normal probability plot shown in Fig. 8.10 based on the data provided in Table 8.1. Data that approximates the normal distribution will fall along the line, with only small deviations from the line. To remove the subjective nature of trying to determine if a point is "too far" from the line to approximate normality, analysts should use the Kolmogorov-Smirnov (K-S) statistic, which tests the null hypothesis that the data follows the specified distribution. K-S values vary between 0 and 1, with values less than 0.05 providing evidence that the null hypothesis of good-

**Figure 8.10** Normal probability plot.

ness of fit should be rejected (i.e., there is evidence the data does not fit the distribution). Values close to 1 indicate a high unlikelihood that the null hypothesis can be rejected, and a higher degree of fit to the distribution. The K-S value of 0.9 in the figure indicates the example data is well approximated by the normal distribution.

## *Lognormal Distribution*

A special case of a normal distribution is the lognormal distribution. When data are not normally distributed, if a transformation of the data using the natural log, or LN, function results in a normally distributed data set, then the original data are assumed to have a lognormal distribution.

To investigate if the lognormal transformation is useful for the original non-normal data:

1. Calculate the natural log for each data value.
2. Check for normality using a normal probability plot and a suitable goodness of fit test, such as the K-S statistic.
3. For the set of transformed data, calculate the mean $\mu$ and the standard deviation $\sigma$.
4. Use the normal distribution to predict the transformed data. Convert a *z*-value to original units using the exponential function (i.e., Excel's EXP function), since the exponential of a lognormal value provides the original value [i.e., EXP(LN(x))=x].

## Lognormal Example

For example, we can calculate the probability of obtaining values less than 20 from a given process with sample data shown in the table below.

| Sample Data Value | Transformed LN value |
|---|---|
| 1.2 | 0.18 |
| 0.3 | −1.2 |
| 5.7 | 1.74 |
| 79.3 | 4.37 |

1. The lognormal transformation of each data value is also provided in the table.
2. The transformed data is plotted on a normal probability plot. The K-S value of 0.8 indicates that the null hypothesis cannot be rejected, and the normal distribution may be assumed for the transformed data.
3. The average and sample standard deviation of the transformed data are calculated as 1.27 and 2.39, respectively.
4. The cumulative normal distribution function is used to calculate the percentile associated with the value of 20, which must first be transformed using the natural log function to its comparable value in the lognormal distribution: LN(20) = 3.00. Using Excel's Norm.Dist function (NORMDIST prior to Excel 2013), we can calculate the percentage below the transformed value of 3.00 (raw value of 20) as 77% by using the equation:

$$=\text{NORM.DIST}(3,1.27,2.39,\text{TRUE})$$

The value associated with the 95% percentile, for example, can be calculated using Excel's Norm.Inv function (NORMINV prior to Excel 2013) as:

$$=\text{NORM.INV}(0.95,1.27,2.39)$$

The function returns the value of 5.20, which must be transformed to the original lognormal distribution using Excel's EXP function (e.g. =EXP(5.20)) to obtain the value of 181.

## *Exponential Distribution*

Another distribution encountered often in quality control work is the exponential distribution. The exponential distribution is especially useful in analyzing reliability. The equation for the probability density function of the exponential distribution is

$$f(x) = \frac{1}{\mu}e^{-x/\mu}, x \geq 0 \qquad (8.15)$$

**Figure 8.11** Exponential pdf curve.

Unlike the normal distribution, the shape of the exponential distribution is highly skewed and there is a much greater area below the mean than above it. In fact, over 63% of the exponential distribution falls below the mean. Figure 8.11 shows an exponential pdf.

Unlike the normal distribution, the exponential distribution has a closed form cumulative density function (cdf), that is, there is an equation which gives the cumulative probabilities directly. Since the probabilities can be determined directly from the equation, no tables are necessary. See Eq. (8.16).

$$P(X \leq x) = 1 - e^{-x/\mu} \qquad (8.16)$$

### Example of Using the Exponential cdf

A city water company averages 500 system leaks per year. What is the probability that the weekend crew, which works from 6 p.m. Friday to 6 a.m. Monday, will get no calls?

We have $\mu$ = 500 leaks per year, which we must convert to leaks per hour. There are 365 days of 24 hours each in a year, or 8760 hours. Thus, mean time between failures (MTBF) is 8760/500 = 17.52 hours. There are 60 hours between 6 p.m. Friday and 6 a.m. Monday. Thus $x$ = 60. Using Eq. (8.16) gives

$$P(X \leq 60) = 1 - e^{-60/17.52} = 0.0967 = 96.7\%$$

Thus, the crew will be working all but 3.3% of the weekends.

### Example of Exponential Probability Calculations Using Microsoft Excel

Microsoft Excel has a built-in capability to analyze exponential probabilities. Note that Excel uses "lambda" rather than the average in its calculations; lambda is the reciprocal

**Figure 8.12** Example of finding exponential probability using Microsoft Excel.

of the average. To solve the above problem using Excel, enter the $x$ value, lambda and the cumulative parameter (True) as shown in Fig. 8.12. The formula result near the bottom of the screen gives the desired probability. (Prior to Excel 2013, the Excel Expondist function was used with identical parameters.)

## *Weibull Distribution*

The Weibull distribution is another useful distribution for continuous data, with applications often found in reliability analysis. The exponential distribution discussed previously is a special case of the more general Weibull distribution, which is completely defined by a shape parameter alpha ($\alpha$) and a scale parameter beta ($\beta$). The cumulative probability at a value $x$ may be found using Eq. (8.17):

$$\text{Probability}(X \leq x) = 1 - e[-(x/\beta)^\alpha] \qquad (8.17)$$

Solving for $x$ provides the value at which a stated cumulative probability $p$ occurs:

$$x = \beta \times (-\ln(1-p))^{1/\alpha} \qquad (8.18)$$

### **Weibull Example**

For example, the cumulative distribution function can be used to predict the percentage below the value of 25 for a Weibull process with a shape parameter of 2.2 and a scale parameter of 31 as:

$$\text{Probability}(X \leq 25) = 1 - e[-(25/31)^{2.2}] = 1 - e[-0.623] = .464 = 64.4\%$$

In Excel, the Weibull cumulative distribution function may be calculated directly using the function Weibull.Dist($x$, $\alpha$, $\beta$, True) [use the Weibull($x$, $\alpha$, $\beta$, True) function prior to Excel 2013]:

$$=\text{WEIBULL.DIST}(25, 2.2, 31, \text{TRUE})$$

Using Eq. (8.18), the value that includes 70% of the distribution can be found as:

$$x = 31 \times (-\ln(1 - .70))^{1/2.2} = 33.73$$

Minitab offers a simpler and more complete approach for Weibull analysis, including the ability to fit a Weibull distribution to estimate the parameters. To solve the initial problem of the cumulative probability at a value of 25, select the *Probability Distributions/ Weibull* option from the *Calc* menu, and enter these parameters into Minitab 17:

Select *Cumulative Probability*
Enter *Shape Parameter* 2.2
Enter *Scale Parameter* 31
Enter *Input Constant* 25
Select *OK*

Likewise, the value that includes 70% of the distribution can be found as:

Select *Inverse Cumulative Probability*
Enter *Shape Parameter* 2.2
Enter *Scale Parameter* 31
Enter *Input Constant* 0.70
Select *OK*

# Control Charts for Variables Data

In statistical process control (SPC), the mean, range, and standard deviation are the statistics most often used for analyzing measurement data. Control charts are used to monitor these statistics. An out-of-control point for any of these statistics is an indication that a special cause of variation is present and that an immediate investigation should be made to identify the special cause.

### Averages and Ranges Control Charts

Averages charts are statistical tools used to evaluate the central tendency of a process over time. Ranges charts are statistical tools used to evaluate the dispersion or spread of a process over time.

*Averages charts* answer the question: "Has a special cause of variation caused the central tendency of this process to change over the time period observed?" *Ranges charts* answer the question: "Has a special cause of variation caused the process distribution to become more or less consistent?" Averages and ranges charts can be applied to many continuous variables such as weight, size, response time, etc.

The basis of the control chart is the *rational subgroup*. Rational subgroups (see "Rational Subgroup Sampling") are composed of items that were produced under essentially the same conditions. The average and range are computed for each subgroup separately, then plotted on the control chart. Each subgroup's statistics are compared to the control limits, and patterns of variation between subgroups are analyzed.

## Subgroup Equations for Averages and Ranges Charts

$$\bar{X} = \frac{\text{sum of subgroup measurements}}{\text{subgroup size}} \qquad (8.19)$$

$$R = \text{largest in subgroup} - \text{smallest in subgroup} \qquad (8.20)$$

## Control Limit Equations for Averages and Ranges Charts

Control limits for both the averages and the ranges charts are computed such that it is highly unlikely that a subgroup average or range from a stable process would fall outside of the limits. All control limits are set at plus and minus three standard deviations from the centerline of the chart. Thus, the control limits for subgroup averages are plus and minus three standard deviations of the mean from the grand average; the control limits for the subgroup ranges are plus and minus three standard deviations of the range from the average range. These control limits are quite robust with respect to non-normality in the process distribution.

To facilitate calculations, constants are used in the control limit equations. Appendix 9 provides control chart constants for subgroup size of 25 (data values) or less. The derivation of the various control chart constants is shown in Burr (1976, pp. 97–105) and assumes there are a "large" number of subgroups of the specified size.

## Control Limit Equations for Ranges Charts

$$\bar{R} = \frac{\text{sum of subgroup ranges}}{\text{number of subgroups}} \qquad (8.21)$$

$$\text{LCL} = D_3 \bar{R} \qquad (8.22)$$

$$\text{UCL} = D_4 \bar{R} \qquad (8.23)$$

## Control Limit Equations for Averages Charts Using $\bar{R}$

$$\bar{\bar{X}} = \frac{\text{sum of subgroup averages}}{\text{number of subgroups}} \tag{8.24}$$

$$\text{LCL} = \bar{\bar{X}} - A_2\bar{R} \tag{8.25}$$

$$\text{UCL} = \bar{\bar{X}} + A_2\bar{R} \tag{8.26}$$

Minitab users should note that the default settings for the averages control chart were incorrectly defined (within the *Estimate* tab in the *Options* dialog box) as the *pooled standard deviation* in Minitab prior to version 17. While this error was corrected in version 17, users of the prior versions should verify the proper setting of *Rbar* when using the *Xbar-R Chart* option in Minitab. The pooled standard deviation should never be applied to a control chart, and in fact removes the primary elements (i.e., variation over time) that define the control chart as a process analysis tool. (See also the discussion of control charts and differences between enumerative and analytical studies in Chap. 7 for further details.)

## Example of Averages and Ranges Control Charts

Table 8.3 contains 25 subgroups of five observations each (i.e., the subgroup size is 5).

**Table 8.3** Data for X-Bar, Ranges, and Sigma Control Charts

| Sample 1 | Sample 2 | Sample 3 | Sample 4 | Sample 5 | X-Bar | Range | Sigma |
|---|---|---|---|---|---|---|---|
| 110 | 93 | 99 | 98 | 109 | 101.8 | 17 | 7.396 |
| 103 | 95 | 109 | 95 | 98 | 100.0 | 14 | 6.000 |
| 97 | 110 | 90 | 97 | 100 | 98.8 | 20 | 7.259 |
| 96 | 102 | 105 | 90 | 96 | 97.8 | 15 | 5.848 |
| 105 | 110 | 109 | 93 | 98 | 103.0 | 17 | 7.314 |
| 110 | 91 | 104 | 91 | 101 | 99.4 | 19 | 8.325 |
| 100 | 96 | 104 | 93 | 96 | 97.8 | 11 | 4.266 |
| 93 | 90 | 110 | 109 | 105 | 101.4 | 20 | 9.290 |
| 90 | 105 | 109 | 90 | 108 | 100.4 | 19 | 9.607 |
| 103 | 93 | 93 | 99 | 96 | 96.8 | 10 | 4.266 |
| 97 | 97 | 104 | 103 | 92 | 98.6 | 12 | 4.930 |
| 103 | 100 | 91 | 103 | 105 | 100.4 | 14 | 5.550 |
| 90 | 101 | 96 | 104 | 108 | 99.8 | 18 | 7.014 |
| 97 | 106 | 97 | 105 | 96 | 100.2 | 10 | 4.868 |
| 99 | 94 | 96 | 98 | 90 | 95.4 | 9 | 3.578 |
| 106 | 93 | 104 | 93 | 99 | 99.0 | 13 | 6.042 |
| 90 | 95 | 98 | 109 | 110 | 100.4 | 20 | 8.792 |

**Table 8.3** Data for X-Bar, Ranges, and Sigma Control Charts (*continued*)

| Sample 1 | Sample 2 | Sample 3 | Sample 4 | Sample 5 | X-Bar | Range | Sigma |
|---|---|---|---|---|---|---|---|
| 96 | 96 | 108 | 97 | 103 | 100.0 | 12 | 5.339 |
| 109 | 96 | 91 | 98 | 109 | 100.6 | 18 | 8.081 |
| 90 | 95 | 94 | 107 | 99 | 97.0 | 17 | 6.442 |
| 91 | 101 | 96 | 96 | 109 | 98.6 | 18 | 6.804 |
| 108 | 97 | 101 | 103 | 94 | 100.6 | 14 | 5.413 |
| 96 | 97 | 106 | 96 | 98 | 98.6 | 10 | 4.219 |
| 101 | 107 | 104 | 109 | 104 | 105.0 | 8 | 3.082 |
| 96 | 91 | 96 | 91 | 105 | 95.8 | 14 | 5.718 |
| | | | | SUM | 2487.2 | 369 | 155.4 |

The control limits are calculated from these data as follows:

*Ranges control chart example*

$$\bar{R} = \frac{\text{sum of subgroup ranges}}{\text{number of subgroups}} = \frac{369}{25} = 14.76$$

$$\text{LCL}_R = D_3 \bar{R} = 0 \times 14.76 = 0$$

$$\text{UCL}_R = D_4 \bar{R} = 2.114 \times 14.76 = 31.2$$

The values of the constants $D_3$ and $D_4$ are obtained from Appendix 9 from the row where the number of observations in the sample (*n*) equals the subgroup size of five. Since it is not possible to have a subgroup range less than zero, the LCL is not shown on the control chart for ranges.

*Averages control chart example*

$$\bar{\bar{X}} = \frac{\text{sum of subgroup averages}}{\text{number of subgroups}} = \frac{2,487.2}{25} = 99.49$$

$$\text{LCL}_X = \bar{\bar{X}} - A_2 \bar{R} = 99.49 - 0.577 \times 14.76 = 91.0$$

$$\text{UCL}_X = \bar{\bar{X}} + A_2 \bar{R} = 99.49 + 0.577 \times 14.76 = 108.0$$

The value of the constant $A^2$ is obtained from Appendix 9 from the row where the number of observations in the sample (*n*) equals the subgroup size of five. The completed averages and ranges control charts are shown in Fig. 8.13.

The charts shown in Fig. 8.13 show a process in statistical control. This merely means that we can predict the limits of variability for this process. To determine the capability of the process with respect to requirements, one must use the methods described later in the process capability analysis section.

**Figure 8.13** Completed averages and ranges control charts (Using *Green Belt XL* software, www.qualityamerica.com).

---

**See online video: The Construction and Interpretation of X-Bar Control Charts at www.mhprofessional.com/ssh5.**

---

## *Averages and Standard Deviation (Sigma) Control Charts*

Averages and standard deviation control charts are conceptually identical to averages and ranges control charts. The difference is that the subgroup standard deviation is used to measure dispersion rather than the subgroup range. The subgroup standard deviation is statistically more efficient than the subgroup range for subgroup sizes greater than 2. This efficiency advantage increases as the subgroup size increases. The inefficiency of the range statistic becomes significant if the subgroup size is 10 or larger, so range charts are not recommended for these large subgroup sizes. However, since Six Sigma analysts will invariably use computer software in their analyses, the standard deviation chart is recommended for all subgroup sizes.

### Subgroup Equations for Averages and Sigma Charts

$$\bar{X} = \frac{\text{sum of subgroup measurements}}{\text{subgroup size}} \tag{8.27}$$

$$s = \sqrt{\frac{\sum_{i=1}^{n}(X_i - \bar{X})^2}{n-1}} \tag{8.28}$$

The standard deviation, s, is computed separately for each subgroup, using the subgroup average rather than the grand average. This is an important point; using the grand average would introduce special cause variation if the process were out of control, thereby underestimating the process capability, perhaps significantly.

## Control Limit Equations for Averages and Sigma Charts

Control limits for both the averages and the sigma charts are computed such that it is highly unlikely that a subgroup average or sigma from a stable process would fall outside of the limits. All control limits are set at plus and minus three standard deviations from the centerline of the chart. Thus, the control limits for subgroup averages are plus and minus three standard deviations of the mean from the grand average. The control limits for the subgroup sigma are placed at plus and minus three standard deviations of sigma from the average sigma. These control limits are quite robust with respect to non-normality in the process distribution.

To facilitate calculations, constants are used in the control limit equations. Appendix 9 provides control chart constants for subgroups of 25 or less data values.

## Control Limit Equations for Sigma Charts Based on $\bar{s}$

$$\bar{s} = \frac{\text{sum of subgroup sigmas}}{\text{number of subgroups}} \quad (8.29)$$

$$\text{LCL} = B_3 \bar{s} \quad (8.30)$$

$$\text{UCL} = B_4 \bar{s} \quad (8.31)$$

## Control Limit Equations for Averages Charts Based on $\bar{s}$

$$\bar{\bar{X}} = \frac{\text{sum of subgroup averages}}{\text{number of subgroups}} \quad (8.32)$$

$$\text{LCL} = \bar{\bar{X}} - A_3 \bar{s} \quad (8.33)$$

$$\text{UCL} = \bar{\bar{X}} + A_3 \bar{s} \quad (8.34)$$

Minitab users should note that the default settings for the averages control chart were incorrectly defined (within the *Estimate* tab in the *Options* dialog box) as the *pooled standard deviation* in Minitab prior to version 17. While this error was corrected in version 17, users of the prior versions should verify the proper setting of *Sbar* when using the *Xbar-S Chart* option in Minitab. The pooled standard deviation should never be applied to a control chart, and in fact removes the primary elements that define the control chart as a process analysis tool. (See also the discussion of control charts in Chap. 7 for further detail).

## Example of Averages and Standard Deviation Control Charts

To illustrate the calculations and to compare the range method to the standard deviation results, the data used in the previous example will be reanalyzed using the subgroup standard deviation rather than the subgroup range.

The control limits are calculated from this data as follows:

*Sigma control chart*

$$\bar{\bar{X}} = \frac{\text{sum of subgroup sigmas}}{\text{number of subgroups}} = \frac{155.4}{25} = 6.22$$

$$\text{LCL}_s = B_3\bar{s} = 0 \times 6.22 = 0$$

$$\text{UCL}_s = B_4\bar{s} = 2.089 \times 6.22 = 13.0$$

The values of the constants $B^3$ and $B_4$ are obtained from Appendix 9 from the row where the number of observations in the sample ($n$) equals the subgroup size of five. Since it is not possible to have a subgroup sigma less than zero, the LCL is not shown on the control chart for sigma for this example.

*Averages control chart*

$$\bar{\bar{X}} = \frac{\text{sum of subgroup averages}}{\text{number of subgroups}} = \frac{2{,}487.2}{25} = 99.49$$

$$\text{LCL}_{\bar{X}} = \bar{\bar{X}} - A_3\bar{s} = 99.49 - 1.427 \times 6.22 = 90.6$$

$$\text{UCL}_{\bar{X}} = \bar{\bar{X}} + A_3\bar{s} = 99.49 + 1.427 \times 6.22 = 108.4$$

The values of the constants $D_3$ and $D_4$ are obtained from Appendix 9 from the row where the number of observations in the sample ($n$) equals the subgroup size of five. The completed averages and sigma control charts are shown in Fig. 8.14. Note that the control limits for the averages chart are only slightly different than the limits calculated using ranges.

Note that the conclusions reached are the same as when ranges were used.

**Figure 8.14** Completed averages and sigma control charts (Using *Green Belt XL* software, www.qualityamerica.com).

See online video: XBar Control Chart Calculations at www.mhprofessional.com/ssh5.

Constructing XBar Control Chart in Minitab at www.mhprofessional.com/ssh5.

## Control Charts for Individual Measurements (X Charts)

Individuals control charts are statistical tools used to evaluate the central tendency of a process over time. They are also called *X charts* or *moving range charts.* Individuals control charts are used when it is not feasible to use averages for process control. There are many possible reasons why averages control charts may not be desirable: observations may be expensive to get (e.g., destructive testing), output may be too homogeneous over short time intervals (e.g., pH of a solution), the production rate may be slow and the interval between successive observations long, etc. Control charts for individuals are often used to monitor batch process, such as chemical processes, where the within-batch variation is so small relative to between-batch variation that the control limits on a standard $\bar{X}$ chart would be too close together. Range charts (sometimes called moving range charts in this application) are used to monitor dispersion between the successive individual observations.[1]

### Calculations for Moving Ranges Charts

As with averages and ranges charts, the range is computed as shown in the previous section:

$$R = \text{largest in subgroup} - \text{smallest in subgroup}$$

where the subgroup is a consecutive pair of process measurements. The range control limit is computed as was described for averages and ranges charts, using the $D_4$ constant for subgroups of 2, which is 3.267. That is,

$$\text{LCL} = 0 \text{ (for } n = 2\text{)}$$

$$\text{UCL} = 3.267 \times R\text{-bar}$$

## Control Limit Equations for Individuals Charts

$$\bar{X} = \frac{\text{sum of measurements}}{\text{number of measurements}} \quad (8.35)$$

$$\text{LCL} = \bar{X} - E_2\bar{R} = \bar{X} - 2.66 \times \bar{R} \quad (8.36)$$

$$\text{UCL} = \bar{X} + E_2\bar{R} = \bar{X} + 2.66 \times \bar{R} \quad (8.37)$$

where $E_2 = 2.66$ is the constant used when individual measurements are plotted, and $\bar{R}$ is based on subgroups of $n = 2$.

## Example of Individuals and Moving Ranges Control Charts

Table 8.4 contains 25 measurements. To facilitate comparison, the measurements are the first observations in each subgroup used in the previous average/ranges and average/standard deviation control chart examples.

**Table 8.4** Data for Individuals and Moving Ranges Control Charts

| Sample | Range |
|--------|-------|
| 110 | None |
| 103 | 7 |
| 97 | 6 |
| 96 | 1 |
| 105 | 9 |
| 110 | 5 |
| 100 | 10 |
| 93 | 7 |
| 90 | 3 |
| 103 | 13 |
| 97 | 6 |
| 103 | 6 |

**Table 8.4** Data for Individuals and Moving Ranges Control Charts (*continued*)

| Sample | Range |
|---|---|
| 90 | 13 |
| 97 | 7 |
| 99 | 2 |
| 106 | 7 |
| 90 | 16 |
| 96 | 6 |
| 109 | 13 |
| 90 | 19 |
| 91 | 1 |
| 108 | 17 |
| 96 | 12 |
| 101 | 5 |
| 96 | 5 |
| SUM: 2476 | 196 |

The control limits are calculated from this data as follows:

*Moving ranges control chart control limits*

$$\bar{R} = \frac{\text{sum of ranges}}{\text{number of ranges}} = \frac{196}{24} = 8.17$$

$$\text{LCL}_R = D_3 \bar{R} = 0 \times 8.17 = 0$$

$$\text{UCL}_R = D_4 \bar{R} = 3.267 \times 8.17 = 26.7$$

*Individuals control chart control limits*

$$\bar{X} = \frac{\text{sum of measurements}}{\text{number of measurements}} = \frac{2{,}476}{25} = 99.0$$

$$\text{LCL}_X = \bar{X} - E_2 \bar{R} = 99.0 - 2.66 \times 8.17 = 77.3$$

$$\text{UCL}_X = \bar{X} + E_2 \bar{R} = 99.0 + 2.66 \times 8.17 = 120.7$$

The completed individuals and moving ranges control charts are shown in Fig. 8.15.

In this case, the conclusions are the same as with averages charts. However, averages charts always provide tighter control than Individual X charts and can be used regardless

**Figure 8.15** Completed individuals and moving ranges control charts (Using *Green Belt XL* software, www.qualityamerica.com).

of the process distribution. That is, while the averages chart is insensitive to the process distribution, the Individual chart may provide control limits that do not reflect the process distribution if the process is significantly non-normal.

## Special Consideration for Non-normal Processes

It's not readily apparent from the calculations, but the control limit calculations shown for the Individual-X chart (and above for the $\bar{X}$ chart) place the control limits at plus and minus three standard deviations from the average, the presumed process centerline. For example, the factor $A_2$ [shown in Eqs. (8.25) and (8.26)] is calculated as follows:

$$A_2 = \frac{3}{d_2\sqrt{n}} \tag{8.38}$$

This factor $A_2$ is multiplied by the quantity $\bar{R}$ in Eqs. (8.25) and (8.26); however, it can also be shown that the standard deviation of the subgroup averages is calculated as:

$$\sigma_{\bar{X}} = \frac{\bar{R}}{d_2\sqrt{n}} \tag{8.39}$$

Therefore, Eq. (8.25) may also be written as:

$$\text{UCL}_{\bar{X}} = \bar{\bar{X}} + 3\sigma_{\bar{X}} \tag{8.40}$$

Likewise, the *process standard deviation* can be calculated as:

$$\sigma_X = \frac{\bar{R}}{d_2} \tag{8.41}$$

This reduces Eq. (8.37) to:

$$\text{UCL}_X = \bar{X} + 3\sigma_X \qquad (8.42)$$

It should be apparent that the lower control limit calculations can be written in a similar fashion. These equations highlight several central premises of these control charts:

▲ The expected value of the plotted statistic is well represented by its average.
▲ The limit of the plotted statistic is equidistant from its centerline in either direction.
▲ The limit of the plotted statistic is well defined at three standard deviations of the plotted statistic from its centerline.

In the $\bar{X}$ chart, the plotted statistic is a subgroup average; in the Individual-X chart, the plotted statistic is an observation. The first and second characteristics apply only to symmetric distributions; that is, statistical distributions whose shape above the process centerline mirrors the shape below the centerline. In these cases, the distribution average exactly coincides with the median (the 50th percentile of the distribution, so that half the distribution lies above the value and half below the value). The third characteristic applies well to symmetric distributions whose shape is similar to the normal distribution, where the region between the plus and minus three sigma values from the average (i.e., the $z$ values from −3 to +3) encompasses 99.73% of the distribution. In these cases, the statistical limits provided by the control chart provide a reasonable bound on the expected process behavior when the process mean or standard deviation are stable (i.e., in the absence of special causes of variation).

Yet, there are many examples of processes in nature and business that do not closely resemble a normal distribution. Process designers and improvement teams often take actions that specifically preclude a symmetric distribution with long tails: cutoff operations have built-in stops, for example. Other times, the nature of the variable defines a boundary, such as cycle times or queue times that can't be less than zero. In those cases, as the process is improved to reduce the cycle time the distribution confronts the bound. The frequent use of the normal distribution is not a testament to its prevalence, but rather to the simplicity of its application. The properties of the normal distribution are well documented, and so long as the actual process distribution is "close enough" to the normal distribution, then its use greatly simplifies the statistical analysis.

Consider the process data (Keller, 2011) plotted on the standard Individual-X control chart in Fig. 8.16. The histogram is displayed on its side in the left pane of the figure to align with the control chart display in the right pane of the figure. Note the sharp boundary of the data close to the zero value, below which no data values exist. The data could represent any process bounded at the zero value, such as cycle time, queue time, surface finish, TIR (total indicator reading), and so on.

**Figure 8.16** Standard Individual-X chart with non-normal process data (Using *Green Belt XL* software, www.qualityamerica.com).

The asymmetrical nature of the process data deviates from the first two conditions noted above, and indeed the calculated lower control limit at minus three standard deviations provides a lower limit far from its closest value. The dead-zone between the lowest possible process value of zero and the calculated lower control limit prevents the control chart from detecting process shifts below the centerline. Any improvement in the process mean (i.e., a shift of the mean to a smaller value) would be undetectable with the chart.

Fortunately, non-normal process data can be conveniently transformed to normality, then analyzed for process stability. The Box-Cox transformation, presented in the context of variance stabilization in Chap. 10, is one strategy. A much simpler form of transformation involves subgrouping the data into rational subgroups and analyzing the subgrouped data on the $\bar{X}$ chart. The subgroup averages plotted on the $\bar{X}$ chart will tend to normality, as predicted by the central limit theorem, even for subgroups as small as three to five observations from fairly skewed process distributions. An example of this approach is presented in the "Process Capability Indices" section later in this chapter. The standard EWMA control chart presented in Fig. 8.46, or simpler Moving Average charts, can also be used to analyze non-normal process data for statistical control.

Yet, the Individual-X chart often provides an easier template for process personnel to analyze the process. It's not uncommon for even seasoned practitioners to compare the plotted averages on the $\bar{X}$ chart to the process specifications, as if they were individual observations. This is incorrect, since (as discussed earlier) the distribution of the subgroup averages is much narrower than the distribution of the observations (by a factor of the square root of $n$, the subgroup size), making a direct comparison of the plotted averages to the specifications misleading: While all the subgroup averages may fit within the specifications, it's possible that all of the observations within the subgroups exceed the specifications. For example, the subgroup containing observations with values of 6, 8, 14, and 16 has an average of 11, which lies directly between the specifications of 10 and 12, even though none of the observations lie within the specification limits. Customer specifications do not apply to the averages. Rather, customers expect each and every data observation to fit within the specification.

Once the stability of the process has been verified with the EWMA (or Moving Average or $\bar{X}$) control chart, an Individual-X chart can be constructed with non-normal control limits that better reflect the process distribution. An example is provided in Fig. 8.17. The advantage of the chart is its simplicity of analysis: Process personnel only need to verify the plotted observation does not violate the control limits established for the fitted process distribution. The run test rules described later in this chapter, modified to reflect the properties of the non-normal distribution, can also be automatically applied by the software.

See online video: **Analyzing Individuals Data on Process Control Charts** at www.mhprofessional.com/ssh5.

**Figure 8.17** Modified Individual-X chart with non-normal process data (Using *Green Belt XL* software, www.qualityamerica.com).

# Control Charts for Attributes Data

## Control Charts for Proportion (p Charts)

The $p$ charts are statistical tools used to evaluate the proportion meeting a condition of interest, such as proportion defective, or proportion of births requiring C-section.

The $p$ charts can be applied when the appropriate performance measure is a unit count. Each sample must either meet the condition of interest, or not meet the condition. For example, is the birth by C-section, or not by C-section? The $p$ charts answer the question: "Has a special cause of variation caused the central tendency of this process to produce an abnormally large or small number of units meeting the condition over the time period observed?"

### $p$ Chart Control Limit Equations

The centerline of the $p$ chart is the average proportion defective and the control limits are set at plus and minus three standard deviations based on an assumed binomial distribution. If the process is in statistical control, then virtually all proportions should be between the control limits and they should fluctuate randomly about the centerline.

$$p = \frac{\text{subgroup defective count}}{\text{subgroup size}} \tag{8.43}$$

$$\bar{p} = \frac{\text{sum of subgroup defective counts}}{\text{sum of subgroup sizes}} \tag{8.44}$$

$$\text{LCL} = \bar{p} - 3\sqrt{\frac{\bar{p}(1-\bar{p})}{n}} \tag{8.45}$$

$$\text{UCL} = \bar{p} + 3\sqrt{\frac{\bar{p}(1-\bar{p})}{n}} \tag{8.46}$$

In Eqs. (8.45) and (8.46), $n$ is the subgroup size. If the subgroup size varies, the control limits will also vary, becoming closer together as $n$ increases.

### Example of $p$ Chart Calculations

The data in Table 8.5 were obtained by opening randomly selected crates from each shipment and counting the number of bruised peaches. There are 250 peaches per crate. Normally, samples consist of 1 crate per shipment. However, when part-time help is available, samples of 2 crates are taken.

For example, the $p$-value for the first subgroup is calculated using Eq. (8.43) as:

$$p = 47/250 = 0.19$$

These values are shown in the last column of Table 8.5. The centerline is calculated using Eq. 8.44 as:

$$\bar{p} = \frac{\text{sum of subgroup defective counts}}{\text{sum of subgroup size}} = \frac{1{,}544}{8{,}000} = 0.193$$

**Table 8.5** Raw Data for *p* Chart

| Shipment No. | Crates | Peaches | Bruised | P |
|---|---|---|---|---|
| 1 | 1 | 250 | 47 | 0.19 |
| 2 | 1 | 250 | 42 | 0.17 |
| 3 | 1 | 250 | 55 | 0.22 |
| 4 | 1 | 250 | 51 | 0.20 |
| 5 | 1 | 250 | 46 | 0.18 |
| 6 | 1 | 250 | 61 | 0.24 |
| 7 | 1 | 250 | 39 | 0.16 |
| 8 | 1 | 250 | 44 | 0.18 |
| 9 | 1 | 250 | 41 | 0.16 |
| 10 | 1 | 250 | 51 | 0.20 |
| 11 | 2 | 500 | 88 | 0.18 |
| 12 | 2 | 500 | 101 | 0.20 |
| 13 | 2 | 500 | 101 | 0.20 |
| 14 | 1 | 250 | 40 | 0.16 |
| 15 | 1 | 250 | 48 | 0.19 |
| 16 | 1 | 250 | 47 | 0.19 |
| 17 | 1 | 250 | 50 | 0.20 |
| 18 | 1 | 250 | 48 | 0.19 |
| 19 | 1 | 250 | 57 | 0.23 |
| 20 | 1 | 250 | 45 | 0.18 |
| 21 | 1 | 250 | 43 | 0.17 |
| 22 | 2 | 500 | 105 | 0.21 |
| 23 | 2 | 500 | 98 | 0.20 |
| 24 | 2 | 500 | 100 | 0.20 |
| 25 | 2 | 500 | 96 | 0.19 |
| | **SUM** | 8,000 | 1,544 | 4.8 |

which is constant for all subgroups. The control limits are calculated using Eq. 8.45 and 8.46, dependent on the subgroup size.

$n = 250$ (1 crate):

$$LCL = \bar{p} - 3\sqrt{\frac{\bar{p}(1-\bar{p})}{n}} = 0.193 - 3\sqrt{\frac{0.193 \times (1 - 0.193)}{250}} = 0.12$$

$$UCL = \bar{p} + 3\sqrt{\frac{\bar{p}(1-\bar{p})}{n}} = 0.193 + 3\sqrt{\frac{0.193 \times (1 - 0.193)}{250}} = 0.27$$

$n = 500$ (2 crates):

$$LCL = 0.193 - 3\sqrt{\frac{0.193 \times (1 - 0.193)}{500}} = 0.14$$

$$UCL = 0.193 + 3\sqrt{\frac{0.193 \times (1 - 0.193)}{500}} = 0.25$$

The control limits and the subgroup proportions are shown in Fig. 8.18.

**Figure 8.18** Completed $p$ control chart.

If the variation in sample size is not too large (typically 20% or less), the average sample size may be used to calculate a common set of control limits for the subgroups. (Always use the actual sample size to calculate the $p$-value for each subgroup). If you're using software to generate the charts it's usually just as easy to show the staggered limits based on the actual sample size; however, some practitioners find the charts with a fixed set of control limits easier to interpret. In our example, the sample size doubled from 250 peaches to 500 peaches, but the difference in the calculated control limits is barely perceptible, as shown in Table 8.6.

**Table 8.6** Effect of Using Average Sample Size

| Sample Size | Lower Control Limit | Upper Control Limit |
|---|---|---|
| 250 | 0.12 | 0.27 |
| 500 | 0.14 | 0.25 |
| 320 | 0.13 | 0.26 |

Notice that the conclusions regarding process performance are the same when using the average sample size as they are using the exact sample sizes, which is often the case if the variation in sample size isn't too great. SPC is all about improved decision-making, so use the simplest method that leads to correct decisions.

It's most important to have adequate data resolution, even with counts data. When counting defective units from a process with high quality levels, a small sample size may reveal zero defective units in many of the samples. The more samples there are with zero defects, the higher likelihood that any sample with one or more defects will be out of control. This chart would not be very useful in practice. A simple alternative is to increase the sample size so most or all samples have a count greater than zero, but if samples are then taken over a long time period this is not desirable, as special causes may consistently appear over these longer periods. A better alternative is to instead monitor the time between defects on a control chart for individuals data.

## *Control Charts for Count of Items* (np *Charts*)

The *np* charts are statistical tools used to evaluate the count of items meeting a condition of interest, such as the count of items nonconforming. *np* charts can be applied to any variable where the appropriate performance measure is a unit count and the subgroup size is held constant. Note that wherever an *np* chart can be used, a *p* chart can be used instead.

### Control Limit Equations for *np* Charts

The centerline of the *np* chart is the average count of defectives-per-subgroup and the control limits are set at plus and minus three standard deviations based on the assumed binomial dstribution. If the process is in statistical control, then virtually all subgroup counts will lie between the control limits, and they will fluctuate randomly about the centerline.

$$np = \text{subgroup defect count} \tag{8.47}$$

$$n\bar{p} = \frac{\text{sum of subgroup defect counts}}{\text{number of subgroups}} \tag{8.48}$$

$$\text{LCL} = n\bar{p} - 3\sqrt{n\bar{p}(1-\bar{p})} \tag{8.49}$$

$$\text{UCL} = n\bar{p} + 3\sqrt{n\bar{p}(1-\bar{p})} \tag{8.50}$$

Note that:

$$\bar{p} = \frac{n\bar{p}}{n} \tag{8.51}$$

## Example of *np* Chart Calculation

The data in Table 8.7 were obtained by opening randomly selected crates from each shipment and counting the number of bruised peaches. There are 250 peaches per crate (constant *n* is required for *np* charts).

**Table 8.7** Raw Data for *np* Chart

| Shipment No. | Bruised Peaches |
|---|---|
| 1 | 20 |
| 2 | 28 |
| 3 | 24 |
| 4 | 21 |
| 5 | 32 |
| 6 | 33 |
| 7 | 31 |
| 8 | 29 |
| 9 | 30 |
| 10 | 34 |
| 11 | 32 |
| 12 | 24 |
| 13 | 29 |
| 14 | 27 |
| 15 | 37 |
| 16 | 23 |
| 17 | 27 |
| 18 | 28 |
| 19 | 31 |
| 20 | 27 |
| 21 | 30 |
| 22 | 23 |
| 23 | 23 |
| 24 | 27 |
| 25 | 35 |
| 26 | 29 |
| 27 | 23 |
| 28 | 23 |
| 29 | 30 |
| 30 | 28 |
| SUM | 838 |

Using the above data the centerline and control limits are found as follows:

$$n\bar{p} = \frac{\text{sum of subgroup defective counts}}{\text{number of subgroups}} = \frac{838}{30} = 27.9$$

$$\text{LCL} = n\bar{p} - 3\sqrt{n\bar{p}(1-\bar{p})} = 27.93 - 3\sqrt{27.9 \times \left(1 - \frac{27.9}{250}\right)} = 13$$

$$\text{UCL} = n\bar{p} + 3\sqrt{n\bar{p}(1-\bar{p})} = 27.93 + 3\sqrt{27.9 \times \left(1 - \frac{27.9}{250}\right)} = 43$$

The control limits and the subgroup defective counts are shown in Fig. 8.19.

**Figure 8.19** Completed *np* control chart.

## Control Charts for Average Occurrences-Per-Unit (u Charts)

*u* charts are statistical tools used to evaluate the average number of occurrences-per-unit produced by a process. Unlike *p* or *np* charts, *u* charts do not necessarily involve counting physical items. Rather, they involve counting of *events*. In the examples above, the *p* and *np* charts were based on counts of peaches that were bruised. When using a *u* chart, one would count the *number of bruises* on each peach. Note the key difference is consideration of the sample unit (e.g., a single peach): For the *p* and *np* charts, based on the binomial distribution, each sample unit (e.g., a single peach) either meets the condition of interest or does not meet the condition. For the *u* chart, we are concerned with the number of occurrences of the condition within each sample unit.

## Control Limit Equations for *u* Charts

The centerline of the *u* chart is the average number of occurrences-per-unit and the control limits are set at plus and minus three standard deviations. If the process is in statistical control then the subgroup occurrences-per-unit should be between the control limits and they should fluctuate randomly about the centerline.

$$u = \frac{\text{subgroup count of occurrences}}{\text{subgroup size in units}} \quad (8.52)$$

$$\bar{u} = \frac{\text{sum of subgroup occurrences}}{\text{sum of subgroup sizes in units}} \quad (8.53)$$

$$\text{LCL} = \bar{u} - 3\sqrt{\frac{\bar{u}}{n}} \quad (8.54)$$

$$\text{UCL} = \bar{u} + 3\sqrt{\frac{\bar{u}}{n}} \quad (8.55)$$

In Eqs. (8.54) and (8.55), *n* is the subgroup size in units. If the subgroup size varies, the control limits will also vary.

One way to determine if a particular set of data is suitable for a *u* chart or a *p* chart is to examine the equation used to compute the centerline for the control chart. If the unit of measure is the same in both the numerator and the denominator, then a *p* chart is indicated, otherwise a *u* chart is indicated. For example, if:

$$\text{Center Line} = \frac{\text{bruises per crate}}{\text{number of crates}} \quad (8.56)$$

then the numerator is in terms of bruises while the denominator is in terms of crates, indicating a *u* chart.

The unit size is arbitrary, but once determined it cannot be changed without recomputing all subgroup occurrences-per-unit and control limits. For example, if the occurrences were accidents and a unit was 100,000 hours worked, then a month with 250,000 hours worked would be 2.5 units and a month with 50,000 hours worked would be 0.5 unit. If the unit size were 200,000 hours then the 2 months would have 1.25 and 0.25 units, respectively. The equations for the centerline and control limits would "automatically" take into account the unit size, so the control charts would give identical results regardless of which unit size is used.

## Example of *u* Chart

The data in Table 8.8 were obtained by opening randomly selected crates from each shipment and counting the number of bruises on peaches. There are 250 peaches per crate. Our unit size will be taken as one full crate, that is, we will be counting crates rather than

**Table 8.8**  Raw Data for *u* Chart

| Shipment No. | Units (Crates) | Flaws | Flaws-Per-Unit |
|---|---|---|---|
| 1 | 1 | 47 | 47 |
| 2 | 1 | 42 | 42 |
| 3 | 1 | 55 | 55 |
| 4 | 1 | 51 | 51 |
| 5 | 1 | 46 | 46 |
| 6 | 1 | 61 | 61 |
| 7 | 1 | 39 | 39 |
| 8 | 1 | 44 | 44 |
| 9 | 1 | 41 | 41 |
| 10 | 1 | 51 | 51 |
| 11 | 2 | 88 | 44 |
| 12 | 2 | 101 | 50.5 |
| 13 | 2 | 101 | 50.5 |
| 14 | 1 | 40 | 40 |
| 15 | 1 | 48 | 48 |
| 16 | 1 | 47 | 47 |
| 17 | 1 | 50 | 50 |
| 18 | 1 | 48 | 48 |
| 19 | 1 | 57 | 57 |
| 20 | 1 | 45 | 45 |
| 21 | 1 | 43 | 43 |
| 22 | 2 | 105 | 52.5 |
| 23 | 2 | 98 | 49 |
| 24 | 2 | 100 | 50 |
| 25 | 2 | 96 | 48 |
| **SUM** | **32** | **1,544** | |

the peaches themselves. Normally, samples consist of one crate per shipment. However, when part-time help is available, samples of two crates are taken.

Using the above data, the number of occurrences in each subgroup is calculated as follows, and shown in the first column of Table 8.8.

$$u = \frac{\text{subgroup count of occurrences}}{\text{subgroup size in units}}$$

The $u$ chart centerline and control limits are found as follows:

$$\bar{u} = \frac{\text{sum of subgroup count of occurrences}}{\text{sum of subgroup unit sizes}} = \frac{1{,}544}{32} = 48.25$$

which is constant for all subgroups.

$n = 1$ unit:

$$\text{LCL} = \bar{u} - 3\sqrt{\frac{\bar{u}}{n}} = 48.25 - 3\sqrt{\frac{48.25}{1}} = 27$$

$$\text{UCL} = \bar{u} + 3\sqrt{\frac{\bar{u}}{n}} = 48.25 + 3\sqrt{\frac{48.25}{1}} = 69$$

$n = 2$ units:

$$\text{LCL} = 48.25 - 3\sqrt{\frac{48.25}{2}} = 34$$

$$\text{UCL} = 48.25 + 3\sqrt{\frac{48.25}{2}} = 63$$

The control limits and the subgroup occurrences-per-unit are shown in Fig. 8.20.

The reader may note that the data used to construct the $u$ chart were the same as those used for the $p$ chart, except that we considered the counts as being counts of occurrences (bruises) instead of counts of physical items (bruised peaches). The practical implications of using a $u$ chart when a $p$ chart should have been used, or vice versa, are usually not serious. The decisions based on the control charts will be quite similar in most cases encountered in Six Sigma regardless of whether a $u$ or a $p$ chart is used.

## Control Charts for Counts of Occurrences-Per-Unit (c Charts)

The $c$ charts are statistical tools used to evaluate the number of occurrences-per-unit produced by a process. The $c$ charts can be applied to any variable where the appropriate

Occurrences-per-unit control chart
(*u* chart)

69.089
62.986
48.25
33.514
27.411

**Figure 8.20** Completed *u* control chart.

performance measure is a count of how often a particular event occurs and samples of constant size are used. The *c* charts answer the question: "Has a special cause of variation caused the central tendency of this process to produce an abnormally large or small number of occurrences over the time period observed?" Note that, unlike *p* or *np* charts, *c* charts do not involve counting physical items. Rather, they involve counting of *events*. For example, when using an *np* chart one would count bruised peaches. When using a *c* chart, one would count the *bruises*.

### Control Limit Equations for *c* Charts

The centerline of the *c* chart is the average number of occurrences-per-unit and the control limits are set at plus and minus three standard deviations. If the process is in statistical control then the subgroup occurrences-per-unit should be between the control limits and they should fluctuate randomly about the centerline.

$$\bar{c} = \frac{\text{sum of subgroup occurrences}}{\text{number of subgroups}} \tag{8.57}$$

$$\text{LCL} = \bar{c} - 3\sqrt{\bar{c}} \tag{8.58}$$

$$\text{UCL} = \bar{c} + 3\sqrt{\bar{c}} \tag{8.59}$$

One way of helping determine whether or not a particular set of data is suitable for a *c* chart or an *np* chart is to examine the equation used to compute the centerline for the control chart. If the unit of measure is the same in both the numerator and the denominator, then a *p* chart is indicated, otherwise a *c* chart is indicated. For example, if:

$$\text{Center Line} = \frac{\text{bruises}}{\text{number of crates}}$$

then the numerator is in terms of bruises while the denominator is in terms of crates, indicating a *c* chart.

The unit size is arbitrary but, once determined, it cannot be changed without recomputing all subgroup occurrences-per-unit and control limits.

## Analysis of *c* Charts

As with all control charts, a special cause is probably present if there are any points beyond either the upper or the lower control limit. Analysis of *c* chart patterns between the control limits is shown later in this chapter.

## Example of *c* Chart

The data in Table 8.9 were obtained by opening randomly selected crates from each shipment and counting the number of bruises. There are 250 peaches per crate. Our unit size will be taken as one full crate, that is, we will be counting crates rather than the peaches themselves. Every subgroup consists of one crate. If the subgroup size varied, a *u* chart would be used.

**Table 8.9**  Raw Data for *c* Chart

| Shipment No. | Flaws |
|---|---|
| 1 | 27 |
| 2 | 32 |
| 3 | 24 |
| 4 | 31 |
| 5 | 42 |
| 6 | 38 |
| 7 | 33 |
| 8 | 35 |
| 9 | 35 |
| 10 | 39 |
| 11 | 41 |

**Table 8.9** Raw Data for c Chart (*continued*)

| Shipment No. | Flaws |
|---|---|
| 12 | 29 |
| 13 | 34 |
| 14 | 34 |
| 15 | 43 |
| 16 | 29 |
| 17 | 33 |
| 18 | 33 |
| 19 | 38 |
| 20 | 32 |
| 21 | 37 |
| 22 | 30 |
| 23 | 31 |
| 24 | 32 |
| 25 | 42 |
| 26 | 40 |
| 27 | 21 |
| 28 | 23 |
| 29 | 39 |
| 30 | 29 |
| SUM | 1,006 |

Using the above data, the centerline and control limits are found as follows:

$$\bar{c} = \frac{\text{sum of subgroup occurrences}}{\text{number of subgroups}} = \frac{1,006}{30} = 33.53$$

$$\text{LCL} = \bar{c} - 3\sqrt{\bar{c}} = 33.53 - 3\sqrt{33.53} = 16.158$$
$$\text{UCL} = \bar{c} - 3\sqrt{8\bar{c}} = 33.53 + 3\sqrt{33.53} = 50.902$$

The control limits and the occurrence counts are shown in Fig. 8.21.

Figure 8.21  Completed c control chart.

See online video: Effect of Multistream Processes on Attribute Control Charts at www.mhprofessional.com/ssh5.

# Control Chart Selection

Selecting the proper control chart for a particular data set is a simple matter if approached properly. The proper approach is illustrated in Fig. 8.22.

To use the decision tree, begin at the leftmost node and determine if the data are measurements or counts. If measurements, then select the control chart based on the subgroup size. If the data are counts, then determine if the counts are of occurrences or pieces. An aid in making this determination is to examine the equation for the process average. If the numerator and denominator involve the same units, then a $p$ or $np$ chart is indicated. If different units of measure are involved, then a $u$ or $c$ chart is indicated. For example, if the average is in accidents-per-month, then a $c$ or $u$ chart is indicated because the numerator is in terms of accidents but the denominator is in terms of time.

## *Rational Subgroup Sampling*

The basis of all control charts is the *rational subgroup*. Rational subgroups are composed of items that were produced under essentially the same conditions. The statistics, for example, the average and range, are computed for each subgroup separately, then plotted on the control chart. When possible, rational subgroups are formed by using consecutive units. Each subgroup's statistics are compared to the control limits, and patterns of variation between subgroups are analyzed. Note the sharp contrast between this approach and the *random sampling* approach used for enumerative statistical methods.

**Figure 8.22** Control chart selection decision tree.

To understand the concept of rational subgrouping, it's useful to understand the role of the subgroup in the control chart. The subgroup provides a snapshot of the process at that moment in time. On an averages ($\bar{X}$) chart, for example, each subgroup provides an estimate of the process location, while the range (or sigma) chart provides the subgroup estimate of the short-term variation. The control limits on the averages chart are defined based on the average within-subgroup variation (estimated using the average within-subgroup range, for example), and when the short-term within-subgroup variation predicts the longer-term between-subgroup variation, then the process is in statistical control. When the longer-term variation is not predicted by the short-term within-subgroup variation, then a special cause has been identified, or more precisely, operationally defined.

A fundamental aspect of the subgroup is thus to estimate the common cause variation within the process, since the within-subgroup variation is used to define the width of the control limits. Therefore, it is critical that the causes of within-subgroup variation be representative of the causes of variation between subgroups. This is not to say that each and every subgroup must contain an identical subset of causes of process variation. Rather, the system that defines the causes is consistent. Consider the following cases:

- Measure the fill volume of each of the six bottles of beer filled by a multiple-head filling station, grouping the six measurements into a subgroup. Take data for a new subgroup (i.e., a different set of bottles from the same six heads) every hour.

▲ Measure the thickness of an extruded plate. Since the plate thickness can vary across its width, measure at three locations to form the subgroup. Take data for a new subgroup (i.e., at a different location along the length) every 10 minutes.

In each of these examples, the subgroup is not rational, since a different system of causes influences the within-subgroup variation than influences the between-subgroup variation. In the beer filling operation, the variation in fill volume is influenced by head-to-head differences. A consistent difference in nozzle size or line pressure in one head versus the others will cause the within-subgroup variation to be excessively large, masking any longer-term variation in fill volume for any of the individual heads. This produces the characteristic behavior noted in Fig. 8.23: many subgroups hugging the centerline of the averages chart. (The circled values are violations of Run Test 7, discussed in the next section, which is used to detect this error in subgrouping.)

**Figure 8.23** Centerline hugging evidence of irrational subgrouping.

A solution to the centerline hugging is offered with a special version of a $\bar{X}$ chart sometimes referred to as a Batch Means chart. (Minitab refers to this chart as the Between-Within Chart.) The range (or sigma) chart is used to monitor the process variation, as usual, but the averages chart is replaced with an Individual-$X$ chart, whose control limits are based on the moving range between the subgroup averages. The data from Fig. 8.23 is plotted with the Batch Means option as shown in Fig. 8.24.

**Figure 8.24** Batch means analysis of Fig. 8.23 data.

The idea of rational subgrouping becomes a bit fuzzy when dealing with $X$ charts, or individuals control charts. The reader may well wonder about the meaning of the term subgrouping when the "subgroup" is a single measurement. The basic idea underlying control charts of all types is to identify the capability of the process. The mechanism by which this is accomplished is careful formation of rational subgroups as defined above. When possible, rational subgroups are formed by using consecutive units. The measure of process variability, either the subgroup standard deviation or the subgroup range, is the basis of the control limits for averages. Conceptually, this is akin to basing the control limits on short-term variation. These control limits are used to monitor variation over time.

As far as possible, this approach also forms the basis of establishing control limits for individual measurements. This is done by forming quasi-subgroups using pairs of consecutive measurements. These "subgroups of 2" are used to compute ranges, or more appropriately, a moving range between, for example, subgroups 1 and 2, 2 and 3, 3 and 4, and so on. These moving ranges form the basis of the estimate of short-term variation, and are used to compute the control limits for the individual measurements. As with larger subgroup sizes, it is important that the observations used to estimate the short-term variation be independent of one another. When they are not independent, we say the process is *autocorrelated* or has serial correlation. (These terms are equivalent.) Autocorrelation may occur in varied processes (Keller, 2011):

- **Service processes**—When customers are in a service queue, where each customer is serviced in turn, the wait time, or time in queue, of any individual customer is influenced by the wait time of the customer immediately prior. The second customer

cannot be serviced until the first customer's service is complete, so the observations are not independent. They are correlated.
- **Chemical processes**—Samples drawn from liquids in batches are influenced by prior samples if the time between samples is small. The batch is essentially homogenous, and its properties change slowly, so measurements at any point in time are correlated with the earlier samples.
- **Discrete part manufacturing**—Feedback control systems to continually adjust the process based on past measurements create autocorrelation in data collected over small time intervals.

The impact of autocorrelation is a bias in the estimate of the process standard deviation. In the examples above, the relatively small short-term standard deviation estimate would create tighter control limits than otherwise indicated by the process, such as shown in Fig. 8.25.

An autocorrelation chart, shown in Fig. 8.26 for the same data in Fig. 8.25, is useful to understand the length of the autocorrelation effects in units of sample period (or lag). The figure shows significant autocorrelation for lags 1 through 4, so independent samples could be taken from a subset of the data, using every fifth sample. (The PACF, or *partial autocorrelation function*, shown in the bottom of the figure indicates significant first order effects, so it's possible lag 1 effect may be dominant, and overwhelming a less significant impact at lags 2 through 4.)

**Figure 8.25** Control chart with autocorrelation present.

**Figure 8.26** Autocorrelation chart.

See online video: Control Chart Interpretation: Autocorrelation at www.mhprofessional.com/ssh5.

## Control Chart Interpretation

Control charts provide the operational definition of the term *special cause*. A special cause is simply anything that leads to an observation beyond a control limit. However, this simplistic use of control charts does not do justice to their power. Control charts are running records of the performance of the process and, as such, they contain a vast store of information on potential improvements. While some guidelines are presented here, control chart interpretation is an art that can only be developed by looking at many control charts and probing the patterns to identify the underlying system of causes present. Freak patterns are the classical special cause situation (Fig. 8.27). Freaks result from causes that have a large effect but that occur infrequently. When investigating freak values, look at the cause-and-effect diagram for items that meet these criteria. The key to identifying freak causes is timelines in collecting and recording the data. If you have difficulty, try sampling more frequently.

Drift is generally seen in processes where the current process value is partly determined by the previous process state. For example, if the process is a plating bath, the content of the tank cannot change instantaneously, instead it will change gradually (Fig. 8.28). Another common example is tool wear: the size of the tool is related to its previ-

**Figure 8.27** Control chart patterns: freaks.

**Figure 8.28** Control chart patterns: drift.

ous size. Once the cause of the drift has been determined, the appropriate action can be taken. Whenever economically feasible, the drift should be eliminated; for example, install an automatic chemical dispenser for the plating bath, or make automatic compensating adjustments to correct for tool wear. Note that the total process variability increases when drift is allowed, which adds cost. When drift elimination is not possible, the control chart can be modified in one of two ways:

▲ Make the slope of the centerline and control limits match the natural process drift. The control chart will then detect departures from the natural drift.
▲ Plot *deviations* from the natural or expected drift.

Cycles often occur due to the nature of the process. Common cycles include hour of the day, day of the week, month of the year, quarter of the year, week of the accounting cycle, and so on (Fig. 8.29). Cycles are caused by modifying the process inputs or methods according to a regular schedule. The existence of this schedule and its effect on the process may or may not be known in advance. Once the cycle has been discovered, action can be taken. The action might be to adjust the control chart by plotting the control measure against a variable base. For example, if a day-of-the-week cycle exists for shipping errors because of the workload, you might plot shipping errors per 100 orders shipped instead of shipping errors per day. Alternatively, it may be worthwhile to change the system to smooth out the cycle. Most processes operate more efficiently when the inputs are relatively stable and when methods are changed as little as possible.

A controlled process will exhibit only "random looking" variation. A pattern where every *n*th item is different is, obviously, nonrandom (Fig. 8.30). These patterns are sometimes quite subtle and difficult to identify. It is sometimes helpful to see if the average fraction defective is close to some multiple of a known number of process streams. For example, if the machine is a filler with 40 stations, look for problems that occur 1/40, 2/40, 3/40, and so on, of the time.

When plotting measurement data the assumption is that the numbers exist on a continuum, that is, there will be many different values in the data set. In the real world, the data are never completely continuous (Fig. 8.31). It usually doesn't matter much if there

**Figure 8.29** Control chart patterns: cycles.

**Figure 8.30** Control chart patterns: repeating patterns.

**Figure 8.31** Control chart patterns: discrete data.

are, say, 10 or more different numbers. However, when there are only a few numbers that appear over and over it can cause problems with the analysis. A common problem is that the *R* chart will underestimate the average range, causing the control limits on both the average and range charts to be too close together. The result will be too many "false alarms" and a general loss of confidence in SPC.

The usual cause of this situation is inadequate measurement resolution. The ideal solution is to obtain measurements with greater resolution (i.e., additional significant decimals). Sometimes the problem occurs because operators, inspectors, or computers are rounding the numbers. The solution here is to record additional digits.

---

**See online video: Control Chart Interpretation: Data Resolution at www.mhprofessional.com/ssh5.**

The reason SPC is done is to accelerate the learning process and to eventually produce an improvement. Control charts serve as historical records of the learning process, and they can be used by others to improve other processes. When an improvement is realized the change should be written on the old control chart; its effect will show up as a less variable process. These charts are also useful in communicating the results to leaders, suppliers, customers, and others interested in quality improvement (Fig. 8.32).

**Figure 8.32**   Control chart patterns: planned changes.

---

**See online video: Control Chart Interpretation: Historical Data at www.mhprofessional.com/ssh5.**

---

Seemingly random patterns on a control chart are evidence of unknown causes of variation, which is not the same as *uncaused* variation. There should be an ongoing effort to reduce the variation from these so-called common causes. Doing so requires that the unknown causes of variation be identified. One way of doing this is a retrospective evaluation of control charts. This involves brainstorming and preparing cause and effect diagrams, then relating the control chart patterns to the causes listed on the diagram. For example, if "operator" is a suspected cause of variation, place a label on the control chart points produced by each operator (Fig. 8.33). If the labels exhibit a pattern, there is evidence to suggest a problem. Conduct an investigation into the reasons and set up controlled experiments (prospective studies) to test any theories proposed. If the experiments indicate a true cause and effect relationship, make the appropriate process improvements. Keep in mind that a statistical *association* is not the same thing as a causal *correlation*. The observed association must be backed up with solid subject-matter expertise and experimental data.

Mixture exists when the data from two different cause systems are plotted on a single control chart (Fig. 8.34). It indicates a failure in creating rational subgroups. The underly-

**Figure 8.33** Control chart patterns: suspected differences.

**Figure 8.34** Control chart patterns: mixture.

ing differences should be identified and corrective action taken. The nature of the corrective action will determine how the control chart should be modified.

*Mixture Example 1.* The mixture represents two different operators who can be made more consistent. A single control chart can be used to monitor the new, consistent process.

*Mixture Example 2.* The mixture is in the number of emergency room cases received on Saturday evening, versus the number received during a normal week. Separate control charts should be used to monitor patient-load during the two different time periods.

---

See online video: Control Chart Interpretation: Multiple Stream Processes at www.mhprofessional.com/ssh5.

## Run Tests

If the process is stable, then the distribution of subgroup averages will be approximately normal. With this in mind, we can also analyze the *patterns* on the control charts to see if they might be attributed to a special cause of variation. To do this, we divide a normal distribution into zones, with each zone one standard deviation wide. Figure 8.35 shows the approximate percentage we expect to find in each zone from a stable process.

Zone C is the area from the mean to the mean plus or minus one sigma, zone B is from plus or minus one sigma to plus or minus two sigma, and zone A is from plus or minus two sigma to plus or minus three sigma. Of course, any point beyond three sigma (i.e., outside of the control limit) is an indication of an out-of-control process.

Since the control limits are at plus and minus three standard deviations, finding the one and two sigma lines on a control chart is as simple as dividing the distance between the grand average and either control limit into thirds, which can be done using a ruler. This divides each half of the control chart into three zones. The three zones are labeled A, B, and C, as shown in Fig. 8.36.

Based on the expected percentages in each zone, sensitive run tests can be developed for analyzing the patterns of variation in the various zones. Remember, the existence of a nonrandom pattern means that a special cause of variation was (or is) probably present.

| A | B | C | C | B | A |
|---|---|---|---|---|---|
| 2.14% | 13.59% | 34.13% | 34.13% | 13.59% | 2.14% |

**Figure 8.35** Percentiles for a normal distribution.

**Figure 8.36** Zones on a control chart.

The run tests shown in Fig. 8.37 are applicable to the averages and Individual-X control charts. Tests 1 through 4 are also applicable to *np* and *c* charts.

Run tests 1, 2, 5, and 6 indicate a general shift in the process. Test 3 indicates a trend toward a process shift. Test 4 is indicative of mixture, or sampling from multiple process streams, on a every-other-one basis, while test 7 indicates sampling from multiple streams within a subgroup (as described in the *Rational Subgroup Sampling* section above). Run test 8 also shows the presence of multiple process streams, where the groups above the mean are from one process stream and the groups below another stream.

Note that, when a point responds to an out-of-control test it is marked as such to make the interpretation of the chart easier. Using this convention, the patterns on the control charts can be used as an aid in troubleshooting. Analyzing control charts with SPC software greatly simplifies this process.

## Short-Run Statistical Process Control Techniques

Short production runs are a way of life with many manufacturing companies. In the future, this will be the case even more often. The trend in manufacturing has been toward smaller production runs with product tailored to the specific needs of individual customers. This is a fundamental aspect of Lean thinking: A move toward producing only what has been ordered by a given customer. Henry Ford's days of "the customer can have any color, as long as it's black" have long since passed.

Classical SPC methods, such as $\bar{X}$ and *R* charts, were developed in the era of mass production of identical parts. Production runs often lasted for weeks, months, or even

**Figure 8.37** Tests for out-of-control patterns on control charts (from Nelson, 1984, pp. 237–239).

years. Many of the "SPC rules of thumb" currently in use were created for this situation. For example, the rule that control limits are not to be calculated until data are available from at least 25 subgroups of 5. This may not have been a problem in 1930, but it certainly is today. In fact, many entire production runs involve fewer parts than required to start a standard control chart!

## Variables Data

Variables data, sometimes called continuous data, involve measurements such as size, weight, pH, temperature, etc. In theory data are variables data if no two values are exactly the same. In practice this is seldom the case. As a rough rule of thumb you can consider data to be variables data if at least ten different values occur and repeat values make up no more than 20% of the data set. If this is not the case, your data may be too discrete to use standard control charts. Consider trying an attribute procedure such as the demerit charts described later in this chapter. We will discuss the following approaches to SPC for short or small runs:

- ▲ **Exact method**—Tables of special control chart constants are used to create $X$, $\bar{X}$ and $R$ charts that compensate for the fact that a limited number of subgroups are available for computing control limits. The exact method is also used to compute control limits when using a code value chart or stabilized $X$ or $\bar{X}$ and $R$ charts (see below). The exact method allows the calculation of control limits that are correct when only a small amount of data is available. As more data become available the exact method updates control limits until, finally, no further updates are required and standard control chart factors can be used (Pyzdek, 1992a).
- ▲ **Code value charts**—Control charts created by subtracting nominal or other target values from actual measurements. These charts are often standardized so that measurement units are converted to whole numbers. For example, if measurements are in thousandths of an inch a reading of 0.011 inch above nominal would be recorded simply as "11." Code value charts enable the user to plot several parts from a given process on a single chart, or to plot several features from a single part on the same control chart. The exact method can be used to adjust the control limits when code value charts are created with limited data.
- ▲ **Stabilized control charts for variables**—Statisticians have known about normalizing transformations for many years. This approach can be used to create control charts that are independent of the unit of measure and scaled in such a way that several different characteristics can be plotted on the same control chart. Since stabilized control charts are independent of the unit of measure, they can be thought of as true *process control charts*. The exact method adjusts the control limits for stabilized charts created with limited data.

### Exact Method of Computing Control Limits for Short and Small Runs

This procedure, adapted from Hillier (1969) and Proschan and Savage (1960), applies to short runs or any situation where a small number of subgroups will be used to set up a control chart. It consists of three stages:

1. Finding the process (establishing statistical control)
2. Setting limits for the remainder of the initial run
3. Setting limits for future runs

The procedure correctly compensates for the uncertainties involved when computing control limits with small amounts of data.

*Stage One: Find the Process*

1. Collect an initial sample of subgroups (*g*). The factors for the recommended minimum number of subgroups are shown in Appendix 12 enclosed in a dark box. If it is not possible to get the minimum number of subgroups, use the appropriate control chart constant for the number of subgroups you actually have.
2. Using Appendix 12 compute the Range chart control limits using the equation upper control limit for ranges $(UCL_R) = D_{4F} \times \bar{R}$. Compare the subgroup ranges to the $UCL_R$ and drop any out-of-control groups. Repeat the process until all remaining subgroup ranges are smaller than $UCL_R$.
3. Using the $\bar{R}$ value found in step 2, compute the control limits for the averages or individuals chart. The control limits are found by adding and subtracting $A_{2F} \times \bar{R}$ from the overall average. Drop any subgroups that have out-of-control averages and recompute. Continue until all remaining values are within the control limits. Go to stage two.

*Stage Two: Set Limits for Remainder of the Initial Run*

1. Using Appendix 12 compute the control limits for the remainder of the run. Use the $A_{2S}$ factors for the $\bar{X}$ and the $D_{4S}$ factors for the R chart; *g* = the number of groups used to compute stage one control limits.

*Stage Three: Set Limits for a Future Run*

1. After the run is complete, combine the raw data from the entire run and perform the analysis as described in stage one above. Use the results of this analysis to set limits for the next run, following the stage two procedure. If more than 25 groups are available, use a standard table of control chart constants.

*Notes*

1. Stage three assumes that there are no special causes of variation between runs. If there are, the process may go out of control when using the stage three control limits. In these cases, remove the special causes. If this isn't possible, apply this procedure to each run separately (i.e., start over each time).
2. This approach will lead to the use of standard control chart tables when enough data are accumulated.

3. The control chart constants for the first stage are $A_{2F}$ and $D_{4F}$ (the $F$ subscript stands for first stage); for the second stage use $A_{2S}$ and $D_{4S}$. These factors correspond to the $A_2$ and $D_4$ factors usually used, except that they are adjusted for the small number of subgroups actually available.

**Setup Approval Procedure**

The following procedure can be used to determine if a setup is acceptable using a relatively small number of sample units.

1. After the initial setup, run 3 to 10 pieces without adjusting the process.
2. Compute the average and the range of the sample.
3. Compute $T = \left[\dfrac{\text{average} - \text{target}}{\text{range}}\right]$

   Use absolute values (i.e., ignore any minus signs). The target value is usually the specification midpoint or nominal.
4. If $T$ is less than the critical $T$ in Table 8.10, accept the setup. Otherwise, adjust the setup to bring it closer to the target. *Note:* There is approximately 1 chance in 20 that an on-target process will fail this test.

Table 8.10  Critical Value for Setup Acceptance

| n | 3 | 4 | 5 | 6 | 7 | 8 | 9 | 10 |
|---|---|---|---|---|---|---|---|---|
| Critical T | 0.885 | 0.529 | 0.388 | 0.312 | 0.263 | 0.230 | 0.205 | 0.186 |

**Example**

Assume we wish to use SPC for a process that involves producing a part in lots of 30 parts each. The parts are produced approximately once each month. The control feature on the part is the depth of a groove and we will be measuring every piece. We decide to use subgroups of size three and to compute the stage one control limits after the first five groups. The measurements obtained are shown in Table 8.11.

Table 8.11  Raw Data for Example of Exact Method

| Subgroup Number | Sample Number 1 | Sample Number 2 | Sample Number 3 | $\bar{X}$ | R |
|---|---|---|---|---|---|
| 1 | 0.0989 | 0.0986 | 0.1031 | 0.1002 | 0.0045 |
| 2 | 0.0986 | 0.0985 | 0.1059 | 0.1010 | 0.0074 |
| 3 | 0.1012 | 0.1004 | 0.1000 | 0.1005 | 0.0012 |
| 4 | 0.1023 | 0.1027 | 0.1000 | 0.1017 | 0.0027 |
| 5 | 0.0992 | 0.0997 | 0.0988 | 0.0992 | 0.0009 |

Using the data in Table 8.11, we can compute the grand average and average range as

Grand average = 0.10053

Average range ($\bar{R}$) = 0.00334

From Appendix 12 we obtain the first stage constant for the range chart of $D_{4F} = 2.4$ in the row for $g = 5$ groups and a subgroup size of 3. Thus:

$$\text{UCL}_R = D_{4F} \times \bar{R} = 2.4 \times 0.00334 = 0.0080$$

All of the ranges are below this control limit, so we can proceed to the analysis of the averages chart. If any $R$ was above the control limit, we would try to determine why before proceeding.

For the averages chart we get:

$$\text{LCL}_{\bar{X}} = \text{grand average} + A_{2F} \times \bar{R}$$

$$= 0.10053 - 1.20 \times 0.00334 = 0.09652 \text{ (rounded)}$$

$$\text{UCL}_{\bar{X}} = \text{grand average} + A_{2F} \times \bar{R}$$

$$= 0.10053 + 1.20 \times 0.00334 = 0.10454 \text{ (rounded)}$$

All of the subgroup averages are between these limits. Now setting limits for the remainder of the run we use $D_{4S} = 3.4$ and $A_{2S} = 1.47$. This gives, after rounding:

$$\text{UCL}_R = 0.01136$$

$$\text{LCL}_{\bar{X}} = 0.09562$$

$$\text{UCL}_{\bar{X}} = 0.10544$$

If desired, this procedure can be repeated when a larger number of subgroups becomes available, say ten groups. This would provide somewhat better estimates of the control limits, but it involves considerable administrative overhead. When the entire run is finished you will have ten subgroups of three per subgroup. The data from all of these subgroups should be used to compute stage one and stage two control limits. The resulting stage two control limits would then be applied to the *next run* of this part number.

By applying this method in conjunction with the code value charts or stabilized charts described below, the control limits can be applied to the next parts produced on this process (assuming the part-to-part difference can be made negligible). Note that if the standard control chart factors were used, the limits for *both* stages would be (values are rounded).

$$UCL_R = 0.00860$$
$$LCL_{\bar{X}} = 0.09711$$
$$UCL_{\bar{X}} = 0.10395$$

As the number of subgroups available for computing the control limits increases, the "short-run" control limits approach the standard control limits. However, if the standard control limits are used when only small amounts of data are available, there is a greater chance of erroneously rejecting a process that is actually in control (Hillier, 1969).

## Code Value Charts

This procedure allows the control of multiple features with a single control chart. It consists of making a simple transformation to the data, namely:

$$\hat{x} = \frac{X - \text{Target}}{\text{unit of measure}} \quad (8.60)$$

The resulting $\hat{x}$ values are used to compute the control limits and as plotted points on the $\bar{X}$ and $R$ charts. This makes the target dimension irrelevant for the purposes of SPC and makes it possible to use a single control chart for several different features or part numbers.

## Example

A lathe is used to produce several different sizes of gear blanks, as is indicated in Fig. 8.38.

Product engineering wants all of the gear blanks to be produced as near as possible to their nominal size. Process engineering believes that the process will have as little deviation for larger sizes as it does for smaller sizes. Quality engineering believes that the inspection system will produce approximately the same amount of measurement error for larger sizes as for smaller sizes. Process capability studies and measurement error studies confirm these assumptions. (I hope you are starting to get the idea that a number of assumptions are being made and that they must be valid before using code value charts.)

Based on these conclusions, the code value chart is recommended. By using the code value chart the amount of paperwork will be reduced and more data will be available for setting control limits. Also, the process history will be easier to follow since the information won't be fragmented among several different charts. The data in Table 8.12 show some of the early results.

Chapter 8 Process Behavior Charts | 355

Process μ = +0.0001
σ = 0.0001

1 ± 0.0010   A   A: C$_p$ = 3.33, C$_{pk}$ = 3.00

0.5 ± 0.0005   B   B: C$_p$ = 1.67, C$_{pk}$ = 1.33

2 ± 0.0015   C   C: C$_p$ = 5.00, C$_{pk}$ = 4.67

**Figure 8.38**  Some of the gear blanks to be machined.

**Table 8.12**  Deviation from Target in Hundred-Thousandths

| Part | Nominal | No. | Sample Number 1 | 2 | 3 | $\bar{X}$ | R |
|---|---|---|---|---|---|---|---|
| A | 1.0000 | 1 | 4 | 3 | 25 | 10.7 | 22 |
|  |  | 2 | 3 | 3 | 39 | 15.0 | 36 |
|  |  | 3 | 16 | 12 | 10 | 12.7 | 6 |
| B | 0.5000 | 4 | 21 | 24 | 10 | 18.3 | 14 |
|  |  | 5 | 6 | 8 | 4 | 6.0 | 4 |
|  |  | 6 | 19 | 7 | 21 | 15.7 | 14 |
| C | 2.0000 | 7 | 1 | 11 | 4 | 5.3 | 10 |
|  |  | 8 | 1 | 25 | 8 | 11.3 | 24 |
|  |  | 9 | 6 | 8 | 7 | 7.0 | 2 |

Note that the process must be able to produce the *tightest tolerance* of ±0.0005 inch. The capability analysis should indicate its ability to do this; that is, $C_{PK}$ should be at least 1.33 based on the tightest tolerance. It will not be allowed to drift or deteriorate when the less stringently toleranced parts are produced. Process control is independent of the product requirements. Permitting the process to degrade to its worst acceptable level (from the product perspective) creates engineering nightmares when the more tightly toleranced parts come along again. It also confuses and demoralizes operators and others trying to maintain high levels of quality. In fact, it may be best to publish only the process performance requirements and to keep the product requirements secret.

The control chart of the data in Table 8.12 is shown in Fig. 8.39. Since only nine groups were available, the exact method was used to compute the control limits. Note that the control chart shows the *deviations* on the $\bar{X}$ and $R$ chart axes, not the actual measured dimensions, For example, the value of Part A, subgroup 1, sample 1 was +0.00004 inch from the target value of 1.0000 inch and it is shown as a deviation of +4 hundred-thousandths; that is, the part checked 1.00004 inches. The stage one control chart shows that the process is obviously in statistical control, but it is producing parts that are consistently too large regardless of the nominal dimension. If the process were on target, the grand average would be very close to 0. The setup problem would have been detected by the second subgroup if the setup approval procedure described earlier in this chapter had been followed.

**Figure 8.39** Code value chart of Table 8.12 data.

This ability to see process performance across different part numbers is one of the advantages of code value charts. It is good practice to actually identify the changes in part numbers on the charts, as is done in Fig. 8.39.

## Stabilized Control Charts for Variables

All control limits, for standard sized runs or short and small runs, are based on methods that determine if a process statistic falls within limits that might be expected from chance variation (common causes) alone. In most cases, the statistic is based on actual measurements from the process and it is in the same unit of measure as the process measurements. As we saw with code value charts, it is sometimes useful to transform the data in some way. With code value charts we used a simple transformation that removed the effect of changing nominal and target dimensions. While useful, this approach still requires that all measurements be in the same units of measurement, for example, all inches, all grams, etc. For example, all of the variables on the control chart for the different gear blanks had to be in units of hundred-thousandths of an inch. If we had also wanted to plot, for example, the perpendicularity of two surfaces on the gear blank we would have needed a separate control chart because the units would be in degrees instead of inches.

Stabilized control charts for variables overcome the units of measure problem by converting all measurements into standard, nondimensional units. Such "standardizing transformations" are not new; they have been around for many years and they are commonly used in all types of statistical analyses. The two transformations we will be using here are shown in Eqs. (8.61) and (8.62).

$$\frac{(\bar{X} - \text{grand average})}{\bar{R}} \qquad (8.61)$$

$$\frac{R}{\bar{R}} \qquad (8.62)$$

As you can see, Eq. (8.61) subtracts the grand average from each subgroup average (or from each individual measurement if the subgroup size is one) and divides the result by $\bar{R}$. Note that this is not the usual statistical transformation where the denominator is $\sigma$. By using $\bar{R}$ as our denominator instead of $s$, we are sacrificing some desirable statistical properties such as normality and independence to gain simplicity. However, the resulting control charts remain valid and the false alarm risk based on points beyond the control limits is identical to standard control charts. Also, as with all transformations, this approach suffers in that it involves plotting numbers that are not in the usual engineering units people are accustomed to working with. This makes it more difficult to interpret the results and spot data entry errors.

Equation (8.62) divides each subgroup range by the average range. Since the numerator and denominator are both in the same unit of measurement, the unit of measurement cancels and we are left with a number that is in terms of the number of average ranges, $R$'s. It turns out that control limits are also in the same units, that is, to compute standard control limits we simply multiply $\bar{R}$ by the appropriate table constant to determine the width between the control limits.

Hillier (1969) noted that this is equivalent to using the transformations shown in Eqs. (8.61) and (8.62) with control limits set at:

$$-A_2 \leq \frac{(\bar{X} - \text{grand average})}{\bar{R}} \leq A_2 \qquad (8.63)$$

for the individuals or averages chart. Control limits are

$$D_3 \leq \frac{R}{\bar{R}} \leq D_4 \qquad (8.64)$$

for the range chart. Duncan (1974) described a similar transformation for attribute charts, $p$ charts in particular (see below), and called the resulting chart a "stabilized $p$ chart." We will call charts of the transformed variables data stabilized charts as well.

Stabilized charts allow you to plot multiple units of measurement on the same control chart. The procedure described in this chapter for stabilized variables charts requires that all subgroups be of the same size. The procedure for stabilized attribute charts, described later in this chapter allows varying subgroup sizes. When using stabilized charts the control limits are always fixed. The raw data are "transformed" to match the scale determined by the control limits. When only limited amounts of data are available, the constants in Appendix 12 should be used for computing control limits for stabilized variables charts. As more data become available, the Appendix 9 constants approach the constants in standard tables of control chart factors. Table 8.13 summarizes the control limits for stabilized averages, stabilized ranges, and stabilized individuals control charts. The values for $A_2$, $D_3$, and $D_4$ can be found in standard control chart factor tables.

### Example

A circuit board is produced on an electroplating line. Three parameters are considered important for SPC purposes: lead concentration of the solder plating bath, plating thickness, and resistance. Process capability studies have been done using more than 25 groups; thus, based on Table 8.13 the control limits are

$$-A_2 \leq \bar{X} \leq A_2$$

**Table 8.13** Control Limits for Stabilized Charts

| Stage | Available Groups | | $\bar{X}$ | R | X | Appendix Table |
|---|---|---|---|---|---|---|
| One | 25 or less | LCL | $-A_{2F}$ | None | $-A_{2F}$ | 12 |
| | | Average | 0 | 1 | 0 | |
| | | UCL | $+A_{2F}$ | $D_{4F}$ | $+A_{2F}$ | |
| Two | 25 or less | LCL | $-A_{2S}$ | None | $-A_{2S}$ | 12 |
| | | Average | 0 | 1 | 0 | |
| | | UCL | $+A_{2S}$ | $D_{4S}$ | $+A_{2S}$ | |
| One or two | More than 25 | LCL | $-A_2$ | $D_3$ | −2.66 | 9 |
| | | Average | 0 | 1 | 0 | |
| | | UCL | $+A_2$ | $D_4$ | +2.66 | |

for the averages control chart, and:

$$-D_3 \leq R \leq D_4$$

for the ranges control chart. The actual values of the constants $A_2$, $D_3$, and $D_4$ depend on the subgroup size; for subgroups of three $A_2 = 1.023$, $D_3 = 0$ and $D_4 = 2.574$.

The capabilities are shown in Table 8.14.

**Table 8.14** Process Capabilities for Example

| Feature Code | Feature | Grand Avg. | Avg. Range |
|---|---|---|---|
| A | Lead % | 10% | 1% |
| B | Plating thickness | 0.005 in | 0.0005 in |
| C | Resistance | 0.1 Ω | 0.0005 Ω |

A sample of three will be taken for each feature. The three lead concentration samples are taken at three different locations in the tank. The results of one such set of sample measurements are shown in Table 8.15, along with their stabilized values.

Table 8.15 Sample Data for Example

| Number | Lead % (A) | Thickness (B) | Resistance (C) |
|---|---|---|---|
| 1 | 11 | 0.0050 in | 0.1000 Ω |
| 2 | 11 | 0.0055 in | 0.1010 Ω |
| 3 | 8 | 0.0060 in | 0.1020 Ω |
| $\bar{X}$ | 10 | 0.0055 in | 0.1010 Ω |
| R | 3 | 0.0010 in | 0.0020 Ω |
| $(\bar{X} - \bar{\bar{X}})/\bar{R}$ | 0 | 1 | 2 |
| $R/\bar{R}$ | 3 | 2 | 4 |

On the control chart *only the extreme values are plotted*. Figure 8.40 shows a stabilized control chart for several subgroups. Observe that the feature responsible for the plotted point is written on the control chart. If a long series of largest or smallest values comes from the same feature it is an indication that the feature has changed. If the process is in statistical control for all features, the feature responsible for the extreme values will vary randomly.

Figure 8.40 Stabilized control chart for variables.

When using stabilized charts it is possible to have a single control chart accompany a particular part or lot of parts through the entire production sequence. For example, the circuit boards described above could have a control chart that shows the results of process and product measurement for characteristics at all stages of production. The chart would then show the "processing history" for the part or lot. The advantage would be a coherent log of the production of a given part. Table 8.16 illustrates a process control plan that could possibly use this approach.

**Table 8.16** PWB Fab Process Capabilities and SPC Plan

| Operation | Feature | $\bar{X}$ | $\bar{R}$ | n |
|---|---|---|---|---|
| Clean | Bath pH | 7.5 | 0.1 | 3/h |
| | Rinse contamination | 100 ppm | 5 ppm | 3/h |
| | Cleanliness quality rating | 78 | 4 | 3 pcs/h |
| Laminate | Riston thickness | 1.5 min | 0.1 mm | 3 pcs/h |
| | Adhesion | 7 in–lb | 0.2 in–lb | 3 pcs/h |
| Plating | Bath lead % | 10% | 1% | 3/h |
| | Thickness | 0.005 in | 0.0005 in | 3 pcs/h |
| | Resistance | 0.1 Ω | 0.0005 Ω | 3 pcs/h |

A caution is in order if the processing history approach is used. When small and short runs are common, the history of a given process can be lost among the charts of many different parts. This can be avoided by keeping a separate chart for each distinct process; additional paperwork is involved, but it might be worth the effort. If the additional paperwork burden becomes large, computerized solutions may be worth investigating.

## *Attribute SPC for Small and Short Runs*

When data are difficult to obtain, as is usual when small or short runs are involved, variables SPC should be used if at all possible. A variables measurement on a continuous scale contains more information than a discrete attributes classification provides. For example, a machine is cutting a piece of metal tubing to length. The specifications call for the length to be between 0.990 and 1.010 inches with the preferred length being 1.000 inch exactly. There are two methods available for checking the process. Method 1 involves measuring the length of the tube with a micrometer and recording the result to the nearest 0.001 inch. Method 2 involves placing the finished part into a "go/no-go gage." With method 2 a part that is shorter than 0.990 inch will go into the "no-go" portion of the gage, while a part that is longer than 1.010 inches will fail to go into the "go" portion of the gage. With method 1 we can determine the size of the part to within 0.001 inch. With method 2 we can only determine the size of the part to within 0.020 inch; that is, either it is within the size tolerance, it's too short, or it's too long. If the process could hold a tolerance of *less than*

0.020 inch, method 1 would provide the necessary information to hold the process to the variability it is capable of holding. Method 2 would not detect a process drift until out of tolerance parts were actually produced.

Another way of looking at the two different methods is to consider each part as belonging to a distinct category, determined by the part's length. Method 1 allows any part that is within tolerance to be placed into one of twenty categories. When out of tolerance parts are considered, method 1 is able to place parts into even more than twenty different categories. Method 1 also tells us if the part is in the best category, namely within ±0.001 inch of 1.000 inch; if not, we know how far the part is from the best category. With method 2 we can place a given part into only three categories: too short, within tolerance, or too long. A part that is far too short will be placed in the same category as a part that is only slightly short. A part that is barely within tolerance will be placed in the same category as a part that is exactly 1.000 inch long.

In spite of the disadvantages, it is sometimes necessary to use attributes data. Special methods must be used for attributes data used to control short run processes. We will describe two such methods:

- Stabilized attribute control charts
- Demerit control charts

## Stabilized Attribute Control Charts

When plotting attribute data statistics from short run processes, two difficulties are typically encountered:

- Varying subgroup sizes
- A small number of subgroups per production run

Item 1 results in messy charts with different control limits for each subgroup, distorted chart scales that mask significant variations, and chart patterns that are difficult to interpret because they are affected by both sample size changes and true process changes. Item 2 makes it difficult to track long-term process trends because the trends are broken up among many different control charts for individual parts. Because of these things, many people believe that SPC is not practical unless large and long runs are involved. This is not the case. In many cases stabilized attribute charts can be used to eliminate these problems. Although somewhat more complicated than classical control charts, stabilized attribute control charts offer a way of realizing the benefits of SPC with processes that are difficult to control any other way.

Stabilized attribute charts may be used if a process is producing part features that are essentially the same from one part number to the next. Production lot sizes and sample sizes can vary without visibly affecting the chart.

*Example One.* A lathe is being used to machine terminals of different sizes. Samples (of different sizes) are taken periodically and inspected for burrs, nicks, tool marks, and other visual defects.

*Example Two.* A printed circuit board hand assembly operation involves placing electrical components into a large number of different circuit boards. Although the boards differ markedly from one another, the hand assembly operation is similar for all of the different boards.

*Example Three.* A job-shop welding operation produces small quantities of "one order only" items. However, the operation always involves joining parts of similar material and similar size. The process control statistic is weld imperfections per 100 inches of weld.

The techniques used to create stabilized attribute control charts are all based on corresponding classical attribute control chart methods. There are four basic types of control charts involved:

- Stabilized $p$ charts for proportion of defective units per sample
- Stabilized $np$ charts for the number of defective units per sample
- Stabilized $c$ charts for the number of defects per unit
- Stabilized $u$ charts for the average number of defects per unit

All of these charts are based on the transformation

$$Z = \frac{\text{sample statistic} - \text{process average}}{\text{process standard deviation}} \tag{8.65}$$

In other words, stabilized charts are plots of the number of standard deviations (plus or minus) between the sample statistic and the long-term process average. Since control limits are conventionally set at ±3 standard deviations, stabilized control charts always have the lower control limit at −3 and the upper control limit at +3. Table 8.17 summarizes the control limit equations for stabilized control charts for attributes.

**Table 8.17** Stabilized Attribute Chart Statistics

| Attribute | Chart | Sample Statistic | Process Average | Process $\sigma$ | Z |
|---|---|---|---|---|---|
| Proportion of defective units | $p$ chart | $p$ | $\bar{p}$ | $\sqrt{\bar{p}(1-\bar{p})/n}$ | $(p-\bar{p})/\sigma$ |
| Number of defective units | $np$ chart | $np$ | $\overline{np}$ | $\sqrt{n\bar{p}(1-\bar{p})}$ | $(np-n\bar{p})/\sigma$ |
| Defects per unit | $c$ chart | $c$ | $\bar{c}$ | $\sqrt{\bar{c}}$ | $(c-\bar{c})/\sigma$ |
| Average defects per unit | $u$ chart | $u$ | $\bar{u}$ | $\sqrt{\bar{u}/n}$ | $(u-\bar{u})/\sigma$ |

When applied to long runs, stabilized attribute charts are used to compensate for varying sample sizes; process averages are assumed to be constant. However, stabilized attribute charts can be created even if the process average varies. This is often done when applying this technique to short runs of parts that vary a great deal in average quality. For example, a wave soldering process used for several missiles had boards that varied in complexity from less than 100 solder joints to over 1,500 solder joints. Tables 8.18 and 8.19 show how the situation is handled to create a stabilized $u$ chart. The unit size is 1,000 leads, set arbitrarily. It doesn't matter what the unit size is set to, the calculations will still produce the correct result since the actual number of leads is divided by the unit size selected. $\bar{u}$ is the average number of defects per 1,000 leads.

***Example Four.*** From the process described in Table 8.18, ten TOW missile boards of type E are sampled. Three defects were observed in the sample. Using Tables 8.17 and 8.18, $Z$ is computed for the subgroup as follows:

**Table 8.18** Data from a Wave Solder Process

| Missile | Board | Leads | Units/Board | $\bar{u}$ |
|---|---|---|---|---|
| Phoenix | A | 1,650 | 1.65 | 16 |
|  | B | 800 | 0.80 | 9 |
|  | C | 1,200 | 1.20 | 9 |
| TOW | D | 80 | 0.08 | 4 |
|  | E | 50 | 0.05 | 2 |
|  | F | 100 | 0.10 | 1 |

$\sigma = \sqrt{\bar{u}/n}$, we get $\bar{u} = 2$ from Table 8.18.

$$n = \frac{50 \times 10}{1{,}000} = 0.5 \text{ units}$$

$$\sigma = \sqrt{2/0.5} = \sqrt{4} = 2$$

$$u = \frac{\text{number of defects}}{\text{number of units}} = \frac{3}{0.5} = 6 \text{ defects per unit}$$

$$Z = \frac{u - \bar{u}}{\sigma} = \frac{6-2}{2} = \frac{4}{2} = 2$$

Since $Z$ is between −3 and +3, we conclude that the process has not gone out of control; that is, it is not being influenced by a special cause of variation.

Table 8.19 shows the data for several samples from this process. The resulting control chart is shown in Fig. 8.41. Note that the control chart indicates that the process was better than average when it produced subgroups 2 and 3 and perhaps 4. Negative $Z$ values mean that the defect rate is below (better than) the long-term process average. Groups 7 and 8 show an apparent deterioration in the process with both groups out of control. Positive $Z$ values indicate a defect rate above (worse than) the long-term process average.

**Table 8.19** Stabilized $u$ Chart Data for Wave Solder

| No. | Board | $\bar{u}$ | Units | Sampled | n | σ | Defects | u | Z |
|---|---|---|---|---|---|---|---|---|---|
| 1 | E | 2 | 0.05 | 10 | 0.50 | 2.00 | 3 | 6.00 | 2.00 |
| 2 | A | 16 | 1.65 | 1 | 1.65 | 3.11 | 8 | 4.85 | −3.58 |
| 3 | A | 16 | 1.65 | 1 | 1.65 | 3.11 | 11 | 6.67 | −3.00 |
| 4 | B | 9 | 0.80 | 1 | 0.80 | 3.35 | 0 | 0.00 | −2.68 |
| 5 | F | 1 | 0.10 | 2 | 0.20 | 2.24 | 1 | 5.00 | 1.79 |
| 6 | E | 2 | 0.05 | 5 | 0.25 | 2.83 | 2 | 8.00 | 2.12 |
| 7 | C | 9 | 1.20 | 1 | 1.20 | 2.74 | 25 | 20.83 | 4.32 |
| 8 | D | 4 | 0.08 | 5 | 0.40 | 3.16 | 6 | 15.00 | 3.48 |
| 9 | B | 9 | 0.80 | 1 | 0.80 | 3.35 | 7 | 8.75 | −0.07 |
| 10 | B | 9 | 0.80 | 1 | 0.80 | 3.35 | 7 | 8.75 | −0.07 |

**Figure 8.41** Control chart of Z values from Table 8.19.

The ability to easily see process trends and changes like these in spite of changing part numbers and sample sizes is the big advantage of stabilized control charts. The disadvantages of stabilized control charts are

- They convert a number that is easy to understand, the number of defects or defectives, into a confusing statistic with no intuitive meaning.
- They involve tedious calculation.

Item 1 can only be corrected by training and experience applying the technique. Item 2 can be handled with computers; the calculations are simple to perform with a spreadsheet. Table 8.19 can be used as a guide to setting up the spreadsheet. Inexpensive handheld computers can be used to perform the calculations right at the process, thus making the results available immediately.

## Demerit Control Charts

As described above, there are two kinds of data commonly used to perform SPC: variables data and attributes data. When short runs are involved, we can seldom afford the information loss that results from using attribute data. However, the following are ways of extracting additional information from attribute data:

- Making the attribute data "less discrete" by adding more classification categories.
- Assigning weights to the categories to accentuate different levels of quality.

Consider a process that involves fabricating a substrate for a hybrid microcircuit. The surface characteristics of the substrate are extremely important. The "ideal part" will have a smooth surface, completely free of any visible flaws or blemishes. However, parts are sometimes produced with stains, pits, voids, cracks, and other surface defects. Although undesirable, most of the less than ideal parts are still acceptable to the customer.

If we were to apply conventional attribute SPC methods to this process the results would probably be disappointing. Since very few parts are actually rejected as unacceptable, a standard $p$ chart or stabilized $p$ chart would probably show a flat line at "zero defects" most of the time, even though the quality level might be less than the target ideal part. Variables SPC methods can't be used because attributes data such as "stains" are not easily measured on a variables scale. Demerit control charts offer an effective method of applying SPC in this situation.

To use demerit control charts we must determine how many imperfections of each type are found in the parts. Weights are assigned to the different categories. The quality score for a given sample is the sum of the weights times the frequencies of each category. Table 8.20 illustrates this approach for the substrate example.

**Table 8.20** Demerit Scores for Substrates

| Subgroup Number → | | 1 | | 2 | | 3 | |
|---|---|---|---|---|---|---|---|
| Attribute | Weight | Freq. | Score | Freq. | Score | Freq. | Score |
| Light stain | 1 | 3 | 3 | | | | |
| Dark stain | 5 | | | 1 | 5 | 1 | 5 |
| Small blister | 1 | | | 2 | 2 | 1 | 1 |
| Medium blister | 5 | 1 | 5 | | | | |
| Pit: 0.01–0.05 mm | 1 | | | | | 3 | 3 |
| Pit: 0.06–0.10 mm | 5 | | | 2 | 10 | | |
| Pit: larger than 0.10 mm | 10 | 1 | 10 | | | | |
| Total demerits → | | | 18 | | 17 | | 9 |

If the subgroup size is kept constant, the average for the demerit control chart is computed as follows (Burr, 1976):

$$\text{Average} = \bar{D} = \frac{\text{sum of subgroup demerits}}{\text{number of subgroups}} \tag{8.66}$$

Control limits are computed in two steps. First compute the weighted average defect rate for each category. For example, there might be the following categories and weights:

| Category | Weight |
|---|---|
| Major | 10 |
| Minor | 5 |
| Incidental | 1 |

Three average defect rates, one each for major, minor, and incidental, could be computed using the following designations:

$\bar{c}_1$ = average number of major defects per subgroup

$\bar{c}_2$ = average number of minor defects per subgroup

$\bar{c}_3$ = average number of incidental defects per subgroup

The corresponding weights might be $W_1 = 10$, $W_2 = 5$, $W_3 = 1$. Using this notation, we compute the demerit standard deviation for this three category example as:

$$\sigma_D = \sqrt{W_1^2 \bar{c}_1 + W_2^2 \bar{c}_2 + W_3^2 \bar{c}_3} \tag{8.67}$$

For the general case the standard deviation is:

$$\sigma_D = \sqrt{\sum_{i=1}^{k} W_i^2 \bar{c}_i} \qquad (8.68)$$

The control limits are:

$$\text{LCL} = \bar{D} - 3\sigma_D \qquad (8.69)$$

$$\text{UCL} = \bar{D} + 3\sigma_D \qquad (8.70)$$

If the lower control limit is negative, it is set to zero.

### Simplified Quality Score Charts

The above procedure, while correct, may sometimes be too burdensome to implement effectively. When this is the case, a simplified approach may be used. The simplified approach is summarized as follows:

1. Classify each part in the subgroup into the following classes (weights are arbitrary).

| Class | Description | Points |
|---|---|---|
| A | Preferred quality. All product features at or very near targets. | 10 |
| B | Acceptable quality. Some product features have departed significantly from target quality levels, but they are a safe distance from the reject limits. | 5 |
| C | Marginal quality. One or more product features are in imminent danger of exceeding reject limits. | 1 |
| D | Reject quality. One or more product features fail to meet minimum acceptability requirements. | 0 |

2. Plot the total scores for each subgroup, keeping the subgroup sizes constant.
3. Treat the total scores as if they were variables data and prepare an individuals and moving range control chart or an $\bar{X}$ and $R$ chart. These charts are described in Pyzdek (1989) and in most texts on SPC.

## *Summary of Short-Run SPC*

Small runs and short runs are common in modern business environments. Different strategies are needed to deal with these situations. Advance planning is essential. Special variables techniques were introduced that compensate for small sample sizes and short runs by using special tables or mathematically transforming the statistics and charts. Attribute short run SPC methods were introduced that make process patterns more evident when small runs are produced. Demerit and scoring systems were introduced that extract more information from attribute data.

## SPC Techniques for Automated Manufacturing

Many people erroneously believe that statistics are not needed when automated manufacturing processes are involved. Since we have measurements from every unit produced, they reason, sampling methods are inappropriate. We will simply correct the process when the characteristic is not on target.

This attitude reflects a fundamental misunderstanding of the relationship between a process and the output of a process. It also shows a lack of appreciation for the intrinsic variability of processes and of measurements. The fact is, even if you have a "complete" data record of every feature of every part produced, you still have only a sample of the output of the process. The process is future-oriented in time, while the record of measurements is past-oriented. Unless statistical control is attained, you will be unable to use the data from past production to predict the variability from the process in the future. (See "Common and Special Causes of Variation" in Chap. 7 for the definition of control.) And without statistical tools you have no sound basis for the belief that statistical control exists.

Another reason process control should be based on an understanding and correct use of statistical methods is the effect of making changes without this understanding. Consider, for example, the following process adjustment rule:

> Measure the diameter of the gear shaft. If the diameter is above the nominal size, adjust the process to reduce the diameter. If the diameter is below the nominal size, adjust the process to increase the diameter.

The problem with this approach is described by Deming's "funnel rules" (see the "Tampering Effects and Diagnosis" section in Chap. 7). This approach to process control will increase the variability of a statistically controlled process by 141%, certainly not what the process control analyst had in mind. The root of the problem is a failure to realize that the part measurement is a sample from the process and, although it provides information about the state of the process, the information is incomplete. Only through using proper statistical methods can the information be extracted, analyzed and understood.

### *Problems with Traditional SPC Techniques*

A fundamental assumption underlying traditional SPC techniques is that the observed values are independent of one another. Although the SPC tools are quite insensitive to moderate violations of this assumption (Wheeler, 1991), automated manufacturing processes often breach the assumption by enough to make traditional methods fail (Alwan and Roberts, 1989). By using scatter diagrams, as described in Chap. 10, you can determine if the assumption of independence is satisfied for your data. If not, you should consider using the methods described below instead of the traditional SPC methods.

A common complaint about nonstandard SPC methods is that they are usually more complex than the traditional methods (Wheeler, 1991). This is often true. However, when dealing with automated manufacturing processes the analysis is usually handled by a computer. Since the complexity of the analysis is totally invisible to the human operator, it makes little difference. Of course, if the operator will be required to act based on the results, he or she must understand how the results are to be used. The techniques described in this chapter which require human action are interpreted in much the same way as traditional SPC techniques.

## *Special and Common Cause Charts*

When using traditional SPC techniques the rules are always the same, namely

- As long as the variation in the statistic being plotted remains within the control limits, leave the process alone.
- If a plotted point exceeds a control limit, look for the cause.

This approach works fine as long as the process remains static. However, the means of many automated manufacturing processes often drift because of inherent process factors. In other words, the drift is produced by *common causes*. In spite of this, there may be known ways of intervening in the process to compensate for the drift. Traditionalists would say that the intervention should be taken in such a way that the control chart exhibits only random variation. However, this may involve additional cost. Mindlessly applying arbitrary rules to achieve some abstract result, like a stable control chart, is poor practice. All of the options should be considered.

One alternative is to allow the drift to continue until the cost of intervention equals the cost of running off-target. This alternative can be implemented through the use of a "common cause chart." This approach, described in Alwan and Roberts (1989) and Abraham and Whitney (1990), involves creating a chart of the process mean. However, unlike traditional $\bar{X}$ charts, there are no control limits. Instead, *action limits* are placed on the chart. Action limits differ from control limits in that they are computed based on costs rather than on statistical theory.

Since the chart shows variation from common causes, violating an action limit does not result in a search for a special cause. Instead, a prescribed action is taken to bring the process closer to the target value.

These charts are called "common cause charts" because the changing level of the process is due to built-in process characteristics. The process mean is tracked by using exponentially weighted moving averages (EWMA). While somewhat more complicated than $\bar{X}$ traditional charts, EWMA charts have a number of advantages for automated manufacturing:

- They can be used when processes have inherent drift.
- EWMA charts provide a forecast of where the next process measurement will be. This allows feed-forward control.
- EWMA models can be used to develop procedures for dynamic process control, as described later in this section.

## *EWMA Common Cause Charts*

When dealing with a process that is essentially static, the predicted value of the average of every sample is simply the grand average. EWMA charts, on the other hand, use the actual process data to determine the predicted process value for processes that may be drifting. If the process has trend or cyclical components, the EWMA will reflect the effect of these components. Also, the EWMA chart produces a forecast of what the *next* sample mean will be; the traditional $\bar{X}$ chart merely shows what the process was doing at the time the sample was taken. Thus, the EWMA chart can be used to take preemptive action to prevent a process from going too far from the target.

If the process has inherent nonrandom components, an EWMA common cause chart should be used. This is an EWMA chart with economic action limits instead of control limits. EWMA control charts, which are described in the next section, can be used to monitor processes that vary within the action limits.

The equation for computing the EWMA is

$$\text{EWMA} = \hat{y}_t + \lambda(y_t - \hat{y}_t) \tag{8.71}$$

In this equation $\hat{y}_t$ is the predicted value of $y$ at time period $t$, $y_t$ is the actual value at time period $t$, and $\lambda$ is a constant between 0 and 1. If $\lambda$ is close to 1, Eq. (8.71) will give little weight to historic data; if $\lambda$ is close to 0 then current observations will be given little weight. EWMA can also be thought of as the forecasted process value at time period $t + 1$, in other words, EWMA = $\hat{y}_{t+1}$.

Since most people already understand the traditional $\bar{X}$ chart, thinking about the relationship between $\bar{X}$ charts and EWMA charts can help you understand the EWMA chart. It is interesting to note that traditional $\bar{X}$ charts give 100% of the weight to the current sample and 0% to past data. This is roughly equivalent to setting $\lambda = 1$ on an EWMA chart. In other words, the traditional $\bar{X}$ chart can be thought of as a special type of EWMA chart where past data are considered to be unimportant (assuming run tests are not applied to the Shewhart chart). This is equivalent to saying that the data points are all independent of one another. In contrast, the EWMA chart uses the information from all previous samples. Although Eq. (8.71) may look as though it is only using the results of the most recent data point, in reality the EWMA weighting scheme applies progressively less weight to each sample result as time passes. Figure 8.42 compares the weighting schemes of EWMA and $\bar{X}$ charts.

In contrast, as $\lambda$ approaches 0 the EWMA chart begins to behave like a cusum chart. With a cusum chart all previous points are given equal weight. Between the two extremes the EWMA chart weights historical data in importance somewhere between the traditional Shewhart chart and the cusum chart. By changing the value of $\lambda$ the chart's behavior can be "adjusted" to the process being monitored.

In addition to the weighting, there are other differences between the EWMA chart and the $\bar{X}$ chart. The "forecast" from the $\bar{X}$ chart is always the same: the next data point will be equal to the historical grand average. In other words, the $\bar{X}$ chart treats all data points as coming from a process that doesn't change its central tendency (implied when the forecast is always the grand average).[2]

**Figure 8.42** $\bar{X}$ versus EWMA weighting.

When using an $\bar{X}$ chart it is not essential that the sampling interval be kept constant. After all, the process is supposed to behave as if it were static. However, the EWMA chart is designed to account for process drift and, therefore, the sampling interval should be kept constant when using EWMA charts. This is usually not a problem with automated manufacturing.

**Example**

Krisnamoorthi (1991) describes a mold line that produces green sand molds at the rate of about one per minute. The molds are used to pour cylinder blocks for large size engines. Application of SPC to the process revealed that the process had an assignable cause that could not be eliminated from the process. The mold sand, which was partly recycled, tended to increase and decrease in temperature based on the size of the block being produced and the number of blocks in the order. Sand temperature is important because it affects the compactability percent, an important parameter. The sand temperature could not be better controlled without adding an automatic sand cooler, which was not deemed economical. However, the effect of the sand temperature on the compactability percent could be made negligible by modifying the amount of water added to the sand so feed-forward control was feasible.

Although Krishnamoorthi doesn't indicate that EWMA charts were used for this process, it is an excellent application for EWMA common cause charts. The level of the sand temperature doesn't really matter, as long as it is known. The sand temperature tends to drift in cycles because the amount of heated sand depends on the size of the casting and how many are being produced. A traditional control chart for the temperature would indicate that sand temperature is out-of-control, which we already know. What is really needed is a method to predict what the sand temperature will be the next time it is checked, then the operator can add the correct amount of water so the effect on the sand compactability percent can be minimized. This will produce an in-control control chart for compactability percent, which is what really matters.

The data in Table 8.21 show the EWMA calculations for the sand temperature data. Using a spreadsheet program, Microsoft Excel for Windows, the optimal value of $\lambda$, which is the value that provided the "best fit" in the sense that it produced the smallest sum of the squared errors, was found to be close to 0.9. Figure 8.43 shows the EWMA common cause chart for this data, and the raw temperature data as well. The EWMA is a forecast of what the sand temperature will be the next time it is checked. The operator can adjust the rate of water addition based on this forecast.

**Table 8.21** Data for EWMA Chart of Sand Temperature

| Sand Temperature | EWMA | Error |
|---|---|---|
| 125 | 125.00* | 0.00 |
| 123 | 125.00 | −2.00† |
| 118 | 123.20‡ | −5.20s |
| 116 | 118.52 | −2.52 |
| 108 | 116.25 | −8.25 |
| 112 | 108.83 | 3.17 |
| 101 | 111.68 | −10.68 |
| 100 | 102.07 | −2.07 |
| 98 | 100.21 | −2.21 |
| 102 | 98.22 | 3.78 |
| 111 | 101.62 | 9.38 |
| 107 | 110.6 | −3.06 |
| 112 | 107.31 | 4.69 |
| 112 | 111.53 | 0.47 |
| 122 | 111.95 | 10.05 |
| 140 | 121.00 | 19.00 |
| 125 | 138.10 | −13.10 |
| 130 | 126.31 | 3.69 |
| 136 | 129.63 | 6.37 |
| 130 | 135.36 | −5.36 |
| 112 | 130.54 | −18.54 |
| 115 | 113.85 | 1.15 |
| 100 | 114.89 | −14.89 |
| 113 | 101.49 | 11.51 |
| 111 | 111.85 | −0.85 |
| 128 | 111.08 | 16.92 |
| 122 | 126.31 | −4.31 |
| 142 | 122.43 | 19.57 |
| 134 | 140.64 | −6.04 |
| 130 | 134.60 | −4.60 |
| 131 | 130.46 | 0.54 |
| 104 | 130.95 | −26.95 |
| 84 | 106.69 | −22.69 |
| 86 | 86.27 | −0.27 |
| 99 | 86.03 | 12.97 |
| 90 | 97.70 | −7.70 |
| 91 | 90.77 | 0.23 |
| 90 | 90.98 | −0.98 |
| 101 | 90.10 | 10.90 |

*The starting EWMA is either the target, or, if there is no target, the first observation.
†Error = Actual observation − EWMA. E.g., −2 = 123 − 125.
‡Other than the first sample, all EWMAs are computed as EWMA = last EWMA + λ × error. E.g., 123.2 = 125 + 0.9 × (−2).

**Figure 8.43** EWMA chart of sand temperature.

### EWMA Control Limits

Although it is not always necessary to put control limits on the EWMA chart, as shown by the above example, it is possible to do so when the situation calls for it. Three sigma control limits for the EWMA chart are computed based on:

$$\sigma^2_{EWMA} = \sigma^2 \left[ \frac{\lambda}{(2-\lambda)} \right] \tag{8.72}$$

For the sand temperature example above, $\lambda = 0.9$, which gives:

$$\sigma^2_{EWMA} = \sigma^2 \left[ \frac{0.9}{(2-0.9)} \right] = 0.82\,\sigma^2$$

$\sigma^2$ is estimated using all of the data. For the sand temperature data $\sigma = 15.37$ so $\sigma_{EWMA} = 15.37 \times \sqrt{0.82} = 13.92$. The $3\sigma$ control limits for the EWMA chart are placed at the grand average plus and minus 41.75. Figure 8.44 shows the control chart for these data. The EWMA line must remain within the control limits. Since the EWMA accounts for "normal drift" in the process centerline, deviations beyond the control limits imply assignable causes other than those accounted for by normal drift. Again, since the effects of changes in temperature can be ameliorated by adjusting the rate of water input, the EWMA control chart may not be necessary.

**Figure 8.44** EWMA control chart of sand temperature.

### Choosing the Value of λ

The choice of λ is the subject of much literature. A value λ of near 0 provides more "smoothing" by giving greater weight to historic data, while a λ value near 1 gives greater weight to current data. Most authors recommend a value in the range of 0.2 to 0.3. The justification for this range of λ values is probably based on applications of the EWMA technique in the field of economics, where EWMA methods are in widespread use. Industrial applications are less common, although the use of EWMA techniques is growing rapidly.

Hunter (1989) proposes a EWMA control chart scheme where λ = 0.4. This value of λ provides a control chart with approximately the same statistical properties as a traditional $\bar{X}$ chart combined with the run tests described in the *AT&T Statistical Quality Control Handbook* (commonly called the Western Electric Rules). It also has the advantage of providing control limits that are exactly half as wide as the control limits on a traditional $\bar{X}$ chart. Thus, to compute the control limits for an EWMA chart when λ is 0.4, you simply compute the traditional $\bar{X}$ chart (or X chart) control limits and divide the distance between the upper and lower control limits by two. The EWMA should remain within these limits.

As mentioned above, the optimal value of λ can be found using some spreadsheet programs. The sum of the squared errors is minimized by changing the value of λ. If your spreadsheet doesn't automatically find the minimum, it can be approximated manually by changing the cell containing λ or by setting up a range of λ values and watching what happens to the cell containing the sum of the squared errors. A graph of the error sum of the squares versus different λ values can indicate where the optimum λ lies.

## Minitab EWMA Example

Minitab has a built-in EWMA analysis capability. We will repeat our analysis for the sand temperature data. Choose Stat > Control Charts > EWMA and you will see a dialog box similar to the one shown in Fig. 8.45. Entering the weight of 0.9 and a subgroup size of 1, then clicking OK, produces the chart in Fig. 8.46.

**Figure 8.45** Minitab EWMA dialog box.

**Figure 8.46** Minitab EWMA chart.

You may notice that the control limits calculated with Minitab are different than those calculated in the previous example. The reason is that Minitab's estimate of sigma is based on the average moving range. This method gives a sigma value of 7.185517, substantially less than the estimate of 15.37 obtained by simply calculating sigma combining all of the data. Minitab's approach removes the effect of the process drift. Whether or not this effect should be removed from the estimate of sigma is an interesting question. In most situations we probably want to remove it so our control chart will be more sensitive, allowing us to detect more special causes for removal. However, as this example illustrates, the situation isn't always clear cut. In the situation described by the example, we might actually want to include the variation from drift into the control limit calculations to prevent operator tampering.

## *EWMA Control Charts Versus Individuals Charts*

In many cases an individuals control chart (*I* chart) will give results comparable to the EWMA control chart. When this is the case it is usually best to opt for the simpler *I* chart. An *I* chart is shown in Fig. 8.47 for comparison with the EWMA chart. The results are very similar to the EWMA chart from Minitab.

**Figure 8.47** *I* chart for sand temperature.

### Special Cause Charts

Whether using a EWMA common cause chart without control limits or an EWMA control chart, it is a good idea to keep track of the forecast errors using a control chart. The special cause chart is a traditional $X$ chart, created using the difference between the EWMA forecast and the actual observed values. Figure 8.48 shows the special cause chart of the sand temperature data analyzed above. The chart indicates good statistical control.

**Figure 8.48** Special cause control chart of EWMA errors.

## SPC and Automatic Process Control

As SPC has grown in popularity its use has been mandated with more and more processes. When this trend reached automated manufacturing processes there was resistance from process control analysts who were applying a different approach with considerable success (Palm, 1990). Advocates of SPC attempted to force the use of traditional SPC techniques as feedback mechanisms for process control. This inappropriate application of SPC was correctly denounced by process control analysts. SPC is designed to serve a purpose fundamentally different than automatic process control (APC). SPC advocates correctly pointed out that APC was not a cure-all and that many process controllers added variation by making adjustments based on data analysis that was statistically invalid.

Both SPC and APC have their rightful place in Six Sigma. APC attempts to dynamically control a process to minimize variation around a target value. This requires valid statistical analysis, which is the domain of the statistical sciences. SPC makes a distinction between special causes and common causes of variation. If APC responds to all variation as if it were the same it will result in missed opportunities to reduce variation by attacking it at the source. A process that operates closer to the target without correction will produce less variation overall than a process that is frequently returned to the target via APC.

However, at times APC must respond to *common cause variation that can't be economically eliminated*, for example, the mold process described above. Properly used, APC can greatly reduce variability in the output.

Hunter (1986) shows that there is a statistical equivalent to the PID control equation commonly used. The PID equation is

$$u(t) = Ke(t) + \frac{K}{T_1}\int_0^1 e(s)ds + KT_D\left(\frac{d_e}{d_t}\right) \quad (8.73)$$

The "PID" label comes from the fact that the first term is a proportional term, the second an integral term and the third a derivative term. Hunter modified the basic EWMA equation by adding two additional terms. The result is the *empirical control equation*.

$$\hat{y}_{t+1} = \hat{y}_t + \lambda e_t + \lambda_2 \sum e_t + \lambda_3 \nabla e_t \quad (8.74)$$

The term $\nabla e_t$ means the first difference of the errors $e_t$, that is, $\nabla e_t = e_t - e_{t-1}$. Like the PID equation, the empirical control equation has a proportional, an integral and a differential term. It can be used by APC or the results can be plotted on a common cause chart and reacted to by human operators, as described above. A special cause chart can be created to track the errors in the forecast from the empirical control equation. Such an approach may help to bring SPC and APC together to work on process improvement.

## Process Capability Indices

A process capability index is a key metric used to predict the ability of the process to meet the customer requirements. As a prediction tool, the process capability index can only be used when the process is in statistical control, as demonstrated on a suitable control chart. For this reason, it is best to always include the control chart when referencing the capability index.

The statistical control chart provides the estimate of process variation and, where needed, process location. The formulas for estimating the process standard deviation are:

*Range chart method*

$$\hat{\sigma} = \frac{\bar{R}}{d_2} \quad (8.75)$$

*Sigma chart method*

$$\hat{\sigma} = \frac{\bar{s}}{c_4} \quad (8.76)$$

The values $d_2$ and $c_4$ are constants from Appendix 9.

When the process is not in control, the process capability index cannot be used. It is not valid, since it relies on estimates of the process average and process standard deviation that are only valid when the process is in control. Unfortunately, there is no single estimate of process average or process standard deviation when the process is out of control. That is, the fact that the process is out of control is indicative of the presence of special causes of variation, which cause shifts in the process mean or standard deviation.

An alternative to the process capability index is the process performance index. For the set of process capability indices presented below ($C_P$, $C_R$, $C_M$, $C_{PK}$, $C_{PM}$) there is a comparable set of process performance indices ($P_P$, $P_R$, $P_M$, $P_{PK}$, $P_{PM}$) that differ in calculation only by the use of the sample standard deviation (in place of the capability index use of the process standard deviation). While the difference may seem minor, the implications are not.

Stated simply, the expected variation in a process can only be predicted when the process is in statistical control. The predictive ability of the control chart lies in the time-ordered nature of its analysis. The sample standard deviation ignores the time basis of the measurements. (See also the discussions in Chap. 7, including Deming's quote on the differences between enumerative and analytical studies.)

The process performance indices are thus limited in scope. They provide only an estimate of the process output for a given batch from which the sample has been obtained. If the process were out of control during the generation of the batch, then it could be difficult to obtain a representative sample of the batch, depending on the magnitude and duration of the process shift associated with each special cause. The proper calculation of sample standard deviation for a multiple-distribution sample requires assumption on the percentage of items in the batch belonging to each sample distribution. Without the use of a control chart, this is an unknown, leading to the most likely requirement of 100% sampling of the batch to truly gauge its ability to meet requirements.

Perhaps the biggest drawback of using process capability indexes is that they take the analysis a step away from the data. The danger is that the analyst will lose sight of the purpose of the capability analysis, which is to improve quality. To the extent that capability indexes help accomplish this goal, they are worthwhile. To the extent that they distract from the goal, they are harmful. The analyst should continually refer to this principle when interpreting capability indexes.

$C_P$—Historically, this was one of the first capability indexes developed. The "natural tolerance" of the process is computed as $6\sigma$ using the control chart (as described above).

$$C_p = \frac{\text{engineering tolerance}}{6\hat{\sigma}} \qquad (8.77)$$

where engineering tolerance = upper specification limit − lower specification limit. This may also be written as:

$$C_P = \frac{USL - LSL}{6\hat{\sigma}} \qquad (8.78)$$

The index simply makes a direct comparison of the process natural tolerance to the engineering requirements. Assuming the process distribution is normal and the process average is exactly centered between the engineering requirements, a $C_p$ index of 1 would give a "capable process." However, to allow a bit of room for process drift, the generally accepted minimum value for $C_p$ is 1.33. In general, the larger $C_p$ is, the better. For a Six Sigma process, that is, a process that produces 3.4 defects-per-million-opportunities including a 1.5 sigma shift, the value of $C_p$ would be 2.

The $C_p$ index has two major shortcomings. First, it can't be used unless there are both upper and lower specifications. Second, it does not account for process centering. If the process average is not exactly centered relative to the engineering requirements, the $C_p$ index will give misleading results. In recent years, the $C_p$ index has largely been replaced by $C_{PK}$ (see below).

$C_R$—The $C_R$ index is equivalent to the $C_p$ index.

$$C_R = 100 \times \frac{6\hat{\sigma}}{\text{engineering tolerance}} \qquad (8.79)$$

where engineering tolerance = upper specification limit − lower specification limit.

The index simply makes a direct comparison of the process to the engineering requirements. Assuming the process distribution is normal and the process average is exactly centered between the engineering requirements, a $C_R$ index of 100% would give a "capable process." However, to allow a bit of room for process drift, the generally accepted maximum value for $C_R$ is 75%. In general, the smaller $C_R$ is the better. The $C_R$ index suffers from the same shortcomings as the $C_p$ index. For a Six Sigma process, that is, a process that produces 3.4 defects-per-million opportunities including a 1.5 sigma shift, the value of $C_R$ would be 50%.

$C_M$—The $C_M$ index is generally used to evaluate machine capability studies, rather than full-blown process capability studies. Since variation will increase when other sources of process variation are added (e.g., tooling, fixtures, materials, etc.), $C_M$ uses an 8 sigma spread rather than a 6 sigma spread to represent the natural tolerance of the process. For a machine to be used on a Six Sigma process, a 10 sigma spread would be used.

$$C_M = \frac{\text{engineering tolerance}}{8\hat{\sigma}} \qquad (8.80)$$

where engineering tolerance = upper specification limit − lower specification limit.

$Z_{MIN}$—The value of $Z_{MIN}$ is simply the smaller of the $Z_L$ or the $Z_U$ values.

$$Z_U = \frac{\text{upper specification} - \bar{\bar{X}}}{\hat{\sigma}} \quad (8.81)$$

$$Z_L = \frac{\bar{\bar{X}} - \text{lower specification}}{\hat{\sigma}} \quad (8.82)$$

$\bar{\bar{X}}$ is the process centerline estimated using the statistical control chart.

$$Z_{MIN} = \text{Minimum}\{Z_L, Z_U\} \quad (8.83)$$

The $Z_U$ and $Z_L$ indices measure the process location (central tendency) relative to its standard deviation and the upper and lower requirements (respectively). If the distribution is normal, the values of $Z_U$ (and $Z_L$) can be used to determine the percentage above the upper requirement (and below the lower requirement) by using Appendix 2. The method is the same as described in the "Normal Distribution" section earlier in this chapter (using $Z_U$ instead of using $Z$).

In general, a larger $Z$ value is better. A value of at least +3 is required to ensure that 0.1% or less defective will be produced. A value of +4 is generally desired to allow some room for process drift. For a Six Sigma process $Z_{MIN}$ would be +6.

$C_{PK}$—The value of $C_{PK}$ is simply $Z_{MIN}$ divided by 3.

$$C_{PK} = \frac{Z_{MIN}}{3} \quad (8.84)$$

Since the smallest value represents the nearest specification, the value of $C_{PK}$ tells you if the process is truly capable of meeting requirements. A $C_{PK}$ of at least +1 is required, and +1.33 is preferred. Note that $C_{PK}$ is closely related to $C_p$, the difference between $C_{PK}$ and $C_p$ represents the potential gain to be had from centering the process. For a Six Sigma process $C_{PK}$ would be 2.

For example, assume that a process is in statistical control based on an X-bar and R chart with subgroups of 5. The grand average (or centerline of the X-bar chart) is calculated as 0.99832, and the average range (or centerline of the R chart) is calculated as 0.02205. From the table of $d_2$ values (Appendix 9), we find $d_2$ is 2.326 for subgroups of 5. Thus, using the equation above for calculating the process standard deviation using the Range chart method:

$$\hat{\sigma} = \frac{0.02205}{2.326} = 0.00948$$

If the process requirements are a lower specification of 0.980 and an upper specification of 1.020 (i.e., 1.000 ± 0.020), the Z values are calculated as:

$$Z_U = \frac{\text{upper specification} - \bar{\bar{X}}}{\hat{\sigma}} = \frac{1.020 - 0.99832}{0.00948} = 2.3$$

$$Z_L = \frac{\bar{\bar{X}} - \text{lower specification}}{\hat{\sigma}} = \frac{0.99832 - 0.980}{0.00948} = 1.9$$

Referring to Appendix 2, we find that approximately 1.1% will be oversized (based on the $Z_u$ value of 2.3) and approximately 2.9% will be undersized (based on the $Z_L$ value of 1.9). Adding the percents finds a total reject rate of 4.0%, which can be equivalently expressed as a DPMO of 40,000 or a yield of 96.0%.

Extending the example shown in the DPMO section above:

$$C_p = \frac{\text{engineering tolerance}}{6\hat{\sigma}} = \frac{1.020 - 0.9800}{6 \times 0.00948} = 0.703$$

$$C_R = 100 \times \frac{6\hat{\sigma}}{\text{engineering tolerance}} = 100 \times \frac{6 \times 0.00948}{0.04} = 142.2\%$$

$$C_M = \frac{\text{engineering tolerance}}{8\hat{\sigma}} = \frac{0.04}{8 \times 0.00948} = 0.527$$

$$Z_{MIN} = \text{Minimum}\{1.9, 2.3\} = 1.9$$

$$C_{PK} = \frac{Z_{MIN}}{3} = \frac{1.9}{3} = 0.63$$

Assuming that the target is precisely 1.000, we compute:

$$C_{pm} = \frac{C_p}{\sqrt{1 + \frac{(\bar{\bar{X}} - T)^2}{\hat{\sigma}^2}}} = \frac{0.703}{\sqrt{1 + \frac{(0.99832 - 1.000)^2}{0.00948^2}}} = 0.692$$

Since the minimum acceptable value for $C_p$ is 1, the 0.703 result indicates that this process cannot meet the requirements. Furthermore, since the $C_p$ index doesn't consider the centering process, we know that the process can't be made acceptable by merely adjusting the process closer to the center of the requirements. Thus, we would expect the $Z_L$, $Z_U$, and $Z_{MIN}$ values to be unacceptable as well.

The $C_R$ value always provides the same conclusions as the $C_p$ index. The number implies that the "natural tolerance" of the process uses 142.2% of the engineering requirement, which is, of course, unacceptable.

The $C_M$ index should be 1.33 or greater. Obviously, it is not. If this were a machine capability study the value of the $C_M$ index would indicate that the machine was incapable of meeting the requirement.

The value of $C_{PK}$ is only slightly smaller than that of $C_p$. This indicates that we will not gain much by centering the process. The actual amount we would gain can be calculated by assuming the process is exactly centered at 1.000 and recalculating $Z_{MIN}$. This gives a predicted total reject rate of 3.6% instead of 4.0%.

Minitab has a built-in capability analysis feature, called a "Six Pack," providing a compact display of the most important statistics and analysis. The control charts indicate if the process is in statistical control. If it's out of control, stop and find out why. The histogram and normal probability plot indicate if the normality assumption is justified. If not, you can't trust the capability indices, and should consider using Minitab's non-normal capability analysis. (An example is provided below.) The "within" capability indices $C_p$ and $C_{PK}$ are based on within-subgroup variation only, called short-term variability. The "overall" capability indices $P_p$ and $P_{PK}$ are based on total variation, sometimes called long-term variability, which includes variation within subgroups and variation between subgroups. A Capability Plot graphically compares within variability (short-term) and overall variability (long-term) to the specifications. Ideally, for a Six Sigma process, the process variability (process tolerance) will be about half of the specifications. However, the capability plot for the example shows that the process tolerance is actually wider than the specifications.

What is missing in the six pack is an estimate of the process yield. There is an option in the six pack to have this information (and a great deal more) stored in the worksheet. Alternatively, you can run Minitab's Capability Analysis (Normal) procedure and get the information along with a larger histogram. The PPM levels confirm the capability and performance indices calculations.

## *Example of Non-Normal Capability Analysis Using Minitab*

Minitab has a built-in capability to perform process capability analysis for non-normal data, which will be demonstrated with an example. The process involved is technical support by telephone. A call center has recorded the total time it takes to "handle" 500 technical support calls. Handle time is a total cycle time metric, which includes gathering preliminary information, addressing the customer's issues, and performing post-call tasks. It is a CTQ metric that also impacts the shareholder. It has been determined that the upper limit on handle time is 45 minutes. Once the data has been collected, it can be analyzed as follows:

*Phase 1—Check for Special Causes.* To begin we must determine if special causes of variation were present during our study. A special cause is operationally defined as points beyond one of the control limits. Some authors recommend that individuals control charts be used for all analysis, so we'll try this first, see Fig. 8.49.

**Figure 8.49** Individuals process behavior chart for handle time.

There are 12 out-of-control points in the chart shown in Fig. 8.49, indicating that special causes are present. However, a closer look will show that there's something odd about the chart. Note that the lower control limit (LCL) is—18:32. Since we are talking about handle time, it is impossible to obtain any result that is less than zero. A reasonable process owner might argue that if the LCL is in the wrong place (which it obviously is), then the upper control limit (UCL) may be as well. Also, the data appear to be strangely cut off near the bottom. Apparently the individuals chart is not the best way to analyze data like these.

But what can be done? Since we don't know if special causes were present, we can't determine the proper distribution for the data. Likewise, if we don't know the distribution of the data we can't determine if special causes are present because the control limits may be in the wrong place. This may seem to be a classic case of "which came first, the chicken or the egg?" Fortunately, there is a way out. The central limit theorem tells us that stable distributions produce normally distributed averages, even when the individuals' data are not normally distributed. Since "stable" means no special causes, then a process with non-normal *averages* would be one that is influenced by special causes, which is precisely what we are looking for. We created subgroups of 10 in Minitab (i.e., observations 1 to 10 are in subgroup 1, observations 11 to 20 are in subgroup 2, etc.) and tested the normality of the averages. The probability plot in Fig. 8.50 indicates that the averages are normally distributed. (Note that EWMA charts and moving average charts are both valid alternatives to the use of $X$-bar charts in this example, since the central limit theorem applies equally well to their plotted groups).

**Figure 8.50** Normality test of subgroups of $n = 10$.

Figure 8.51 shows the control chart for the process using averages instead of individuals. The chart indicates that the process is in statistical control. The process average is stable at 18.79 minutes. The LCL is comfortably above zero at 5.9 minutes; any average below this is an indication that things are better than normal and we'd want to know why in case we can do it all of the time. Any average above 31.67 minutes indicates worse than normal behavior and we'd like to find the reason and fix it. Averages between these two limits are normal for this process.

**Figure 8.51** Averages of handle time (*n* = 10 per subgroup).

***Phase 2—Examine the Distribution.*** Now that stability has been determined, we can trust the histogram of individual observations to give us an accurate display of the distribution of handle times. The histogram shows the distribution of actual handle times, which we can compare to the upper specification limit of 45 minutes. This couldn't be done with the control chart in Fig. 8.51 because it shows averages, not individual times. Figure 8.52 shows the histogram of handle time with the management upper requirement

**Figure 8.52** Histogram of handle time.

of 45 minutes drawn in. Obviously, a lot of calls exceed the 45 minute requirement. Since the control chart is stable, we know that this is what we can expect from this process. There is no point in asking why a particular call took longer than 45 minutes. The answer is "It's to be expected for this process, unless we make a fundamental change to the process." If management doesn't like the answer they'll need to sponsor one or more Six Sigma projects to improve the process.

*Phase 3—Predicting the Long-Term Defect Rate for the Process.* The histogram makes it visually clear that the process distribution is non-normal. This conclusion can be tested statistically with Minitab by going to Stats > Basic Statistics > Normality test. Minitab presents the data in a chart specially scaled so that normally distributed data will plot as a straight line (Fig. 8.53). The vertical axis is scaled in cumulative probability and the horizontal in actual measurement values. The plot shows that the data are not even close to falling on the straight line, and the *P*-value of 0 confirms that the data are not normal.[3] To make a prediction about the defect rate we need to find a distribution that fits the data reasonably well. Minitab offers an option that performs capability analysis using the Weibull rather than the normal distribution. Choose Stat > Quality Tools > Capability Analysis (Weibull) and enter the column name for the handle time data. The output is shown in Fig. 8.54.

**Figure 8.53** Normality test of handle time.

**Process data**

| | |
|---|---|
| USL | 45.0000 |
| Target | * |
| LSL | 0.0000 |
| Mean | 18.9618 |
| Sample N | 500 |
| Shape | 1.5237 |
| Scale | 21.0433 |

**Overall (LT) capability**

| | |
|---|---|
| Pp | 0.62 |
| PPU | 0.58 |
| PPL | 2.40 |
| Ppk | 0.58 |

**Observed LT performance**

| | |
|---|---|
| PPM < LSL | 0.00 |
| PPM > USL | 56000.00 |
| PPM total | 56000.00 |

**Expected LT performance**

| | |
|---|---|
| PPM < LSL | 0.00 |
| PPM > USL | 41422.03 |
| PPM total | 41422.03 |

**Figure 8.54** Capability analysis of handle times based on the Weibull distribution.

Minitab calculates process performance indices rather than process capability indices (i.e., $P_{PK}$ instead of $C_{PK}$). This means that the denominator for the indices is the overall standard deviation rather than the standard deviation based on the control chart's within-subgroup variability. This is called the long-term process capability, which Minitab labels as "Overall (LT) Capability." When the process is in statistical control, as this one is, there will be little difference in the estimates of the standard deviation. When the process is not in statistical control, the short-term capability estimates have no meaning, and the long-term estimates are of dubious value as well. Process performance indices are interpreted in exactly the same way as their process capability counterparts. Minitab's analysis indicates that the process is not capable ($P_{PK} < 1$). The estimated long-term performance of the process is 41,422 defects per million calls. The observed performance is even worse, 56,000 defects per million calls. The difference is a reflection of lack of fit. The part of the Weibull curve we're most interested in is the tail area above 45, and the curve appears to drop off more quickly than the actual data. When this is the case it is better to estimate the long-term performance using the actual defect count rather than Minitab's estimates.

See online video: **Process Capability in Minitab** at www.mhprofessional.com/ssh5.

# CHAPTER 9

# Measurement Systems Evaluation

A good measurement system possesses certain properties. First, it should produce a number that is "close" to the actual property being measured, that is, it should be *accurate*. Second, if the measurement system is applied repeatedly to the same object, the measurements produced should be close to one another, that is, it should be *repeatable*. Third, the measurement system should be able to produce accurate and consistent results over the entire range of concern, that is, it should be *linear*. Fourth, the measurement system should produce the same results when used by any properly trained individual, that is, the results should be *reproducible*. Finally, when applied to the same items the measurement system should produce the same results in the future as it did in the past, that is, it should be *stable*. The remainder of this section is devoted to discussing ways to ascertain these properties for particular measurement systems.

## Definitions

**Bias**—The difference between the average measured value and a reference value is referred to as *bias*. The reference value is an agreed-upon standard, such as a standard traceable to a national standards body (see below). When applied to attribute inspection, bias refers to the ability of the attribute inspection system to produce agreement on inspection standards. Bias is controlled by *calibration*, which is the process of comparing measurements to standards. The concept of bias is illustrated in Fig. 9.1.

**Repeatability**—AIAG defines repeatability as the variation in measurements obtained with one measurement instrument when used several times by one appraiser, while measuring the identical characteristic on the same part. Variation obtained when the measurement system is applied repeatedly under the same conditions is usually caused by conditions inherent in the measurement system.

ASQ defines *precision* as "The closeness of agreement between randomly selected individual measurements or test results. *Note:* The standard deviation of the error of

**Figure 9.1** Bias illustrated.

measurement is sometimes called *imprecision*. This definition is similar to repeatability. Repeatability is illustrated in Fig. 9.2.

**Reproducibility**—Reproducibility is the variation in the average of the measurements made by different appraisers using the same measuring instrument when measuring the identical characteristic on the same part. Reproducibility is illustrated in Fig. 9.3.

**Stability**—Stability is the total variation in the measurements obtained with a measurement system on the same master or parts when measuring a single characteristic over an extended time period. A system is said to be stable if the results are the same at different points in time. Stability is illustrated in Fig. 9.4.

**Linearity**—The difference in the bias values through the expected operating range of the gage. Linearity is illustrated in Fig. 9.5.

Historically, calibration has been the standard approach to limit the effects of bias, long considered the fundamental source of measurement error. Modern measurement system analysis goes well beyond calibration. A gage can be perfectly accurate when checking a standard and still be entirely unacceptable for measuring a product or controlling a process. This section illustrates techniques for quantifying discrimination, stability, bias, repeatability, reproducibility, and variation for a measurement system. Control charts are used to provide graphical portrayals of the measurement processes, enabling the analyst to detect special causes that numerical methods alone would not detect.

**Figure 9.2** Repeatability illustrated.

**Figure 9.3** Reproducibility illustrated.

**Figure 9.4** Stability illustrated.

**Figure 9.5** Linearity illustrated.

## Measurement System Discrimination

Discrimination, sometimes called resolution, refers to the ability of the measurement system to divide measurements into "data categories." All parts within a particular data category will measure the same. For example, if a measurement system has a resolution of 0.001 inch, then items measuring 1.0002, 1.0003, 0.9997 would all be placed in the data category 1.000, that is, they would all measure 1.000 inch with this particular measurement system. A measurement system's discrimination should enable it to divide the region of interest into many data categories. In Six Sigma, the region of interest is the smaller of the tolerance (the high specification minus the low specification) or six standard deviations. A measurement system should be able to divide the region of interest into at least five data categories. For example, if a process was capable (i.e., the tolerance is greater than six standard deviations) and $\sigma = 0.0005$, then a gage with a discrimination of 0.0005 would be acceptable (six data categories), but one with a discrimination of 0.001 would not (three data categories). When unacceptable discrimination exists, the range chart shows discrete "jumps" or "steps." This situation is illustrated in Fig. 9.6.

Note that on the control charts shown in Fig. 9.6, the data plotted are the same, except that the data on the bottom X-bar and Range charts were rounded. The effect is most easily seen on the *Range* chart, which appears highly stratified. As sometimes happens (but not always), the result is to make the *X*-bar chart go out of control, even though the process is in control, as shown by the control charts at the top of the figure, created with unrounded data. The remedy is to use a measurement system capable of additional discrimination, resulting in more significant digits in the measurements.

## Stability

*Measurement system stability* is the change in bias over time when using a measurement system to measure a given master part or standard. *Statistical stability* is a broader term that refers to the overall consistency of measurements over time, including variation from *all causes*, including bias, repeatability, reproducibility, etc. A system's statistical stability is determined through the use of control charts. Averages and range charts are typically used to plot measurements of a standard or a master part repeatedly over a short time interval, such as an hour; then the measurements are repeated at longer predetermined intervals, such as weekly. Expertise of the specific measurement system will dictate the subgroup size, sampling intervals, and measurement procedures to be followed. A (statistically) stable system will show no out-of-control signals on an *X*-control chart of the averages' readings.

**Figure 9.6** Inadequate gage discrimination on a control chart.

Once statistical stability has been achieved, measurement system stability can be estimated from the process standard deviation based on the control chart used for process control.

## Bias

Bias is the difference between an observed average measurement result and a reference value. Estimating bias involves identifying a standard to represent the reference value, then obtaining multiple measurements on the standard. The standard might be a master part whose value has been determined by a measurement system with much less error than the system under study, or by a standard traceable to NIST. Since parts and processes vary over a range, bias is measured at a point within the range. If the gage is nonlinear, bias will not be the same at each point in the range (see the definition of linearity earlier).

Bias can be determined by selecting a single appraiser and a single reference part or standard. The appraiser then obtains a number of repeated measurements on the reference part. Bias is then estimated as the difference between the average of the repeated measurement and the known value of the reference part or standard.

### Example of Computing Bias

A standard with a known value of 25.4 mm is checked 10 times by one mechanical inspector using a dial caliper with a resolution of 0.025 mm. The readings obtained are:

| 25.425 | 25.425 | 25.400 | 25.400 | 25.375 |
|--------|--------|--------|--------|--------|
| 25.400 | 25.425 | 25.400 | 25.425 | 25.375 |

The average is found by adding the 10 measurements together and dividing by 10:

$$\bar{X} = \frac{254.051}{10} = 25.4051 \text{ mm}$$

The bias is the average minus the reference value, that is:

$$\text{Bias} = \text{average} - \text{reference value}$$

$$= 25.4051 \text{ mm} - 25.400 \text{ mm} = 0.0051 \text{ mm}$$

The bias of the measurement system can be stated as a percentage of the tolerance or as a percentage of the process variation. For example, if this measurement system were to be used on a process with a tolerance of ±0.25 mm, then:

$$\% \text{ bias} = 100 \times |\text{bias}|/\text{tolerance}$$

$$= 100 \times 0.0051/0.5 = 1\%$$

This is interpreted as follows: this measurement system will, on average, produce results that are 0.0051 mm larger than the actual value. This difference represents 1% of the allowable product variation. The situation is illustrated in Fig. 9.7.

## *Repeatability*

Measurement repeatability is estimated by comparing the repeated measurements of multiple units. It's important to use real units from the process, rather than the pristine samples used in calibration or bias estimates. Actual units from the process, spanning the full width of the tolerance or process variation, will likely present the real-world challenges in

**Figure 9.7** Bias example illustrated.

measurement that we seek to quantify with the study. Machined parts may have burrs or rough surfaces that make the units difficult to measure, just as in a real production environment. It's also important to include all the steps (i.e., sources of error) usually taken to obtain a measurement. In a chemical laboratory, this would include sample preparation, for example.

A measurement system is repeatable if its variability is consistent, as evidenced by constructing a range or sigma chart based on repeated measurements of parts that cover a significant portion of the process variation or the tolerance, whichever is greater. If the range or sigma chart is out of control, then special causes influence the measurement system repeatability. If the range or sigma chart is in control, then repeatability can be estimated.

## Example of Estimating Repeatability

The data in Table 9.1 are from a measurement study involving three appraisers. Each appraiser measured the length of ten parts; each part was checked three times by each appraiser.

We can construct a special purpose X-bar/Range chart (discussed in Ch. 8) to analyze the repeatability error, as displayed in Fig. 9.8. The range chart for repeatability indicates that all the range values are less than the upper control limit. This indicates that the variability due to repeatability is consistent, that is, there are no special causes of variation influencing the measurement system repeatability.

**Table 9.1** Measurement System Repeatability & Reproducibility (R&R) Study Data

| Appraiser | Piece | Trial 1 | Trial 2 | Trial 3 |
|---|---|---|---|---|
| Joe | 1 | 5.65 | 5.65 | 5.7 |
| | 2 | 5.15 | 5.2 | 5.2 |
| | 3 | 6.25 | 6.35 | 6.3 |
| | 4 | 6.15 | 6.05 | 6.05 |
| | 5 | 5.8 | 5.85 | 5.75 |
| | 6 | 5.75 | 5.8 | 5.6 |
| | 7 | 5.85 | 5.8 | 5.6 |
| | 8 | 5.7 | 5.7 | 5.65 |
| | 9 | 6.55 | 6.5 | 6.45 |
| | 10 | 5.85 | 5.7 | 5.75 |
| Karen | 1 | 5.7 | 5.75 | 5.65 |
| | 2 | 5.1 | 5.15 | 5.2 |
| | 3 | 6.3 | 6.25 | 6.25 |
| | 4 | 6.05 | 6.1 | 6.15 |
| | 5 | 5.9 | 5.8 | 5.85 |
| | 6 | 6.05 | 5.95 | 5.75 |
| | 7 | 5.8 | 5.6 | 5.6 |
| | 8 | 5.7 | 5.85 | 5.75 |
| | 9 | 6.45 | 6.45 | 6.55 |
| | 10 | 5.8 | 5.85 | 5.9 |
| Robert | 1 | 5.5 | 5.6 | 5.45 |
| | 2 | 5.3 | 5.25 | 5.35 |
| | 3 | 5.95 | 6.3 | 6 |
| | 4 | 5.9 | 6 | 6.1 |
| | 5 | 5.7 | 5.95 | 5.9 |
| | 6 | 5.4 | 5.6 | 5.45 |
| | 7 | 5.35 | 5.45 | 5.6 |
| | 8 | 5.6 | 5.65 | 5.6 |
| | 9 | 6.4 | 6.25 | 6.4 |
| | 10 | 5.7 | 5.75 | 5.65 |

**Figure 9.8** Repeatability control charts (Green Belt XL).

Note that many of the averages are outside of the control limits. This is good! Consider that the spread of the *X*-bar chart's control limits is based on the average range, which is based on the repeatability error. If the averages were within the control limits it would mean that the part-to-part variation was less than the variation due to gage repeatability error, an undesirable condition. That the part-to-part variation is larger than the repeatability error means the part-to-part variation can be measured in spite of the repeatability error.

## *Reproducibility*

A measurement system is reproducible when there is consistency between the measurements of different appraisers. Appraiser-to-appraiser variation represents a bias due to appraisers. The appraiser bias, or reproducibility, can be estimated by comparing each appraiser's average with that of the other appraisers.

## Reproducibility Example

A control chart for reproducibility can also be constructed by comparing the appraisers' average for each part, as shown in Fig. 9.9. The range chart at bottom shows that all of the range values are less than the upper control limit, indicating that the variability due to reproducibility is consistent, that is, there are no special causes of variation evident from reproducibility error across the ten parts. Where instability is noted on this chart, it often points to an opportunity for improvement in the measurement process. For example, a burr on one of the machined parts could cause that part to have an increased measure of variation, as noticed in the range chart for reproducibility. Detecting this instability in measurement error could lead to an improved measurement process, such as by removing all burrs prior to measurement.

**Figure 9.9** Reproducibility control charts (Green Belt XL).

The averages chart for reproducibility, however, indicates significant variation amongst the appraiser averages, evidence of reproducibility error, as demonstrated by the points plotted beyond the statistical limits.

Figure 9.10 shows the Repeatability chart, with separate limits calculated for each appraiser. The repeatability range chart suggests that Robert's repeatability error is larger than the other appraisers, who used the same measurement equipment, pointing perhaps to a measurement technique unique to Robert that produced higher levels of repeatability error. The averages chart for repeatability suggests a potential bias in Robert's measurements. The software allows us to easily exclude Robert from the analysis, and doing so results in a reduction in R&R from 12% to 5% of the total variance. These would be reasonable targets if we are able to understand and correct the root causes of the increased variation in Robert's trial to trial (repeatability) error and measurement bias (reproducibility error). Often, it's a matter of technique, so the analyst is wise to conduct the study

**Figure 9.10** Repeatability control charts segmented by Appraiser (Green Belt XL).

and observe (without comment) the appraisers' methods, which may direct us towards new best practices that recognize requirements for sample preparation or reconcile within piece variation. For example, when measuring a diameter where an out-of-round condition may exist, is the maximum or minimum diameter reported?

## *Part-to-Part Variation*

The *X*-bar chart for repeatability plots the part-to-part variation. As stated earlier, if the measurement system is adequate, *most of the parts will fall outside of the X-bar chart control limits*. If fewer than half of the parts are beyond the control limits, then the measurement system is not capable of detecting normal part-to-part variation for this process.

Part-to-part variation can be estimated once the measurement process is shown to have adequate discrimination and to be stable, accurate, linear (see below), and consistent with respect to repeatability and reproducibility. If control charts are used on the process, then they will likely provide a better estimate of part-to-part variation than the limited number of parts available in this study (in this case, ten parts). In that case, the process sigma values reported in your SPC software can be input as an option in the R&R analysis to improve its estimates.

$$\sigma_p = 1.23/3.18 = 0.388$$

## *Summary Reporting*

Notice in Figures 9.8 through 9.10 that the Green Belt XL software lists each component of variation, presenting the calculated standard deviation for the component as well as its percent relative to the overall variance, the tolerance, and the total process variation. Generally, the variances components reported will sum to 100% of the parameter; the variation as a percent of tolerance or process variation will not. For example, the variance due to Repeatability (5.34%) plus that due to Reproducibility (7.09%) sum to the Repeatability & Reproducibility variance reported (12.43%). Likewise, the Repeatability & Reproducibility variance (12.43%) plus the Part variance (87.57%) equals 100% of the estimated process variance.

A comparison of the measurement system variation to the tolerance provides insight into whether the system is well designed for product acceptance decisions. Likewise, a comparison to the process variation ensures the measurement system can properly estimate process variation via the control chart. General recommendations for the measurement system error (i.e., R&R) as a percent of process variation or tolerance is 10% or less.

The analyses in Figures 9.8 through 9.10 were done using the ANOVA method (discussed in Chapter 10), which offers some improvement over the historical range method of analysis. In addition to providing a more accurate solution, the ANOVA method also provides some clarity in the estimates of reproducibility.

Most statistical software will also provide a statistic for the number of distinct categories defined by the measurement system, which attempts to describe the discrimination of the measurement system. (The Green Belt XL software used in the previous analysis allows the user to turn off the display of that statistic, so it is not shown in the above figures). The calculation of this statistic was historically referenced in AIAG publications, so it was widely used; however, the author cited as the originator of the statistic has since disavowed the use of the statistic for that purpose, insisting that was never the intent (Wheeler, 2010). As such, it is no longer recommended as a means the quantify the discrimination.

See online video: Interpreting R&R Analysis at www.mhprofessional.com/ssh5.

## Gage R&R Analysis Using Minitab

Minitab offers several methods for performing gage R&R studies. Use nested studies when each part can be measured by only one operator, as with destructive testing; otherwise, choose crossed studies from the Stat > Quality Tools menu. In addition to crossed or nested options, Minitab offers both ANOVA and $X$ bar and $R$ methods of analysis, so it is recommended to use the ANOVA technique as mentioned earlier.

The Minitab ANOVA for the example R&R study above is shown in Table 9.2. The $P$ values show the probability that the observed variation for the source row might be due to chance. By convention, a $P$ value less than 0.05 is the critical value for deciding that a source of variation is "significant," that is, greater than zero. In this example, all terms are statistically significant, since their $p$ values are less than 0.05. Because the Appraiser term is significant, we conclude there are statistically significant differences between operators, prompting an investigation into underlying causes. Since the interaction term is significant, we conclude that one operator has obtained different results with some, but not all, parts.

**Table 9.2**  Two-Way ANOVA Table with Interaction

| Source | DF | SS | MS | F | P |
|---|---|---|---|---|---|
| Piece | 9 | 9.7451 | 1.08279 | 48.9226 | 0.000 |
| Appraiser | 2 | 0.3177 | 0.15886 | 7.1777 | 0.005 |
| Piece*Appraiser | 18 | 0.3984 | 0.02213 | 3.0764 | 0.001 |
| Repeatability | 60 | 0.4317 | 0.00719 | | |
| Total | 89 | 10.8929 | | | |

$\alpha$ to remove interaction term = 0.05

The Appraiser*Piece interaction represents 3.7% of total variance, or about half of the reproducibility variance. Returning to Figure 9.9, we see that the range between appraiser averages for part 6 is larger than the other parts, approaching (yet within) the statistical limit. When the interaction is limited to one or a few identifiable parts, it's useful to look for measurement challenges on those pieces. For this data, excluding part 6 from the analysis does not remove the significance of the Appraiser*Piece interaction, so the interaction is not limited to part 6; however, excluding Robert's data from the analysis removed the statistical significance of the interaction, again confirming the significance of his measurements to the Reproducibility error. A review of Figure 9.10 shows a general inconsistency between Robert and the other appraisers on multiple parts. While looking for measurement challenges on part 6 is warranted given its increased Reproducibility error, a comparative review of the appraisers' techniques, not limited to part 6, would also be appropriate.

**Table 9.3** Variance Components

| Source | VarComp | % Contribution (of VarComp) |
|---|---|---|
| Total Gage R&R | 0.016731 | 12.43 |
|   Repeatability | 0.007194 | 5.35 |
|   Reproducibility | 0.009537 | 7.09 |
|     Appraiser | 0.004558 | 3.39 |
|     Appraiser*Piece | 0.004979 | 3.70 |
| Part-To-Part | 0.117851 | 87.57 |
| Total Variation | 0.134582 | 100.00 |

Minitab decomposes the total variance into components in Table 9.3. The VarComp column shows the variance attributed to each source, while the % Contribution shows the percentage of the total variance accounted for by each source. The analysis indicates that nearly all of the variation is between parts, as one would hope.

**Table 9.4** Gage Evaluation

| Source | StdDev (SD) | Study Var (6 × SD) | %Study Var (%SV) | %Tolerance (SV/Toler) |
|---|---|---|---|---|
| Total Gage R&R | 0.129350 | 0.77610 | 35.26 | 38.81 |
|   Repeatability | 0.084820 | 0.50892 | 23.12 | 25.45 |
|   Reproducibility | 0.097658 | 0.58595 | 26.62 | 29.30 |
|     Appraiser | 0.067510 | 0.40506 | 18.40 | 20.25 |
|     Appraiser*Piece | 0.070565 | 0.42339 | 19.24 | 21.17 |
| Part-To-Part | 0.343294 | 2.05976 | 93.58 | 102.99 |
| Total Variation | 0.366855 | 2.20113 | 100.00 | 110.06 |

The variance analysis shown in Table 9.3, while accurate, is not in original units. (Variances are the squares of measurements.) Technically, this is the correct way to analyze information on dispersion because variances are additive, while dispersion measurements expressed in original units are not. However, there is a natural interest in seeing an analysis of dispersion in the original units, so Minitab provides this in Table 9.4. The StdDev column is the standard deviation, or the square root of the VarComp column in Table 9.3. The Study Var column shows (in this example) the variation represented by six times the StdDev. The % Study Var column is the Study Var column divided by the total variation due to all sources. The % Tolerance column is calculated as the Study Var column result divided by the tolerance (in this case, upper specification of 7; lower specification of 5). It is interesting that the percent Tolerance column total is greater than 100%. This indicates that the measured process spread exceeds the tolerance. Although this isn't a process capability analysis, the data do indicate a possible problem meeting tolerances.

The information in Table 9.3 is presented graphically in Fig. 9.11, with other useful charts similar to what was previously discussed.

---

See online video: R&R Analysis with ANOVA at www.mhprofessional.com/ssh5.

---

**Figure 9.11** Gage R&R Report (Minitab 18).

## *Linearity*

Linearity can be determined by choosing parts or standards that cover all or most of the operating range of the measurement instrument. Bias is determined at each point in the range and a linear regression analysis is performed.

Linearity is defined as the slope times the process variance or the slope times the tolerance, whichever is greater. A scatter diagram should also be plotted from the data.

### Linearity Example

The following example is taken from *Measurement Systems Analysis*, published by the Automotive Industry Action Group.

A plant foreman was interested in determining the linearity of a measurement system. Five parts were chosen throughout the operating range of the measurement system based upon the process variation. Each part was measured by a layout inspection to determine its reference value. Each part was then measured 12 times by a single appraiser. The parts were selected at random. The part average and bias average were calculated for each part as shown in Fig. 9.12. The part bias was calculated by subtracting the part reference value from the part average.

| Part → | 1 | 2 | 3 | 4 | 5 |
|---|---|---|---|---|---|
| Average | 2.49 | 4.13 | 6.03 | 7.71 | 9.38 |
| Ref. Value | 2.00 | 4.00 | 6.00 | 8.00 | 10.00 |
| Bias | +0.49 | +0.13 | +0.03 | -0.29 | -0.62 |

**Figure 9.12** Gage data summary.

A linear regression analysis was performed. In the regression, $x$ is the reference value and $y$ is the bias. The results are shown in Fig. 9.13.

The *P*-values indicate that the result is statistically significant, that is, there is actually a bias in the gage. The slope of the line is –0.132, and the intercept is 0.74. $R^2 = 0.98$, indicating that the straight line explains about 98% of the variation in the bias readings. The results can be summarized as follows:

| Bias | $b + ax$ = 0.74 – 0.132 (Reference value) |
|---|---|
| Linearity | |slope| x process variation = 0.132 x 6 = 0.79, where 6 is the tolerance |
| % Linearity | 100% x |slope| = 13.2% |

**408** | The Six Sigma Handbook

| Summary output | | | | | |
|---|---|---|---|---|---|
| *Regression statistics* | | | | | |
| Multiple R | 0.98877098 | | | | |
| R Square | 0.97766805 | | | | |
| Adjusted R Square | 0.97022407 | | | | |
| Standard Error | 0.07284687 | | | | |
| Observations | 5 | | | | |
| ANOVA | | | | | |
| | df | ss | ms | F | Significance F |
| Regression | 1 | 0.69696 | 0.69696 | 131.336683 | 0.00142598 |
| Residual | 3 | 0.01592 | 0.00530667 | | |
| Total | 4 | 0.71288 | | | |
| | Coefficients | Standard error | t Stat | P-value | |
| Intercept | 0.74 | 0.07640244 | 9.68555413 | 0.0023371 | |
| Ref. Value | -0.132 | 0.0115181 | -11.460222 | 0.00142598 | |

**Figure 9.13** Regression analysis of linearity summary data.

Note that the zero bias point is found at:

$$x = -\left(\frac{\text{intercept}}{\text{slope}}\right) = -\left(\frac{0.74}{-0.132}\right) = 5.61$$

In this case, this is the point of least bias. Greater bias exists as you move further from this value.

This information is summarized graphically in Fig. 9.14.

**Figure 9.14** Graphical analysis of linearity.

## Linearity Analysis Using Minitab

Minitab has a built-in capability to perform gage linearity analysis. Figure 9.15 shows the data layout and dialog box. Figure 9.16 shows the Minitab output.

**Figure 9.15** Minitab gage linearity dialog box.

**Figure 9.16** Minitab gage linearity output.

Note that Minitab doesn't show the *P*-values for the analysis so it is necessary to perform a supplementary regression analysis anyway to determine the statistical significance of the results. For this example, it is obvious from the scatter plot that the slope of the line isn't zero, so a *P*-value isn't required to conclude that non-linearity exists. The results aren't so clear for bias, which is only 0.867%. In fact, if we perform a one-sample *t*-test of the hypothesis that the mean bias is 0, we get the results shown in Fig. 9.17, which indicate the bias could be 0 (*P* = 0.797).[1]

```
One-Sample T: bias

Test of mu = 0 vs mu not = 0

Variable      N       Mean      StDev    SE Mean
bias          5      -0.052     0.422      0.189

Variable           95.0% CI              T        P
bias          ( -0.576,  0.472)       -0.28    0.797
```

**Figure 9.17** One-sample *t*-test of bias.

## Attribute Measurement Error Analysis

Attribute data consist of classifications rather than measurements. Attribute inspection involves determining the classification of an item, for example, is it "good" or "bad"? The principles of good measurement for attribute inspection are the same as for measurement inspection (Table 9.5). Thus, it is possible to evaluate attribute measurement systems in much the same way as we evaluate variable measurement systems. Much less work has been done on evaluating attribute measurement systems. The proposals provided in this book are those I've found to be useful for my employers and clients. The ideas are not part of any standard, and you are encouraged to think about them critically before adopting them. I also include an example of Minitab's attribute gage R&R analysis.

**Table 9.5**  Attribute Measurement Concepts

| Measurement Concept | Interpretation for Attribute Data | Suggested Metrics and Comments |
|---|---|---|
| Accuracy | Items are correctly categorized. | $$\frac{\text{Number of times correctly classified by all}}{\text{Total number of evaluations by all}}$$ Requires knowledge of the "true" value. |
| Bias | The proportion of items in a given category is correct. | Overall average proportion in a given category (for all inspectors) minus correct proportion in a given category. Averaged over all categories. Requires knowledge of the "true" value. |
| Repeatability | When an inspector evaluates the same item multiple times in a short time interval, she assigns it to the same category every time. | For a given inspector: $$\frac{\text{Total number of times repeat classications agree}}{\text{Total number of repeat classifications}}$$ Overall: Average of repeatabilities |
| Reproducibility | When all inspectors evaluate the same item, they all assign it to the same category. | $$\frac{\text{Total number of times classifications for all concur}}{\text{Total number of classification}}$$ |
| Stability | The variability between attribute R&R studies at different times. | <table><tr><th>Metric</th><th>Stability Measure for Metric</th></tr><tr><td>Repeatability</td><td>Standard deviation of repeatabilities</td></tr><tr><td>Reproducibility</td><td>Standard deviation of reproducibilities</td></tr><tr><td>Accuracy</td><td>Standard deviation of accuracies</td></tr><tr><td>Bias</td><td>Average bias</td></tr></table> |
| Linearity | When an inspector evaluates items covering the full set of categories, her classifications are consistent across the categories. | Range of inaccuracy and bias across all categories. Requires knowledge of the "true" value. Note: Because there is no natural ordering for nominal data, the concept of linearity doesn't really have a precise analog for attribute data on this scale. However, the suggested metrics will highlight interactions between inspectors and specific categories. |

## Operational Definitions

An operational definition is defined as a requirement that includes a means of measurement. "High quality solder" is a requirement that must be operationalized by a clear definition of what "high quality solder" means. This might include verbal descriptions, magnification power, photographs, physical comparison specimens, and many more criteria.

### Examples of Operational Definitions

1. Operational definition of the Ozone Transport Assessment Group's (OTAG) goal.
   **Goal**—To identify reductions and recommend transported ozone and its precursors which, in combination with other measures, will enable attainment and maintenance of the ozone standard in the OTAG region.
   Suggested operational definition of the goal:
   - A general modeled reduction in ozone and ozone precursors aloft throughout the OTAG region; and
   - A reduction of ozone and ozone precursors both aloft and at ground level at the boundaries of non-attainment area modeling domains in the OTAG region; and
   - A minimization of increases in peak ground level ozone concentrations in the OTAG region (This component of the operational definition is in review.).
2. Wellesley College Child Care Policy Research Partnership operational definition of unmet need:
   - Standard of comparison to judge the adequacy of neighborhood services: the median availability of services in the larger region (Hampden County).
   - Thus, our definition of unmet need: The difference between the care available in the neighborhood and the median level of care in the surrounding region (stated in terms of child care slots indexed to the age-appropriate child population—"slots-per-tots").
3. Operational definitions of acids and bases:
   - An *acid* is any substance that increases the concentration of the $H^+$ ion when it dissolves in water.
   - A *base* is any substance that increases the concentration of the $OH^-$ ion when it dissolves in water.
4. Operational definition of "intelligence":
   - Administer the Stanford-Binet IQ test to a person and score the result. The person's intelligence is the score on the test.
5. Operational definition of "dark blue carpet."
   - A carpet will be deemed to be dark blue, if judged by an inspector medically certified as having passed the U.S. Air Force test for color-blindness.

▼ It matches the PANTONE color card 7462 C when both carpet and card are illuminated by GE "cool white" fluorescent tubes.
▼ Card and carpet are viewed at a distance between 16 and 24 inches.

## *How to Conduct Attribute Inspection Studies*

Some commonly used approaches to attribute inspection analysis are shown in Table 9.6.

**Table 9.6** Methods of Evaluating Attribute Inspection

| True Value | Method of Evaluation | Comments |
|---|---|---|
| Known | **Expert Judgment:** An expert looks at the classifications after the operator makes normal classifications and decides which are correct and which are incorrect. | • Metrics:<br>　1. Percent correct<br>• Quantifies the accuracy of the classifications<br>• Simple to evaluate<br>• Who says the expert is correct?<br>• Care must be taken to include all types of attributes<br>• Difficult to compare operators since different units are classified by different people<br>• Acceptable level of performance must be decided upon. Consider cost, impact on customers, etc. |
| Known | **Round Robin Study:** A set of carefully identified objects is chosen to represent the full range of attributes.<br>1. Each item is evaluated by an expert and its condition recorded.<br>2. Each item is evaluated by every inspector at least twice. | • Metrics:<br>　1. Percent correct by inspector<br>　2. Inspector repeatability<br>　3. Inspector reproducibility<br>　4. Stability<br>　5. Inspector "linearity"<br>• Full range of attributes included<br>• All aspects of measurement error quantified<br>• People know they're being watched, may affect performance<br>• Not routine conditions<br>• Special care must be taken to ensure rigor<br>• Acceptable level of performance must be decided upon for each type of error. Consider cost, impact on customers, etc. |
| Unknown | **Inspector Concurrence Study:** A set of carefully identified objects is chosen to represent the full range of attributes, to the extent possible.<br>1. Each item is evaluated by every inspector at least twice. | • Metrics:<br>　1. Inspector repeatability<br>　2. Inspector reproducibility<br>　3. Stability<br>　4. Inspector "linearity"<br>• Like a round robin, except true value isn't known<br>• No measures of accuracy or bias are possible. Can only measure agreement between equally qualified people<br>• Full range of attributes included<br>• People know they're being watched, may affect performance<br>• Not routine conditions<br>• Special care must be taken to ensure rigor<br>• Acceptable level of performance must be decided upon for each type of error. Consider cost, impact on customers, etc. |

### Example of Attribute Inspection Error Analysis

Two sheets with identical lithographed patterns are to be inspected under carefully controlled conditions by each of the three inspectors. Each sheet has been carefully examined multiple times by journeymen lithographers and they have determined that one of the sheets should be classified as acceptable, the other as unacceptable. The inspectors sit on a stool at a large table where the sheet will be mounted for inspection. The inspector can adjust the height of the stool and the angle of the table. A lighted magnifying glass is mounted to the table with an adjustable arm that lets the inspector move it to any part of the sheet (see Fig. 9.18).

**Figure 9.18** Lithography inspection station table, stool, and magnifying glass.

Each inspector checks each sheet once in the morning and again in the afternoon. After each inspection, the inspector classifies the sheet as either acceptable or unacceptable. The entire study is repeated the following week. The results are shown in Table 9.7.

In Table 9.7 the part column identifies which sheet is being inspected, and the standard column is the classification for the sheet based on the journeymen's evaluations. A 1 indicates that the sheet is acceptable, a 0 that it is unacceptable. The columns labeled InspA, InspB, and InspC show the classifications assigned by the three inspectors, respectively. The reproducible column is a 1 if all three inspectors agree on the classification, whether their classification agrees with the standard or not. The accurate column is a 1 if *all three* inspectors classify the sheet correctly as shown in the standard column.

### Individual Inspector Accuracy

Individual inspector accuracy is determined by comparing each inspector's classification with the standard. For example, in cell C2 of Table 9.7 inspector A classified the unit as acceptable, and the standard column in the same row indicates that the classification is

**Table 9.7** Results of Lithography Attribute Inspection Study

|   | A | B | C | D | E | F | G | H | I |
|---|---|---|---|---|---|---|---|---|---|
| 1 | Part | Standard | InspA | InspB | InspC | Date | Time | Reproducible | Accurate |
| 2 | 1 | 1 | 1 | 1 | 1 | Today | Morning | 1 | 1 |
| 3 | 1 | 1 | 0 | 1 | 1 | Today | Afternoon | 0 | 0 |
| 4 | 2 | 0 | 0 | 0 | 0 | Today | Morning | 1 | 1 |
| 5 | 2 | 0 | 0 | 0 | 1 | Today | Afternoon | 0 | 0 |
| 6 | 1 | 1 | 1 | 1 | 1 | Last Week | Morning | 1 | 1 |
| 7 | 1 | 1 | 1 | 1 | 0 | Last Week | Afternoon | 0 | 0 |
| 8 | 2 | 0 | 0 | 0 | 1 | Last Week | Morning | 0 | 0 |
| 9 | 2 | 0 | 0 | 0 | 0 | Last Week | Afternoon | 1 | 1 |

correct. However, in cell C3 the unit is classified as unacceptable when it actually is acceptable. Continuing this evaluation shows that inspector A made the correct assessment 7 out of 8 times, for an accuracy of 0.875 or 87.5%. The results for all inspectors are given in Table 9.8.

**Table 9.8** Inspector Accuracies

| Inspector | A | B | C |
|---|---|---|---|
| Accuracy | 87.5% | 100.0% | 62.5% |

## Repeatability and Pairwise Reproducibility

Repeatability is defined in Table 9.5 as the same inspector getting the same result when evaluating the same item more than once within a short time interval. Looking at InspA we see that when she evaluated Part 1 in the morning of "Today" she classified it as acceptable (1), but in the afternoon she said it was unacceptable (0). The other three morning/afternoon classifications matched each other. Thus, her repeatability is 3/4 or 75%.

Pairwise reproducibility is the comparison of each inspector with every other inspector when checking the same part at the same time on the same day. For example, on Part 1/Morning/Today, InspA's classification matched that of InspB. However, for Part 1/*Afternoon*/Today InspA's classification was different than that of InspB. There are eight such comparisons for each pair of inspectors. Looking at InspA versus InspB we see that they agreed 7 of the 8 times, for a pairwise repeatability of 7/8 = 0.875.

In Table 9.9 the diagonal values are the repeatability scores and the off-diagonal elements are the pairwise reproducibility scores. The results are shown for "Today," "Last Week" and both combined.

**Table 9.9** Repeatability and Pairwise Reproducibility for Both Days Combined

| | Overall | | | | Today | | | | Last Week | | |
|---|---|---|---|---|---|---|---|---|---|---|---|
| | A | B | C | | A | B | C | | A | B | C |
| A | 0.75 | 0.88 | 0.50 | A | 0.50 | 0.75 | 0.50 | A | 1.00 | 1.00 | 0.50 |
| B | | 1.00 | 0.50 | B | | 1.00 | 0.75 | B | | 1.00 | 0.50 |
| C | | | 0.25 | C | | | 0.50 | C | | | 0.00 |

## Overall Repeatability, Reproducibility, Accuracy, and Bias

Information is always lost when summary statistics are used, but the data reduction often makes the tradeoff worthwhile. The calculations for the overall statistics are operationally defined as follows:

- *Repeatability* is the average of the repeatability scores for the two days combined; that is, $(0.75 + 1.00 + 0.25)/3 = 0.67$.
- *Reproducibility* is the average of the reproducibility scores for the two days combined (see Table 9.7); that is:

$$\left(\frac{1+0+1+0}{4}\right) + \left(\frac{1+0+0+1}{4}\right) \bigg/ 2 = 0.50$$

- *Accuracy* is the average of the accuracy scores for the two days combined (see Table 9.7); that is:

$$\left(\frac{1+0+1+0}{4}\right) + \left(\frac{1+0+0+1}{4}\right) \bigg/ 2 = 0.50$$

- *Bias* is the estimated proportion in a category minus the true proportion in the category. In this example the true percent defective is 50% (1 part in 2). Of the 24 evaluations, 12 evaluations classified the item as defective. Thus, the bias is $0.5 - 0.5 = 0$.

## Overall Stability

Stability is calculated for each of the above metrics separately, as shown in Table 9.10.

## Interpretation of Results

- The system *overall* appears to be unbiased and accurate. However, the evaluation of individual inspectors indicates that there is room for improvement.
- The results of the individual accuracy analysis indicate that inspector C has a problem with accuracy, see Table 9.8.

**Table 9.10** Stability Analysis

| Stability of... | Operational Definition of Stability | Stability Result |
|---|---|---|
| Repeatability | Standard deviation of the six repeatabilities (0.5, 1, 0.5, 1, 1, 0) | 0.41 |
| Reproducibility | Standard deviation of the average reproducibilities. For data in Table 9.9, STDEV [(AVERAGE (H2:H5), AVERAGE (H6:H9)] | 0.00 |
| Accuracy | Standard deviation of the average accuracies. For data in Table 9.9, = STDEV [AVERAGE (I2:I5), AVERAGE (I6:I9)] | 0.00 |
| Bias | Average of bias over the 2 weeks | 0.0 |

▲ The results of the R&R (pairwise) indicate that inspector C has a problem with both repeatability and reproducibility; see Table 9.9.

▲ The repeatability numbers are not very stable (Table 9.10). Comparing the diagonal elements for Today with those of Last Week in Table 9.9, we see that inspectors A and C tended to get different results for the different weeks. Otherwise, the system appears to be relatively stable.

▲ Reproducibility of inspectors A and B is not perfect. Some benefit might be obtained from looking at reasons for the difference.

▲ Since inspector B's results are more accurate and repeatable, studying her might lead to the discovery of best practices.

## *Minitab Attribute Gage R&R Example*

Minitab includes a built-in capability to analyze attribute measurement systems, known as "attribute gage R&R." We will repeat the above analysis using Minitab.

Minitab can't work with the data as shown in Table 9.7, it must be rearranged. Once the data are in a format acceptable to Minitab, we enter the Attribute Gage R&R Study dialog box by choosing Stat > Quality Tools > Attribute Gage R&R Study (see Fig. 9.19). Note the checkbox "Categories of the attribute data are ordered." Check this box if the data are ordinal and have more than two levels. Ordinal data means, for example, a 1 is in some sense "bigger" or "better" than a 0. For example, if we ask raters in a taste test a question like the following: "Rate the flavor as 0 (awful), 1 (OK), or 2 (delicious)." Our data are ordinal (acceptable is better than unacceptable), but there are only two levels, so we will not check this box.

Minitab evaluates the repeatability of appraisers by examining how often the appraiser "agrees with him/herself across trials." It does this by looking at all of the classifications for each part and counting the number of *parts* where *all* classifications agreed. For our example each appraiser looked at two parts four times each. Minitab's output, shown in

**418** | The Six Sigma Handbook

**Figure 9.19** Attribute gage R&R dialog box and data layout within appraiser analysis.

Fig. 9.20, indicates that InspA rated 50% of the parts consistently, InspB 100%, and InspC 0%. The 95% confidence interval on the percentage agreement is also shown. The results are displayed graphically in Fig. 9.21.

```
Within Appraiser
Assessment Agreement

Appraiser  # Inspected  # Matched  Percent (%)      95.0% CI
InspA               2          1         50.0 (  1.3,  98.7)
InspB               2          2        100.0 ( 22.4, 100.0)
InspC               2          0          0.0 (  0.0,  77.6)

# Matched: Appraiser agrees with him/herself across trials.
```

**Figure 9.20** Minitab within appraiser assessment agreement.

## Accuracy Analysis

Minitab evaluates accuracy by looking at how often all of an appraiser's classifications for a given part agree with the standard. Figure 9.22 shows the results for our example. As before, Minitab combines the results for both days. The plot of these results is shown in Fig. 9.23.

**Figure 9.21** Plot of appraiser assessment agreement.

```
Each Appraiser vs Standard

Assessment Agreement

Appraiser  # Inspected  # Matched  Percent (%)      95.0% CI
InspA                2         1         50.0  (  1.3,  98.7)
InspB                2         2        100.0  ( 22.4, 100.0)
InspC                2         0          0.0  (  0.0,  77.6)

# Matched: Appraiser's assessment across trials agrees with standard
```

**Figure 9.22** Minitab appraiser versus standard agreement.

**Figure 9.23** Plot of appraiser versus standard assessment agreement.

Minitab also looks at whether or not there is a distinct pattern in the disagreements with the standard. It does this by counting the number of times the appraiser classified an item as a 1 when the standard said it was a 0 (the 1/0 Percent column), how often the appraiser classified an item as a 0 when it was a 1 (the 0/1 Percent column), and how often the appraiser's classifications were mixed, i.e., is not repeatable (the # Mixed Percent column). The results are shown in Fig. 9.24. The results indicate that there is no consistent bias, defined as consistently putting a unit into the same wrong category. The problem, as was shown in the previous analysis, is that appraisers A and C are not repeatable.

```
Assessment Disagreement

Appraiser     # 1/0 Percent (%)    # 0/1 Percent (%)    # Mixed Percent (%)
InspA              0      0.0           0      0.0           1      50.0
InspB              0      0.0           0      0.0           0       0.0
InspC              0      0.0           0      0.0           2     100.0

# 1/0: Assessments across trials = 1 / standard = 0.
# 0/1: Assessments across trials = 0 / standard = 1.
# Mixed: Assessments across trials are not identical.
```

**Figure 9.24**  Minitab appraiser assessment disagreement analysis.

## Between Appraiser Assessments

Next, Minitab looks at all of the appraiser assessments for each part and counts how often every appraiser agrees on the classification of the part. The results, shown in Fig. 9.25, indicate that this never happened during our experiment. The 95% confidence interval is also shown.

```
Between Appraisers
Assessment Agreement

# Inspected # Matched Percent (%)       95.0% CI
     2           0          0.0  (  0.0,  77.6)

# Matched: All appraisers' assessments agree with each other.
```

**Figure 9.25**  Minitab between appraisers assessment agreement.

## All Appraisers Versus Standard

Finally, Minitab looks at all of the appraiser assessments for each part and counts how often every appraiser agrees on the classification of the part and their classification agrees with the standard. This can't be any better than the between appraiser assessment agreement shown in Fig. 9.25. Unsurprisingly, the results, shown in Fig. 9.26, indicate that this never happened during our experiment. The 95% confidence interval is also shown.

```
All Appraisers vs Standard
Assessment Agreement

# Inspected # Matched Percent (%)     95.0% CI
         2         0     0.0 (  0.0,  77.6)

# Matched: All appraisers' assessments agree with standard.
```

**Figure 9.26**  Minitab assessment versus standard agreement across all appraisers.

# CHAPTER 10

# Analyze Phase

The key objectives of the Analyze phase include

- For existing processes, analyze the value stream to identify ways to eliminate the gap between the current performance and the desired performance.
- Analyze the sources of variation that contribute to the gap (for DMAIC) or that will contribute to the design performance (for DMADV).
- Determine the drivers, the little $x$'s that correlate to the customer requirements (CTQ, CTS, CTC) and significantly influence the process or design.
- Use benchmarking techniques described in Chap. 3 to evaluate best in class for similar products or services.

## Value Stream Analysis

A value stream consists of all activities, both value added and non-value added, required to bring a product from raw material into the hands of the customer, a customer requirement from order to delivery, and a design from concept to launch. Value stream improvement usually begins at the door-to-door level within a facility (that is, from the door in to the door out), and then expands outward to eventually encompass the full value stream including suppliers and customers (Womack and Jones, 1996, p. 311). A value stream consists of product and service flows, as well as information flows.

Lean principles are used to analyze the value stream. Lean, also known as the Lean Production System, has its origins in the post–World War II era in Japan. It was developed by Taiichi Ohno, a Toyota production executive, in response to a number of problems that plagued Japanese industry. The main problem was that of high-variety production required to serve the domestic Japanese market. Mass production techniques, which were developed by Henry Ford to economically produce long runs of identical product, were ill-suited to the situation faced by Toyota. Today the conditions faced by Toyota in the late 1940s are common throughout industry and Lean is being adopted by businesses all over the world as a way to improve efficiency and to better serve customers.

The Lean approach (the term Lean was coined in the early 1990s by MIT researchers) systematically minimizes waste—sometimes referred to by its Japanese name *muda*—in the value stream. *Muda* includes all types of defective work, not just defective products. Wasted time, motion, and materials are all *muda*. Ohno (1988) identified the following types of waste in business:

- Defects
- Overproduction
- Inventories (in process or finished goods)
- Unnecessary processing
- Unnecessary movement of people
- Unnecessary transport of goods
- Waiting

Womack and Jones (1996) added another type of waste:

- Designing goods and services that don't meet customers' needs

Ohno has since clarified (2013):

I don't know who came up with it but people often talk about "the seven types of waste." This might have started when the book came out, but waste is not limited to seven types. There's an old expression: "He without bad habits has seven," meaning even if you think there's no waste you will find at least seven types. So I came up with overproduction, waiting, etc., but that doesn't mean there are only seven types. So don't bother thinking about "what type of waste is this?" Just get on with it and do *kaizen*.

Clearly, the focus is on value. Value is what customers want or need, and are willing and able to pay for. Waste is any activity that consumes resources but creates no value for the customer, thus waste activities are called "non-value added." Differentiating between the two may not be easy, especially for new products or services, but it must be done. For existing products use focus groups, surveys, and other methods described in this text. For new products, consider the DFSS methods. Most importantly, **do not rely on internal sources!** Most companies start with what they already know and go from there, tweaking their existing offering in some way. Customer input involves asking customers what they like or don't like about the existing offering, or what they'd like to see added or changed. The result is incremental change that may or may not address what the customers are really after. The definition of value must begin with the producer and customer jointly analyzing value and challenging old beliefs.

In *Lean Thinking*, authors Womack and Jones consider a team tasked with reducing defects on supermarket shelving. The number one problem identified was "weld dents," a condition caused when brackets were welded to the shelves. A great deal of effort went into inspecting shelves for this condition, running laboratory tests to determine the impact of weld dents on the durability of the shelves, reworking shelves that had weld dents, etc. Scrap costs were very high. When the team met with customers to try to operationally define unacceptable weld dents they made an amazing discovery: customers didn't know what weld dents were! Even more strange, when shown shelves with no weld dents and those with "extreme" weld dents, customers couldn't care less. However, customers *did* care about the shape of the front of the shelves. They wanted nice, straight looking shelf fronts that looked streamlined when lined up in long supermarket aisles. They were not happy at all with what was being delivered. No one inside the company knew that this was important to customers, and no efforts were underway to improve this aspect of the product.

If the supermarket manager was asked to define value, chances are he wouldn't say "Shelves that have straight fronts that line up." Instead he might say "Shelves that look good to my customers when they look down the aisle." The importance of obtaining the voice of the customer, and using this voice to drive business processes, was discussed in Chaps. 2 and 3. Those vital Six Sigma lessons need to be integrated into Lean as well.

With your definition of value in hand, you can now begin to evaluate which activities add value and which activities are waste. The results are often surprising. In some cases, most process activities are not value added. For example, one Six Sigma team working on improving purchase order (PO) cycle time (defined as the time from receiving a request for a PO to the time the requestor received the PO) conducted a little test. They made a list of all the people whose signature was needed for PO approval. Then the team members (with the approval of the director of purchasing) hand-carried 10 POs through the process. Each purchasing agent was to treat the team member's request as their number 1 priority, dropping every other activity until it was completed. The team discovered that it took an average of about 6 hours to process a PO in this manner. The average processing time in the real world was 6 weeks. Assuming a 40-hour work-week, the value-added time accounted for only 2.5% of the total time a PO was in the system. The remaining 97.5% was waste.

Yet, the full extent of the waste had not been uncovered. During the walk-throughs, the team questioned the need for each approval, and discovered that, in some cases, such as POs for standard hardware or basic supplies, the requestor could be empowered to directly place the order. Many POs could be eliminated completely with automatic *pull* ordering systems (see Chap. 11 for more on pull systems). The value-added portion of the purchase order approval process was tiny indeed.

The immediate impact of such discoveries is often fear. Imagine yourself as the director of purchasing or a purchasing agent. Along comes a team with data that indicate that most of your department is non-value added. Is it any wonder that change agents so often talk about "resistance to change"? Who wouldn't resist change when that change is you losing your job?

Yet this is often the case and the leadership needs to face up to this reality and to plan for it. They have a responsibility to the shareholders that dictates that they reduce waste. They have a responsibility to customers to produce value. But they also have a responsibility to the employees to treat them fairly. Unless all the leadership makes it clear that fair treatment is guaranteed, you can expect strong resistance from people threatened by the change.

In this example, the purchasing department needs to rethink the value they add, that is, their mission. Perhaps their focus is better defined as improving the integration of the supply chain with the rest of the value stream (see the following section for a discussion of the value stream). This might involve looking at how suppliers can help design easier-to-produce parts, how they can deliver to precisely the right place and at precisely the right time, what they can do to help your customers succeed, etc. This is easier to do in the process enterprise, where core business processes control the definition of work (see Chap. 2). In the end, the transformed "purchasing department" will probably look much different than it did at the beginning.

The good news is that when Lean organizations redefine value, they often discover the key to more customers (and more sales). The increased demand often outpaces the rate at which resources are converted from waste to value. Although this isn't guaranteed, it happens often enough to provide a measure of comfort to employees, especially if they see it happening in their own organization. They may still need to acquire new skills to do a different kind of work, but they are usually able to adapt.

When trying to identify waste it may be helpful to think of certain categories of waste. One handy mnemonic is CLOSEDMITTS (Spencer, 1999), as described in Table 10.1.

**Table 10.1** CLOSEDMITTS

| Type of Waste | Example |
| --- | --- |
| Complexity | Unnecessary steps, excessive documentation, too many permissions needed |
| Labor | Inefficient operations, excess headcount |
| Overproduction | Producing more than the customer demands. Producing before the customer needs it |
| Space | Storage for inventory, parts awaiting disposition, parts awaiting rework and scrap storage. Excessively wide aisles. Other wasted space |
| Energy | Wasted power or human energy |
| Defects | Repair, rework, repeated service, multiple calls to resolve problems |
| Materials | Scrap, ordering more than is needed |
| Idle materials | Material that just sits, inventory |
| Time | Waste of time |
| Transportation | Movement that adds no value |
| Safety hazards | Unsafe or accident-prone environments |

## Value Stream Mapping

Value stream mapping, also known as material and information flow mapping, is a variation of process mapping that looks at how value flows into and through a process and to the customer, and how information flow facilitates the work flow. One way to view a process is the logical flow of work. Another view is the physical flow of work. Figure 10.1 shows the logical flow of work for a technical support process. The team determined that the value-added steps were the ones shown with a drop-shadow box. The process map makes it obvious that the bulk of the work in the process is not part of the value stream; it's waste. Based on the times shown for each activity and the path for a given support issue, somewhere between 38 and 49% of the total time is used for value-added activities.

**Figure 10.1** Flow of work through email technical support.

Unfortunately, in many cases some of the non-value-added steps cannot be immediately eliminated. The non-value-added and unnecessary steps that can be eliminated without consequence to the business or the customer are sometimes referred to as Type II waste. A few of the steps involve recording information that can be used in the future to make it faster and easier to find the right answers. This is an example of a non-value-added task that is necessary, based on the internal policies. Sometimes referred to as business value added (BVA) tasks, since they could be justified as necessary for business operations, these tasks cannot be eliminated immediately without consequence. Other examples of typical BVA activities include most quality functions, such as inspections, audits, and SPC, as well as related management approvals. These functions only exist because of poor quality levels: customers would not be willing to pay for these if they had alternate suppliers that could guarantee perfect quality without the cost and risk of these activities. In some cases, BVA activities are related to regulations, such as in the production of pharmaceuticals. In these cases, at least some of the BVA activities can be reduced when justified as unnecessary by sufficient analysis. Many times, BVA activities are excellent targets for significant cost reduction through process improvement and redesign. Of course, in the absence of process improvement and process redesign, the business need still exists.

More elaborate value stream maps can also be developed using somewhat standardized symbols shown in Fig. 10.2. An example value stream map for a manufacturing oper-

**Figure 10.2** Common symbols used in value stream maps (from *MS Visio 2013*).

ation is shown in Fig. 10.3. It should be clear the primary focus on the value stream map is the identification of waste. In that regard, simple is sometimes better, and the use of all the symbols or inclusion of all data should not overwhelm this focus. In Fig. 10.3, the timeline segments provide quick reference for viewers to differentiate between the value-added processing times and the non-value-added lead times for each activity in the system, which are then summarized in the upper-left corner of the figure. Potential targets for improvement are noted by the *kaizen* bursts.

The lead times noted for each activity in Fig. 10.3 are obvious symptoms of waste that prevent continuous flow of the material through the process steps. These non-value-added wait times are key concerns for most lean practitioners, causing non-value-added in-process inventory, ultimately due to the lack of balance in the processing (i.e., flow) rate between preceding and subsequent process activities. In the Improve stage, we seek to balance the flow between the process activities, also known as *level loading* the process, to remove the work in process (WIP) inventories. Obviously, higher levels of WIP inventory are associated with higher costs of material and labor that goes into the WIP, so the improvement in cash flow in reduced WIP inventories is one source of value, Lower WIP inventories also allow flexibility, such as in in response to a change in customer demand.

WIP inventories are prevalent when batch processing is defined for one or more of the process steps. It is a fallacy that processes are more efficient when process work is carried out by specialized departments or equipment that process work in batches. While these specific departments or equipment may be most efficient when run at economic lot quantities designed to minimize set-up time or material delivery costs, they lack efficiency relative to the overall value streams. From a systems perspective, time is lost (i.e., waste is created) when each single item awaits the completion of other items needed to form the batch. Likewise, waste is additionally created when each single item, for which customers are waiting, must wait for the remainder of the batch to be processed. Batch processing is yet another example of process optimization at the expense of system optimization, which Deming often criticized.

Batches are often considered cost effective when one or more of the following situations occur:

- Cost of movement of material is significant
- Process is designed for multiple items
- Setup time dominates the per item cycle time

With the prevalence of the global supply chain in today's economy, the time and costs associated with shipping materials is sometimes considered a necessary burden. Yet in a perfect world, if your supplier was next door and you only needed one replacement part to repair a machine, you'd order the single part needed and be done with it until the next

**Figure 10.3** Example value stream map (Wilson, 2010).

## Process Flow Diagram

**Information Flow:**
- 30/90 Day Forecasts → Home Office Planning
- Weekly Updates often modified daily via telecons
- Customer → Weekly Update, 30/90 Day Forecasts
- Delay: 2 Dys — *Eliminate Delay*
- Weekly / Daily with Shift Updates (to process steps)
- Daily / Weekly (shipping)

### Rework
| | | |
|---|---|---|
| CT | 18 | Min |
| Recovery | 72 | % |
| Operators | 3 | Stf |

17 H / 0.00 Min

*S/D Not needed* | *SMED*

### Bake
| | | |
|---|---|---|
| CT | 210 | Min |
| Batch Size | 24-48 | Unt |
| Avail | 99 | % |
| Operators | 0 | Stf |

### Polish
| | | |
|---|---|---|
| CT | 3.5 | Min |
| CO | 0 | Min |
| Operators | 1 | Stf |

### CNC Lathe
| | | |
|---|---|---|
| CT Lg | 9.5 | Min |
| CT Sml | 8.5 | Min |
| CT | 9.00 | Min |
| CO | 45 | Min |
| Avail | 99 | % |
| Scrap | 0 | % |
| Operators | 2 | Stf |

### Final Assembly
| | | |
|---|---|---|
| CT Lg | 6.5 | Min |
| CT Sml | 6.0 | Min |
| CT | 6.25 | Min |
| CO | 0.2 | Min |
| Scrap | 0.2 | % |
| Cells | 2 | Off |
| Staff per cell | 2 | Stf |
| Operators | 4 | Stf |

### Final Test
| | | |
|---|---|---|
| CT | 6.0 | Min |
| CO | 0 | Min |
| Scrap | 4.8 | % |
| Operators | 4 | Stf |

### Pack & Ship
| | | |
|---|---|---|
| Monthly Demand | 3550 | Unt |
| CT | 2.5 | Min |
| Operators | 1 | Stf |

**Timeline:**
| 5.0 Hrs | 28 Hrs | 24 Hrs | 1 Hrs | 0.2 Hrs |
|---|---|---|---|---|
| 3.50 Min | 9.00 Min | 6.25 Min | 6.00 Min | 2.50 Min |

### Summary
| | | |
|---|---|---|
| Total Lead Time | 210.20 | Hrs |
| Total Processing Time | 1.02 | Hrs |

time you needed the part, since it was so easy to replace. If your supplier is on the other side of the planet, and it takes weeks or months to receive a replacement part, you're likely to order several to have a few on hand, to avoid waiting again. This is waste, since you've spent money before you needed to, and you can't spend that money elsewhere. If the machine is eventually replaced, you're stuck with replacement parts you'll never use. Your supplier's other customers may have had to wait to receive their parts, since you bought more than you needed. In the Improve stage, we would recognize these inefficiencies and reduce the space between supplier and customer to reduce costs of movement. Offshore production efficiencies may be less than perceived if true costs of consumer shipments, consumer holding unused inventory, and consumers waiting for delayed shipments are considered.

When processes are designed to produce multiple items, they may be inefficient for small batches. Consider baking a batch of cookies when you really only want to eat a few fresh ones today, and have no interest in stale ones tomorrow. In Lean, you sometimes hear about so-called monuments, which refer to large and expensive fixtures or technologies created for a specific task around which all related processes were designed, similar to how traffic is routed around a statue (i.e., a monument) in the town square. The monuments were usually designed to efficiently process product at high quantity to achieve "efficiencies of scale," and often purchased at great initial cost with a long lifecycle planned to achieve its eventual cost savings. The monuments are often an obstacle in our drive to small batch sizes.

Setup time is the time to change from the last item of the previous order to the first good item of the next order. When setup time dominates, we claim "efficiency" in spreading the setup costs across the batch. In the automotive industry, circa pre-1980, dedicated lines for each product was commonplace, requiring week-long shutdowns to switch from product to product. Likewise, printing presses required elaborate setup procedures, causing book publishers to order large quantities to spread the setup costs over many items. This resulted in large inventories, a disincentive to revise a book with new material. With the advent of digital printing, setup time was reduced, smaller batch sizes became affordable, and a revolution occurred in book publishing, allowing even unknown authors to self-publish with minimal cost.

Setup includes preparation, replacement, location and adjustment activities.

▲ Preparation refers to the tasks associated with getting or storing tools or WIP needed for the process. For example, retrieving printer paper from closet; downloading the process instructions on the computer; moving completed items to the next process step; starting-up software that we need to process the order, and so on.

▲ Replacement refers to the tasks associated with adding or removing items or tools. For example, the movement of test fixtures; loading of new material into the hopper; loading paper in the copy machine.

▲ Location tasks are those associated with positioning or placement during setup. Examples include setting temperature profiles for heating; adjusting cutoff length for specific product; placing the chunk of deli meat in the slicer.

▲ Adjustment refers to tasks associated with ensuring correct process settings. Examples include monitoring temperature of furnace; checking cutoff length; proofing copy before printing.

In the Analyze stage, we want to differentiate between these different setup activities, since our response in the Improve stage will vary based on the type. Additionally, in the Analyze stage, we should note whether each setup activity is *internal* or *external*. Internal setup activities require an inactive (i.e., shutdown) process, meaning that no items can be run while the setup activity is taking place. For example, loading a standard dishwasher is an internal process activity: no items can be cleaned while it is being loaded. External setup activities may be performed while the process is operational (i.e., they are offline activities). In the Improve stage, we will seek to convert internal process activities to external process activities to achieve overall process cycle time reduction.

It should also be obvious that any other disruptions to the process flow, notably process rework and defects, will also prevent continuous flow of the process material. Most lean experts consider low defect rates and consistent process stability precursors to establishing flow through the lean methodologies (Wilson, 2010). Inventories often exist specifically to mask those process deficiencies, and removing the inventories without first improving the quality and repeatability of the process through the statistical process control methods of Ch. 8 will cause great hardship for your customer. This is the fallacy of stand-alone Lean programs that neglect the fundamental quality levels critical to success.

## *Spaghetti Charts*

Current state physical work flow is often depicted on *spaghetti charts*. A spaghetti chart is a map of the path taken by a specific product as it travels down the value stream in a mass-production organization, so-called because the product's route typically looks like a plate of spaghetti. To create a spaghetti chart, like the one shown on the left in Fig. 10.4, tell a person to "be the part" and to physically walk through the process as the part would go through it. Sometimes a part travels *miles* in the original process configuration, and only a few feet in the Lean layout. The Lean layout is shown on the right in Fig. 10.4. The difference between the current state layout and the Lean layout demonstrates waste due to unnecessary movement of product or personnel.

When setting goals for a future state process, it is often helpful to stretch the mind. One way to become inspired is to identify the absolute best in class performance for a particular activity. For example, the quick lube joints' claim to exceptional value is that they

**Figure 10.4** Spaghetti chart versus Lean flow.

can get you in and out in 15 minutes or less, much quicker than the corner service station, which often took a couple of hours or more. But consider the pit crew of a NASCAR racing team, which can perform maintenance on a car so fast (14 seconds or less) they make your local Quickie Lube look like they're working at a crawl. And during those 14 seconds they do a great deal more than change the car's fluids. They gas it up, wash the windows, change all of the tires, etc. There are many published examples of Lean achievements that can serve to educate and inspire. At the CAMI factory operated by GM and Suzuki, machine changeover time was reduced from 36 hours to 6 minutes. Methods for achieving improved flow are discussed in Chap. 11.

## Analyzing the Sources of Variation

Whether the focus of the Six Sigma project relates directly to the mechanics of the value stream, such as cycle time and cost/capacity-related resource allocation, or indirectly, as in the relative quality of its output, a critical piece of the analysis is the understanding of the relative contributions of the sources of variation impacting the value stream. A review of the statistical control chart constructed in the Measure phase will provide input as to the type of variation: common cause intrinsic to the process or special causes that occur sporadically under specific conditions. As discussed in the Measure stage, the response to each of these types of variation differs significantly.

The potential sources of process variation may be brainstormed by the Six Sigma team using a cause and effect diagram. These potential causes must then be analyzed for their significance using more advanced statistical tools, including designed experiments and their associated enumerative methods.

The basic statistical methods, including confidence intervals and hypothesis tests, are discussed in the sections that follow. While these rather simple statistical methods may be used to directly compare a sample to its desired properties, or one sample to another, designed experiments will build on these concepts in applying ANOVA (analysis of variance) techniques to multiple sources of variation, allowing quantification of the relative contribution of each source to the total error. General regression and correlation analysis is presented as a precursor to the designed experiments, to aid in the understanding of the experimental analysis techniques.

## *Cause and Effect Diagrams*

With most practical applications, the number of possible causes for any given problem can be huge. Dr. Kaoru Ishikawa developed a simple method of graphically displaying the causes of any given quality problem. His method is called by several names: the Ishikawa diagram, the fishbone diagram, and the cause and effect diagram.

Cause and effect diagrams are tools that are used to organize and graphically display all of the knowledge a group has relating to a particular problem.

Constructing the cause and effect diagram is very simple. The steps are

1. Draw a box on the far right-hand side of a large sheet of paper and draw a horizontal arrow that points to the box. Inside of the box, write the description of the problem you are trying to solve.
2. Write the names of the categories above and below the horizontal line. Think of these as branches from the main trunk of the tree.
3. Draw in the detailed cause data for each category. Think of these as limbs and twigs on the branches.

A good cause and effect diagram will have many "twigs," as shown in Fig. 10.5. If your cause and effect diagram doesn't have a lot of smaller branches and twigs, it shows that the understanding of the problem is superficial. Chances are that you need the help of someone outside of your group to aid in the understanding, perhaps someone more closely associated with the problem.

Cause and effect diagrams come in several basic types. The dispersion analysis type is created by repeatedly asking "why does this dispersion occur?" For example, we might want to know why all of our fresh peaches don't have the same color. A production process class cause and effect diagram uses production processes as the main categories, or branches of the diagram, with processes joined by the horizontal line.

**Figure 10.5** Cause and effect diagram.

The cause enumeration cause and effect diagram simply displays all possible causes of a given problem grouped according to rational categories. This type of cause and effect diagram lends itself readily to the brainstorming approach we are using.

A variation of the basic cause and effect diagram, developed by Dr. Ryuji Fukuda of Japan, is cause and effect diagrams with the addition of cards, or CEDAC. The main difference is that the group gathers ideas outside of the meeting room on small cards, as well as in group meetings. The cards also serve as a vehicle for gathering input from people who are not in the group; they can be distributed to anyone involved with the process. Often the cards provide more information than the brief entries on a standard cause and effect diagram. The cause and effect diagram is built by actually placing the cards on the branches.

## Boxplots

A *boxplot* displays summary statistics for a set of distributions. It is a plot of the 25th, 50th, and 75th percentiles, as well as values far removed from the rest.

Figure 10.6 shows an annotated sketch of a boxplot. The lower boundary of the box is the 25th percentile. Tukey refers to the 25th and 75th percentile "hinges." Note that the

**Figure 10.6** Annotated boxplot.

50th percentile is the median of the overall data set, the 25th percentile is the median of those values below the median, and the 75th percentile is the median of those values above the median. The horizontal line inside the box represents the median. Fifty percent of the cases are included within the box. The box length corresponds to the interquartile range, which is the difference between the 25th and 75th percentiles.

The boxplot includes two categories of cases with outlying values. Cases with values that are more than 3 box-lengths from the upper or lower edge of the box are called extreme values. On the boxplot, these are designated with an asterisk (*). Cases with values that are between 1.5 and 3 box-lengths from the upper or lower edge of the box are called outliers and are designated with a circle. The largest and smallest observed values that aren't outliers are also shown. Lines are drawn from the ends of the box to these values. (These lines are sometimes called whiskers and the plot is then called a box-and-whiskers plot.)

Despite its simplicity, the boxplot contains an impressive amount of information. From the median you can determine the central tendency, or location. From the length of the box, you can determine the spread, or variability, of your observations. If the median is not in the center of the box, you know that the observed values are skewed. If the median is closer to the bottom of the box than to the top, the data are positively skewed. If the median is closer to the top of the box than to the bottom, the opposite is true: the distribution is negatively skewed. The length of the tail is shown by the whiskers and the outlying and extreme points.

Boxplots are particularly useful for comparing the distribution of values in several groups. Figure 10.7 shows boxplots for the salaries for several different job titles.

**Figure 10.7** Boxplots of salary by job category.

The boxplot makes it easy to see the different properties of the distributions. The location, variability, and shapes of the distributions are obvious at a glance. This ease of interpretation is something that statistics alone cannot provide.

## *Statistical Inference*

This section discusses the basic concept of statistical inference. The reader should also consult the glossary in Appendix 1 for additional information. Inferential statistics belong to the enumerative class of statistical methods. All statements made in this section are valid only for stable processes, that is, processes in statistical control. Although most applications of Six Sigma are analytic, there are times when enumerative statistics prove useful. The term *inference* is defined as (1) the act or process of deriving logical conclusions from premises known or assumed to be true, or (2) the act of reasoning from factual knowledge or evidence. Inferential statistics provide information that is used in the process of inference. As can be seen from the definitions, inference involves two domains: the premises and the evidence or factual knowledge. Additionally, there are two conceptual frameworks for addressing premises questions in inference: the design-based approach and the model-based approach.

As discussed by Koch and Gillings (1983), a statistical analysis whose only assumptions are random selection of units or random allocation of units to experimental conditions results in *design-based inferences*; or, equivalently, randomization-based inferences. The objective is to structure sampling such that the sampled population has the same characteristics as the target population. If this is accomplished, then inferences from the sample are said to have internal validity. A limitation on design-based inferences for experimental studies is that formal conclusions are restricted to the finite population of subjects that actually received treatment, that is, they lack *external validity*. However, if sites and subjects are selected at random from larger eligible sets, then models with random effects provide one possible way of addressing both internal and external validity considerations. One important consideration for external validity is that the sample coverage includes all relevant subpopulations; another is that treatment differences be homogeneous across subpopulations. A common application of design-based inference is the survey.

Alternatively, if assumptions external to the study design are required to extend inferences to the target population, then statistical analyses based on postulated probability distributional forms (e.g., binomial, normal, etc.) or other stochastic processes yield *model-based inferences*. A focus of distinction between design-based and model-based studies is the population to which the results are generalized rather than the nature of the statistical methods applied. When using a model-based approach, external validity requires substantive justification for the model's assumptions, as well as statistical evaluation of the assumptions.

Statistical inference is used to provide probabilistic statements regarding a scientific inference. Science attempts to provide answers to basic questions, such as can this machine meet our requirements? Is the quality of this lot within the terms of our contract? Does the new method of processing produce better results than the old? These questions are answered by conducting an experiment, which produces data. If the data vary, then statistical inference is necessary to interpret the answers to the questions posed. A statistical model is developed to describe the probabilistic structure relating the observed data to the quantity of interest (the *parameters*), that is, a scientific hypothesis is formulated. Rules are applied to the data and the scientific hypothesis is either rejected or not. In formal tests of a hypothesis, there are usually two mutually exclusive and exhaustive hypotheses formulated: a *null hypothesis* and an *alternate hypothesis*.

## Chi-Square, Student's t, and f Distributions

In addition to the distributions present earlier in the Measure phase, these three distributions are used in Six Sigma to test hypotheses, construct confidence intervals, and compute control limits.

### Chi-Square Distribution

Many characteristics encountered in Six Sigma have normal or approximately normal distributions. It can be shown that in these instances the distribution of sample variances has the form (except for a constant) of a chi-square distribution, symbolized $\chi^2$. Tables have been constructed giving abscissa values for selected ordinates of the cumulative $\chi^2$ distribution. One such table is given in Appendix 4.

The $\chi^2$ statistic is calculated for a sample of $n$ items with a sample standard deviation $s$ and a population standard deviation $\sigma$ as:

$$\chi^2 = \frac{(n-1)s^2}{\sigma^2} \tag{10.1}$$

The $\chi^2$ distribution varies with the degrees of freedom $\upsilon$, which for our purposes is equal to the sample size minus 1. For each value of $\upsilon$ there is a different $\chi^2$ distribution. Equation (10.2) gives the pdf for the $\chi^2$.

$$f(\chi^2) = \frac{e^{-\chi^2/2}(\chi^2)^{(\upsilon-2)/2}}{2^{\upsilon/2}\left(\frac{\upsilon-2}{2}\right)!} \tag{10.2}$$

Figure 10.8 shows the pdf for $\upsilon = 4$.

**Figure 10.8** $\chi^2$ pdf for $\upsilon = 4$.

*Example of Chi-Square.* The use of $\chi^2$ is illustrated in this example to find the probability that the variance of a sample of $n$ items from a specified normal universe will equal or exceed a given value $s^2$; we compute $\chi^2 = (n-1)\,s^2/\sigma^2$. Now, let's suppose that we sample $n = 10$ items from a process with $\sigma^2 = 25$ and wish to determine the probability that the sample variance will exceed 50. Then:

$$\frac{(n-1)s^2}{\sigma^2} = \frac{9(50)}{25} = 18$$

We look in Appendix 4 ($\chi^2$) at the line for $\upsilon = 10 - 1 = 9$ and note that 18 falls between the columns for the percentage points of 0.025 and 0.05. Thus, the probability of getting a sample variance in excess of 50 is about 3%.

It is also possible to determine the sample variance that would be exceeded only a stated percentage of the time. For example, we might want to be alerted when the sample variance exceeded a value that should occur only once in 100 times. Then we set up the $\chi^2$ equation, find the critical value from Appendix 4, and solve for the sample variance. Using the same values as above, the value of $s^2$ that would be exceeded only once in 100 times is found as follows:

$$\frac{9s^2}{\sigma^2} = \frac{9s^2}{25} = 21.7 \rightarrow s^2 = \frac{21.7 \times 25}{9} = 60.278$$

In other words, the variance of samples of size 10, taken from this process, should be less than 60.278, 99% of the time.

*Example of Chi-Squared Probability Calculations Using Microsoft Excel.* Microsoft Excel has a built-in capability to calculate chi-squared probabilities. To solve the above problem using Excel, use the CHISQ.DIST.RT function and enter the calculated $\chi^2$ value and the degrees of freedom as shown in Fig. 10.9. The formula result near the bottom of the figure gives the desired probability.

**Figure 10.9** Example of finding chi-squared probability using Microsoft Excel.

The CHISQ.DIST.RT function provides the probability of exceeding the specified chi-square value (i.e., the percentage associated with the right tail of the distribution). In addition, the CHISQ.DIST function is also available, providing the probability the sample variance will be less than the stated value (i.e., the probability associated with the left tail of the distribution) when its CUMULATIVE parameter is set to TRUE (otherwise, the pdf is provided). Prior to Excel 2013, the CHIDIST function was used to calculate the right tail probability. The left tail probability is calculated by subtracting the right tail probability from one.

*Example of Inverse Chi-Squared Probability Calculations Using Microsoft Excel.* Microsoft Excel has a built-in capability to calculate chi-squared probabilities, making it unnecessary to look up the probabilities in tables. To find the critical chi-squared value for the above problem using Excel, use the CHISQ.INV.RT function and enter the desired probability (0.01) and degrees of freedom (10-1), as shown in Fig. 10.10. The formula result near the bottom of the figure gives the desired critical value of 21.7.

The CHISQ.INV function is also available, providing the chi-square value associated with the left tail of the distribution at the stated alpha value. Prior to Excel 2013, the CHIINV function provided the chi-square value associated with the right-tail alpha value entered.

**Figure 10.10** Example of finding inverse chi-squared probability using Microsoft Excel.

## Student's *t* Distribution

The *t* statistic is commonly used to test hypotheses regarding means, regression coefficients, and a wide variety of other statistics used in quality engineering. "Student" was the pseudonym of W. S. Gosset, whose need to quantify the results of small-scale experiments motivated him to develop and tabulate the probability integral of the ratio that is now known as the *t* statistic and is shown in Eq. (10.3).

$$t = \frac{\bar{X} - \mu}{s/\sqrt{n}} \tag{10.3}$$

In Eq. (10.3), the denominator is the standard deviation of the sample mean. Percentage points of the corresponding distribution function of *t* may be found in Appendix 3. There is a *t* distribution for each sample size of $n > 1$. As the sample size increases, the *t* distribution approaches the shape of the normal distribution, as shown in Fig. 10.11.

One of the simplest (and most common) applications of the student's *t* test involves using a sample from a normal population with mean $\mu$ and variance $\sigma^2$. This is demonstrated in the hypothesis testing section later in this chapter.

## *F* Distribution

Suppose we have two random samples drawn from a normal population. Let $s_1^2$ be the variance of the first sample and $s_1^2$ be the variance of the second sample. The two samples need not have the same sample size. The statistic *F* given by:

$$F = \frac{s_1^2}{s_2^2} \tag{10.4}$$

**Figure 10.11** Student's t distributions.

has a sampling distribution called the *F distribution*. There are two sample variances involved and two sets of degrees of freedom, $n_1 - 1$ in the numerator and $n_2 - 1$ in the denominator. Appendixes 5 and 6 provide values for the 1 and 5 percentage points for the *F* distribution. The percentages refer to the areas to the right of the values given in the tables. Figure 10.12 illustrates two *F* distributions.

## *Point and Interval Estimation*

So far, we have introduced a number of important statistics including the sample mean, the sample standard deviation, and the sample variance. These sample statistics are called *point estimators* because they are single values used to represent population parameters. It is also possible to construct an interval about the statistics that has a predetermined probability of including the true population parameter. This interval is called a *confidence interval*. Interval estimation is an alternative to point estimation that gives us a better idea of the magnitude of the sampling error. Confidence intervals can be either one-sided or two-sided. A one-sided or confidence interval places an upper or lower bound on the value of a parameter with a specified level of confidence. A two-sided confidence interval places both upper and lower bounds.

In almost all practical applications of enumerative statistics, including Six Sigma applications, we make *inferences* about *populations* based on data from *samples*. In this chapter, we have talked about sample averages and standard deviations; we have even used these numbers to make statements about future performance, such as long-term yields or potential failures. A problem arises that is of considerable practical importance: any

**Figure 10.12** F distributions.

estimate that is based on a sample has some amount of sampling error. This is true even though the sample estimates are the "best estimates" in the sense that they are (usually) unbiased estimators of the population parameters.

### Estimates of the Mean When σ Is Known

For random samples with replacement, the sampling distribution of $\bar{X}$ has a mean $\mu$ and a standard deviation equal to $\sigma/\sqrt{n}$. For large samples the sampling distribution of $\bar{X}$ is approximately normal, and normal tables can be used to find the probability that a sample mean will be within a given distance of $\mu$.

A confidence interval can be defined such that 1-α percent of the samples will include $\mu$:

$$\bar{X} - Z_{\alpha/2} \frac{\sigma}{\sqrt{n}} < \mu < \bar{X} + Z_{\alpha/2} \frac{\sigma}{\sqrt{n}} \qquad (10.5)$$

For example, in 95% of the samples we will observe a mean within ±1.96 standard deviations of $\mu$. In other words, in 95% of the samples the interval from $\bar{X} - 1.96\sigma/\sqrt{n}$ to $\bar{X} + 1.96\sigma/\sqrt{n}$ will include $\mu$. This interval is called a "95% confidence interval for estimating $\mu$." It is usually shown using inequality symbols:

$$\bar{X} - 1.96\sigma/\sqrt{n} < \mu < \bar{X} + 1.96\sigma/\sqrt{n}$$

The factor 1.96 is the Z value obtained from the normal table in Appendix 2. It corresponds to the Z value beyond which 2.5% of the population lie. Since the normal distribution is symmetric, 2.5% of the distribution lies above Z and 2.5% below −Z. The notation commonly used to denote Z values for confidence interval construction or hypothesis testing is $Z_{\alpha/2}$ where $100(1 - \alpha)$ is the desired confidence level in percent. For example, if we want 95% confidence, $\alpha = 0.05$, $100(1 - \alpha) = 95\%$, and $Z_{0.025} = 1.96$. In hypothesis testing the value of $\alpha$ is known as the *significance level*.

***Example: Estimating μ When σ Is Known.*** Suppose that σ is known to be 2.8. Assume that we collect a sample of $n = 16$ and compute $\bar{X} = 15.7$. Using the equation mentioned in previous section, we find the 95% confidence interval for μ as follows:

$$\bar{X} - 1.96\sigma/\sqrt{n} < \mu < \bar{X} + 1.96\sigma/\sqrt{n}$$

$$15.7 - 1.96(2.8/\sqrt{16}) < \mu < 15.7 + 1.96(2.8/\sqrt{16})$$

$$14.33 < \mu < 17.07$$

There is a 95% *level of confidence* associated with this interval. The numbers 14.33 and 17.07 are sometimes referred to as the *confidence limits*.

Note that this is a two-sided confidence interval. There is a 2.5% probability that 17.07 is lower than μ and a 2.5% probability that 14.33 is greater than μ. (Stated differently, 95% of the confidence intervals created from samples from this population would include the population mean μ). If we were only interested in, say, the probability that μ were greater than 14.33, then the one-sided confidence interval is μ > 14.33 at s 97.5% confidence level.

***Example of Using Microsoft Excel to Calculate the Confidence Interval for the Mean When Sigma Is Known.*** Microsoft Excel has a built-in capability to calculate confidence intervals for the mean. The dialog box in Fig. 10.13 shows the input for the above example. (Prior to Excel 2013, the CONFIDENCE function was used with the same parameters). The formula result near the bottom of the figure provides the interval width as approximately 1.37. To find the lower confidence limit, subtract the calculated width from the mean. To find the upper confidence limit, add the width to the mean.

The confidence functions in Excel assume the two-sided case, and automatically adjust the alpha value entered to alpha/2 to provide the two-sided confidence interval. To find the one-sided confidence interval on the mean in Excel, replace the specified alpha value with twice the alpha value. For example, to obtain the 95% one-sided confidence level, enter 0.10 in place of the 0.05 shown in the figure. This result (1.15) can then be added to the mean to obtain the 95% one-sided upper confidence limit, or subtracted from the mean to obtain the 95% one-sided lower confidence limit.

**Figure 10.13** Example of finding the confidence interval when sigma is known using Microsoft Excel.

### Example of Estimating μ When σ Is Unknown

When σ is not known, as is common in practice, σ is replaced with $s$ in calculating confidence intervals for μ. In addition, $Z_{\alpha/2}$ is replaced with $t_{\alpha/2,n-1}$ and the percentiles are obtained from tables for student's $t$ distribution instead of the normal tables. Let's revisit the example above and assume that instead of knowing σ, it was estimated from the sample. Using the data within the sample of $n = 16$, we computed $s = 2.8$ and $\bar{X} = 15.7$. The $t$ value for $n - 1 = 15$ degrees of freedom at the $\alpha/2 = 0.025$ level is shown in Appendix 3 as 2.131. The 95% confidence interval becomes

$$\bar{X} - 2.131s/\sqrt{n} < \mu < \bar{X} + 2.131s/\sqrt{n}$$

$$15.7 - 2.131(2.8/\sqrt{16}) < \mu < 15.7 + 2.131(2.8/\sqrt{16})$$

$$14.21 < \mu < 17.19$$

It can be seen that this interval is wider than the one obtained for known σ. The $t_{\alpha/2}$ value found for 15 df is 2.131 (see Appendix 3), which is greater than $Z_{\alpha/2} = 1.96$ above.

*Example of Using Microsoft Excel to Calculate the Confidence Interval for the Mean When Sigma Is Unknown.* The CONFIDENCE.T function is used to obtain the two-sided confidence interval for the mean when sigma is calculated based on the sample data, as shown in Figure 10.14.

This function can also be used for one-sided limits, replacing the specified alpha value with twice the alpha value to calculate the one-sided limit. (See explanation in earlier example).

**Figure 10.14** Example of finding the confidence interval when sigma is unknown using Microsoft Excel.

Prior to Excel 2013, Microsoft Excel had no built-in capability to calculate confidence intervals for the mean when sigma is not known. Instead, users calculate t-values using its TINV function. This information can be entered into an equation and used to find the desired confidence limits. Note the TINV function assumes a two-sided case, so that the specified probability (i.e., alpha) value is split equally as $\alpha/2$ between the upper and lower side of the interval. For a one-sided case, the user would enter $2\alpha$ into the TINV function's Probability field to obtain the $t$-value associated with $\alpha$. For example, a one-sided 95% upper confidence limit on the mean is calculated as:

$$= 15.7 + TINV(0.10, 15) \times (2.8/SQRT(16))$$

This quantity evaluates to 16.93, so $\mu < 16.93$.

Beginning in Excel 2013, the TINV function, designed as noted above for the two-sided case, is replaced with the T.INV.2T function specific to two-sided applications. In the example above, the 95% two-sided $t$-value is obtained in MS Excel as:

$$= T.INV.2T(0.05, 15)$$

The quantity evaluates to 2.131 (identical to the older TINV function result noted above).

The newly added T.INV function is available for one-sided applications, and directly provides the $t$-value for the entered $\alpha$ probability. Note that while the Appendix 3 table in this book assumes the right tail alpha, so that all tabulated values are positive, the MS Excel function provides negative values for alpha values less than 0.5. This impacts the formula for the one-sided lower confidence limit, in that the quantity is added to the calculated mean to obtain both the upper and lower confidence limit (rather than subtracting the quantity to obtain the lower limit). For example, a one-sided 95% lower confidence limit on the mean is calculated as:

$$= 15.7 + \text{T.INV}(0.05, 15) \times (2.8/\text{SQRT}(16))$$

This quantity evaluates to 14.47, so $\mu > 14.47$.

The one-sided 95% upper confidence limit on the mean is calculated as:

$$= 15.7 + \text{T.INV}(0.95, 15) \times (2.8/\text{SQRT}(16))$$

This quantity evaluates to 16.93, so $\mu < 16.93$.

### Estimates of the Standard Deviation and Variance

Confidence intervals can be constructed for the standard deviation $\sigma$ and the variance $\sigma^2$ using the chi-square distribution.

A $(1 - \alpha)$% one-sided upper confidence limit on the variance is defined as:

$$\sigma^2 < \frac{(n-1)s^2}{\chi^2_{1-\alpha, n-1}} \tag{10.6}$$

A $(1 - \alpha)$% one-sided upper confidence limit on the standard deviation is defined as:

$$\sigma < \sqrt{\frac{(n-1)s^2}{\chi^2_{1-\alpha, n-1}}} \tag{10.7}$$

A $(1 - \alpha)$% one-sided lower confidence limit on the variance is defined as:

$$\sigma^2 > \frac{(n-1)s^2}{\chi^2_{\alpha, n-1}} \tag{10.8}$$

A $(1 - \alpha)$% one-sided lower confidence limit on the standard deviation is defined as:

$$\sigma < \sqrt{\frac{(n-1)s^2}{\chi^2_{\alpha, n-1}}} \tag{10.9}$$

*Examples of Standard Deviation and Variance Estimates Using Microsoft Excel.* For example, 98% one-sided confidence levels on the variance and standard deviation can be computed as follows from a sample of 16 items with a sample standard deviation of 5.73.

A 98% one-sided upper confidence limit on the variance is calculated within MS Excel as:

$$=((16-1)*(5.73^2))/\text{CHIINV}(0.98,16-1) \quad \text{(Prior to Excel 2013)}$$

$$=((16-1)*(5.73^2))/\text{CHISQ.INV.RT}(0.98,16-1)$$
(Beginning Excel 2013)

So, $\sigma^2 < 82.29$

A 98% one-sided upper confidence limit on the standard deviation is calculated within MS Excel as:

$$=\text{SQRT}(((16-1)*(5.73^2))/\text{CHIINV}(0.98, 16-1)) \quad \text{(Prior to Excel 2013)}$$

$$=\text{SQRT}(((16-1)*(5.73^2))/\text{CHISQ.INV.RT}(0.98, 16-1))$$
(Beginning Excel 2013)

So, $\sigma < 9.07$

A 98% one-sided lower confidence limit on the variance is calculated within MS Excel as:

$$=((16-1)*(5.73^2))/\text{CHIINV}(1-0.98,16-1) \quad \text{(Prior to Excel 2013)}$$

$$=((16-1)*(5.73^2))/\text{CHISQ.INV}(0.98, 16-1)$$
(Beginning Excel 2013)

So, $\sigma^2 > 17.43$

A 98% one-sided lower confidence limit on the standard deviation is calculated within MS Excel as:

$$=\text{SQRT}(((16-1)*(5.73^2))/\text{CHIINV}(1-0.98,16-1))$$
(Prior to Excel 2013)

$$=\text{SQRT}(((16-1)*(5.73^2))/\text{CHISQ.INV}(0.98,16-1))$$
(Beginning Excel 2013)

So, $\sigma > 4.17$

## Estimates of Occurrence Intervals

The chi-square distribution can also be used to define confidence intervals on the number of occurrences.

A $(1 - \alpha)$% two-sided confidence interval on a sample of $r$ occurrences is calculated as:

$$\frac{\chi^2_{1-\frac{\alpha}{2},2r}}{2} < \mu < \frac{\chi^2_{\frac{\alpha}{2},2r+2}}{2}$$

A $(1 - \alpha)$% one-sided upper confidence limit on a sample of $r$ occurrences is calculated as:

$$\mu < \frac{\chi^2_{\alpha, 2r+2}}{2}$$

A $(1 - \alpha)$% one-sided lower confidence limit on a sample of $r$ occurrences is calculated as:

$$\mu < \frac{\chi^2_{1-\alpha, 2r}}{2}$$

*Example of Confidence Interval on Occurrence Using Microsoft Excel.* Seven complaints were received in the last fiscal quarter. The 95% confidence limits on the average number of complaints in a fiscal quarter are calculated as follows.

The 95% two-sided upper confidence limit on the number of occurrences is calculated within MS Excel as:

=CHIINV(0.025,(2*7)+2)/2    (Prior to Excel 2013)

= CHISQ.INV.RT(0.025,(2*7)+2)/2    (Beginning Excel 2013)

Each of these evaluates to 14.4.

The 95% two-sided lower confidence limit on the number of occurrences is calculated within MS Excel as:

=CHIINV(0.975,(2*7))/2    (Prior to Excel 2013)

=CHISQ.INV.RT(0.975,(2*7))/2    (Beginning Excel 2013)

Each of these evaluates to 2.8, so the 95% two-sided confidence interval on the number of occurrences is $2.8 < \mu < 14.4$.

A 95% one-sided upper confidence limit on the number of occurrences is calculated within MS Excel as:

=CHIINV(0.05,(2*7)+2)/2    (Prior to Excel 2013)

=CHISQ.INV.RT(0.05,(2*7)+2)/2    (Beginning Excel 2013)

Each of these evaluate to 13.1, so $\mu < 13.1$.

The 95% two-sided lower confidence limit on the number of occurrences is calculated within MS Excel as:

$$=\text{CHIINV}(0.95,(2*7))/2 \quad \text{(Prior to Excel 2013)}$$

$$=\text{CHISQ.INV.RT}(0.95,(2*7))/2 \quad \text{(Beginning Excel 2013)}$$

Each of these evaluates to 3.3, so $\mu > 3.3$.

## Hypothesis Testing

Statistical inference generally involves four steps:

1. Formulating a hypothesis about the population or "state of nature"
2. Collecting a sample of observations from the population
3. Calculating statistics based on the sample
4. Either accepting or rejecting the hypothesis based on a predetermined acceptance criterion

There are two types of error associated with statistical inference:

- Type I error ($\alpha$ error)—The probability that a hypothesis that is actually true will be rejected. The value of $\alpha$ is known as the significance level of the test.
- Type II error ($\beta$ error)—The probability that a hypothesis that is actually false will be accepted. Type II errors are often plotted in what is known as an operating characteristics curve.

Confidence intervals are usually constructed as part of a *statistical test of hypotheses*. The hypothesis test is designed to help us make an inference about the true population value at a desired level of confidence. We will look at a few examples of how hypothesis testing can be used in Six Sigma applications.

### Example: Two-Sided Hypothesis Test of Sample Mean

Experiment: The nominal specification for filling a bottle with a test chemical is 30 cc. The plan is to draw a sample of $n = 25$ units from a stable process and, using the sample mean and standard deviation, construct a two-sided confidence interval (an interval that extends on either side of the sample average) that has a 95% probability of including the true population mean. If the interval includes 30, conclude that the lot mean is 30, otherwise conclude that the lot mean is not 30.

Result: A sample of 25 bottles was measured and the following statistics computed

$$\bar{X} = 28 \text{ cc}$$

$$s = 6 \text{ cc}$$

The appropriate test statistic is *t*, given by the formula

$$t = \frac{\bar{X} - \mu}{s/\sqrt{n}} = \frac{28 - 30}{6/\sqrt{25}} = -1.67$$

Appendix 3 provides values for the *t* statistic at various degrees of freedom (df), $n - 1$. For a 95% two-sided test, $\alpha/2 = 0.025$, so looking in the $t_{.025}$ column and the 24 df row provides a *t* value of 2.064. Since the absolute value of this *t* value is greater than our test statistic, we fail to reject the hypothesis that the lot mean is 30 cc. Using statistical notation, this is shown as:

$$H_0: \mu = 30 \text{ cc (the } null\ hypothesis)$$

$$H_1: \mu \neq 30 \text{ cc (the } alternate\ hypothesis)$$

$$\alpha = .05\ (Type\ I\ error\ or\ level\ of\ significance)$$

Critical region: $-2.064 \leq t_0 \leq +2.064$

Test statistic: $t = -1.67$

Since *t* lies inside the critical region, fail to reject $H_0$, and accept the hypothesis that the lot mean is 30 cc for the data at hand. Failing to reject the null hypothesis is sometimes referred to as the weak conclusion, since we cannot disprove the assertion of the null hypothesis.

Note that the critical value of *t* in the above example can also be calculated in MS Excel using its TINV function as TINV(0.05, 25-1). Beginning Excel 2013, the comparable two-sided T.INV.2T function may be used as T.INV.2T(0.05, 25-1). In both cases, the quantity evaluates to 2.064, since Excel apportions the alpha value equally to the two sides.

## Example: One-Sided Hypothesis Test of Sample Mean

In some cases, rather than testing whether the mean equals an assumed value, we are more interested in whether the mean is greater than or less than a specified value. For example, we might be interested in whether the fill volume in the last example exceeded the minimum requirement of 26 cc. The one-sided hypothesis test on the mean takes the following form:

$$H_0: \mu < 26 \text{ cc (the } null\ hypothesis)$$

$$H_1: \mu \geq 26 \text{ cc (the } alternate\ hypothesis)$$

$$\alpha = .05\ (Type\ I\ error\ or\ level\ of\ significance)$$

Notice that the null hypothesis is stated as less than 26 cc in this example to take advantage of the potential strong conclusion possible in rejecting the null hypothesis. That is, there are two possibilities resulting from the hypothesis test:

- Given sufficient statistical evidence (as determined by the statistics of the sample), we can reject the null hypothesis. This is known as the strong conclusion. In rejecting the null hypothesis, we can assert the alternative hypothesis, so it's useful to set your alternative hypothesis to what you would like to prove.
- Lacking sufficient statistical evidence, we can only fail to reject the null hypothesis. This is known as the weak conclusion. We are not proving the null hypothesis; rather there is insufficient evidence to reject it. In these cases, it's often wise to gather additional data, since failing to reject may merely result from the lack of a sufficient quantity of data. It's useful to think of this in terms of the American judicial system, in which defendants are presumed innocent until proven guilty. When insufficient evidence is presented to prove guilt beyond a reasonable doubt, the defendant is declared not guilty, which does not necessarily mean innocent of the charges. It simply means there is insufficient evidence to prove guilt. So it is with hypothesis tests.

The test statistic for this example is calculated as:

$$t = \frac{\bar{X} - \mu}{s/\sqrt{n}} = \frac{28 - 26}{6/\sqrt{25}} = 1.67$$

The null hypothesis for this one-sided test, where the alternative is stated as $\mu > \mu_0$, is rejected if the calculated $t$-value exceeds the critical value of the test statistic at the right tail of the distribution. (In the case of the alternative hypothesis of $\mu < \mu_0$, the null is rejected if it is less than the critical value in the left tail of the distribution.)

For this one-sided case, the critical value of $t$ at a 95% level may be calculated using MS Excel's TINV function as TINV(0.10,25-1), where the specified alpha value of 0.05 is doubled to compensate for the TINV function's assumption of a two-sided interval. (See discussion in the sections above.) Beginning in Excel 2013, the T.INV function may be used as:

$$\text{T.INV}(0.95, 25\text{-}1)$$

to achieve the same result of 1.71. (Note that both functions have provided the right-tail value; the left-tail value is −1.71.)

Since the calculated value of $t$ of 1.67 is not greater than the critical value of 1.71, we fail to reject the null hypothesis that the mean is less than 26. It may be; we can't disprove that.

A related problem is to determine the level of significance in the one-sided test, that is, the percentile of the distribution corresponding to the calculated *t*-value. This is usually a more interesting (and practical) problem than the rejection of the null based on a pre-defined alpha value, since (as in this example) one might wonder if our calculated value is "close enough" to the critical value that we should collect some additional data. Excel's TDIST function is useful in answering these questions.

For this example, we use TDIST(1.67, 25-1, 1), where the third property (the numeral 1) specifies a one-tail test. (Beginning in Excel 2013, use the T.DIST.RT(1.67, 25-1) to obtain the same result, which is specifically the one-tailed test and hence has only two arguments.) The formula evaluates to 0.054, which indicates the probability of data this extreme (5.4%) if the null hypothesis were accepted. This value is very close to our specified alpha value of 0.05, so additional data may prove helpful to reduce the uncertainty. Alternatively, for most situations we would likely decide that 5.4% is a sufficiently small level of uncertainty and reject the null hypothesis at this level. In Minitab, the test is easily performed using the *1-Sample t* option from the *Stat/Basic Statistics* menu. Select the *Summarized Data* option to enter data from questions posed as above, with the *Options* button allowing specification of the confidence level and the form of the alternative hypothesis (i.e., a one-sided or two-sided test).

### Example: Hypothesis Test of Two Sample Means

A modified *t* statistic can be used to test if two samples are likely to come from populations with the same mean. Excel and Minitab provide easy-to-use tools for this analysis. Although both offer options for assuming either equal variance of the populations or unequal variance of the populations, the latter choice is recommended as a more robust solution.

The two-sided hypothesis test for equality of the two sample means takes the following form:

$$H_0: \mu_1 = \mu_2 \text{ (the null hypothesis)}$$

$$H_1: \mu_1 \neq \mu_2 \text{ (the alternate hypothesis)}$$

$$\alpha = .05 \text{ (Type I error or level of significance)}$$

For example, consider the sample output from two machines, shown next.

| Machine A | Machine B |
|---|---|
| 223.14 | 221.44 |
| 192.18 | 201.69 |
| 216.95 | 246.18 |
| 228.36 | 201.98 |
| 224.3 | 208.26 |
| 247.38 | 198.83 |
| 240.81 | |
| 192.32 | |
| 223.1 | |
| 221.91 | |
| 177.67 | |
| 193.01 | |
| 229.59 | |
| 275.35 | |
| 223.29 | |
| 227.5 | |
| 200.85 | |

In Minitab, use the *2-Sample t* option from the *Stat/Basic Statistics* menu. Select the *Summarized Data* option to enter data from questions posed as above, with the *Options* button allowing specification of the confidence level and the form of the alternative hypothesis (i.e., a one-sided or two-sided test). A hypothesized difference may also be stated but is less common. The output is shown in Fig. 10.15.

**Two-Sample *t*-Test and CI: Machine A, Machine B**

```
Two-sample T for Machine A vs Machine B

            N    Mean   StDev   SE Mean
Machine A  17   219.9   23.7      5.7
Machine B   6   213.1   18.1      7.4
Difference = μ (Machine A) - μ (Machine B)

Estimate for difference:   6.80
95% CI for difference:    (-13.81, 27.42)
T-Test of difference = 0 (vs ≠): T-Value = 0.73   P-Value = 0.483   DF = 11
```

**Figure 10.15** Minitab output for two-sample *t*-test.

The high *P*-value provided in the result indicates the null hypothesis of equal means cannot be rejected at the specified 95% confidence. Similarly, the confidence interval for the difference in population means includes the value 0, so it is possible the means are the same.

Excel provides similar functionality (with additional information) in its *T Test: Two Samples with Unequal Variance* option available within its *Data Analysis* add-on, as shown in Fig. 10.16. (This add-on is included in all versions of Excel, but is often not installed by default.) For a simple output of the *P*-value associated with the test, use the T.TEST function (e.g., =T.TEST(D10:D26,E10:E15,2,3) for a two-sided test or =T.TEST(D10:D26,E10:E15,1,3) for a one-sided test, assuming data in ranges indicated) available beginning with Excel 2013.

t-Test: Two-Sample Assuming Unequal Variances

|  | Machine A | Machine B |
|---|---|---|
| Mean | 219.86529 | 213.0633 |
| Variance | 559.56393 | 328.947 |
| Observations | 17 | 6 |
| Hypothesized mean Difference | 0 | |
| df | 12 | |
| t Stat | 0.7261649 | |
| P(T<=t) one-tail | 0.240831 | |
| t Critical one-tail | 1.7822867 | |
| P(T<=t) two-tail | 0.481662 | |
| t Critical two-tail | 2.1788128 | |

**Figure 10.16** Excel output for two-sided *t*-test of two sample means.

In some cases, data from two samples is paired, in that each of the sample observations is from the same physical item. For example, we might wish to compare the measurements of 10 items, using two difference measurement techniques. This analysis is offered via Minitab's *Paired t* option from the *Stat/Basic Statistics* menu. In Excel, it is an option within the functions noted above.

## Example: Hypothesis Test of Two-Sample Variances

The variance of machine X's output, based on a sample of $n = 25$ taken from a stable process, is 100. Machine Y's variance, based on a sample of 10, is 50. The manufacturing representative from the supplier of machine X contends that the result is a mere "statistical fluke." Assuming that a "statistical fluke" is something that has less than 1 chance in 100, test the hypothesis that both variances are actually equal.

The test statistic used to test for equality of two sample variances is the $F$ statistic, which, for this example, is given by the equation:

$$F = \frac{s_1^2}{s_2^2} = \frac{100}{50} = 2, \text{ numerator df} = 24, \text{ denominator df} = 9$$

Using Appendix 5 for $F_{.99}$, we find that for 24 df in the numerator and 9 df in the denominator $F = 4.73$. Based on this we conclude that the manufacturer of machine X could be right, the result could be a statistical fluke. This example demonstrates the volatile nature of the sampling error of sample variances and standard deviations.

Minitab's *2 Variances* test (from the *Stat/Basic Statistics* menu) may be used for this example. Select the *Sample variances* option from the drop-down list to enable entry of the summarized data, and the *Options* button to specify the confidence level and the form of the alternative hypothesis (i.e., a one-sided or two-sided test); a hypothesized ratio other than 1 may also be stated but is less common. The 99% confidence interval noted in Fig. 10.17 includes the ratio 1, so the variances may be equal at that confidence level (i.e., the null hypothesis of equal variance cannot be rejected).

**Figure 10.17** Minitab output of hypothesis test on variance of two samples.

Excel's *F-Test Two-Sample for Variances* option within the *Data Analysis* add-on can be used when raw data is available, but does not provide an option for summarized data such as shown here.

## Example: Hypothesis Test of a Standard Deviation Compared to a Standard Value

A machine is supposed to produce parts in the range of 0.500 inch plus or minus 0.006 inch. Based on this, your statistician computes that the absolute worst standard deviation tolerable is 0.002 inch. In looking over your capability charts you find that the best machine in the shop has a standard deviation of 0.0022, based on a sample of 25 units. In discussing the situation with the statistician and management, it is agreed that the machine will be used if a one-sided 95% confidence interval on sigma includes 0.002.

The correct statistic for comparing a sample standard deviation with a standard value is the chi-square statistic. For our data we have $s = 0.0022$, $n = 25$, and $\sigma_0 = 0.002$. The $\chi^2$ statistic has $n - 1 = 24$ degrees of freedom. Thus:

$$\chi^2 = \frac{(n-1)s^2}{\sigma^2} = \frac{24 \times (0.0022)^2}{(0.002)^2} = 29.04$$

Appendix 4 gives, in the 0.05 column (since we are constructing a one-sided confidence interval) and the df = 24 row, the critical value $\chi^2 = 36.42$. Since our computed value of $\chi^2$ is less than 36.42, we use the machine. The reader should recognize that all of these exercises involved a number of assumptions, for example, that we "know" that the best machine has a standard deviation of 0.0022. In reality, this knowledge must be confirmed by a stable control chart.

## *Resampling (Bootstrapping)*

A number of criticisms have been raised regarding the methods used for estimation and hypothesis testing:

- They are not intuitive.
- They are based on strong assumptions (e.g., normality) that are often not met in practice.
- They are difficult to learn and to apply.
- They are error-prone.

In recent years a new method of performing these analyses has been developed. It is known as resampling or bootstrapping. The new methods are conceptually quite simple: using the data from a sample, calculate the statistic of interest repeatedly and examine the distribution of the statistic. For example, say you obtained a sample of $n = 25$ measurements from a lot and you wished to determine a confidence interval on the statistic $C_{PK}$.[1] Using resampling, you would tell the computer to select a sample of $n = 25$ *from the sample results*, compute $C_{PK}$, and repeat the process many times, say 10,000 times. You would then determine whatever percentage point value you wished by simply looking at the results. The samples would be taken "with replacement," that is, a particular value from the original sample might appear several times (or not at all) in a resample.

Resampling has many advantages, especially in the era of easily available, low-cost computer power. Spreadsheets can be programmed to resample and calculate the statistics of interest. Compared with traditional statistical methods, resampling is easier for most people to understand. It works without strong assumptions, and it is simple. Resampling doesn't impose as much baggage between the engineering problem and the statistical result as conventional methods. It can also be used for more advanced problems, such as modeling, design of experiments, etc.

For a discussion of the theory behind resampling, see Efron (1982). For a presentation of numerous examples using a resampling computer program, see Simon (1992).

## Regression and Correlation Analysis

The simplest tool used in regression analysis is often called a *scatter plot* or *scatter diagram*. A scatter diagram is a plot of one variable versus another. One variable is called the *independent variable* and it is usually shown on the horizontal (bottom) axis. The other variable is called the *dependent variable* and it is shown on the vertical (side) axis.

Scatter diagrams are used to evaluate cause and effect relationships. The assumption is that the independent variable is causing a change in the dependent variable. Scatter plots are used to answer such questions as "Does the length of training have anything to do with the amount of scrap an operator makes?"

For example, an orchard manager has been keeping track of the weight of peaches on a day by day basis. The data are provided in Table 10.2.

**Table 10.2** Raw Data for Scatter Diagram

| Number | Days on Tree | Weight (Ounces) |
|--------|--------------|-----------------|
| 1      | 75           | 4.5             |
| 2      | 76           | 4.5             |
| 3      | 77           | 4.4             |
| 4      | 78           | 4.6             |
| 5      | 79           | 5.0             |
| 6      | 80           | 4.8             |
| 7      | 80           | 4.9             |
| 8      | 81           | 5.1             |
| 9      | 82           | 5.2             |
| 10     | 82           | 5.2             |

**Table 10.2** Raw Data for Scatter Diagram (*continued*)

| Number | Days on Tree | Weight (Ounces) |
|--------|--------------|-----------------|
| 11 | 83 | 5.5 |
| 12 | 84 | 5.4 |
| 13 | 85 | 5.5 |
| 14 | 85 | 5.5 |
| 15 | 86 | 5.6 |
| 16 | 87 | 5.7 |
| 17 | 88 | 5.8 |
| 18 | 89 | 5.8 |
| 19 | 90 | 6.0 |
| 20 | 90 | 6.1 |

The scatter diagram is shown in Fig. 10.18.

**Figure 10.18** Completed scatter diagram. (From Pyzdek, 1990, p. 68. Copyright © 1990 by Quality Publishing.)

## Pointers for Using Scatter Diagrams

- Scatter diagrams display different patterns that must be interpreted; Fig. 10.19 provides a scatter diagram interpretation guide.
- Be sure that the independent variable, $X$, is varied over a sufficiently large range. When $X$ is changed only a small amount, you may not see a correlation with $Y$, even though the correlation really does exist.
- If you make a prediction for $Y$, for an $X$ value that lies outside of the range you tested, be advised that the prediction is highly questionable and should be tested thoroughly. Predicting a $Y$ value beyond the $X$ range actually tested is called *extrapolation*.

**Figure 10.19** Scatter diagram interpretation guide. (From Pyzdek, 1990, p. 69)

- ▲ Keep an eye out for the effect of variables you didn't evaluate. Often, an uncontrolled variable will wipe out the effect of your $X$ variable. It is also possible that an uncontrolled variable will be causing the effect and you will mistake the $X$ variable you are controlling as the true cause. This problem is much less likely to occur if you choose $X$ levels at random. An example of this is our peaches. It is possible that any number of variables changed steadily over the time period investigated. It is possible that these variables, and not the independent variable, are responsible for the weight gain (e.g., was fertilizer added periodically during the time period investigated?).
- ▲ Beware of "happenstance" data! Happenstance data are data that were collected in the past for a purpose different than constructing a scatter diagram. Since little or no control was exercised over important variables, you may find nearly anything. Happenstance data should be used only to get ideas for further investigation, never for reaching final conclusions. One common problem with happenstance data is that the variable that is truly important is not recorded. For example, records might show a correlation between the defect rate and the shift. However, perhaps the real cause of defects is the ambient temperature, which also changes with the shift.
- ▲ If there is more than one possible source for the dependent variable, try using different plotting symbols for each source. For example, if the orchard manager knew that some peaches were taken from trees near a busy highway, he could use a different symbol for

those peaches. He might find an interaction, that is, perhaps the peaches from trees near the highway have a different growth rate than those from trees deep within the orchard.
- Although it is possible to do advanced analysis without plotting the scatter diagram, this is generally bad practice. This misses the enormous learning opportunity provided by the graphical analysis of the data.

See online video: Scatter Diagrams at www.mhprofessional.com/ssh5.

### Correlation and Regression

Correlation analysis (the study of the strength of the linear relationships among variables) and regression analysis (modeling the relationship between one or more independent variables and a dependent variable) are activities of considerable importance in Six Sigma. A regression problem considers the frequency distributions of one variable when another is held fixed at each of several levels. A correlation problem considers the joint variation of two variables, neither of which is restricted by the experimenter. Correlation and regression analyses are designed to assist the analyst in studying cause and effect. Of course, statistics cannot by themselves establish cause and effect. Proving cause and effect requires sound scientific understanding of the situation at hand. The statistical methods described in this section assist the analyst in performing this task.

### *Linear Models*

A linear model is simply an expression of a type of association between two variables, $x$ and $y$. A *linear relationship* simply means that a change of a given size in $x$ produces a proportionate change in $y$. Linear models have the form:

$$y = a + bx \qquad (10.10)$$

where $a$ and $b$ are constants. The equation simply says that when $x$ changes by 1 unit, $y$ will change by $b$ units. This relationship can be shown graphically.

In Fig. 10.20, $a = 1$ and $b = 2$. The term $a$ is called the intercept and $b$ is called the slope. When $x = 0$, $y$ is equal to the intercept. Fig. 10.20 depicts a perfect linear fit, for example, if $x$ is known we can determine $y$ exactly. Of course, perfect fits are virtually unknown when real data are used. In practice we must deal with error in $x$ and $y$. These issues are discussed below.

Many types of associations are nonlinear, but can be converted to linear for ease of analysis, as described later in this chapter.

When conducting regression and correlation analysis we can distinguish two main types of variables. One type we call *predictor variables* or *independent variables*; the other,

**Figure 10.20**   Scatter diagram of a linear relationship.

*response variables* or *dependent variables*. By predictor or independent variables we usually mean variables that can either be set to a desired variable (e.g., oven temperature) or else take values that can be observed but not controlled (e.g., outdoors ambient humidity). As a result of changes that are deliberately made, or simply take place in the predictor variables, an effect is transmitted to the response variables (e.g., the grain size of a composite material). We are usually interested in discovering how changes in the predictor variables affect the values of the response variables. Ideally, we hope that a small number of predictor variables will "explain" nearly all of the variation in the response variables.

In practice, it is sometimes difficult to draw a clear distinction between independent and dependent variables. In many cases it depends on the objective of the investigator. For example, an analyst may treat ambient temperature as a predictor variable in the study of paint quality, and as the response variable in a study of clean room particulates. However, the above definitions are useful in planning Six Sigma studies.

Another idea important to studying cause and effect is that of the *data space* of the study. The data space of a study refers to the region bounded by the range of the independent variables under study. In general, predictions based on values outside the data space studied, called *extrapolations*, are little more than speculation and not advised. Figure 10.21 illustrates the concept of data space for two independent variables. Defining the data space can be quite tricky when large numbers of independent variables are involved.

While the numerical analysis of data provides valuable information, it should always be supplemented with graphical analysis as well. Scatter diagrams are one very useful supplement to regression and correlation analysis. Figure 10.22 illustrates the value of supplementing numerical analysis with scatter diagrams.

Chapter 10　Analyze Phase | **465**

**Figure 10.21**　Data space.

**Figure 10.22**　Illustration of the value of scatter diagrams. (From Tufte, 2001.)

In other words, although the scatter diagrams clearly show four distinct processes, the statistical analysis does not. In Six Sigma, numerical analysis alone is not enough.

## Least-Squares Fit

If all data fell on a perfectly straight line it would be easy to compute the slope and intercept given any two points. However, the situation becomes more complicated when there is "scatter" around the line. That is, for a given value of $x$, more than one value of $y$ appears. When this occurs, we have error in the model. Figure 10.23 illustrates the concept of error.

**Figure 10.23** Error in the linear model.

The model for a simple linear regression with error is

$$y = a + bx + \epsilon \qquad (10.11)$$

where $\epsilon$ represents error. Generally, assuming the model adequately fits the data, errors are assumed to follow a normal distribution with a mean of 0 and a constant standard deviation. The standard deviation of the errors is known as the *standard error*. We discuss ways of verifying our assumptions below.

When error occurs, as it does in nearly all "real-world" situations, there are many possible lines that might be used to model the data. Some method must be found that provides, in some sense, a "best-fit" equation in these everyday situations. Statisticians have developed a large number of such methods. The method most commonly used in Six

Sigma finds the straight line that minimizes the sum of the squares of the errors for all of the data points. This method is known as the "least-squares" best-fit line. In other words, the least-squares best-fit line equation is $y'_i = a + bx_i$ where $a$ and $b$ are found so that the sum of the squared deviations from the line is minimized. The best-fit equations for $a$ and $b$ are

$$b = \frac{\Sigma(X_i - \bar{X})(Y_i - \bar{Y})}{\Sigma(X_i - \bar{X})^2} \qquad (10.12)$$

$$a = \bar{Y} - b\bar{X} \qquad (10.13)$$

where the sum is taken over all $n$ values. Most spreadsheets and scientific calculators have a built-in capability to compute $a$ and $b$. As stated above, there are many other ways to compute the slope and intercept (e.g., minimize the sum of the absolute deviations, minimize the maximum deviation, etc.); in certain situations one of the alternatives may be preferred. The reader is advised to consult books devoted to regression analysis for additional information [see, for example, Draper and Smith (1981)].

The reader should note that the fit obtained by regressing $x$ on $y$ will not in general produce the same line as would be obtained by regressing $y$ on $x$. This is illustrated in Fig. 10.24.

**Figure 10.24** Least-squares lines of weight versus height and height versus weight.

When weight is regressed on height the equation indicates the average weight (in pounds) for a given height (in inches). When height is regressed on weight the equation indicates the average height for a given weight. The two lines intersect at the average height and weight.

These examples show how a single independent variable is used to model the response of a dependent variable. This is known as *simple linear regression*. It is also possible to model the dependent variable in terms of two or more independent variables; this is known as *multiple linear regression*. The mathematical model for multiple linear regression has additional terms for the additional independent variables. Equation (10.14) shows a linear model when there are two independent variables.

$$\hat{y} = a + b_1 x_1 + b_2 x_2 + \epsilon \tag{10.14}$$

where $x_1, x_2$ are independent variables, $b_1$ is the coefficient for $x_1$ and $b_2$ is the coefficient for $x_2$.

## Example of Regression Analysis

A restaurant conducted surveys of 42 customers, obtaining customer ratings on staff service, food quality, and overall satisfaction with their visit to the restaurant. Figure 10.25 shows the regression analysis output from a spreadsheet regression function (Microsoft Excel).

**Summary output**

Regression statistics

| | |
|---|---|
| Multiple R | 0.847 |
| R square | 0.717 |
| Adjusted R square | 0.703 |
| Standard error | 0.541 |
| Observations | 42 |

ANOVA

| | df | ss | ms | F | Significance F |
|---|---|---|---|---|---|
| Regression | 2 | 28.97 | 14.49 | 49.43 | 0.00 |
| Residual | 39 | 11.43 | 0.29 | | |
| Total | 41 | 40.40 | | | |

| | Coefficients | Standard error | t Stat | P-value | Lower 95% | Upper 95% |
|---|---|---|---|---|---|---|
| Intercept | -1.188 | 0.565 | -2.102 | 0.042 | -2.331 | -0.045 |
| Staff | 0.902 | 0.144 | 6.283 | 0.000 | 0.611 | 1.192 |
| Food | 0.379 | 0.163 | 2.325 | 0.025 | 0.049 | 0.710 |

**Figure 10.25** Regression analysis output.

The data consist of two independent variables, staff and food quality, and a single dependent variable, overall satisfaction. The basic idea is that the quality of staff service and the food are causes and the overall satisfaction score is an *effect*. The regression output is interpreted as follows:

- **Multiple R**—The multiple correlation coefficient. It is the correlation between $y$ and $\hat{y}$. For the example: multiple $R = 0.847$, which indicates that $y$ and $\hat{y}$ are highly correlated, which implies that there is an association between overall satisfaction and the quality of the food and service.
- **R square**—The square of multiple R, it measures the proportion of total variation about the mean $\bar{Y}$ explained by the regression. For the example: $R^2 = 0.717$, which indicates that the fitted equation explains 71.7% of the total variation about the average satisfaction level.
- **Adjusted R squared**—A measure of $R^2$ "adjusted for degrees of freedom." The equation is

$$\text{Adjusted } R^2 = 1 - (1 - R^2)\left(\frac{n-1}{n-p}\right) \qquad (10.15)$$

where $p$ is the number of parameters (coefficients for the $x$'s) estimated in the model. For the example: $p = 2$, since there are two $x$ terms. Some experimenters prefer the adjusted $R^2$ to the unadjusted $R^2$, while others see little advantage to it (e.g., Draper and Smith, 1981, p. 92).
- **Standard error**—The standard deviation of the residuals. The *residual* is the difference between the observed values of $y$ and the predicted values based on the regression equation.
- **Observations**—Refer to the number of cases in the regression analysis, or $n$.
- **ANOVA or analysis of variance**—A table examining the hypothesis that the variation explained by the regression is zero. If this is so, then the observed association could be explained by chance alone. The rows and columns are those of a standard one-factor ANOVA table (discussed in more detail later in this chapter). For this example, the important item is the column labeled "Significance F." The value shown, 0.00, indicates that the probability of getting these results due to chance alone is less than 0.01; that is, the association is probably not due to chance alone. Note that the ANOVA applies to the entire *model*, not to the individual variables.
- The next table in the output examines each of the terms in the linear model separately. The *intercept* is as described above, and corresponds to our term $a$ in the linear equation. Our model uses two independent variables. In our terminology staff = $b_1$, food = $b_2$.

Thus, reading from the coefficients column, the linear model is: $\bar{y} = -1.188 + 0.902 \times$ staff score $+ 0.379 \times$ food score. The remaining columns test the hypotheses that each coefficient in the model is actually zero.

- ▲ **Standard error column**—Gives the standard deviations of each term, i.e., the standard deviation of the intercept = 0.565, etc.
- ▲ **$t$ Stat column**—the coefficient divided by the standard error, i.e., it shows how many standard deviations the observed coefficient is from zero.
- ▲ **$P$-value**—Shows the area in the tail of a $t$ distribution beyond the computed $t$ value. For most experimental work, a $P$-value less than 0.05 is accepted as an indication that the coefficient is significantly different than zero. All of the terms in our model have significant $P$-values.
- ▲ **Lower 95 and upper 95% columns**—A 95% confidence interval on the coefficient. If the confidence interval does not include zero, we will reject the hypothesis that the coefficient is zero. None of the intervals in our example include zero.

See online video: Regression Analysis at www.mhprofessional.com/ssh5.

## Correlation Analysis

As mentioned earlier, a correlation problem considers the joint variation of two variables, neither of which is restricted by the experimenter. Unlike regression analysis, which considers the effect of the independent variable(s) on a dependent variable, correlation analysis is concerned with the joint variation of one independent variable with another. In a correlation problem, the analyst has two measurements for each individual item in the sample. Unlike a regression study where the analyst controls the values of the $x$ variables, correlation studies usually involve spontaneous variation in the variables being studied. Correlation methods for determining the strength of the linear relationship between two or more variables are among the most widely applied statistical techniques. More advanced methods exist for studying situations with more than two variables (e.g., canonical analysis, factor analysis, principal components analysis, etc.), however, with the exception of multiple regression, our discussion will focus on the linear association of two variables at a time.

In most cases, the measure of correlation used by analysts is the statistic $r$, sometimes referred to as *Pearson's product-moment correlation*. Usually, $x$ and $y$ are assumed to have a bivariate normal distribution. Under this assumption $r$ is a sample statistic that estimates the population correlation parameter $r$. One interpretation of $r$ is based on the linear regression model described earlier, namely, that $r^2$ is the proportion of the total variability in the $y$ dat,a which can be explained by the linear regression model. The equation for $r$ is

$$r = \frac{s_{xy}}{s_x s_y} = \frac{n\Sigma xy - \Sigma x \Sigma y}{\sqrt{[n\Sigma x^2 - (\Sigma x)^2][n\Sigma y^2 - (\Sigma y)^2]}} \qquad (10.16)$$

and, of course, $r^2$ is simply the square of $r$. $r$ is bounded at −1 and +1. When the assumptions hold, the significance of $r$ is tested by the regression ANOVA.

Interpreting $r$ can become quite tricky, so scatter plots should always be used (see above). When the relationship between $x$ and $y$ is nonlinear, the "explanatory power" of $r$ is difficult to interpret in precise terms and should be discussed with great care. While it is easy to see the value of very high correlations such as $r = 0.99$, it is not so easy to draw conclusions from lower values of $r$, even when they are statistically significant (i.e., they are significantly different than 0.0). For example, $r = 0.5$ does *not* mean the data show half as much clustering as a perfect straight-line fit. In fact, $r = 0$ does *not* mean that there is no relationship between the $x$ and $y$ data, as Fig. 10.26 shows. When $r > 0$, $y$ tends to increase when $x$ increases. When $r < 0$, $y$ tends to decrease when $x$ increases.

**Figure 10.26** Interpreting $r = 0$ for curvilinear data.

Although $r = 0$, the relationship between $x$ and $y$ is perfect, albeit nonlinear. At the other extreme, $r = 1$, a "perfect correlation," does not mean that there is a cause and effect relationship between $x$ and $y$. For example, both $x$ and $y$ might be determined by a third variable, $z$. In such situations, $z$ is described as a *lurking variable*, which "hides" in the background, unknown to the experimenter. Lurking variables are behind some of the infamous silly associations, such as the association between teacher's pay and liquor sales (the lurking variable is general prosperity).[2]

Establishing causation requires solid scientific understanding. Causation cannot be "proven" by statistics alone. Some statistical techniques, such as path analysis, can help determine if the correlations between a number of variables are consistent with causal assumptions. However, these methods are beyond the scope of this book.

## Designed Experiments

Designed experiments play an important role in quality improvement. While the confidence intervals and hypothesis tests previously discussed are limited to rather simple comparisons between one sample and requirements or between two samples, the designed experiments will use ANOVA (analysis of variance) techniques to partition the variation in a response amongst the potential sources of variation. This section will introduce the basic concepts involved and it will contrast the statistically designed experiment with the "one factor at a time" (OFAT) approach used traditionally. Also briefly discussed are the concepts involved in Taguchi methods, statistical methods named after their creator, Dr. Genichi Taguchi.

### The Traditional Approach Versus Statistically Designed Experiments

The traditional approach, which most of us learned in high school science class, is to hold all factors constant except one. When this approach is used we can be sure that the variation is due to a cause and effect relationship, or so we are told. However, this approach suffers from a number of problems:

- It usually isn't possible to hold all other variables constant.
- There is no way to account for the effect of joint variation of independent variables, such as interaction.
- There is no way to account for experimental error, including measurement variation.

The statistically designed experiment usually involves varying two or more variables simultaneously and obtaining multiple measurements under the same experimental conditions. The advantage of the statistical approach is threefold:

- Interactions can be detected and measured. Failure to detect interactions is a major flaw in the OFAT approach.
- Each value does the work of several values. A properly designed experiment allows you to use the same observation to estimate several different effects. This translates directly to cost savings when using the statistical approach.
- Experimental error is quantified and used to determine the confidence the experimenter has in his conclusions.

### Terminology

Much of the early work on the design of experiments involved agricultural studies. The language of experimental design still reflects these origins. The experimental area was literally a piece of ground. A block was a smaller piece of ground with fairly uniform properties. A plot was smaller still, and it served as the basic unit of the design. As the plot was planted, fertilized, and harvested, it could be split simply by drawing a line. A treatment was actually a treatment, such as the application of fertilizer. Unfortunately for the Six Sigma analyst, these terms are still part of the language of experiments. The analyst must do his or her best to understand quality improvement experimenting using these terms. Natrella (1963) recommends the following:

Experimental area can be thought of as the scope of the planned experiment. For us, a block can be a group of results from a particular operator, or from a particular machine, or on a particular day—any planned natural grouping that should serve to make results from one block more alike than results from different blocks. For us, a treatment is the factor being investigated (material, environmental condition, etc.) in a single factor experiment. In factorial experiments (where several variables are being investigated at the same time) we speak of a treatment combination and we mean the prescribed levels of the factors to be applied to an experimental unit. For us, a yield is a measured result and, happily enough, in chemistry it will sometimes be a yield.

## Definitions

A designed experiment is an experiment where one or more factors, called independent variables, believed to have an effect on the experimental outcome are identified and manipulated according to a predetermined plan. Data collected from a designed experiment can be analyzed statistically to determine the effect of the independent variables, or combinations of more than one independent variable. An experimental plan must also include provisions for dealing with extraneous variables, that is, variables not explicitly identified as independent variables.

- **Response variable**—The variable being investigated, also called the *dependent variable*, sometimes called simply *response*.
- **Primary variables**—The controllable variables believed most likely to have an effect. These may be quantitative, such as temperature, pressure, or speed, or they may be qualitative such as vendor, production method, and operator.
- **Background variables**—Variables, identified by the designers of the experiment, which may have an effect but either cannot or should not be deliberately manipulated or held constant. The effect of background variables can contaminate primary variable effects unless they are properly handled. The most common method of handling background variables is blocking (blocking is described later in this chapter).
- **Experimental error**—In any given experimental situation, a great many variables may be potential sources of variation. So many, in fact, that no experiment could be designed that deals with every possible source of variation explicitly. Those variables

that are not considered explicitly are analogous to common causes of variation. They represent the "noise level" of the process, and their effects are kept from contaminating the primary variable effects by *randomization*. Randomization is a term meant to describe a procedure that assigns test units to test conditions in such a way that any given unit has an equal probability of being processed under a given set of test conditions.

▲ **Interaction**—A condition where the effect of one factor depends on the level of another factor. Interaction is illustrated in Fig. 10.27.

**Figure 10.27** Illustration of interaction.

## *Design Characteristics*

Good experiments don't just happen, they are a result of careful planning. A good experimental plan depends on (Natrella, 1963):

▲ The purpose of the experiment
▲ Physical restrictions on the process of taking measurements
▲ Restrictions imposed by limitations of time, money, material, and personnel

The analyst must explain clearly why the experiment is being done, why the experimental treatments were selected, and how the completed experiment will accomplish the stated objectives. The experimental plan should be in writing and it should be endorsed by all key participants. The plan will include a statement of the objectives of the experiment, the experimental treatments to be applied, the size of the experiment, the time frame, and a brief discussion of the methods to be used to analyze the results. Two concepts are of particular interest to the Six Sigma analyst: replication and randomization.

▲ **Replication**—The collection of more than one observation for the same set of experimental conditions. Replication allows the experimenter to estimate experimental error. If variation exists when all experimental conditions are held constant, the cause must be something other than the variables being controlled by the experimenter. Experimental error can be estimated without replicating the entire experiment. If a process has been in statistical control for a period of time, experimental error can be estimated from the control chart. Replication also serves to decrease bias due to uncontrolled factors.

- **Randomization**—In order to eliminate bias from the experiment, variables not specifically controlled as factors should be randomized. This means that allocations of specimens to treatments should be made using some mechanical method of randomization, such as a random numbers table. Randomization also ensures valid estimates of experimental error.

## *Types of Design*

Experiments can be designed to meet a wide variety of experimental objectives. A few of the more common types of experimental designs are defined here.

- **Fixed-effects model**—An experimental model where all possible factor levels are studied. For example, if there are three different materials, all three are included in the experiment.
- **Random-effects model**—An experimental model where the levels of factors evaluated by the experiment represent a sample of all possible levels. For example, if we have three different materials but only use two materials in the experiment.
- **Mixed model**—An experimental model with both fixed and random effects.
- **Completely randomized design**—An experimental plan where the order in which the experiment is performed is completely random, for example,

| Level | Test Sequence Number |
|---|---|
| A | 7, 1, 5 |
| B | 2, 3, 6 |
| C | 8, 4 |

- **Randomized-block design**—An experimental design is one where the experimental observations are divided into "blocks" according to some criterion. The blocks are filled sequentially, but the order within the block is filled randomly. For example, assume we are conducting a painting test with different materials, material A and material B. We have four test pieces of each material. Ideally, we would like to clean all of the pieces at the same time to ensure that the cleaning process doesn't have an effect on our results; but what if our test requires that we use a cleaning tank that cleans two test pieces at a time? The tank load then becomes a "blocking factor." We will have four blocks, which might look like this:

| Material | Tank Load | Test Piece Number |
|----------|-----------|-------------------|
| A        | 1         | 7                 |
| B        |           | 1                 |
| B        | 2         | 5                 |
| A        |           | 2                 |
| B        | 3         | 3                 |
| A        |           | 6                 |
| B        | 4         | 4                 |
| A        |           | 8                 |

Since each material appears exactly once per cleaning tank load we say the design is *balanced*. The material totals or averages can be compared directly. The reader should be aware that statistical designs exist to handle more complicated "unbalanced designs."

▲ **Latin-square designs**—Designs where each treatment appears once and only once in each row and column. A Latin-square plan is useful when it is necessary or desirable to allow for two specific sources of nonhomogeneity in the conditions affecting test results. Such designs were originally applied in agricultural experimentation when the two sources of nonhomogeneity were the two directions on the field and the "square" was literally a square piece of ground. Its usage has been extended to many other applications where there are two sources of nonhomogeneity that may affect experimental results—for example, machines, positions, operators, runs, and days. A third variable is then associated with the other two in a prescribed fashion. The use of Latin squares is restricted by two conditions:

▲ The number of rows, columns, and treatments must all be the same.
▲ There must be no interactions between row and column factors.

Natrella (1963, pp. 13–30) provides the following example of a Latin square. Suppose we wish to compare four materials with regard to their wearing qualities. Suppose further that we have a wear-testing machine that can handle four samples simultaneously. Two sources of inhomogeneity might be the variations from run to run, and the variation among the four positions on the wear machine. In this situation, a 4 × 4 Latin square will enable us to allow for both sources of inhomogeneity if we can make four runs. The Latin square plan is as in Fig. 10.28 (the four materials are labeled A, B, C, D).

|     | Position number |     |     |     |
| --- | --- | --- | --- | --- |
| Run | (1) | (2) | (3) | (4) |
| 1 | A | B | C | D |
| 2 | B | C | D | A |
| 3 | C | D | A | B |
| 4 | D | A | B | C |

**Figure 10.28**  A 4 × 4 Latin square.

The procedure to be followed in using a given Latin square is as follows:

1. Permute the columns at random.
2. Permute the rows at random.
3. Assign letters randomly to the treatments.

## One-Factor ANOVA

The following example will be used to illustrate the interpretation of a single factor analysis of variance. With the widespread availability of computers, few people actually perform such complex calculations by hand. The analysis below was performed using Microsoft Excel. Commonly used statistical methods such as regression and ANOVA are included in most high-end spreadsheets.

The coded results in Table 10.3 were obtained from a single-factor, completely randomized experiment, in which the production outputs of three machines (A, B, and C) were to be compared.

**Table 10.3**  Experimental Raw Data (Coded)

| A | B | C |
| --- | --- | --- |
| 4 | 2 | −3 |
| 8 | 0 | 1 |
| 5 | 1 | −2 |
| 7 | 2 | −1 |
| 6 | 4 | 0 |

An ANOVA of these results produced the results shown in Table 10.4.

The first part of Table 10.4 shows descriptive statistics for the data; the analyst should always look carefully at these easily understood results to check for obvious errors. The results show that the means vary from a low of −1 for machine C to a high of 6 for machine A.

Table 10.4 Results of the Analysis

**ANOVA: SINGLE FACTOR**

SUMMARY

| Groups | Count | Sum | Average | Variance |
|---|---|---|---|---|
| A | 5 | 30.000 | 6.000 | 2.500 |
| B | 5 | 9.000 | 1.800 | 2.200 |
| C | 5 | −5.000 | −1.000 | 2.500 |

ANOVA

| Source of Variation | SS | df | MS | F | P-value | F crit |
|---|---|---|---|---|---|---|
| Between groups | 124.133 | 2 | 62.067 | 25.861 | 0.000 | 3.885 |
| Within groups | 28.800 | 12 | 2.400 | | | |
| Total | 152.933 | 14 | | | | |

## ANOVA Procedure

ANOVA proceeds as follows:

1. State the null and alternative hypotheses: the ANOVA table tests the hypotheses: $H_0$ (all means are equal) versus $H_a$ (at least two of the means are different).
2. Choose the level of significance. For this analysis a significance level $\alpha = 0.05$ was selected.
3. Compute the $F$ statistic, the ratio of the mean square between groups to the mean square within groups.
4. Assuming that the observations are random samples from normally distributed populations with equal variances, and that the hypothesis is true, the critical value of $F$ is found in Appendix 5 or 6. The numerator will have the degrees of freedom shown in the degrees of freedom column for the between groups row. The denominator will have the degrees of freedom shown in the degrees of freedom column for the within groups row.
5. If the computed $F > F_{1-\alpha}$ then reject the null hypothesis and conclude the alternate hypothesis. Otherwise fail to reject the null hypothesis.

The ANOVA table shows that for these data $F$ computed is 62.067/2.4 = 25.861 and $F$ critical at $\alpha = 0.05$ with numerator df = 2 and denominator df = 12 is 3.885.[3] Since 25.861 > 3.885 we reject the null hypothesis and conclude that the machines produce different results. Note that all we know is that at least the two extreme machines (A and C) are different. The ANOVA does *not* tell us if A and B or B and C are significantly different. There are methods that can make this determination, such as *contrasts*. The reader is referred to a text on design of experiments, for example, Montgomery (1984) for additional information.

## Performing ANOVA Manually

On rare occasions (such as taking a Black Belt exam), the analyst may find that computers are not available and the analysis must be performed "by hand." The analysis is illustrated below.

|  |  | Total | N | Sum of Squares |
|---|---|---|---|---|
| Treatment A | 4, 8, 5, 7, 6 | 30 | 5 | 190 |
| Treatment B | 2, 0, 1, 2, 4 | 9 | 5 | 25 |
| Treatment C | −3, 1, −2, −1, 0 | −5 | 5 | 15 |
| Totals |  | 34 | 15 | 230 |

$$\text{Total sum of square} = 230 = \frac{(34)^2}{15} = 152.933$$

$$\text{Treatment sum of square} = \frac{(30)^2}{5} + \frac{(9)^2}{5} + \frac{(-5)^2}{5} - \frac{(34)^2}{15} = 124.133$$

Error sum of squares = total sum of squares − treatment sum of squares

$$= 152.933 - 124.133 = 28.8$$

These values are placed in the sum of squares (SS) column in the ANOVA table (Table 10.4). The remainder of the ANOVA table is obtained through simple division.

## *Two-Way ANOVA with No Replicates*

When experiments are conducted that involve two factors, and it is not possible to obtain repeat readings for a given set of experimental conditions, a two-way analysis of variance may be used. The following example assumes that experimental treatments are assigned at random. Note that if the factors involved are each tested at only two levels, the full factorial analysis method described below could also be used.

### Example of Two-Way ANOVA with No Replicates

An experiment was conducted to evaluate the effect of different detergents and water temperatures on the cleanliness of ceramic substrates. The experimenter selected three different detergents based on their pH levels, and conducted a series of experiments at four different water temperatures. Cleanliness was quantified by measuring the contamination of a distilled water beaker after rinsing the parts cleaned using each treatment combination. The coded data are shown in Table 10.5.

**Table 10.5** Cleaning Experiment Raw Data

|  | Detergent A | Detergent B | Detergent C |
|---|---|---|---|
| Cold | 15 | 18 | 10 |
| Cool | 12 | 14 | 9 |
| Warm | 10 | 18 | 7 |
| Hot | 6 | 12 | 5 |

Part one of the Excel output (Table 10.6) provides descriptive statistics on the different treatment levels. The ANOVA table is shown in part two. Note that in the previously presented raw data table the rows represent the different temperatures and the columns the different detergents. Because there are no replicates, Excel is not able to provide an estimate of the interaction of detergent and water temperature. If you suspect that an interaction may be present, then you should try to replicate the experiment to estimate this effect. For this experiment, any $P$-value less than 0.05 would indicate a significant effect. The ANOVA table indicates that there are significant differences between the different detergents and the different water temperatures. To identify *which* differences are significant the experimenter can examine the means of the different detergents and water temperatures using $t$-tests. (Excel's data analysis tools add-in includes these tests.) Be aware that the Type I error is affected by conducting multiple $t$-tests. If the Type I error on a single $t$-test is $\alpha$, then the overall Type I error for $k$ such tests is $1 - (1 - \alpha)^k$. For example, if $\alpha = 0.01$ and three pairs of means are examined, then the combined Type I error for all three $t$-tests is $1 - (1 - 0.01)^3 = 1 - (0.99)^3 = 0.03$. Statistical methods exist that guarantee an overall level of Type I error for simultaneous comparisons (Hicks, 1973, pp. 31–38).

## *Two-Way ANOVA with Replicates*

If you are investigating two factors that might interact with one another, and you can obtain more than one result for each combination of experimental treatments, then two-way analysis of variance with replicates may be used for the analysis. Spreadsheets such as Microsoft Excel include functions that perform this analysis.

### **Example of Two-Way ANOVA with Replicates**

An investigator is interested in improving a process for bonding photoresist to copper clad printed circuit boards. Two factors are to be evaluated: the pressure used to apply the photoresist material and the preheat temperature of the photoresist. Three different pressures and three different temperatures are to be evaluated; the number of levels need

**Table 10.6** Cleaning Experiment Two-Way ANOVA Output from Microsoft Excel (Two-Factor without Replication)

**SUMMARY OUTPUT**

|  | Count | Sum | Average | Variance |
|---|---|---|---|---|
| Cold water | 3 | 43 | 14.333333 | 16.33333 |
| Cool water | 3 | 35 | 11.666667 | 6.333333 |
| Warm water | 3 | 35 | 11.666667 | 32.33333 |
| Hot water | 3 | 23 | 7.6666667 | 14.33333 |
| Detergent A | 4 | 43 | 10.75 | 14.25 |
| Detergent B | 4 | 62 | 15.5 | 9 |
| Detergent C | 4 | 31 | 7.75 | 4.916667 |

**ANOVA**

| Source of variation | SS | df | MS | F | P-value | F crit |
|---|---|---|---|---|---|---|
| Rows | 68 | 3 | 22.666667 | 8.242424 | 0.015043179 | 4.757055 |
| Columns | 122.1666667 | 2 | 61.083333 | 22.21212 | 0.001684751 | 5.143249 |
| Error | 16.5 | 6 | 2.75 |  |  |  |
| Total | 206.6666667 | 11 |  |  |  |  |

not be the same for each factor and there is no restriction on the total number of levels. Each experimental combination of variables is repeated five times. Note that while Excel requires equal numbers of replicates for each combination of treatments, most statistical analysis packages allow different sample sizes to be used. The experimenter recorded the number of photoresist defects per batch of printed wiring boards. The coded data are shown in Table 10.7.

These data were analyzed using Excel's two-way ANOVA with replicates function. The results are shown in Table 10.8.

As before, part one of the Excel output provides descriptive statistics on the different treatment levels. The ANOVA table is shown in part two. Because there are now replicates, Excel is able to provide an estimate of the interaction of pressure and temperature. For this experiment, the experimenter decided that any $P$-value less than 0.05 would indicate a significant effect. The ANOVA table $P$-value of less than 0.001 indicates that there are significant differences between the different columns (pressure), but the $P$-value of 0.6363 indicates that there is not a significant difference between the rows (temperature). The interaction of pressure and temperature is also not significant, as indicated by the $P$-value of 0.267501.

Table 10.7  Photoresist Experiment Raw Data ANOVA Results

|  | High Pressure | Medium Pressure | Low Pressure |
|---|---|---|---|
| High temp | 39 | 32 | 18 |
|  | 30 | 31 | 20 |
|  | 35 | 28 | 21 |
|  | 43 | 28 | 25 |
|  | 25 | 29 | 26 |
| Med temp | 38 | 10 | 22 |
|  | 31 | 15 | 28 |
|  | 31 | 25 | 29 |
|  | 30 | 31 | 26 |
|  | 35 | 36 | 20 |
| Low temp | 30 | 21 | 25 |
|  | 35 | 22 | 24 |
|  | 36 | 25 | 20 |
|  | 37 | 24 | 21 |
|  | 39 | 27 | 21 |

Since the *P*-value indicates that at least one difference is significant, we know that the largest difference of 34.26666667 − 23.06666667 = 11.2 is significant. To identify *which other* differences are significant the experimenter can examine the means of the different pressures using *t*-tests. (Excel's data analysis tools add-in includes these tests.) Be aware that the Type I error is affected by conducting multiple *t*-tests. If the Type I error on a single *t*-test is $\alpha$, then the overall Type I error for $k$ such tests is $1 - (1 - \alpha)^k$. For example, if $\alpha = 0.01$ and three pairs of means are examined, then the combined Type I error for all three *t*-tests is $1 - (1 - 0.01)^3 = 1 - (0.99)^3 = 0.03$.

## *Full and Fractional Factorial*

Full factorial experiments are those where at least one observation is obtained for every possible combination of experimental variables. For example, if A has 2 levels, B has 3 levels, and C has 5 levels, a full factorial experiment would have at least 2 × 3 × 5 = 30 observations.

*Fractional factorial or fractional replicate* are experiments where there are some combinations of experimental variables where observations were not obtained. Such experi-

**Table 10.8** Photoresist Experiment Two-Way ANOVA Output from Microsoft Excel (Two-Factor with Replication)

**SUMMARY OUTPUT**

|  | High pressure | Medium pressure | Low pressure | Total |
|---|---|---|---|---|
| **High temperature** | | | | |
| Count | 5 | 5 | 5 | 15 |
| Sum | 172 | 148 | 110 | 430 |
| Average | 34.4 | 29.6 | 22 | 28.66667 |
| Variance | 50.8 | 3.3 | 11.5 | 46.66667 |
| **Medium temperature** | | | | |
| Count | 5 | 5 | 5 | 15 |
| Sum | 165 | 117 | 125 | 407 |
| Average | 33 | 23.4 | 25 | 27.13333 |
| Variance | 11.5 | 117.3 | 15 | 59.98095 |
| **Low temperature** | | | | |
| Count | 5 | 5 | 5 | 15 |
| Sum | 177 | 119 | 111 | 407 |
| Average | 35.4 | 23.8 | 22.2 | 27.13333 |
| Variance | 11.3 | 5.7 | 4.7 | 43.26667 |
| **Total** | | | | |
| Count | 15 | 15 | 15 | |
| Sum | 514 | 384 | 346 | |
| Average | 34.26666667 | 25.6 | 23.06666667 | |
| Variance | 22.06666667 | 44.68571429 | 10.92380952 | |

**ANOVA**

| Source of variation | SS | df | MS | F | P-value | F crit |
|---|---|---|---|---|---|---|
| Sample | 23.5111111 | 2 | 11.7555556 | 0.45781 | 0.6363 | 3.259444 |
| Columns | 1034.84444 | 2 | 517.422222 | 20.1506 | 1.34E-06 | 3.259444 |
| Interaction | 139.555556 | 4 | 34.8888889 | 1.35872 | 0.267501 | 2.633534 |
| Within | 924.4 | 36 | 25.6777778 | | | |
| Total | 2122.31111 | 44 | | | | |

ments may not allow the estimation of every interaction. However, when carefully planned, the experimenter can often obtain all of the information needed at a significant saving.

See online video: Design of Experiments Setup at www.mhprofessional.com/ssh5.

## Analyzing Factorial Experiments

A simple method exists for analyzing the common $2^n$ experiment. The method, known as the Yates method, can be easily performed with a pocket calculator or programmed into a spreadsheet. It can be used with any properly designed $2^n$ experiment, regardless of the number of factors being studied.

To use the Yates algorithm, the data are first arranged in standard order (of course, the actual running order is random). The concept of standard order is easier to understand if demonstrated. Assume that we have conducted an experiment with three factors, A, B, and C. Each of the three factors is evaluated at two levels, which we will call low and high. A factor held at a low level will be identified with a "−" sign, one held at a high level will be identified with a "+" sign. The eight possible combinations of the three factors are identified using the scheme shown in the table below.

| ID | A | B | C |
|---|---|---|---|
| (1) | − | − | − |
| a | + | − | − |
| b | − | + | − |
| ab | + | + | − |
| c | − | − | + |
| ac | + | − | + |
| bc | − | + | + |
| abc | + | + | + |

Note that the table begins with all factors at their low level. Next, the first factor is high and all others are low. When a factor is high, it is shown in the ID column, otherwise it is not. For example, whenever *a* appears it indicates that factor A is at its high level. To complete the table you simply note that as each factor is added to the table it is "multiplied" by each preceding row. Thus, when *b* is added it is multiplied by *a*, giving the row *ab*. When *c* is added it is multiplied by, in order, *a*, *b*, and *ab*, giving the remaining rows in the table. (As an exercise, the reader should add a fourth factor D to the above table. Hint: the result will be a table with eight more rows.) Once the data are in standard order, add a column for the data and one additional column for each variable, for example, for our three variables we will add four columns.

| ID | A | B | C | Data | 1 | 2 | 3 |
|---|---|---|---|---|---|---|---|
| (1) | − | − | − |  |  |  |  |
| a | + | − | − |  |  |  |  |
| b | − | + | − |  |  |  |  |
| ab | + | + | − |  |  |  |  |
| c | − | − | + |  |  |  |  |
| ac | + | − | + |  |  |  |  |
| bc | − | + | + |  |  |  |  |
| abc | + | + | + |  |  |  |  |

Record the observations in the data column (if the experiment has been replicated, record the totals). Now record the sum of the data values in the first two rows—that is, (1) + a in the first cell of the column labeled column 1. Record the sum of the next two rows in the second cell (i.e., b + ab). Continue until the top half of column 1 is completed. The lower half of column 1 is completed by subtracting one row from the next, for example, the fifth value in column 1 is found by subtracting −5 − 2 = −3. After completing column 1 the same process is completed for column 2, using the values in column 1. Column 3 is created using the values in column 2. The result is shown below.

| ID | A | B | C | Data | 1 | 2 | 3 |
|---|---|---|---|---|---|---|---|
| (1) | − | − | − | −2 | −7 | 21 | −17 |
| a | + | − | − | −5 | 28 | -38 | −15 |
| b | − | + | − | 15 | −29 | −5 | 55 |
| ab | + | + | − | 13 | −9 | −10 | 1 |
| c | − | − | + | −12 | −3 | 35 | −59 |
| ac | + | − | + | −17 | −2 | 20 | −5 |
| bc | − | + | + | −2 | −5 | 1 | −15 |
| abc | + | + | + | −7 | −5 | 0 | −1 |

### Example of Yates Method

The table below shows sample data from an actual experiment. The experiment involved a target shooter trying to improve the number of targets hit per box of 25 shots. Three variables were involved: $a$ = the gauge of the shotgun (12-gauge and 20-gauge), $b$ = the shot *size* (6 shot and 8 shot), and $c$ = the length of the handle on the target thrower (short or long). The shooter ran the experiment twice. The column labeled "1st" is the number of hits the first time the combination was tried. The column labeled "2nd" is the number of hits the second time the combination was tried. The Yates analysis begins with the sums shown in the column labeled Sum.

| ID | 1st | 2nd | Sum | 1 | 2 | 3 | Effect | df | SS | MS | F Ratio |
|---|---|---|---|---|---|---|---|---|---|---|---|
| 1 | 22 | 19 | 41 | 86 | 167 | 288 | 18 | Avg. | | | |
| a | 21 | 24 | 45 | 81 | 121 | 20 | 2.5 | 1 | 25.00 | 25.00 | 3.64 |
| b | 20 | 18 | 38 | 58 | 9 | 0 | 0 | 1 | 0.00 | 0.00 | 0.00 |
| ab | 21 | 22 | 43 | 63 | 11 | 4 | 0.5 | 1 | 1.00 | 1.00 | 0.15 |
| c | 12 | 15 | 27 | 4 | −5 | −46 | −5.75 | 1 | 132.25 | 132.25 | 19.24 |
| ac | 12 | 19 | 31 | 5 | 5 | 2 | 0.25 | 1 | 0.25 | 0.25 | 0.04 |
| bc | 13 | 15 | 28 | 4 | 1 | 10 | 1.25 | 1 | 6.25 | 6.25 | 0.91 |
| abc | 20 | 15 | 35 | 7 | 3 | 2 | 0.25 | 1 | 0.25 | 0.25 | 0.04 |
| Error | | | | | | | | 8 | 55.00 | 6.88 | |
| Total | 141 | 147 | | | | | | 15 | 220.00 | | |

The first row in the Effect column is simply the first row of column 3 (288) divided by the count ($r \times 2^n$); this is simply the average. Subsequent rows in the Effect column are found by dividing the numbers in column 3 by $r \times 2^{n-1}$. The Effect column provides the impact of the given factor on the response; thus, the shooter hit, on average, 2.5 more targets per box when shooting a 12-gauge than he did when shooting a 20-gauge.

The next question is whether or not these differences are statistically significant, that is, could they be due to chance alone? To answer this question we will use the F ratio of the effect MS for each factor to the error MS. The degrees of freedom (df) for each effect is simply 1 (the number of factor levels minus 1), the total df is $N - 1$, and the error df is the total df minus the sum of the factor degrees of freedom. The sum of squares (SS) for each factor is the column 3 value squared divided by $r \times 2^n$; for example., $SS_A = 20^2/16 = 25$. The total SS is the sum of the individual values squared minus the first row in column 3 squared divided by $r \times 2^n$; for example:

$$(22^2 + 21^2 + \ldots 15^2) - \frac{288^2}{16} = 220$$

The error SS is the total SS minus the factor SS. The MS and F columns are computed using the same approach as shown above for one-way ANOVA. For the example the F ratio for factor c (thrower) is significant at $\alpha < 0.01$ and the F ratio for factor a (gauge) is significant at $\alpha < 0.10$; no other F ratios are significant.

***Screening Experiments.*** In many cases, there are a sufficient number of variables that even a fractional factorial design, with interactions, would be quite large. In those cases, a screening design is useful as an initial pass to filter out factors that are highly insignificant.

Consider this example of reducing order processing time (adapted from Keller, 2011) with the following list of factors and their experimental conditions:

| Variable | Low Level (−) | High Level (+) |
|---|---|---|
| A: Number of line items in order | 2 | 14 |
| B: Product family | A | B |
| C: Person processing order | Charlie | Andrew |
| D: Method of order processing | Method 1 | Method 2 |
| E: Background music | No | Yes |
| F: Input Type | Paper | Online |

The choice of product and person were made with some consideration toward best and worst case. Of the multiple operators working the process, Andrew is the most experienced; Charlie the least experienced. Information gleaned from the Analyze stage box plot indicated that Product A has the lowest median but highest variation in cycle time, while product B has a similar level of variation as Product C, but a higher median. Since it is a rare case when there is only a single line item on an order, the low level for line items was chosen as 2; the upper level of 14 was chosen based on a review of the available data. Levels for other factors were limited by the factor itself.

As a screening experiment, the team was not interested in estimating factor interactions, only the main effects. A replicated $2^{6-3}$ fractional factorial design was chosen, requiring $2^{6-3}$ = 8 experimental runs. Each run will be replicated (i.e., repeated in a random order) for a total of 16 runs. Generally, it's better to replicate a small design than run an extended design to estimate all combinations of factor interactions.

The data is shown in Table 10.9. The run order for each replicate of eight conditions has been randomized. The cycle time results for each condition are shown in the "response" column. The design is of resolution III: Main effects are not confounded with each other, but may be confounded with two-factor (and higher) interactions. The initial analysis of the data will determine if these main factors are significant.

In Table 10.9 the "−" indicates that the variable is run at its low level, while a "+" sign indicates that it is to be run at its high level. For example, run 16 was processed as follows:

- Number of line items in order = 14
- Product family = A
- Person processing order = Charlie
- Method of order processing = Method 2

**488** | The Six Sigma Handbook

**Table 10.9** Screening Experiment Data (Adapted from Keller, 2011)

| Run | A | B | C | D | E | F | Response |
|---|---|---|---|---|---|---|---|
| 1 | + | + | − | + | + | − | 19.2 |
| 2 | − | − | + | − | + | − | 6.5 |
| 3 | − | + | − | − | − | + | 5.7 |
| 4 | − | + | + | + | − | − | 9.3 |
| 5 | + | + | + | − | + | + | 11.8 |
| 6 | − | − | − | + | + | + | 8.2 |
| 7 | + | − | + | + | − | + | 17.6 |
| 8 | + | − | − | − | − | − | 10.8 |
| 9 | − | + | − | + | − | + | 10.3 |
| 10 | + | + | + | + | + | + | 17.8 |
| 11 | − | − | − | − | + | + | 6.2 |
| 12 | + | − | + | − | − | + | 10.8 |
| 13 | − | − | + | + | + | − | 9.3 |
| 14 | − | + | + | − | − | − | 6.4 |
| 15 | + | + | − | − | + | − | 10.8 |
| 16 | + | − | − | + | − | − | 17 |

▲ Background Music = No
▲ Input Type = Paper

The *p*-values associated with each of the main factors in Table 10.10 indicate the significance of the factor: Factors with *p*-values larger than 0.05 (or perhaps 0.10 in this early screening) are deemed insignificant. The analysis indicates that factors A (number of line items in order) and D (method of order processing) produce significant effects. Graphically, the effects of A and D are shown to be further from the line than can be accounted for by random variation, as shown in Fig. 10.29.

The data from this screening design is analyzed again, with only Factors A, D and the AD interaction included. Terms should generally be removed from the model one at a time (based on the highest p-value); the *Stepwise* regression option using *Backwards Elimination* in Minitab automates this process. Note also that the AD interaction can now be analyzed with the limited number of data runs since four of the original six main factors were eliminated in the first-pass analysis. The analysis provides the regression equation useful for modeling the process as:

cycle time = 11.11 + 3.37 × (line items) + 2.48 × (method) + 0.94 × (line items) × (method)

where the line items and method are both expressed in coded (−1,1) terms.

**Table 10.10** First-Pass Results of Screening Experiment; Coded Coefficients (Minitab 17)

**Analysis of Variance**

| Source | DF | Adj SS | Adj MS | F-Value | P-Value |
|---|---|---|---|---|---|
| Model | 6 | 281.964 | 46.994 | 23.10 | 0.000 |
| Linear | 6 | 281.964 | 46.994 | 23.10 | 0.000 |
| A | 1 | 181.576 | 181.576 | 89.27 | 0.000 |
| B | 1 | 1.501 | 1.501 | 0.74 | 0.413 |
| C | 1 | 0.106 | 0.106 | 0.05 | 0.825 |
| D | 1 | 98.506 | 98.506 | 48.43 | 0.000 |
| E | 1 | 0.226 | 0.226 | 0.11 | 0.747 |
| F | 1 | 0.051 | 0.051 | 0.02 | 0.878 |
| Error | 9 | 18.306 | 2.034 | | |
| Total | 15 | 300.269 | | | |

| Term | Effect | Coef | SE Coef | T-Value | P-Value |
|---|---|---|---|---|---|
| Constant | | 11.106 | 0.357 | 31.15 | 0.000 |
| A | 6.738 | 3.369 | 0.357 | 9.45 | 0.000 |
| B | 0.613 | 0.306 | 0.357 | 0.86 | 0.413 |
| C | 0.163 | 0.081 | 0.357 | 0.23 | 0.825 |
| D | 4.963 | 2.481 | 0.357 | 6.96 | 0.000 |
| E | 0.237 | 0.119 | 0.357 | 0.33 | 0.747 |
| F | −0.113 | −0.056 | 0.357 | −0.16 | 0.878 |

**Figure 10.29** Significant factor effects.

**Figure 10.30**  Residuals analysis of final model (Minitab 17).

The model is further analyzed by reviewing its residuals using Minitab's four-in-one plot in Fig.10.30. The residuals fit well along a normal probability line (upper-left figure), suggesting the model fits the data well at each observation (possible exception is observation 1, with a standardized residual exceeding 2.0). No obvious patterns of systematic error in the model are identified in the two right-hand figures.

A correlation analysis indicates the approximate percent of total variation in the order processing time associated with each term in the model:

- Number of line items: 60%
- Processing method: 33%
- Line items/method interaction: 5%

The majority of the variation in order processing time is associated with the differences in the number of items ordered. The relative impact of the person processing the order as well as the product family had no significance, contrary to earlier results using box whisker and data stratification. As is often the case, analysis of happenstance data observed in the process is unduly influenced by the nature of the data: orders for certain product families had higher propensity for more line items, so the influence of the number of line items on processing time appeared to be an influence due to product family. Similarly, specific persons tended to process orders depending on the product family, so this also appeared significant. Product family and person were both correlated with the number of line items, so it was impossible to distinguish their effects individually with the happenstance data.

Furthermore, the experimental data analysis identified the processing method as a significant factor, which was not detected in the happenstance data. With these experimental results in hand, the team considered ways to reduce the impact of line items in the order, such as methods to quickly add line items. They standardized on Method 2 to reduce order processing time by over 30%. Once the changes are validated by additional experimental runs, they can be implemented to reduce both the median processing time as well as the variation in processing time (adapted from Keller, 2011).

See online video: Design of Experiments Analysis at www.mhprofessional.com/ssh5.

See online video: Regression Analysis in Excel at www.mhprofessional.com/ssh5.

See online video: Regression Analysis in Excel with Interactions at www.mhprofessional.com/ssh5.

## *Power and Sample Size*

The term *power* of a statistical test refers to the probability that the test will lead to correctly rejecting a false null hypothesis, that is, $1 - \beta$, where beta is the probability of failing to reject the false null hypothesis. Generally, the power of a statistical test is improved when:

- There is a large difference between the null and alternative conditions.
- The population sigma is small.
- The sample size is large.
- The significance ($\alpha$) is large.

Many statistical software packages provide Power and Sample Size calculations. Minitab's *Power and Sample Size* option in the *Stat* menu can estimate these for a variety of test formats.

### Example

Consider a one-way ANOVA test of the hypothesis that four populations have equal means. A sample of $n = 5$ is taken from each population whose historical standard deviation is 2.0. If we are interested in detecting a difference of 3 units in the means, the software can estimate the power of the test after completing the *Power and Sample Size for one-way ANOVA* dialog box as:

- Number of levels: 4
- Sample sizes: 5
- Values of the maximum difference between means: 3
- Standard deviation: 2
- Significance level (in the *Options dialog*): 0.05

The probability the assumption of equal means is rejected is found to be about 39% in this case. Note that if the sample size is increased to 10 the power is improved to 77%.

## Testing Common Assumptions

Many statistical tests are only valid if certain underlying assumptions are met. In most cases, these assumptions are stated in the statistical textbooks along with the descriptions of the particular statistical technique. This chapter describes some of the more common assumptions encountered in Six Sigma project work and how to test for them. However, the subject of testing underlying assumptions is a big one and you might wish to explore it further with a Master Black Belt.

### Continuous Versus Discrete Data

Data come in two basic flavors: continuous and discrete, as discussed in Chap. 7. To review the basic idea, continuous data are numbers that can be expressed to any desired level of precision, at least in theory. For example, using a mercury thermometer I can say that the temperature is 75 degrees Fahrenheit. With a home digital thermometer I could say it's 75.4 degrees. A weather bureau instrument could add additional decimal places. Discrete data can only assume certain values. For example, the counting numbers can only be integers. Some survey responses force the respondent to choose a particular number from a list (pick a rating on a scale from 1 to 10).

Some statistical tests assume that you are working with either continuous or discrete data. For example, ANOVA assumes that continuous data are being analyzed, while chi-square and correspondence analysis assume that your data are counts. In many cases the tests are insensitive to departures from the data-type assumption. For example, expenditures can only be expressed to two decimal places (dollars and cents), but they can be treated as if they are continuous data. Counts can usually be treated as continuous data if there are many different counts in the data set. For example, if the data are defect counts ranging from 10 to 30 defects with all 21 counts showing up in the data (10, 11, 12, 28, 29, 30).

*You Have Discrete Data But Need Continuous Data.* In some cases, however, the data type matters. For example, if discrete data are plotted on control charts intended for continuous data the control limit calculations will be incorrect. Run tests and other non-parametric tests will also be affected by this. The problem of "discretized" data is often caused by rounding the data to too few decimal places when they are recorded. This rounding can be human caused, or it might be a computer program not recording or displaying enough digits. The simple solution is to record more digits. The problem may be caused by an inadequate measurement system. This situation can be identified by a measurement system analysis (see Chap. 9). The problem can be readily detected by creating a dot plot of the data.

*You Have Continuous Data But Need Discrete Data.* Let's say you want to determine if operator experience has an impact on the defects. One way to analyze this is to use a technique such as regression analysis to regress $X$ = years of experience on $Y$ = defects.

Another would be to perform a chi-square analysis on the defects by experience level. To do this you need to put the operators into discrete categories, then analyze the defects in each category. This can be accomplished by "discretizing" the experience variable. For example, you might create the following discrete categories:

| Experience (years) | Experience Category |
| --- | --- |
| Less than 1 | New |
| 1 to 2 | Moderately experienced |
| 3 to 5 | Experienced |
| More than 5 | Very experienced |

The newly classified data are now suitable for chi-square analysis or other techniques that require discrete data.

## Independence Assumption

Statistical independence means that two values are not related to one another. In other words, knowing one value provides no information as to what the other value is. If you throw two dice and I tell you that one of them is a 4, that information doesn't help you predict the value on the other die. Many statistical techniques assume that the data are independent. For example, if a regression model fits the data adequately, then the residuals will be independent. Control charts assume that the individual data values are independent; that is, knowing the diameter of piston 100 doesn't help me predict the diameter of piston 101, nor does it tell me what the diameter of piston 99 was. If I don't have independence, the results of my analysis will be wrong. I will believe that the model fits the data when it does not. I will tamper with controlled processes.

Independence can be tested in a variety of ways. If the data are normal (testing the normality assumption is discussed below) then the run tests described for control charts can be used.

A scatter plot can also be used. Let $y = X_{t-1}$ and plot $X$ versus $Y$. You will see random patterns if the data are independent. Software such as Minitab offer several ways of examining independence in time series data. Note: lack of independence in time series data is called *autocorrelation*.

If you don't have independence you have several options. In many cases the best course of action is to identify the reason why the data are not independent and fix the underlying cause. If the residuals are not independent, add terms to the model. If the process is drifting, add compensating adjustments.

If fixing the root cause is not a viable option, an alternative is to use a statistical technique that accounts for the lack of independence—for example, the EWMA control chart or a time series analysis that can model autocorrelated data. Another is to modify the technique to work with your autocorrelated data, such as using sloped control limits on the control chart. If data are cyclical you can create uncorrelated data by using a sampling interval equal to the cycle length. For example, you can create a control chart comparing performance on Monday mornings.

**Normality Assumption**

Statistical techniques such as *t*-tests, *Z*-tests, ANOVA, and many others assume that the data are at least approximately normal. This assumption is easily tested using software. There are two approaches to testing normality: graphical and statistical.

*Graphical Evaluation of Normality.* One graphical approach involves plotting a histogram of the data, then superimposing a normal curve over the histogram. This approach works best if you have at least 200 data points, and the more the merrier. For small data sets the interpretation of the histogram is difficult; the usual problem is seeing a lack of fit when none exists. In any case, the interpretation is subjective and two people often reach different conclusions when viewing the same data. Figure 10.31 shows four histograms for normally distributed data with mean = 10, sigma = 1 and sample sizes ranging from 30 to 500.

**Figure 10.31** Histograms with normal curves for different sample sizes.

An alternative to the histogram/normal curve approach is to calculate a "goodness-of-fit" statistic and a *P*-value. This gives an unambiguous acceptance criterion; usually the researcher rejects the assumption of normality if $P < 0.05$. However, it has the disadvantage of being nongraphical. This violates the three rules of data analysis:

1. Plot the data.
2. Plot the data.
3. Plot the data.

To avoid violating these important rules, the usual approach is to supplement the statistical analysis with a probability plot. The probability plot is scaled so that normally distributed data will plot as a straight line. Figure 10.32 shows the probability plots that correspond to the histograms and normal curves in Fig. 10.31. The following table shows that the *P*-values are all comfortably above 0.05, leading us to conclude that the data are reasonably close to the normal distribution.

**Figure 10.32** Normal probability plots and goodness of fit tests.

| N | 30 | 100 | 200 | 500 |
|---|---|---|---|---|
| P-value | 0.139 | 0.452 | 0.816 | 0.345 |

*What to Do If the Data Aren't Normal.* When data are not normal, the following steps are usually pursued:

- **Do nothing**—Often the histogram or probability plot shows that the normal model fits the data well "where it counts." If the primary interest is in the tails, for example, and the curve fits the data well there, then proceed to use the normal model despite the fact that the *P*-value is less than 0.05. Or if the model fits the middle of the distribution well and that's your focus, go with it. Likewise, if you have a very large sample you may get *P*-values greater than 0.05 even though the model appears to fit well *everywhere*. I work with clients who routinely analyze data sets of 100,000+ records. Samples this large will flag functionally and economically unimportant departures from normality as "statistically significant," but it isn't worth the time or the expense to do anything about it.
- **Transform the data**—It is often possible to make the data normal by performing a mathematical operation on the data. For example, if the data distribution has very long tails to the high side, taking the logarithm often creates data that are normally distributed. Minitab's control chart feature offers the Box-Cox normalizing power transformation that works with many data distributions encountered in Six Sigma work. The downside to transforming is that data have to be returned to the original measurement scale before being presented to nontechnical personnel. Some statistics can't be directly returned to their original units; for example, if you use the log transform then you can't find the mean of the original data by taking the inverse log of the mean of the transformed data.
- **Use averages**—Averages are a special type of transformation because averages of subgroups always tend to be normally distributed, even if the underlying data are not. Sometimes the subgroup sizes required to achieve normality can be quite small.
- **Fit another statistical distribution**—The normal distribution isn't the only game in town. Try fitting other curves to the data, such as the Weibull or the exponential. Most statistics packages, such as Minitab, have the ability to do this. If you have a knack for programming spreadsheets, you can use Excel's solver add-in to evaluate the fit of several distributions.
- **Use a non-parametric technique**—There are statistical methods, called non-parametric methods, that don't make any assumptions about the underlying distribution of the data. Rather than evaluating the differences of parameters such as the mean or variance, non-parametric methods use other comparisons. For example, if the observations are paired they may be compared directly to see if the after is different than the before. Or the method might examine the pattern of points above and below the median to see if the before and after values are randomly scattered in the

two regions. Or ranks might be analyzed. Non-parametric statistical methods are discussed later in this chapter.

## Equal-Variance Assumption

Many statistical techniques assume equal variances. ANOVA tests the hypothesis that the *means* are equal, not that *variances* are equal. In addition to assuming normality, ANOVA assumes that variances are equal for each treatment. Models fitted by regression analysis are evaluated partly by looking for equal variances of residuals for different levels of *Xs* and *Y*.

Minitab's test for equal variances is found in Stat > ANOVA > Test for Equal Variances. You need a column containing the data and one or more columns specifying the factor level for each data point. If the data have already passed the normality test, use the *P*-value from Bartlett's test to test the equal variances assumption. Otherwise, use the *P*-value from Levene's test. The test shown in Fig. 10.33 involved five factor levels and Minitab shows a confidence interval bar for sigma of each of the five samples; the tick mark in the center of the bar represents the sample sigma. These are the data from the sample of 100 analyzed earlier and found to be normally distributed, so Bartlett's test can be used. The *P*-value from Bartlett's test is 0.182, indicating that we can expect this much variability from populations with equal variances 18.2% of the time. Since this is greater than 5%, we fail to reject the null hypothesis of equal variances. Had the data not been normally distributed we would've used Levene's test, which has a *P*-value of 0.243 and leads to the same conclusion.

**Figure 10.33** Output from Minitab's test for equal variances.

## Linear Model Assumption

Many types of associations are nonlinear. For example, over a given range of $x$ values, $y$ might increase, and for other $x$ values, $y$ might decrease. This *curvilinear relationship* is shown in Fig. 10.34.

**Figure 10.34** Scatter diagram of a curvilinear relationship.

Here we see that $y$ increases when $x$ increases and is less than 1, and decreases as $x$ increases when $x$ is greater than 1. Curvilinear relationships are valuable in the design of robust systems. A wide variety of processes produce such relationships.

It is often helpful to convert these nonlinear forms to linear form for analysis using standard computer programs or scientific calculators. Several such transformations are shown in Table 10.11.

Fit the straight line $Y_T = b_0 + b_1 X_T$ using the usual linear regression procedures (see below). In all formulas, substitute $Y_T$ for $Y$ and $X_T$ for $X$. A simple method for selecting a transformation is to simply program the transformation into a spreadsheet and run regressions using every transformation. Then select the transformation that gives the largest value for the statistic $R^2$.

There are other ways of analyzing nonlinear responses. One common method is to break the response into segments that are piecewise linear, and then to analyze each piece separately. For example, in Fig. 10.34 $y$ is roughly linear and increasing over the range $0 < x < 1$ and linear and decreasing over the range $x > 1$. Of course, if the analyst has access to powerful statistical software, nonlinear forms can be analyzed directly.

**Table 10.11** Some Linearizing Transformations

| If the Relationship Is of the Form: | Plot the Transformed Variables $Y_T$ | $X_T$ | Convert Straight-Line Constants ($B_0$ and $B_1$) to Original Constants $b_0$ | $b_1$ |
|---|---|---|---|---|
| $Y = a + \dfrac{b}{X}$ | $Y$ | $\dfrac{1}{X}$ | $a$ | $b$ |
| $\dfrac{1}{Y} = a + bX$ | $\dfrac{1}{Y}$ | $X$ | $a$ | $b$ |
| $Y = \dfrac{X}{a + bX}$ | $\dfrac{X}{Y}$ | $X$ | $a$ | $b$ |
| $Y = ab^X$ | $\log Y$ | $X$ | $\log a$ | $\log b$ |
| $Y = ae^{bx}$ | $\log Y$ | $X$ | $\log a$ | $b \log e$ |
| $Y = aX^b$ | $\log Y$ | $\log X$ | $\log a$ | $b$ |
| $Y = a + bX^n$ where $n$ is known | $Y$ | $X^n$ | $a$ | $b$ |

Source: Natrella, 1963, pp. 5–31.

# Analysis of Categorical Data

## Making Comparisons Using Chi-Square Tests

In Six Sigma, there are many instances when the analyst wants to compare the percentage of items distributed among several categories. The things might be operators, methods, materials, or any other grouping of interest. From each of the groups a sample is taken, evaluated, and placed into one of several categories (e.g., high quality, marginal quality, reject quality). The results can be presented as a table with $m$ rows representing the groups of interest and $k$ columns representing the categories. Such tables can be analyzed to answer the question "Do the groups *differ* with regard to the proportion of items in the categories?" The chi-square statistic can be used for this purpose.

### Example of Chi-Square Test

The following example is from Natrella (1963):

Rejects of metal castings were classified by cause of rejection for three different weeks, as given in the following tabulation. The question to be answered is: Does the distribution of rejects differ from week to week?

|  | Cause of Rejection |||||||  |
| --- | --- | --- | --- | --- | --- | --- | --- | --- |
|  | Sand | Misrun | Shift | Drop | Corebreak | Broken | Other | Total |
| Week 1 | 97 | 8 | 18 | 8 | 23 | 21 | 5 | 180 |
| Week 2 | 120 | 15 | 12 | 13 | 21 | 17 | 15 | 213 |
| Week 3 | 82 | 4 | 0 | 12 | 38 | 25 | 19 | 180 |
| Total | 299 | 27 | 30 | 33 | 82 | 63 | 39 | 573 |

Chi-square ($X^2$) is computed by first finding the expected frequencies in each cell. This is done using the equation:

$$\text{Frequency expected} = f_e = \frac{\text{Row sum} \times \text{column sum}}{\text{overall sum}}$$

For example, for week 1, the frequency expected of sand rejects is $(180 \times 299)/573 = 93.93$. The table below shows the frequency expected for the remainder of the cells.

|  | Sand | Misrun | Shift | Drop | Corebreak | Broken | Other |
| --- | --- | --- | --- | --- | --- | --- | --- |
| Week 1 | 93.93 | 8.48 | 9.42 | 10.37 | 25.76 | 19.79 | 12.25 |
| Week 2 | 111.15 | 10.04 | 11.15 | 12.27 | 30.48 | 23.42 | 14.50 |
| Week 3 | 93.93 | 8.48 | 9.42 | 10.37 | 25.76 | 19.79 | 12.25 |

The next step is to compute $X^2$ as follows:

$$X^2 = \frac{\Sigma}{\text{over all cells}} \frac{(\text{frequency expected} - \text{frequency observed})^2}{\text{frequency expected}}$$

$$= \frac{(93.93 - 97)^2}{93.93} + \ldots + \frac{(12.25 - 19)^2}{12.25} = 45.60$$

Next choose a value for $\alpha$; we will use $\alpha = 0.10$ for this example. The degrees of freedom for the $X^2$ test are $(k - 1)(m - 1) = 12$. Referring to Appendix 4 we find the critical value of $X^2 = 18.55$ for our values. Since our computed value of $X^2$ exceeds the critical value, we conclude that the weeks differ with regard to proportions of various types of defectives.

## Logistic Regression

Logistic regression, like least squares regression, investigates the relationship between a response variable and one or more predictors. However, linear regression is used when response variables are continuous, while logistic regression techniques are used with categorical response variables. We will look at three different types of logistic regression, based on the type of response variable being analyzed (see Table 10.12).

**Table 10.12** Types of Logistic Regression Analysis

| Response Variable and Logistic Regression Type | Number of Response Categories | Response Characteristics | Examples |
| --- | --- | --- | --- |
| Binary | 2 | Two levels | Go/not-go, pass/fail, buy/doesn't buy, yes/no, recovers/dies, male/female |
| Ordinal | 3 or more | Natural ordering of the levels | Dissatisfied/neutral/satisfied, none/mild/severe, fine/medium/coarse |
| Nominal | 3 or more | No natural ordering of the levels | Black/white/Hispanic, black hair/brown hair/blonde hair, sunny/rainy/cloudy |

The basic idea behind logistic regression is very simple, as shown in Fig. 10.35. $X$ is a hypothetical "cause" of a response. $X$ can be either continuous or categorical. $Y$ is an event that we are interested in, and it must be categorical. A model can have multiple $X$s, but only one response variable. For example, $Y$ might be whether a prospect purchased a magazine or not, and $X$s might be the age and race of the prospect. The model would produce a prediction of the probability of a magazine being purchased based on the age and race of the prospect, which might be used to prioritize a list for telemarketing purposes.

**Figure 10.35** Logistic regression.

## The Logit

Figure 10.35 illustrates a direct modeling of the proportion responding versus a predictor variable. The problem is that in the real world the response pattern can take on a wide variety of forms and a simple model of the proportion responding as a function of predictors isn't flexible enough to take on all of the various shapes. The solution to this is to use a mathematical function, called the logit, that makes it possible to develop versatile models. The formula for the logit is shown in Eq. (10.17). Although it looks intimidating, it is really very similar to the equation for a linear regression. Notice that $e$ is raised to a power that is just a linear function of the $X$'s. In fact, the power term is just the multiple linear regression model. However, where linear regression can only model straight-line functions, the logit takes on a wide variety of curve shapes as the estimates of the parameters vary. Figure 10.36 shows logit curves for a few values of $\beta$, with $\alpha$ held constant at 0 (changing $\alpha$ would result in shifting the curves left or right).

$$P(x) = \frac{e^{\alpha + \beta_1 x_1 + \beta_2 x_2 + \ldots + \beta_n x_n}}{1 + e^{\alpha + \beta_1 x_1 + \beta_2 x_2 + \ldots + \beta_n x_n}} \qquad (10.17)$$

**Figure 10.36** Plot of the logit for $\alpha = 0$, $\beta$ varies.

## Odds Ratios

When the logit link is used (it's the default in most software packages, including Minitab), logistic regression evaluates the odds of some event of interest happening versus the odds of it not happening. This is done via *odds ratios*. "Odds" and probabilities are similar, but not identical. In a standard deck of cards there are 13 different card values, ace, king, queen, and so on. The *odds* of a randomly selected card being an ace is 12-to-1, i.e., there are 12 nonaces to 1 ace. The *probability* of selecting an ace is 1-in-13, that is, there are 13 choices of which 1 is an ace. In most statistical analyses used in Six Sigma work we use probabilities, but logistic regression uses odds for its calculations.

Consider a Six Sigma project involving a Web site. The goal of the project is to make it easier for customers to find what they are looking for. A survey was administered to people who visited the Web site and the results in Table 10.13 were obtained. The Black Belt wants to know if the design change had an impact on the customer's ability to find an answer to their question.

**Table 10.13** Odds Ratio Example

| Web Site Design | Found Answer | Didn't Find Answer |
|---|---|---|
| Old | 50 | 169 |
| New | 26 | 46 |

The odds ratio for these data is calculated as follows:

Odds of finding answer with old design = 50/169 = 0.0296

Odds of finding answer with new design = 26/46 = 0.565

Odds ratio = 0.565/0.296 = 1.91

It can be seen that the odds of the customer finding the answer appears to be 91% better with the new design than with the old design. However, to interpret this result properly we must know if this improvement is statistically significant. We can determine this by using binary logistic regression.

Note: Another way to analyze these data is to use chi-square. Logistic regression, in addition to providing a predictive model, will sometimes work when chi-square analysis will not.

## *Binary Logistic Regression*

Minitab's binary logistic regression function is located in the Stat > Regression menu. The data must be arranged in one of the formats Minitab accepts. Minitab's Binary Logistic Regression dialog box (Fig. 10.37) shows the input for this problem in columns C1, C2, C3, and C4. Column C4 is a code value that is 0 if the customer visited after the change, 1 otherwise.

**Figure 10.37** Minitab's Binary Logistic Regression dialog box.

## Interpreting Minitab's Binary Logistic Regression Output

There is a great deal of information displayed in Fig. 10.38; let's take a closer look at it. At the top we see that Minitab used the logit link in the analysis, which is its default. Next Minitab summarizes the response information, which matches the input in Table 10.13 (odds ratio example). Next we see the predictive model coefficients. The coefficient labeled "Constant" (0.5705) is the value for $\alpha$ in Eq. (10.17), and the coefficient labeled "WhenCode" is the coefficient for $\beta$. The $P$ column is the test for significance and $P < 0.05$ is the critical value. Since $P < 0.05$ for both the constant and the WhenCode, we conclude that the constant is not zero and that when the data were taken (before or after the design change) made a difference.

In the WhenCode row we have three additional columns: odds ratio, 95% confidence interval lower limit and 95% confidence interval upper limit. The odds ratio is the 1.91 we calculated directly earlier. The 95% confidence interval on the odds ratio goes from 1.07 to 3.40. If the design change made no difference, the expected value of the odds ratio would be 1.00. Since the interval doesn't include 1.00 we conclude (at 95% confidence) that the design change made a difference. This conclusion is confirmed by the $P$-value of 0.029 for the test that all slopes are equal (testing for equal slopes is equivalent to testing the null hypothesis that the design change had no effect).

```
Binary Logistic Regression: Outcome versus WhenCode

Link Function: Logit

Response Information

Variable  Value
Outcome   Not found    215   (Event)
          Found         76
          Total        291
Frequency: HowMany

Logistic Regression Table
                                         Odds      95% CI
Predictor   Coef    SE Coef    Z     P   Ratio   Lower   Upper
Constant    0.5705  0.2454   2.33  0.020
WhenCode    0.6473  0.2935   2.21  0.027  1.91   1.07    3.40

Log-Likelihood = -164.745
Test that all slopes are zero: G = 4.739, DF = 1, P-Value = 0.029

* NOTE * No goodness of fit tests performed.
*        The model uses all degrees of freedom.

Measures of Association:
(Between the Response Variable and Predicted Probabilities)
Pairs       Number  Percent   Summary Measures
Concordant    4394   26.9%    Somers' D              0.13
Discordant    2300   14.1%    Goodman-Kruskal Gamma  0.31
Ties          9646   59.0%    Kendall's Tau-a        0.05
Total        16340  100.0%
```

- Here's our odds-ratio
- Interval does not include 1
- The impact is statistically significant.
- 76 found *215 not found = 16340 pairs. A pair is *concordant* if actual "found" also has higher predicted probability of "found," *discordant* if the opposite is true, *tied* if probabilities are equal.
- Values range from 0 to 1. Closer to 1 = better predictability. The model isn't very good at prediction, but it's significantly better than random guessing.

**Figure 10.38** Output from Minitab binary logistic regression.

Had we had a covariate term (an *X* on a continuous scale) Minitab would've performed a goodness of fit test by dividing the data into ten groups and performing a chi-square analysis of the resulting table.

Next Minitab compares the predicted probabilities with the actual responses. The data are compared pairwise, predicted: found and not found versus actual: found and not found. A pair is "concordant" if actual and predicted categories are the same, "discordant" if they are different, and "tied" otherwise. Table 10.14 shows the classifications for our example.

**Table 10.14** Concordant and Discordant Results

| Design | Correct Result | Incorrect Result | Actual Count | Result |
|---|---|---|---|---|
| Old | Not found |  | 169 | Concordant |
|  |  | Found | 50 | Discordant |
| New | Found |  | 26 | Concordant |
|  |  | Not Found | 46 | Discordant |

The total number of found times not found pairs is 76 × 215 = 16340. The total number of concordant pairs is 169 × 26 = 4394. The total number of discordant pairs is 50 × 46 = 2300. The remaining 16340 − 4394 − 2300 = 9646 pairs are ties. The model correctly discriminated between and classified the concordant pairs, or 27%. It incorrectly classified the discordant pairs, or 14%.

Somers' D, Goodman-Kruskal Gamma, and Kendall's Tau-a are summaries of the table of concordant and discordant pairs. The numbers have the same numerator: the number of concordant pairs minus the number of discordant pairs. The denominators are the total number of pairs with Somers' D, the total number of pairs excepting ties with Goodman-Kruskal Gamma, and the number of all possible observation pairs for Kendall's Tau-a. These measures most likely lie between 0 and 1 where larger values indicate a better predictive ability of the model. The three summary measures of fit range between 0.05 and 0.31. This isn't especially impressive, but the *P*-value and the concordance/discordance analysis indicate that it's better than randomly guessing.

### Conclusion

The main conclusion is found in the odds ratio and *P*-value. The new design is better than the original design. The mediocre predictability of the model indicates that there's more to finding the correct answer than the different web designs. In this case it would probably pay to continue looking for ways to improve the process, only 36% of the customers find the correct answer (a process sigma that is less than zero!).

## *Ordinal Logistic Regression*

If the response variable has more than two categories, and if the categories have a natural order, then use ordinal logistic regression. Minitab's procedure for performing this analysis assumes parallel logistic regression lines. You may also want to perform a nominal logistic regression, which doesn't assume parallel regression lines, and compare the results. An advantage to using ordinal logistic regression is that the output includes estimated probabilities for the response variables as a function of the factors and covariates.

### Example of Ordinal Logistic Regression

A call center conducted a survey of its customers to determine the impact of various call center variables on overall customer satisfaction. Customers were asked to read a statement, then to respond by indicating the extent of their agreement with the statement. The two survey items we will analyze are

Q3: The technical support representative was professional. (*X*)
Q17: I plan to use XXX in the future, should the need arise. (*Y*)

Customers were asked to choose one of the following responses to each question:

- I strongly disagree with the statement.
- I disagree with the statement.

▲ I neither agree nor disagree with the statement.
▲ I agree with the statement.
▲ I strongly agree with the statement.

The results are shown in Table 10.15. Table 10.16 presents the first part of the Minitab worksheet for the data—note that this is the same information as in Table 10.15, just rearranged. There is one row for each combination of responses to Q3 and Q17.

**Table 10.15**  Survey Response Cross-Tabulation

**Frequency Table**

|             | Q17 Response |   |    |     |     |
|-------------|---|---|----|-----|-----|
| Q3 Response | 1 | 2 | 3  | 4   | 5   |
| 1           | 7 | 6 | 7  | 12  | 9   |
| 2           | 5 | 2 | 8  | 18  | 3   |
| 3           | 4 | 2 | 20 | 42  | 10  |
| 4           | 7 | 5 | 24 | 231 | 119 |
| 5           | 0 | 2 | 14 | 136 | 303 |

**Table 10.16**  Table 10.15 Data Reformatted for Minitab

| Q3 Response | Freq | Q17 Response |
|---|---|---|
| 1 | 7 | 1 |
| 2 | 5 | 1 |
| 3 | 4 | 1 |
| 4 | 7 | 1 |
| 5 | 0 | 1 |
| 1 | 6 | 2 |
| 2 | 2 | 2 |
| Etc. | Etc. | Etc. |

Minitab's dialog box for this example is shown in Fig. 10.39. The storage dialog box allows you to tell Minitab to calculate the probabilities for the various responses. I also recommend telling Minitab to calculate the number of occurrences so that you can cross-check your frequencies with Minitab's to ensure that you have the data in the correct format. When you tell Minitab to store results, the information is placed in new columns in your active worksheet, not in the session window. Note the data entries for the response, frequency, model, and factors.

Figure 10.39  Ordinal Logistic Regression Minitab dialog boxes.

Minitab's session window output is shown in Fig. 10.40. For simplicity only part of the output is shown. The goodness-of-fit statistics (concordance, discordance, etc.) have been omitted, but the interpretation is the same as for binary logistic regression. Minitab needs to designate one of the response values as the reference event. Unless you specifically choose a reference event, Minitab defines the reference event based on the data type:

```
Ordinal Logistic Regression: Q17RESPONSE versus Q3RESPONSE

Link Function: Logit

Response Information

Variable   Value    Count
Q17RESPO   1           23
           2           17
           3           73
           4          439
           5          444
           Total      996
Frequency: FREQ

    24 cases were used
     1 cases contained missing values
       or was a case with zero frequency.

Logistic Regression Table
                                            Odds       95% CI
Predictor     Coef   SE Coef      Z     P   Ratio   Lower   Upper
Const(1)   -2.0630    0.3358  -6.14  0.000
Const(2)   -1.4516    0.3094  -4.69  0.000
Const(3)   -0.1995    0.2926  -0.68  0.496
Const(4)    2.5735    0.3121   8.25  0.000
Q3RESPON
  2        -0.0905    0.4253  -0.21  0.831   0.91    0.40    2.10
  3        -0.5898    0.3638  -1.62  0.105   0.55    0.27    1.13
  4        -1.8408    0.3176  -5.80  0.000   0.16    0.09    0.30
  5        -3.2571    0.3254 -10.01  0.000   0.04    0.02    0.07

Log-likelihood = -941.733
Test that all slopes are zero: G = 246.547, DF = 4, P-Value = 0.000
```

Figure 10.40  Minitab ordinal logistic regression session window output.

- For numeric factors, the reference event is the greatest numeric value.
- For date/time factors, the reference event is the most recent date/time.
- For text factors, the reference event is the last in alphabetical order.

A summary of the interpretation follows:

- The odds of a reference event is the ratio of *P*(event) to *P*(not event).
- The estimated coefficient can also be used to calculate the odds ratio, or the ratio between two odds. Exponentiating the parameter estimate of a factor yields the ratio of *P*(event)/*P*(not event) for a certain factor level compared to the reference level.

You can change the default reference event in the *Options* subdialog box. For our example, category 5 (strongly agree) is the reference event. The odds ratios are calculated as the probability of the response being a 5 versus the probability that it is not a 5. For factors, the *smallest* numerical value is the reference event. For the example, this is a Q3 response of 1.

The odds ratios and their confidence intervals are given near the bottom of the table. A negative coefficient and an odds ratio less than 1 indicate that higher responses to Q17 tend to be associated with higher responses to Q3. Odds ratios whose confidence intervals do not include 1.00 are statistically significant. For example, this applies to responses of 4 or 5 to Q3, that is, a customer who chooses a 4 or 5 in response to Q3 is more likely to choose a 5 in response to Q17.

The statistical probabilities stored by Minitab are plotted in Fig. 10.41. The lines for Q3 = 4 and Q3 = 5, the factor categories with significant odds ratios, are shown as bold lines. Note that the gap between these two lines and the other lines is greatest for Q17 = 5.

**Figure 10.41** Minitab stored results.

## Nominal Logistic Regression

Nominal logistic regression, as indicated in Table 10.12, is used when the response is categorical, there are two or more response categories, and there is no natural ordering of the response categories. It can also be used to evaluate whether the parallel line assumption of ordinal logistic regression is reasonable.

### Example of Nominal Logistic Regression

Upon further investigation the Master Black Belt discovered that the Black Belt working on the Web site redesign project described in the binary logistic regression example section above had captured additional categories. Rather than just responding that the answer to their question was found or not found, there were several other response categories (Figs. 10.42 and 10.43). Since the various not found subcategories have no natural order, nominal logistic regression is the correct procedure for analyzing these data.

The result of Minitab's analysis, shown in Fig. 10.44, shows that only the odds ratio for found and worked versus not found is significant. The confidence interval for all other found subcategories compared with found and worked includes 1.00. The family *P*-value is a significance test for all comparisons simultaneously. Since we are making four comparisons, the significance level is higher than that of each separate test.

**Figure 10.42**  Minitab's nominal logistic regression dialog box.

```
Nominal Logistic Regression: Response versus Code
Response Information

Variable  Value
Response  Not Found            215   (Reference Event)
          Found, Unclear        18
          Found, incomplete     17
          Found, Failed         40
          Found and Worked      76
          Total                366
Frequency: Number

Logistic Regression Table
                                                 Odds         95% CI
Predictor     Coef    SE Coef      Z      P      Ratio    Lower    Upper
Logit 1: (Found, Unclear/Not Found)
Constant    -2.5649    0.2878    -8.91  0.000
Code         0.3457    0.5519     0.63  0.531    1.41     0.48     4.17

Logit 2: (Found, incomplete/Not Found)
Constant    -2.8273    0.3254    -8.69  0.000
Code         0.9446    0.5201     1.82  0.069    2.57     0.93     7.13

Logit 3: (Found, Failed/Not Found)
Constant    -1.6959    0.1954    -8.68  0.000
Code         0.0645    0.4136     0.16  0.876    1.07     0.47     2.40

Logit 4: (Found and Worked/Not Found)
Constant    -1.2179    0.1610    -7.57  0.000
Code         0.6473    0.2935     2.21  0.027    1.91     1.07     3.40

Log-likelihood = -425.268
Test that all slopes are zero: G = 7.053, DF = 4, P-Value = 0.133
```

**Figure 10.43** Minitab nominal logistic regression output.

**Figure 10.44** Interpretation of Minitab nominal logistic regression output.

- Interval includes 1, p-value > 0.05. i.e., "found but unclear" isn't significantly different than the reference event.
- "Found and worked" occurs significantly more often than the reference event.
- This is a "family" p-value

## Comparison with Chi-Square

If a chi-square analysis is performed on the web redesign data Minitab produces the output shown in Fig. 10.45. Note that the chi-square procedure prints a warning that there are two cells with less than the recommended minimum expected frequency of 5.0. It also gives a *P*-value of 0.116, which is greater than the critical value of 0.05, leading to a somewhat different conclusion than the logistic regression analysis. The chi-square test only lets us look at the significance of the overall result, which is analogous to the "family *P*-value" test performed in the nominal logistic regression analysis. However, in this case we are primarily concerned with the improved odds of finding the *correct* answer with the new web design versus the old web design, which is provided by logit 4 of the logistic regression.

**Chi-Square Test: Before, After**

Expected counts are printed below observed counts

|   | Before | After | Total |
|---|--------|-------|-------|
| 1 | 50     | 26    | 76    |
|   | 56.69  | 19.31 |       |
| 2 | 10     | 7     | 17    |
|   | 12.68  | 4.32  |       |
| 3 | 13     | 5     | 18    |
|   | 13.43  | 4.57  |       |
| 4 | 31     | 9     | 40    |
|   | 29.84  | 10.16 |       |
| 5 | 169    | 46    | 215   |
|   | 160.37 | 54.63 |       |
| Total | 273 | 93    | 366   |

Chi-Sq = 0.789 + 2.317 +
0.567 + 1.663 +
0.014 + 0.040 +
0.045 + 0.133 +
0.465 + 1.364 = 7.396
DF = 4, P-Value = 0.116
2 cells with expected counts less than 5.0

*Chi-square only gives an overall result*

*Family P-value > 0.05*

*Sample size too small.*

**Figure 10.45** Chi-square analysis of web design data.

## Non-Parametric Methods

The most commonly used statistical tests ($t$-tests, $Z$-tests, ANOVA, etc.) are based on a number of assumptions (see testing assumptions above). Non-parametric tests, while not assumption-free, make no assumption of a specific *distribution* for the population. The qualifiers (assuming) for non-parametric tests are always much less restrictive than for their parametric counterparts. For example, classical ANOVA requires the assumptions of mutually independent random samples drawn from normal distributions that have equal variances, while the non-parametric counterparts require only the assumption that the samples come from any identical continuous distributions. Also, classical statistical methods are strictly valid only for data measured on interval or ratio scales, while non-parametric statistics apply to frequency or count data and to data measured on nominal or ordinal scales. Since interval and ratio data can be transformed to nominal or ordinal data, non-parametric methods are valid in all cases where classical methods are valid; the reverse is not true. Ordinal and nominal data are very common in Six Sigma work. Nearly all customer and employee surveys, product quality ratings, and many other activities produce ordinal and nominal data.

So if non-parametric methods are so great, why do we ever use parametric methods? When the assumptions hold, parametric tests will provide greater power than non-parametric tests. That is, the probability of rejecting $H_0$ when it is false is higher with parametric tests than with a non-parametric test using the same sample size. However, if the assumptions do not hold, then non-parametric tests may have considerably greater power than their parametric counterparts.

It should be noted that non-parametric tests perform comparisons using medians rather than means, ranks rather than measurements, and signs of difference rather than measured differences. In addition to not requiring any distributional assumptions, these statistics are also more robust to outliers and extreme values.

The subject of non-parametric statistics is a big one and there are many entire books written about it. We can't hope to cover the entire subject in a book about Six Sigma. Instead, we briefly describe the non-parametric tests performed by Minitab (Fig. 10.46). Minitab's non-parametric tests cover a reasonably wide range of applications to Six Sigma work, as shown in Table 10.17.

Table 10.17  Applications for Minitab's Non-parametric Tests

| Minitab Non-parametric Test | What It Does | Parametric Analogs |
|---|---|---|
| 1-sample sign | Performs a one-sample sign test of the median and calculates the corresponding point estimate and confidence interval. | 1-sample Z-test<br>1-sample t-test |
| 1-sample Wilcoxon | Performs a one-sample Wilcoxon signed rank test of the median and calculates the corresponding point estimate and confidence interval. | 1-sample Z-test<br>1-sample t-test |
| Mann-Whitney | Performs a hypothesis test of the equality of two population medians and calculates the corresponding point estimate and confidence interval. | 2-sample t-test |
| Kruskal-Wallis | Kruskal-Wallis performs a hypothesis test of the equality of population medians for a one-way design (two or more populations). This test is a generalization of the procedure used by the Mann-Whitney test.<br>See also: Mood's median test. | One-way ANOVA |
| Mood's median test | Performs a hypothesis test of the equality of population medians in a one-way design. Sometimes called a median test or sign scores test.<br>Mood's median test is robust against outliers and errors in data, and is particularly appropriate in the preliminary stages of analysis.<br>Mood's median test is more robust against outliers than the Kruskal-Wallis test, but is less powerful (the confidence interval is wider, on the average) for analyzing data from many distributions, including data from the normal distribution.<br>See also: Kruskal-Wallis test. | One-way ANOVA |
| Friedman | Performs a non-parametric analysis of a randomized block experiment. Randomized block experiments are a generalization of paired experiments. The Friedman test is a generalization of the paired sign test with a null hypothesis of treatments having no effect. This test requires exactly one observation per treatment-block combination. | Two-way ANOVA<br>Paired sign test |
| Runs tests | Test whether or not the data order is random. Use Minitab's Stat > Quality Tools > Run Chart to generate a run chart. | None |
| Pairwise averages | Pairwise averages calculates and stores the average for each possible pair of values in a single column, including each value with itself. Pairwise averages are also called Walsh averages. Pairwise averages are used, for example, for the Wilcoxon method. | None |

| | | |
|---|---|---|
| Pairwise differences | Pairwise differences calculates and stores the differences between all possible pairs of values formed from two columns. These differences are useful for non-parametric tests and confidence intervals. For example, the point estimate given by Mann-Whitney can be computed as the median of the differences. | None |
| Pairwise slopes | Pairwise slopes calculates and stores the slope between all possible pairs of points, where a row in y-x columns defines a point in the plane. This procedure is useful for finding robust estimates of the slope of a line through the data. | Simple linear regression |
| Levene's test | Test for equal variances. This method considers the distances of the observations from their sample median rather than their sample mean. Using the sample median rather than the sample mean makes the test more robust for smaller samples. | Bartlett's test |
| Non-parametric Dist Analysis—Censored Data | Analyzes times-to-failure when no distribution can be found to fit the (censored) data. Tests for the equality of survival curves. | Parametric Dist Analysis—Censored data |
| Hazard plots—non-parametric distribution analysis | If data are right censored, plots empirical hazard function or actuarial estimates. If data are arbitrarily censored, plots actuarial estimates. | Hazard plots—parametric distribution analysis |

© *All Rights Reserved. 2000 Minitab, Inc. Used by permission.*

**Figure 10.46** Minitab's non-parametric tests.

## Guidelines on When to Use Non-parametric Tests

Use non-parametric analysis when *any* of the following are true (Gibbons, 1993):

▲ The data are counts or frequencies of different types of outcomes.
▲ The data are measured on a nominal scale.
▲ The data are measured on an ordinal scale.
▲ The assumptions required for the validity of the corresponding parametric procedure are not met or cannot be verified.
▲ The shape of the distribution from which the sample is drawn is unknown.
▲ The sample size is small.
▲ The measurements are imprecise.
▲ There are outliers and/or extreme values in the data, making the median more representative than the mean.

Use a parametric procedure when *both* of the following are true:

▲ The data are collected and analyzed using an interval or ratio scale of measurement.
▲ All of the assumptions required for the validity of that parametric procedure can be verified.

# CHAPTER 11

# The Improve/Design Phase

The primary objective of the Improve or Design stage of DMAIC/DMADV is to implement the new system. The first consideration is to prioritize the various opportunities, if more than one proposal exists. Once a preferred approach has been determined, the new process or product design is defined and optimal settings established. This new design can then be evaluated for risks and potential failure modes. If any of these steps require changes in prior assumptions, then steps must be repeated to properly evaluate the new proposal.

## Using Customer Demands to Make Design and Improvement Decisions

Customer demands can easily be converted into design requirements and specifications. The "Translating Customer Demands" section in Chap. 2 provided a detailed example of these techniques. The project selection methods of Chap. 4, including the prioritization matrix, Pareto analysis, benefit-cost analysis, and the theory of constraints throughput-based analysis, may be similarly applied to select the optimal improvement opportunities. Another simple comparison method is shown below.

### Pugh Concept Selection Method

The Pugh concept selection method is a simple alternative to the above approach of evaluating competing design concepts. The Pugh approach utilizes a simple matrix diagram to compare alternative concepts (Fig. 11.1). One concept is dubbed the "baseline," and all others are evaluated relative to the baseline in a simple Matrix diagram. In DMAIC the baseline is the current process. In DMADV, where there is no existing process or where the current process is deemed too bad to be salvaged, the baseline process is the one found "best" according to some criterion (e.g., fastest cycle time, lowest cost, fewest errors). If an

| Pugh Concept Selection Matrix Comparison Criteria | Baseline | Concept 1 | Concept 2 | Concept 3 | Concept 4 | Concept 5 | Concept 6 | Concept 7 |
|---|---|---|---|---|---|---|---|---|
| Criterion 1 | S | + | S | | | | | |
| Criterion 2 | S | S | – | | | | | |
| Criterion *n* | S | | | | | | | |
| Total +'s | 0 | 1 | 0 | | | | | |
| Total –'s | 0 | 0 | 1 | | | | | |

Compare current with selected alternatives

+ = Better Alternative, – = Worse Alternative, S = Same as Baseline

**Figure 11.1** Pugh concept selection matrix.

alternative concept is better than the baseline with respect to a given criterion, it is given a "+" for that criterion. If it is worse it is given a "–." Otherwise it is considered to be the same and given an "S." Concept scores are found by summing the plus and minus signs, providing a count of pros and cons. This is only one input into the final choice of concepts, but the structure of the approach stimulates thought and discussion and usually proves to be very worthwhile.

## Lean Techniques for Optimizing Flow

In the Analyze stage, process streams were analyzed to distinguish between value and non-value-added activities. In the Improve stage, we seek to optimize the flow by reducing waste.

Flow focuses on the object of value. The product, design, service, order, etc., that is being created for the customer. The focus is *not* on the department, the supplier, the factory, the procedure, the tooling, the setup, the inventory, or any other facet of the enterprise, or its operation. All work practices are carefully evaluated and rethought to eliminate stoppages of any kind so the object of value proceeds smoothly and continuously to the customer.

Flow requires that the whole process be considered simultaneously. Generally, the process begins with the order and ends with the customer receiving what was ordered. It requires, in effect, a customer-driven organization as described in Chap. 2 to ensure that value is properly specified, designed, produced, and delivered to the customer.

The Analyze stage will uncover a myriad of issues that impact flow and must be suitably addressed.

## *Unnecessary Process Steps*

One of the first steps for cycle time reduction is to eliminate the type 2 waste; the process steps that are simply not necessary. These may include activities such as routine authorizations or approvals, or information and data collection that is not necessary or not used. (It's remarkable how often data is collected but never shared or used by the appropriate personnel).

## *Excessive Movement of Material or Personnel*

When process cycle times are influenced by excessive movement of material and/or personnel between process steps, such as noted on the Spaghetti Diagram, a more efficient design of the process layout or procedures can reduce this NVA physical movement and overall cycle time. For example, if a part is inspected by an offline inspection department in a batch operation far away from the process line, an inline inspection by inspectors or even process personnel at the process location while the process is running would reduce unnecessary movement and provide more immediate results useful for quick prevention of non-conformances. A common outcome of the physical redesign is less departmentalization. Within an organization, you may see re-assignment of individuals from functional departments into multi-functional work cells. This improves the flow of the process, so work is not batched up at each department. The work cell approach, where each process step is close in location to its next step, also increases visibility of problems as the item moves from one step in the process to the next. (This *transparency* is a key lean concept that keeps everyone aware of current status).

The tools of 5S are also useful for reducing unnecessary movement. The term 5S is derived from five Japanese words to orient action: Seiri (organization), Seiton (tidiness), Seiso (purity), Seiketsu (cleanliness), and Shitsuke (discipline). The English translation is often stated as Sort, Set in order (also known as Straighten), Shine, Standardize, and Sustain. These terms are defined as follows:

- **Sort**—Clearly distinguish what is necessary to do the job from what is not. Eliminate the unnecessary. Ohno (2013) notes, "If you are holding on to your parts and stacking them in your warehouse just because you worked hard to make them, this is not sorting."
- **Set in order (straighten)**—Put needed items in their correct place to allow for easy accessibility and retrieval. Ohno (2013) adds, "Marking lines to prevent people from

going in and out from an area or to limit the height for stacking things has nothing to do with sorting and setting in order. I told them, 'We call sort, set in order, sweep and sanitize the 4S, but we do not have an S for lining things up.'"

- ▲ **Shine**—Keep the workplace clean and clear of clutter. This promotes safety as well as efficiency.
- ▲ **Standardized cleanup**—Develop an approach to regularly apply sort, set in order, and shine to maintain a clean and orderly work environment that facilitates productive work.
- ▲ **Sustain**—Incorporate the discipline required to achieve the (above) 4S into your everyday activities.

## Bottleneck or Constraint

Constraints, or bottlenecks, require special attention. A process constraint is that step or part of the process that limits the throughput of the entire process. As such, they determine how much output the process can produce. When a constraint isn't producing, the *process* isn't producing. Every effort needs to be focused on assuring that:

- ▲ The constraint has sufficient resources to keep running.
- ▲ Every unit supplied to the constraint is of acceptable quality.
- ▲ Every unit produced by the constraint is of acceptable quality.
- ▲ The constraint is operated in as efficient a manner as is possible.

These issues require emphasis on process stability, perhaps the most frequently overlooked requirement for achieving continuous flow of the process. It is fundamental that the process output is predictable at each process step to ensure consistent flow from one step to the next. A process that is out of control (on a statistical basis, as described in Chap. 8) is unpredictable, preventing continuous flow. When the upstream process is also incapable of consistently producing quality parts, the constraint will be deprived of resource; when the constraint itself is operating at a poor-quality level, then valuable process time is wasted producing material that is not usable to the customer. The SPC control charts discussed in Chapter 8 allow us to differentiate between the special causes of instability, and the common causes of process variation inherent to the process design. The designed experiments presented in Chapter 10 provide the methods for identifying the root causes of the special cause variation, as well as the main contributors of common cause variation. Techniques for optimizing the process for one or more process outcomes are discussed in the "Using Empirical Model Building to Optimize" section later in this chapter. The "Optimization Using Simulation" section provide a means of evaluating process flow, bottlenecks, staffing levels and cycle times using what-if scenarios.

## Process Errors Requiring Rework

Cycle time improvements can also be achieved through a reduction of errors requiring rework. Obviously, time spent producing error is a form of process waste, as is the resulting disruption in process flow. Practices for reducing errors include standardization of procedures, mistake-proofing, and improvement of process capability (see previous discussion of bottlenecks or constraints).

Standardization is a Lean approach to define a specified method for performing tasks to reduce variation and achieve commonality in the result. Looking forward, it sets the baseline. Ohno (2013) notes, "Where there is no standard, there can be no kaizen." There is an inherent paradox between standardization and kaizen: "But in the beginning, you must perform the standard work, and as you do, you should find things you don't like, and you will think of one kaizen idea after another. Then you should implement these ideas right away and make them the new standard." And he cautions: "But sometimes this can be kaiaku (change for the worse). If it is a change for the worse, just fix it right away. Don't change it back to the way it was, but try something else."

Mistake proofing is discussed within the concept of the *Risk Assessment Tools* later in this chapter. As errors requiring rework are eliminated, the Business Value Added inspections and approvals currently necessary may also be reduced or eliminated. The overall cycle time will be reduced, possibly even beyond the elimination of complaints to a level of customer delight.

## Excess In-process Inventory

As discussed in Chapter 10, inventories are a sign of waste. Simply, you've made a product that does not have an immediate customer. In many cases, the inventory level is specifically defined to ensure that product can be shipped when ordered by the customer simply because the total cycle time required to produce the product is excessive: the customer can't wait that long. In this way, traditional mass production is a *push* system. Push systems can be summarized as "Make a lot of stuff as cheaply as possible and hope people will buy it." Push systems minimize the number of setups and changeovers and use dedicated, specially designed equipment to produce identical units.

An alternative approach is found in pull systems. Pull systems can be summarized as "Don't make anything until it is needed, then make it fast." A pull system controls the flow and quantity produced by replacing items when they are consumed. Pull systems are achieved through level loading and facilitated by implementation of flexible processes.

Flexible processes are lightweight and maneuverable tools, fixtures and equipment positioned to improve safety, ergonomics, quality, and productivity. They are the opposite of the big, heavy, permanently positioned counterparts (aka, "monuments") traditionally used for mass production. A flexible shop can be quickly reconfigured to produce different items to meet varied customer demands. A completely flexible process would allow the factory to be instantly reconfigured to produce an item as soon as an order for it arrived. Reducing setup times is key to this process agility.

Level loading is the process of generating a schedule that is constant (i.e. level), stable and responsive to the market. The goal of level loading is to make the same quantity of an item every day. It is driven by *takt* time. (Takt is the German word for "metronome" and is used to indicate the desired rhythm of the process.) A level loaded schedule for a particular product can be obtained as follows:

$$takt \text{ time} = \frac{\text{daily work time}}{\text{daily quantity needed}} \qquad (11.1)$$

Work time does not include lunches, breaks, or other process downtime. Generally, *takt* time is used to create short-term (daily, weekly) work schedules.

### Example of *Takt* Time Calculation

A satellite manufacturer receives orders for 26 satellites per year.

$$takt \text{ time} = \frac{260 \text{ work days/year}}{26 \text{ satellites/year}} = 10 \text{ days per satellite}$$

This means that every work cell and operation must move one system's worth of work through every ten work days, no less and no more. For example, if an average satellite requires ten batteries, then the battery operation needs to produce one battery per work day, if a satellite needs 1,000 circuit boards, then 100 boards need to be completed every work day. Resources are allocated to each of these activities appropriately to match this *takt* time.

If the historical process average is 20 satellites per year, then the time to produce a satellite is 13 work days, substantially short of the ten-day *takt* time. In this case efforts need to focus on improving cycle time. On the other hand, if the historical average is 30 satellites per year, then production time is only 8.67 days per satellite and focus should be on increasing sales and reducing resources to the level dictated by customer demand.

The *takt* time is thus a standard time to be targeted for each process activity, so that flow or balance is achieved for the process as a whole. A properly balanced process will not suffer from the waste of inventory waiting to be processed by a subsequent process activity. An item will become available for the next process step as soon as it is needed (i.e., just in time), so the subsequent process does not suffer the waste of downtime waiting for the unit, and the prior process does not suffer the waste of producing output that is not yet needed.

Putting all of these things together, the ideal scenario becomes: a customer orders an item or items (pull), the factory has the resources to produce the order (level loading), processes are configured to create the items ordered (flexible process), the order is produced and delivered to the customer exactly when he needs it.

In traditional (i.e., push) systems, an economic lot size is defined. The lot size refers to the quantity of an item that is ordered from the plant or supplier or issued as a standard quantity to the production process. Larger lot sizes lead to larger quality problems due to delayed feedback, excessive inventory, obsolete inventory, etc. (collectively known as *carrying costs*). Of course, there are offsetting benefits such as quantity discounts, fewer setups, lower transportation costs, etc. (collectively known as *setup costs*). In practice, the costs and benefits must be balanced to achieve an optimum.

An economic order quantity (EOQ) can be defined that balances the setup and carrying costs.

$$EOQ = \sqrt{\frac{2do}{i}} \qquad (11.2)$$

where $d$ is the annual demand; $o$ is the setup or order cost; and $i$ is the annual carrying cost.

The following example is provided by the IQF Study Guide. The annual demand for an item is 1231 units. The setup cost for the machinery to produce the product is $27. The cost invested in each unit is $83, and the annual carrying costs are 7.9%. The EOQ is calculated as:

$$EOQ = \sqrt{\frac{2(1231)(27)}{(83)(0.079)}} = 101$$

The number of setups per year is then:

$$\# \text{ Setups per year} = \frac{1231}{101} = 12.2$$

Annual setup costs:

$$\text{Annual Setup Costs} = (\$27)(12.2) = \$329.40$$

The average inventory level is ½ the EOQ, so the annual carrying cost is:

$$\text{Annual Carrying Costs} = 0.5(101)(83)(0.079) = \$331.13$$

The total cost (setup and carrying) is thus $329.40 + $331.12 = $660.53.

Once these total costs are calculated, we can consider the impact of various improvement activities. For example, if the setup costs are reduced by 50%, what is the total annual savings?

The revised EOQ is calculated as:

$$EOQ = \sqrt{\frac{2(1231)(0.5)(27)}{(83)(0.079)}} = 72$$

The revised number of setups per year is then:

$$\text{\# Setups per year} = \frac{1231}{72} = 17.1$$

The revised annual setup costs:

$$\text{Annual Setup Costs} = (0.5)(\$27)(17.1) = \$230.85$$

The revised annual carrying cost is

$$\text{Annual Carrying Costs} = 0.5(72)(83)(0.079) = \$236.05$$

The total cost (setup and carrying) is thus $230.85 + $236.05 = $466.90, resulting in a savings of $193.63.

The EOQ serves as an economic justification for batch processing. So long as setup costs dominate a process, batch processing will offer incentives. Reducing the setup costs lowers the EOQ, effectively reducing carrying costs as well. In the Analyze stage, the setup times were analyzed and classified as *internal* (performed only when the process is stopped) or *external* (can be performed while the process is running), and as *preparation, replacement, location* or *adjustment* activities. In the Improve stage, we can reduce the process cycle time associated with setup activities by replacing internal activities with external activities. For example, loading a lab's dishwasher would typically be an internal activity, since the dishwasher would not be running as the racks are loaded. However, if additional racks are purchased, then they can be loaded offline while the dishwasher is running. This could reduce the total stoppage time for loading the dishwasher, as the current rack of clean dishes is replaced with the pre-loaded rack of dirty dishes.

Recommended practices to reduce the setup time associated with each setup activity are summarized below.

- **Preparation** (i.e., tasks associated with getting or storing tools or WIP needed for the process):
  - Convert from departments to work cells to minimize the time required to move the finished product to the next process step.

- Store tools and materials locally, such as advocated by the 5S principles.
- Convert to Always-Ready-To-Go. Make the software or instructions instantly accessible.

▲ **Replacement** (i.e., tasks associated with adding or removing items or tools, such as the movement of test fixtures; loading of new material into the hopper; loading paper in the copy machine):
- Reduce the number of setup steps required, such as through a redesign of fixtures;.
- Establish commonality of setups for product families. When we establish the same setup procedures for multiple items, we naturally have less instances of change required, reducing the setup time. This is the 5S tool of standardization. The process is simplified by reducing the number of special items that are processed: the higher the process complexity, the longer the cycle time. Henry Ford, in offering the first affordable automobile, realized the efficiency advantages offered by standardization. His motto (*any color you want, so long as it's black*) exemplified the notion of standardization. The modern approach to standardization is not that we need to limit options, but rather that we need to recognize the advantages of simplified processes, and seek to remove special cases where they provide little value at the cost of increased cycle times. Decision points and subsequent parallel paths on flowcharts provide indication of process complexities that can sometimes be avoided. By grouping parts or services into families, we can recognize that there are common methods that can be applied, thus simplifying processes and reducing overall cycle times.
- The 5S tools of sorting and straightening also help to reduce movement and wait times.

▲ **Location** (i.e., tasks associated with positioning or placement during setup, such as setting temperature profiles for heating; adjusting cutoff length for specific product; placing the chunk of deli meat in the slicer):
- Poka yoke (mistake proofing the process), as discussed in the "Risk Assessment" section later in this chapter.
- Commonality of setups as previously mentioned (the 5S tool of standardization).

▲ **Adjustment** (i.e., tasks associated with ensuring correct process settings, such as monitoring temperature of furnace; checking cutoff length; proofing copy before printing):
- Process control. If we can improve the repeatability of the process, then the adjustments would not be necessary. Often this is achieved though robust design methods discussed later in this chapter.

## Understanding Queues to Balance Processes

Once the takt time is determined for a particular product, attention will move to matching the flow or processing rate of each process step to the takt time. We seek to maintain a balanced flow rate across each step of the process to eliminate the in-process inventories created by the unbalanced steps. We can use the statistics of *queuing theory*, sometimes referred to as *waiting line systems*, to better understand these issues.

Queues, or lines, form in a process whenever the arrival rate (the number of customers, orders or items waiting to be processed in a given time period) exceeds the service (or processing) rate (the number of customers, orders or items that can be processed in a given time period). In the real world, the actual arrival times and processing times are variable, so even when the average arrival rate exactly equals the average processing rate, lines will form.

Like most statistical analysis, general queuing theory will rely on a few basic assumptions that simplify the problem. In the real world, human behavior can complicate the statistics. For example, humans arriving at the checkout line in the grocery store may decide that a long line is not worth the wait, and decide not to enter the line. This is commonly referred to as balking. Once in the line, the customer may decide to leave the line (referred to as reneging). When there are multiple servers (i.e., checkout clerks) available, the system may be defined with either a single line feeding all servers or a separate line for each server. (Generally, the single line for all servers provides the minimum average time in line for all customers.) When there are multiple lines, customers can also jockey between lines in an attempt to minimize their time in line. Each of these issues tend to complicate the analysis.

For our purposes, we will assume the following conditions:

- Arrival rate is well described by a Poisson distribution with an average arrival rate of lambda ($\lambda$), meaning that the time between orders follows an exponential distribution with an average $1/\lambda$.
- Processing rate is well described by a Poisson distribution with an average arrival rate of mu ($\mu$), meaning that the processing time for any given order follows an exponential distribution with an average $1/\mu$.
- Processing rate is greater than the arrival rate (i.e., $\mu > \lambda$).
- Orders are processed on a first-come, first-serve basis.
- There is a single processing line (sometimes called a *single server* system).
- There are an infinite number of orders.
- Orders are not removed from the line before processing.

With these basic assumptions, we can easily calculate some key performance metrics:

Average utilization of the system: $p = \dfrac{\lambda}{\mu}$

Average number of orders in the system: $L = \dfrac{\lambda}{\mu - \lambda}$

Average number of orders in queue: $L_Q = pL$

Average time in system (including processing): $W = \dfrac{1}{\mu - \lambda}$

Average time in queue: $W_Q = pW$

Probability that $n$ customers are in the system at any given time: $P_n = (1 - p)p^n$

For example, if there are, on average, 20 purchase orders to process in an 8-hour day, and each order averages 20 minutes to process:

Arrival rate $\lambda$ = 20 orders per day/8 hours per day = 2.5 orders per hour

Processing rate $\mu$ = (1 order/20 minutes) × (60 minutes/hour) = 3 orders per hour

Average utilization of the system: $p = \dfrac{\lambda}{\mu} = \dfrac{2.5}{3} = 0.83 = 83\%$

Average number of orders in the system: $L = \dfrac{\lambda}{\mu - \lambda} = \dfrac{2.5}{3 - 25} = 5$

Average number of orders in queue: $L_Q = pL = 0.83 \times 5 = 4.2$

Average time in system (including processing): $W = \dfrac{1}{\mu - \lambda} = \dfrac{1}{3 - 2.5} = 2$ hours

Average time in queue: $W_Q = pW = 0.83 \times 2 = 1.67$ hours

Probability that $n$ orders are in the system at any given time:
No orders: $P_0 = (1 - p)p^n = (1 - 0.83) \times (0.83)^0 = 0.17 = 17\%$
1 order: $P_1 = (1 - p)p^n = (1 - 0.83) \times (0.83)^1 = 0.14 = 14\%$
2 orders: $P_2 = (1 - p)p^n = (1 - 0.83) \times (0.83)^2 = 0.12 = 12\%$
3 orders: $P_3 = (1 - p)p^n = (1 - 0.83) \times (0.83)^3 = 0.10 = 10\%$

We can expand the system by adding processing lines (sometimes referred to as a *multiple server* systems). In this case, in addition to the earlier assumptions, we'll assume a single queue for orders, and that the orders will go to the first available processing line (i.e., server). When there are $s$ processing lines:

Average utilization of the system: $p = \dfrac{\lambda}{s\mu}$

Probability that no orders are in the system: $P_0 = \left[ \displaystyle\sum_{n=0}^{s-1} \dfrac{(\lambda/\mu)^n}{n!} + \dfrac{(\lambda/\mu)^s}{s!}\left(\dfrac{1}{1-p}\right)\right]^{-1}$

Average number of orders in queue: $LQ = \dfrac{P_0(\lambda/\mu)^s p}{s!(1-p)^2}$

Average time in queue: $W_Q = \dfrac{L_Q}{\lambda}$

Average time in system (including processing): $W = W_Q + \dfrac{1}{\mu}$

Average number of orders in the system: $L = \lambda W$

Probability that $n$ orders are in the system at any given time:

$P_n = \dfrac{(\lambda/\mu)^n P_0}{n!}$ for $n \leq s$

$P_n = \dfrac{(\lambda/\mu)^n P_0}{s! s^{n-s}}$ for $n > s$

For example, let's have the same processing rate (per server), but now the arrival rate is twice the earlier rate and there are 2 servers:

Arrival rate $\lambda$ = 40 orders per day/8 hours per day = 5 orders per hour

Processing rate $\mu$ = (1 order/20 minutes)(60 minutes/hour) = 3 orders per hour

Number of servers $s = 2$

Average utilization of the system: $p = \dfrac{\lambda}{s\mu} = \dfrac{5}{2 \times 3} = 0.83 = 83\%$

Probability that no orders are in the system:

$P_0 = \left[\displaystyle\sum_{n=0}^{s-1} \dfrac{(\lambda/\mu)^n}{n!} + \dfrac{(\lambda/\mu)^s}{s!}\left(\dfrac{1}{1-p}\right)\right]^{-1} = \left[1 + \dfrac{(5/3)^1}{1!} + \dfrac{(5/3)^2}{2!(1-0.83)}\right]^{-1} = 0.09 = 9\%$

Average number of orders in queue: $L_Q = \dfrac{P_0(\lambda/\mu)^s p}{s!(1-p)^2} = \dfrac{0.09(5/3)^2(0.83)}{2!(1-0.83)^2} = 3.8$

Average time in queue: $W_Q = \dfrac{L_Q}{\lambda} = \dfrac{3.8}{5} = 0.76$ hours

Average time in system (including processing): $W = W_Q + \dfrac{1}{\mu} = 0.76 + \dfrac{1}{3} = 1.1$ hours

Average number of orders in the system: $L = \lambda W = 5 \times 1.1 = 5.5$

Probability that $n$ orders are in the system at any given time:

No orders: $P_0 = \dfrac{(\lambda/\mu)^n}{n!} P_0 = \dfrac{(5/3)^0}{0!} \times 0.09 = 0.09 = 9\%$

1 order: $P_1 = \frac{(\lambda/\mu)^n}{n!}P_0 = \frac{(5/3)^1}{1!} \times 0.09 = 0.15 = 15\%$

2 orders: $P_2 = \frac{(\lambda/\mu)^n}{n!}P_0 = \frac{(5/3)^0}{2!} \times 0.09 = 0.13 = 13\%$

3 orders: $P_3 = \frac{(\lambda/\mu)^n}{s!s^{n-s}}P_0 = \frac{(5/3)^0}{2!2^{3-2}} \times 0.09 = 0.11 = 11\%$

This analysis allows us to understand how reductions in processing rate and addition of processing lines (or servers) can effectively meet the demands of the process. For example, by designing a process that minimizes the number of customers in line, we reduce the tendency of customers to renege, balk or jockey. We will also see later in this chapter how simulation can be used for more complicated problems, although it should be clear that the above analysis will be useful for a wide variety of waiting line problems.

## Using Empirical Model Building to Optimize

Empirical model building is a statistical approach for determining optimal process or design settings. It uses a series of experimental designs to reduce the total possible process or product design space to hone in on the optimal settings with regard to one or more requirements.

If you are new to design of experiments and empirical model building, a metaphor may prove helpful. Imagine that you suddenly wake up in a strange wilderness. You don't know where you are, but you'd like to climb to the top of the nearest hill to see if there are any signs of civilization. What would you do?

A first step might be to take a good look around you. Is there anything you should know before starting out? You would probably pay particular attention to things that might be dangerous. If you are in a jungle these might be dangerous animals, quicksand, and other things to avoid. You'd also look for things that could be used for basic survival, such as food, shelter, and clothing. You may wish to establish a "base camp" where you can be ensured that all the basic necessities are available; a safe place to return to if things get a bit too exciting. In empirical modeling we also need to begin by becoming oriented with the way things are before we proceed to change them. We will call this *knowledge discovery* activity Phase 0.

Now that you have a feel for your current situation and you feel confident that you know something about where you are, you may begin planning your trip to the highest hill. Before starting out you will probably try to determine what you will need to make the trip. You are only interested in things that are truly important. However, since you are new to jungle travel, you decide to make a few short trips to be sure that you have what you need. For your first trip, you pack up every conceivable item and set out. In all likelihood, you will discover that you have more than you need. Those things that are not important you will leave at your camp. As part of your short excursions you also learn something about the local terrain close to your camp; not much, of course, but enough to identify which direction is uphill. This phase is equivalent to a *screening experiment*, which we call Phase I.

You now feel that you are ready to begin your journey. You take only those things you will need and head out into the jungle in the uphill direction. From time to time you stop to get your bearings and to be sure that you are still moving in the right direction. We call this hill-climbing *steepest ascent*, or Phase II.

At some point you notice that you are no longer moving uphill. You realize that this doesn't mean that you are at the highest point in your area of the jungle, only that you are no longer moving in the right direction. You decide to stop and make camp. The next morning you begin to explore the local area more carefully, making a few short excursions from your camp. The jungle is dense and you learn that the terrain in the immediate vicinity is irregular, sometimes steep, sometimes less steep. This is in contrast to the smooth and consistent uphill slope you were on during your ascent. We call this phase of your journey the *factorial experiment*, or Phase III.

Now you decide that a more organized approach will be needed to locate the nearby peak. You break out the heavy artillery, the GPS you've been carrying since the beginning (one of those cheap ones that don't have built-in maps)! You take several altitude readings from near your camp, and others at a carefully measured distance on all major compass headings. Each time you carefully record the altitude on a hand-drawn map. You use the map to draw contour lines of equal altitude and eventually a picture emerges that clearly shows the location of the top of the hill. This is the *composite design phase*, which we call Phase IV.

At last you reach the top of the hill. You climb to the top of a tree and are rewarded with a spectacular view, the best for miles around. You decide that you love the view so much, you will build your home on this hill and live there permanently. You make your home sturdy and strong, able to withstand the ravages of wind and weather that are sure to come to your little corner of the jungle. In other words, your home design is *robust*, or impervious to changes in its environment. We call the activity of building products and processes that are insensitive to changes in their operating parameters *robust product and process design*, which is Phase V of the journey.

Now that this little tale has been told, let's go on to the real thing, improving your products, processes, and services.

## Phase 0: Getting Your Bearings

### "Where Are We Anyway?"

Before any experimentation can begin the team should get an idea of what the major problems are, important measures of performance, costs, time, and other resources available for experimentation, etc. Methods and techniques for conducting Phase 0 research are described in Chap. 10.

The central premise of the approach described in this section is that learning is, by its very nature, a sequential process. The experimenter, be it an individual or a team, begins with relatively little specific knowledge and proceeds to gain knowledge by conducting experiments on the process. As new knowledge is acquired, the learner is better able to determine which step is most appropriate to take next. In other words, experimentation always involves guesswork; but guesses become more educated as experimental data become available for analysis.

This approach is in contrast to the classical approach where an effort is made to answer all conceivably relevant questions in one large experiment. The classical approach to experimentation was developed primarily for agricultural experiments. Six Sigma applications are unlike agricultural applications in many ways, especially in that results become available quickly. The approach described here takes advantage of this to accelerate and direct learning.

We will use an example from electronic manufacturing. At the outset, a team of personnel involved in a soldering process received a mission from another team that had been evaluating problems for the factory as a whole. The factory team had learned that a leading reason for customer returns was solder problems. Another team discovered that the solder area spent more resources in terms of floor space than other areas; a major usage of floor space was for the storage of defective circuit boards and the repair of solder defects. Thus, the solder process improvement team was formed and asked to find ways to eliminate solder defects if possible, or to at least reduce them by a factor of ten. Team members included a Six Sigma technical leader, a process engineer, an inspector, a production operator, and a product engineer.

The team spent several meetings reviewing Pareto charts and problem reports. It also performed a process audit that uncovered several obvious problems. When the problems were repaired the team conducted a process capability study, which revealed a number of special causes of variation, which were investigated and corrected. Over a four-month period, this preliminary work resulted in a 50% reduction in the number of solder defects, from about 160 defects per standard unit to the 70 to 80 defect range. The productivity of the solder area nearly doubled as a result of these efforts. While impressive, the results were still well short of the factor-of-ten minimum improvement the team was asked to deliver.

## Phase I: The Screening Experiment

### "What's Important Here?"

At this point the process was stable and the team was ready to move from the process control stage to the process improvement stage. This involved conducting designed experiments to measure important effects. The solder team decided to list as many items as possible that might be causing solder problems. Since many variables had already been studied as part of the Phase 0 work, the list was not unreasonably long. The team looked at ways to control the variables listed and was able to develop methods for eliminating the effects of many variables on their list. The remaining list included the following factors:

| Variable | Low Level (−) | High Level (+) |
| --- | --- | --- |
| A: Prebaking of boards in an oven | No | Yes |
| B: Preheat time | 10 s | 20 s |
| C: Preheat temperature | 150°F | 200°F |
| D: Distance from preheat element to board surface | 25 cm | 50 cm |
| E: Line speed | 3 fpm | 5 fpm |
| F: Solder temperature | 495°F | 505°F |
| G: Circuit density | Low | High |
| H: Was the board in a fixture? | No | Yes |

This information was used to create an experimental design using a statistical software package. There are many packages on the market that perform similar analyses to the one shown here.

Since this is only to be a screening experiment, the team was not interested in obtaining estimates of factor interactions. The focus was to identify important main effects. The software allows selection from among several designs. The Black Belt decided upon the design that would estimate the main effects with the smallest number of test units. This design involved testing 16 units. The data matrix produced by the computer is shown in Table 11.1. The run order has been randomized by the computer, and is shown in the first column. If the experiment cannot be conducted in that particular order, the computer software would allow the data to be run in blocks and it would adjust the analysis accordingly. The design is a standard $2^4$ factorial design of resolution IV, which means that main effects are not confounded with each other or any two-factor interactions.

**Table 11.1** Screening Experiment Layout Data Matrix (Randomized)

| Run | A | B | C | D | E | F | G | H | Response |
|---|---|---|---|---|---|---|---|---|---|
| 1 | + | − | − | − | − | + | + | + | 65 |
| 2 | + | − | + | + | − | + | − | − | 85 |
| 3 | + | + | − | − | + | + | − | − | 58 |
| 4 | − | + | − | − | + | − | + | + | 57 |
| 5 | − | − | − | − | − | − | − | − | 63 |
| 6 | + | + | + | + | + | + | + | + | 75 |
| 7 | − | + | − | + | − | + | + | − | 77 |
| 8 | − | + | + | − | − | + | − | + | 60 |
| 9 | + | − | + | − | + | − | − | + | 67 |
| 10 | + | + | + | − | − | − | + | − | 56 |
| 11 | − | − | + | − | + | + | + | − | 63 |
| 12 | − | − | − | + | + | + | − | + | 81 |
| 13 | + | + | − | + | − | − | − | + | 73 |
| 14 | + | − | − | + | + | − | + | − | 87 |
| 15 | − | + | + | + | + | − | − | − | 75 |
| 16 | − | − | + | + | − | − | + | + | 84 |

In Table 11.1 the minus sign indicates that the variable is run at its low level, while a plus sign indicates that it is to be run at its high level. For example, the unit for run #16 was processed as follows:

- Prebaking = No
- Preheat time = 10 sec
- Preheat temperature = 200°F
- Distance from preheat element to board surface = 50 cm
- Line speed = 3 fpm
- Solder temperature = 495°F
- Circuit density = High
- Fixture used = Yes
- Measured Response: Defects per standard unit = 84

Experimental data were collected using the randomized run order recommended by the software. The "response" column are data that were recorded in terms of defective solder joints per "standard unit," where a standard unit represented a circuit board with a median number of solder joints. The results of the factorial design regression from Minitab 18 are shown in Table 11.2.

**Table 11.2** Results of Experimental Data Analysis, Fractional Factorial Fit (Minitab 18)

**Estimated effects and coefficients for response (coded units)**

| Term | Effect | Coef | SE Coef | T-value | P-value | VIF |
|---|---|---|---|---|---|---|
| Constant |  | 70.375 | 0.660 | 106.67 | 0.000 |  |
| A | 0.750 | 0.375 | 0.660 | 0.57 | 0.588 | 1.00 |
| B | −8.000 | −4.000 | 0.660 | −6.06 | 0.001 | 1.00 |
| C | 0.500 | 0.250 | 0.660 | 0.38 | 0.716 | 1.00 |
| D | 18.500 | 9.250 | 0.660 | 14.02 | 0.000 | 1.00 |
| E | 0.000 | 0.000 | 0.660 | 0.00 | 1.000 | 1.00 |
| F | 0.250 | 0.125 | 0.660 | 0.19 | 0.855 | 1.00 |
| G | 0.250 | 0.125 | 0.660 | 0.19 | 0.855 | 1.00 |
| H | −0.250 | −0.125 | 0.660 | −0.19 | 0.855 | 1.00 |

**ANOVA for defects (coded units)**

| Source of variation | df | Seq. SS | Adj. SS | Adj. MS | F | P-value |
|---|---|---|---|---|---|---|
| Main effects | 8 | 1629.00 | 1629.00 | 203.63 | 29.24 | 0.000 |
| Residual error | 7 | 48.75 | 48.75 | 6.96 |  |  |
| Total | 15 | 1677.75 |  |  |  |  |

**Model Summary**

| S | R-sq | R-sq(adj) | R-sq(pred) |
|---|---|---|---|
| 2.63899 | 97.09% | 93.77% | 84.82% |

A model that fits the data well would produce residuals that fall along a straight line. The Black Belt concluded that the fit of the model was adequate as shown by Minitab 18's four-in-one residuals plot in Figure 11.2. Starting at the upper left and moving clockwise: the normal probability plot show the standardized residuals following a normal distribution (the Anderson-Darling statistic significantly above 0.05 fails to reject the null hypothesis that the residuals are normally distributed); the standardized residuals versus fits plot verifies constant variance of the residuals; the standardized residuals versus order plot shows no obvious correlation of the residuals; the histogram of the standardized residuals shows no obvious skew or outliers.

The analysis indicates that only factors $B$ (preheat time) and $D$ (distance from preheat element to board surface) produce significant effects, as shown by the p-value less than 0.05. Figure 11.3 shows a normal probability plot of the standardized experimental effects. If the factor's effect was due to chance variation it would plot close to the line representing normal variation. In Fig. 11.3 the effects of $B$ and $D$ are shown to be further from the line than can be accounted for by random variation indicating significance of these effects.

Chapter 11 The Improve/Design Phase | 535

**Figure 11.2** Residuals from experimental model.

**Figure 11.3** Significant factor effects.

Since the response is a defect count, the graph indicates that the low level of factor $D$ gives better results, while the high level of factor $B$ gives the better results. This can also be seen by examination of the coefficients for the variables. When $D$ is low the average defect rate is 18.5 defects per unit better (i.e., lower) than when $D$ is high; when $B$ is high the average defect rate is 8 defects per unit better than when $B$ is low.

The team met to discuss these results. They decided to set all factors that were not found to be statistically significant to the levels that cost the least to operate, and factors $B$ and $D$ at their midpoints. The process would be monitored at these settings for a while to determine that the results were similar to what the team expected based on the experimental analysis. While this was done, another series of experiments would be planned to further explore the significant effects uncovered by the screening experiment.

## Phase II: Steepest Ascent (Descent)

### "Which Way Is Up?"

Based on the screening experiment, the linear model for estimating the defect rate was found from the coefficients in Table 11.2 to be

$$\text{Defect rate} = 70.375 - 4B + 9.25D$$

The team wanted to conduct a series of experiments to evaluate factors $B$ and $D$. The Phase I experiment reveals the direction and ratio at which $B$ and $D$ should be changed to get the most rapid improvement in the defect rate, that is, the direction of steepest ascent (where "ascent" means improvement in the measurement of interest). To calculate a series of points along the direction of steepest ascent, start at the center of the design and change the factors in proportion to the coefficients of the fitted equation; that is, for every 4-unit increase in factor $B$ we decrease factor $D$ 9.25 units. For the data at hand, the center of the experiment and unit sizes are shown in Table 11.3.

**Table 11.3** Unit Sizes and Center of Experiment

| Factor | Unit Size | Center |
|--------|-----------|--------|
| B | 5 | 15 s |
| D | 12.5 | 37.5 cm |

A test unit was produced at the center value of $B$ and $D$. The team decided that they would increase the preheat time ($B$) in increments of 5 seconds (1 unit), while lowering the distance from the heating element ($D$) by increments of $(9.25/4) \times 12.5$ cm $= 28.9$ cm. This resulted in a single experiment where $B = 20$ seconds, $D = 8.6$ cm. The result was 52 defects per unit. However, despite the improved solder defect performance, the team noted that at the short distance the board was beginning to scorch. This necessitated that the team abandon the steepest ascent path. They conducted a series of experiments where board scorching was examined at different distances to the preheating element (factor $D$) and determined that a distance of at least 15 cm was required to be confident they would avoid scorching. To allow a margin of safety, the team set the distance D at 20 cm. They then proceeded to increase preheat time in 5-second intervals, producing one board at each preheat setting. The results are shown in Table 11.4.

These data are presented graphically in Fig. 11.4.

**Table 11.4** Data for Experiments on Path of Steepest Descent

| Run | B (s) | D (cm) | Average Defects |
|-----|-------|--------|-----------------|
| 1   | 15    | 37.5   | 70              |
| 2   | 20    | 8.75   | 52              |
| 3   | 25    | 20     | 51              |
| 4   | 30    | 20     | 31              |
| 5   | 35    | 20     | 18              |
| 6   | 40    | 20     | 12              |
| 7   | 45    | 20     | 10              |
| 8   | 50    | 20     | 13              |

**DEFECTS VS B & D SETTINGS**

| D       | 37.5 | 8.75 | 20 | 20 | 20 | 20 | 20 | 20 |
|---------|------|------|----|----|----|----|----|----|
| B       | 15   | 20   | 25 | 30 | 35 | 40 | 45 | 50 |
| DEFECTS | 70   | 52   | 51 | 31 | 18 | 12 | 10 | 13 |

SETTING

**Figure 11.4** Steepest descent results.

With the distance fixed at 20 cm from the preheat element to the board surface, the best results were obtained with a preheat time of 45 seconds. Beyond that the defect rate was greater.

## Phase III: The Factorial Experiment

The team decided to conduct a factorial experiment near the best settings to explore that experimental region more thoroughly to estimate the two-factor BD interaction as well as the main effects. They also wished to determine if there was any curvature in the response, indicating a local maximum or minimum. This required adding a third level (only linear estimates are possible with two-level designs) a center point positioned midway between each factor's high and low settings. Finally, to obtain an estimate of the experimental error in the region the experiment would be replicated.

The coded values used for each factor in the experiment are shown in parentheses in Table 11.5. The runs marked 0, 0 are center points. Note that each combination (i.e., set of plus and minus signs or zeros) is repeated three times to allow an estimate of the error. The design was centered at the $B$ value with the highest response, $B = 45$ seconds. The interval for $D$ was reduced to 2.5 cm and the experiment was centered one interval above $D = 20$ (i.e., at $D = 22.5$).

**Table 11.5** Results of Full Factorial Experiment with Center Points and Replicates

| Run | B | D | Result |
|---|---|---|---|
| 1 | 40 (−1) | 20.0 (−1) | 11 |
| 2 | 45 (0) | 22.5 (0) | 9 |
| 3 | 50 (1) | 25.0 (1) | 11 |
| 4 | 40 (−1) | 25.0 (1) | 15 |
| 5 | 50 (1) | 20.0 (−1) | 12 |
| 6 | 45 (0) | 22.5 (0) | 10 |
| 7 | 40 (−1) | 25.0 (1) | 17 |
| 8 | 40 (−1) | 25.0 (1) | 15 |
| 9 | 50 (1) | 20.0 (−1) | 11 |
| 10 | 50 (1) | 25.0 (1) | 11 |
| 11 | 40 (−1) | 20.0 (−1) | 13 |
| 12 | 40 (−1) | 20.0 (−1) | 13 |
| 13 | 50 (1) | 25.0 (1) | 11 |
| 14 | 50 (1) | 20.0 (−1) | 11 |
| 15 | 45 (0) | 22.5 (0) | 10 |

The results from Minitab 18 are shown in Table 11.6. The *P-value* column indicates all terms in the model are significant (any *P* value below 0.05 indicates a significant effect). This analysis is confirmed by the ANOVA table.

The *P-value* column of the ANOVA table indicates that main effects, the two-way interaction, and curvature are all significant *(P < 0.05)*. Curvature is measured by comparing the average response at the center points with the responses at the corner points of the design. The fact that curvature is significant means that we are no longer experimenting in a linear region of the responses.

This means that our original coefficients, which were based on the linear model, are no longer adequate. Upon seeing these results, the Black Belt decided that it was necessary to proceed to Phase IV to better investigate the response region and to try to locate a stationary optimum.

**Table 11.6** Factorial Regression: Result Versus B, D, CenterPt (Minitab 18)

**Analysis of Variance**

| Source | DF | Adj SS | Adj MS | F-Value | P-Value |
|---|---|---|---|---|---|
| Model | 4 | 61.333 | 15.3333 | 23.00 | 0.000 |
| Linear | 2 | 30.833 | 15.4167 | 23.12 | 0.000 |
| B | 1 | 24.083 | 24.0833 | 36.12 | 0.000 |
| D | 1 | 6.750 | 6.7500 | 10.12 | 0.010 |
| 2-Way Interactions | 1 | 10.083 | 10.0833 | 15.12 | 0.003 |
| B*D | 1 | 10.083 | 10.0833 | 15.12 | 0.003 |
| Curvature | 1 | 20.417 | 20.4167 | 30.62 | 0.000 |
| Error | 10 | 6.667 | 0.6667 | | |
| Total | 14 | 68.000 | | | |

**Model Summary**

| S | R-sq | R-sq(adj) | R-sq(pred) |
|---|---|---|---|
| 0.816497 | 90.20% | 86.27% | 77.94% |

**Coded Coefficients**

| Term | Effect | Coef | SE Coef | T-Value | P-Value | VIF |
|---|---|---|---|---|---|---|
| Constant | | 12.583 | 0.236 | 53.39 | 0.000 | |
| B | −2.833 | −1.417 | 0.236 | −6.01 | 0.000 | 1.00 |
| D | 1.500 | 0.750 | 0.236 | 3.18 | 0.010 | 1.00 |
| B*D | −1.833 | −0.917 | 0.236 | −3.89 | 0.003 | 1.00 |
| Ct Pt | | −2.917 | 0.527 | −5.53 | 0.000 | 1.00 |

Regression Equation in Uncoded Units

Result = 12.583 − 1.417 B + 0.750 D − 0.917 B*D − 2.917 Ct Pt

## Phase IV: The Composite Design

At this point in the empirical model building, it is appropriate to use a *central composite design* to obtain additional information on the region where the process is operating. This design involves augmenting the corner points and center point of the previous factorial experiment with axial points, as shown in Fig. 11.5. The points extend the design beyond the levels previously designed by the high and low values for each factor. The team decided that they could allow the distance to be decreased somewhat below the 20 cm "minimum" distance because they had added a 5 cm margin of safety. They also noted that they were now taking relatively small experimental steps compared to the large jumps they took during steepest ascent. Recommended levels for the axial points are shown in Table 11.7 based on the number of corner points $k$ included in the design to be extended. In general, the level is defined at $k^{1/4}$ (Montgomery, 1991). For example, the axial point of a $2^2$ design with four corner points is recommended at $4^{1/4} = 1.414$; the eight corner points of a standard $2^3$ design places the axial points at 1.68. Using these recommended axial point levels ensure the design is rotatable, meaning that the variance of the predicted response is constant at all points equidistant from the design center (Cornell, 1990), a desirable property for the design.

**Figure 11.5** Central composite design for solder process.

**Table 11.7** Central Composite Design Axial Point Location α

| k | 2 | 3 | 4 | 5 |
|---|---|---|---|---|
| α | 1.414 | 1.682 | 2.00 | 2.378 |

With a central composite design's axial points, the fitted equation is expanded to include higher order (squared) terms:

$$y = \beta_0 + \beta_1 x_1 + \beta_2 x_2 + \beta_{11} x_1^2 + \beta_{22} x_2^2 + \beta_{12} x_1 x_2 + \epsilon \qquad (11.3)$$

The region described by this equation may contain a maximum, a minimum, or a "saddle point." At a maximum any movement away from the stationary point will cause the response to decrease. At the minimum any movement away from the stationary point will cause the response to increase. At a saddle point moving away from the stationary value of one variable will cause a decrease, while moving away from the stationary value of the other variable will cause an increase. Some DOE software will tell you the values of $X$ and $Y$ at the stationary point, and the nature of the stationary point (max, min, or saddle). Other DOE software display two-dimensional and three-dimensional drawings that graphically describe the region of experimentation. It is usually not difficult to interpret the response surface drawings.

The design is augmented using Minitab's *Modify Design* option to add axial points. In this case, there were 12 corner points in the design, so Minitab places the axial points at $12^{1/4} = 1.86$ coded units from the center. The data collected by the team are shown in Table 11.8, and added to the earlier data shown in Table 11.5. Since the new data was collected at a different time period than the original data, it is added as a new block. (The earlier data has a value of 1 in Minitab's *Block* column; this new data for the axial points shown in Table 11.8 is entered with the *Blocks* column set to 2.)

Table 11.8  Central Composite Design Experiment and Data

| B | D | Defects |
|---|---|---|
| −1.86121 | 0.00000 | 14 |
| 0.00000 | −1.86121 | 11 |
| 1.86121 | 0.00000 | 7 |
| 0.00000 | 1.86121 | 15 |

The design with the axial points in now analyzed using Minitab's *Analyze Response Surface Design* option. The analysis, including the blocks and all terms (i.e., B, D, BD, $B^2$, $D^2$), is shown in Table 11.9.

**Table 11.9** Response Surface Regression: Defects Versus B, D

**Analysis of Variance**

| Source | DF | Adj SS | Adj MS | F-Value | P-Value |
|---|---|---|---|---|---|
| Model | 6 | 98.873 | 16.4788 | 24.49 | 0.000 |
|   Blocks | 1 | 12.985 | 12.9848 | 19.30 | 0.001 |
|   Linear | 2 | 61.926 | 30.9628 | 46.02 | 0.000 |
|     B | 1 | 47.638 | 47.6384 | 70.80 | 0.000 |
|     D | 1 | 14.287 | 14.2873 | 21.23 | 0.001 |
|   Square | 2 | 26.667 | 13.3333 | 19.82 | 0.000 |
|     B*B | 1 | 9.636 | 9.6359 | 14.32 | 0.003 |
|     D*D | 1 | 26.475 | 26.4752 | 39.35 | 0.000 |
|   2-Way Interaction | 1 | 10.083 | 10.0833 | 14.99 | 0.002 |
|     B*D | 1 | 10.083 | 10.0833 | 14.99 | 0.002 |
| Error | 12 | 8.074 | 0.6729 | | |
|   Lack-of-Fit | 2 | 1.408 | 0.7038 | 1.06 | 0.384 |
|   Pure Error | 10 | 6.667 | 0.6667 | | |
| Total | 18 | 106.947 | | | |

**Model Summary**

| S | R-sq | R-sq(adj) | R-sq(pred) |
|---|---|---|---|
| 0.820281 | 92.45% | 88.68% | 78.79% |

**Coded Coefficients**

| Term | Coef | SE Co-ef | T-Value | P-Value | VIF |
|---|---|---|---|---|---|
| Constant | 8.182 | 0.709 | 11.54 | 0.000 | |
| Blocks | | | | | |
|   1 | 1.484 | 0.338 | 4.39 | 0.001 | 2.14 |
| B | −1.586 | 0.189 | −8.41 | 0.000 | 1.00 |
| D | 0.869 | 0.189 | 4.61 | 0.001 | 1.00 |
| B*B | 1.097 | 0.290 | 3.78 | 0.003 | 2.14 |
| D*D | 1.819 | 0.290 | 6.27 | 0.000 | 2.14 |
| B*D | −0.917 | 0.237 | −3.87 | 0.002 | 1.00 |

**Regression Equation in Uncoded Units**

Result = 8.182 − 1.586 B + 0.869 D + 1.097 B*B + 1.819 D*D − 0.917 B*D

Equation averaged over blocks.

The *P*-values indicate that all terms are significant, except the Lack of Fit (which is a good thing). The significance of the Block is not so good, and should be investigated with additional experiments.

The response surface 3D and contour plots (in coded units) are shown in Figs. 11.6(a) and (b).

**Figure 11.6a**  Response surface plot for defect data.

**Figure 11.6b**  Contour plot for defect data.

An optimal response can also be predicted in Minitab (*Stat>DOE>Response Surface>Response Optimizer*). In this case, we are interested in minimizing the response (defects), so use a target value of zero for the response. Minitab predicts a minimum response of 7.6 can be attained with (coded) values of $B = 0.696$ and $D = -0.056$. Using the most recent design conditions defined in Table 11.5, this corresponds to real (experimental) units of $B = 48.5$ s and $D = 22.4$ cm.

At this point the team decided that they had reached a point of diminishing returns as far as the current process is concerned. The data indicate that the existing process for wave solder can, if properly controlled, manage to produce 10 or fewer defects per standard unit at the center of the last design. This is about 16 times better than the process was producing at the outset of the project and about 7 times better than the average result of the first experiment.

The team, guided by the Black Belt, decided to set the process at the center point of the last experiment ($B = 0$, $D = 0$) and to implement Evolutionary Operation (EVOP) to pursue further optimization. EVOP involves running a series of designed experiments on production units, with operating personnel making small changes (Box and Draper, 1969). By restricting EVOP to small process changes the risk of producing scrap is reduced. Although the movements in process settings are small, the cumulative improvement in performance can be substantial. The apparent gradual slope of the defect rate in the $+B$ direction also made it unlikely that the process would "fall off of a cliff" during EVOP.

The Black Belt helped set up EVOP on the process and train supervisory and hourly personnel in its use. She also agreed to provide ongoing support in the form of periodic visits and availability should questions arise. The team decided that after turning over process improvement to operating personnel, they would look at ways of maintaining their gains, while simultaneously investigating basic process and product design changes to obtain further improvement.

## *Phase V: Robust Product and Process Design*

Maintaining gains involves, among other things, creating processes and products that operate close to their optimum conditions even when changes occur. Robust design can begin with careful study of the contour plot. Note that if you start at $B = D = 0$ and move from along a line from left to right the response changes relatively slowly. However, if you move from the center along a line from lower to upper, the defect rate increases rapidly. Robust process control planning should take such nonlinearity into account. If there is a need to change factor $B$ or $D$, they should be changed in a way that avoids increasing the defect rate. This does *not* mean that all changes should be forbidden; after all, without change there can be no learning or improvement. However, changes should be monitored (as with EVOP) to provide a filter between the customer and the production of nonconforming product that may occur during the learning process.

More formally, robust design can be integrated into experimental design. The methods described by Genichi Taguchi are a well-known approach to integrating DOE and product and process design. While there has been much criticism of Taguchi's statistical approach, there is a broad consensus that his principles of robust parameter design are both valid and valuable contributions to Six Sigma analysis.

## Taguchi Robustness Concepts

This section will introduce some of the special concepts introduced by Dr. Genichi Taguchi of Japan. A complete discussion of Taguchi's approach to designed experiments is beyond the scope of this book. However, many of Taguchi's ideas are useful, such as his concept of designing processes and products to be robust to the potential sources of variation that could influence them.

*Introduction.* Quality is defined as the loss imparted to the society from the time a product is shipped (Taguchi, 1986). Taguchi divides quality control efforts into two categories: online quality control and off-line quality control. Online quality control involves diagnosis and adjusting of the process, forecasting and correction of problems, inspection and disposition of product, and follow-up on defectives shipped to the customer. Off-line quality control is quality and cost control activities conducted at the product and the process design stages in the product development cycle. There are three major aspects to off-line quality control:

- System design is the process of applying scientific and engineering knowledge to produce a basic functional prototype design. The prototype model defines the initial settings of product or process design characteristics.
- Parameter design is an investigation conducted to identify settings that minimize (or at least reduce) the performance variation. A product or a process can perform its intended function at many settings of its design characteristics. However, variation in the performance characteristics may change with different settings. This variation increases both product manufacturing and lifetime costs. The term *parameter design* comes from an engineering tradition of referring to product characteristics as product parameters. An exercise to identify optimal parameter settings is therefore called *parameter design.*
- Tolerance design is a method for determining tolerances that minimize the sum of product manufacturing and lifetime costs. The final step in specifying product and process designs is to determine tolerances around the nominal settings identified by parameter design. It is still a common practice in industry to assign tolerances by

convention rather than scientifically. Tolerances that are too narrow increase manufacturing costs, and tolerances that are too wide increase performance variation and the lifetime cost of the product.

Expected loss is the monetary losses an arbitrary user of the product is likely to suffer at an arbitrary time during the product's life span due to performance variation. Taguchi advocates modeling the loss function so the issue of parameter design can be made more concrete. The most often-used model of loss is the quadratic loss function illustrated in Fig. 11.7. Note that the loss from operating the process is found by integrating the process pdf over the dollar-loss function. Under this model there is always a benefit to:

▲ Moving the process mean closer to the target value
▲ Reducing variation in the process

Of course, there is often a cost associated with these two activities.

**Figure 11.7** Taguchi's quadratic loss function.

Weighing the cost/benefit ratio is possible when viewed from this perspective.

Note the contrast between the quadratic loss function and the conceptual loss function implicit in the traditional management view. The traditional management approach to loss is illustrated in Fig. 11.8.

Interpretation of Fig. 11.8: there is no loss as long as a product or service meets requirements. There is no "target" or "optimum": just barely meeting requirements is as good as operating anywhere else within the zone of zero loss. Deviating a great deal from

**Figure 11.8** Traditional approach to loss.

requirements incurs the same loss as being just barely outside the prescribed range. The process distribution is irrelevant as long as it meets the requirements.

Note that under this model of loss there is no incentive for improving a process that meets the requirements since there is no benefit, that is, the loss is zero. Thus, cost > benefit for any process that meets requirements. This effectively destroys the idea of continuous improvement and leads to the acceptance of an "acceptable quality level" as an operating standard.

Noise is the term used to describe all those variables, except design parameters, that cause performance variation during a product's life span and across different units of the product. Sources of noise are classified as either external sources or internal sources. External sources of noise are variables external to a product that affect the product's performance. Internal sources of noise are the deviations of the actual characteristics of a manufactured product from the corresponding nominal settings.

Performance statistics estimate the effect of noise factors on the performance characteristics. Performance statistics are chosen so that maximizing the performance measure will minimize expected loss. Many performance statistics used by Taguchi use "signal to noise ratios," which account jointly for the levels of the parameters and the variation of the parameters.

### Summary of the Taguchi Method

The Taguchi method for identifying settings of design parameters that maximize a performance statistic is summarized by Kackar (1985):

- Identify initial and competing settings of the design parameters, and identify important noise factors and their ranges.

▲ Construct the design and noise matrices, and plan the parameter design experiment.
▲ Conduct the parameter design experiment and evaluate the performance statistic for each test run of the design matrix.
▲ Use the values of the performance statistic to predict new settings of the design parameters.
▲ Confirm that the new settings do indeed improve the performance statistic.

## Data Mining, Artificial Neural Networks, and Virtual Process Mapping

As beneficial and productive as design of experiments can be, the process of conducting them has its drawbacks. The workplace, be it a factory, a retail establishment, or an office, is designed around a routine. The routine is the "real work" that must be done to generate the sales, which, in turn, produce the revenues that keep the enterprise in existence. By its very nature, experimenting means disrupting the routine. Important things are changed to determine what effect they have on various important metrics. Often, these effects are unpleasant; that's why they weren't changed in the first place! The routine was often established to steer a comfortable course that avoids the disruption and waste that results from making changes.

The problem is, without change things can never improve. Six Sigma generates as much improvement by changing things as it does by reducing variability.

In this section we present a way of conducting "virtual" experiments using existing data and artificial neural network (neural net) software. Neural nets are popular because they have a proven track record in many data mining and decision-support applications. Neural nets are a class of very powerful, general purpose tools readily applied to prediction, classification, and clustering. They have been applied across a broad range of industries from predicting financial series to diagnosing medical conditions, from identifying clusters of valuable customers to identifying fraudulent credit card transactions, from recognizing numbers written on checks to predicting failure rates of engines (Berry and Linoff, 1997). In this section we explore only the application of neural nets to design of experiments for Six Sigma, but this merely scratches the surface of the potential applications of neural nets for quality and performance improvement.

Neural networks use a digital computer to model the neural connections in human brains. When used in well-defined domains, their ability to generalize and learn from data mimics our ability to learn from experience. However, there is a drawback. Unlike a well-planned and executed DOE, a neural network does not provide a mathematical model of the process.[2] For the most part, neural networks must be approached as black boxes with mysterious internal workings, much like the mystery of the human mind it is designed to imitate.

All companies record important data, some in well-designed data warehouses, some in file drawers. These data represent potential value to the Six Sigma team. They contain information that can be used to evaluate process performance. If the data include information on process settings, for example, they may be matched up to identify possible cause and effect relationships and point the direction for improvement. The activity of sifting through a database for useful information is known as *data mining*. The process works as follows:

1. Create a detailed inventory of data available throughout the organization.
2. Determine the variables that apply to the process being improved.
3. Using a subset of the data that include the most extreme values, train the neural net to recognize relationships between patterns in the independent variables and patterns in the dependent variables.
4. Validate the neural net's predictive capacity with the remaining data.
5. Perform experimental designs as described in the section above entitled "Using Empirical Model Building to Optimize." However, instead of making changes to the actual process, make changes to the "virtual process" as modeled by the neural net.
6. Once Phase IV has been completed, use the settings from the neural net as a starting point for conducting experiments on the actual process. In other words, begin experimenting at Phase I with a screening experiment.

It can be seen that the entire soft experimentation process is part of Phase 0 in the empirical model building process. It helps answer the question "Where are we?" It is important to recognize that neural net experiments are not the same as live experiments. However, the cost of doing them is minimal compared with live experiments and the process of identifying input and output variables, deciding at which levels to test these variable, etc. will bear fruit when the team moves on to the real thing. Also, soft experiments allow a great deal more "what if?" analysis, which may stimulate creative thinking from team members.

## *Example of Neural Net Models*

The data in Table 11.10 are from the solder process described above. Data were not gathered for a designed experiment, but were merely collected during the operation of the process. The data were used to train and validate a neural net.

The neural net model is shown in Fig. 11.9.

The model was trained using the above data, producing the process map shown in Fig. 11.10.

You can see that the surface described by the neural net is similar to the one modeled earlier using DOE. Both models direct the *B* and *D* settings to similar levels and both make similar predictions for the defect rate.

**Table 11.10** Solder Process Data for Virtual Process Mapping

| PH_Time | PH_Distance | Defects |
|---|---|---|
| 38 | 22.5 | 15 |
| 40 | 20 | 13 |
| 40 | 25 | 16 |
| 45 | 17.5 | 15 |
| 45 | 22.5 | 5 |
| 45 | 26 | 11 |
| 50 | 20 | 12 |
| 42 | 22.5 | 10 |
| 50 | 25 | 3 |
| 42 | 22 | 11 |
| 46 | 22 | 4 |
| 55 | 25 | 4 |
| 55 | 21 | 17 |
| 55 | 25 | 15 |
| 50 | 24 | 3 |
| 49 | 25 | 3 |
| 57 | 37 | 10 |
| 35 | 25 | 20 |
| 45 | 37.5 | 17 |
| 30 | 20 | 27 |
| 30 | 22.5 | 33 |
| 30 | 25 | 37 |
| 30 | 27.5 | 50 |
| 30 | 37.5 | 57 |
| 50 | 20 | 13 |
| 50 | 22.5 | 5 |
| 50 | 25 | 3 |
| 50 | 30 | 5 |
| 50 | 14 | 12 |
| 50 | 37.5 | 14 |
| 50 | 45 | 16 |
| 50 | 50 | 40 |
| 60 | 20 | 35 |
| 60 | 25 | 18 |
| 60 | 37.5 | 12 |

Figure 11.9 Neural net model for solder defects.

Figure 11.10 Neural net process map for solder defects.

The neural net software also allows "what if" analysis. Since these data are from the region where the team ran its last phase of experiments, they could be used to conduct virtual DOE. The neural net's What If? contour plot dialog box is shown in Fig. 11.11.

Figure 11.11 What If? contour plot dialog box.

The virtual DOE values are entered in the What If? dialog box, and the neural net's predictions are used in the experimental design just as if they had been obtained using data from a real experiment. If you have data covering the entire region of interest, the neural net may bring you very close to the optimum settings even before you do your first actual experiment.

## Optimization Using Simulation

Simulation is a means of experimenting with a detailed model of a real system to determine how the system will respond to changes in its structure, environment, or underlying assumptions. A system is defined as a combination of elements that interact to accomplish a specific objective. A group of machines performing related manufacturing operations would constitute a system. These machines may be considered, as a group, an element in a larger production system. The production system may be an element in a larger system involving design, delivery, etc.

Simulations allow the system designer to solve problems. To the extent that the computer model behaves as the real world system it models, the simulation can help answer important questions. Care should be taken to prevent the model from becoming the focus of attention. If important questions can be answered more easily without the model, then the model should not be used

The modeler must specify the scope of the model and the level of detail to include in the model. Only those factors that have a significant impact on the model's ability to serve its stated purpose should be included. The level of detail must be consistent with the purpose. The idea is to create, as economically as possible, a replica of the real-world system that can provide answers to important questions. This is usually possible at a reasonable level of detail.

Well-designed simulations provide data on a wide variety of systems metrics, such as throughput, resource utilization, queue times, and production requirements. While useful in modeling and understanding existing systems, they are even better suited to evaluating proposed process *changes*. In essence, simulation is a tool for rapidly generating and evaluating ideas for process improvement. By applying this technology to the creativity process, Six Sigma improvements can be greatly accelerated.

### *Predicting CTQ Performance*

A key consideration for any design concept is the CTQ that would result from deploying the design. It is often very difficult to determine the overall result of a series of process steps, but relatively easy to study each step individually. Software can then be used to simulate the process a number of times and calculate the CTQ at the end of the series.

### Example of Predicting CTQ Performance

An order fulfillment process design concept has been studied by benchmarking each step. Some of the steps were observed internally, while others were operating in companies considered best in class for the particular step. The overall CTQ target was to ship within 24 hours of receiving the customer's call. The distribution of the time to complete each of the individual steps is shown in Fig. 11.12.

An Excel-based computer program[3] was used to simulate 10,000 orders going through the process, producing the results shown in Fig. 11.13. The simulation indicates that the CTQ Total Cycle Time will be met 99.9% of the time, for a process sigma level of 4.6. Since the process goal is Six Sigma, there is a gap of 1.4 sigma (about 997 PPM) to be addressed. The distribution of the individual steps provides valuable clues about where to focus our attention. Analysis of Fig. 11.13 indicates that the problem is a long tail to the right, so we should look at steps where this is also the case. "Ship order" and "enter order" are both prime suspects and candidates for improvement. A new concept design would then be resimulated.

## *Simulation Tools*

Not long ago, computer simulation was the exclusive domain of highly trained systems engineers. These early simulations were written in some general purpose programming language, such as FORTRAN, Pascal, or C. However, modern computer software has greatly simplified the creation of simulation models. With graphical user interfaces and easy drawing-based model creation, it is now almost as easy to create a simulation as it is to draw a flow chart (see Fig. 11.14).

While the user interface may *look* like an ordinary flow chart, it is much more powerful. A closer look at Fig. 11.14 reveals that additional information is recorded on the chart. Next to box 2: Receive Call, we see In: 60. This means that the simulation program looked at 60 simulated phone calls. By following the arrows we can learn that 20 of these calls were sales calls and that this kept the simulated sales representatives active for 2 hours and 2 minutes. Other data are available from the diagram, including the fact that QA checking cost $34.83 and that one order remained to be shipped at the conclusion of the simulation. If the model is based on an existing system, these numbers will be compared to actual process numbers to validate the model. The first simulation should always model the past to ensure the model's forecasts for the future.

**554** | The Six Sigma Handbook

| Process Step | Performance Information |
|---|---|
| Open order | Cell B2: Exponential Distribution<br>Assumption Name: Open Order Time (Hours)<br>Rate 12.05<br>0.04 to +Infinity |
| Enter order into system | Cell C2: Gamma Distribution<br>Assumption Name: Enter Order<br>Loc. 0.17  Scale 1.00  Shape 2<br>0.17 to +Infinity |
| Pick order | Cell D2: Normal Distribution<br>Assumption Name: Pick Order<br>Mean 4.00  Std Dev 0.40<br>−Infinity to +Infinity |

**Figure 11.12** Order fulfillment process.

| Process Step | Performance Information |
|---|---|
| Stage order | Cell E2: Uniform Distribution — Assumption Name: Stage Order; Min 2.70, Max 3.30 |
| Package order | Cell F2: Triangular Distribution — Assumption Name: Package; Min 2.00, Likeliest 3.50, Max 4.00 |
| Ship order | Cell G2: Extreme Value Distribution — Assumption Name: Ship Order; -Infinity, +Infinity; Mode 2.00, Scale 1.00, End: Max |

**Figure 11.12** Order fulfillment process *(continued)*.

**556** | The Six Sigma Handbook

**Figure 11.13** Cycle time for 10,000 orders.

**Figure 11.14** Simulation software interface.

As intriguing as models of existing systems can be, the real power and excitement begins when simulation models are applied to process changes. Refer to the simple model shown in Fig. 11.14. There are many questions that might arise regarding this process, for example:

- Our new promotion is expected to double the number of orders phoned in, what effect will that have on production?
- If the QA check was performed by production personnel, what effect would that have on QA cost? Total cost? Production throughput?

In general, the model lets us determine what happens to $Y_i$ if we change $X$. Changes often create *unanticipated consequences* throughout a complex system due to their effects on interrelated processes. For example, changing the volume of calls might cause an increase in the idle time of the QA checker because it increases the delay time in the "Get Parts" bottleneck process. Once this fact has been revealed by the simulation, the manager can deal with it. Furthermore, the manager's proposed solution can also be tested by simulation before it's tried in the real world. For example, the manager might propose to cross-train the QA person to be able to help Get Parts. This would theoretically reduce the wait at the Get Parts step while simultaneously increasing the utilization of the QA person. The simulation would allow this hypothesis to be tested before it's tried. Perhaps it will show that the result is merely to move the bottleneck from one process step to another, rather than eliminating it. Anyone who has spent any length of time in the working world is familiar with these "hydraulic models" where managers' attempts to fix one problem only result in the creation of new problems. By discovering this before trying it, money is saved and morale improved.

## *Random Number Generators*

The heart of any simulation is the generation of random numbers. Random numbers from specific distributions are generated by transforming random numbers from the unit, uniform distribution. Virtually all programming languages, as well as electronic spreadsheets, include a unit, uniform random number generator. Technically, these unit, uniform random number generators are pseudorandom number generators, as the algorithms used to generate them take away a small portion of the randomness. Nevertheless, these algorithms are extremely efficient and for all practical purposes the result is a set of truly random numbers.

A simple way to generate distribution-specific random numbers is to set the cumulative distribution function equal to a unit, random number and take the inverse. Consider the exponential distribution:

$$F(x) = 1 - e^{-\lambda x} \tag{11.4}$$

By setting $r$, a random variable uniformly distributed from zero to one, equal to $F(x)$ and inverting the function, an exponentially distributed random variable, $x$, with a failure rate of $\lambda$ is created.

$$\begin{aligned} r &= 1 - e^{-\lambda x} \\ 1 - r &= e^{-\lambda x} \\ \ln(1 - r) &= -\lambda x \\ x &= \frac{-\ln(1 - r)}{\lambda} \end{aligned} \tag{11.5}$$

This expression can be further reduced; the term $1 - r$ is also uniformly distributed from zero to one. The result is

$$x = -\frac{\ln r}{\lambda} \tag{11.6}$$

Table 11.11 contains some common random number generators.

After the desired random number generator(s) have been constructed, the next step is to mathematically model the situation under study. After completing the model, it is important to *validate* and *verify* the model. A valid model is a reasonable representation of the situation being studied A model is verified by determining that the mathematical and computer model created represents the intended conceptual model.

Enough iterations should be included in the simulation to provide a *steady-state* solution, which is reached when the output of the simulation from one iteration to the next changes negligibly. When calculating means and variances, 1,000 iterations is usually sufficient. If calculating confidence limits, many more iterations are required; after all, for 99% confidence limits the sample size for the number of random deviates exceeding the confidence limit is 1/100th the number of iterations.

### Example: A Simulation of Receiving Inspection[4]

This example describes a simulation model of a complex inspection operation at a factory of a large unionized defense contractor. The plant receives four types of parts: electrical, simple mechanical, complex mechanical, and parts or materials that require nondestructive testing (NDT). Union regulations required four different inspector grades. The plant

**Table 11.11** Random Number Generators

| Distributors | Probability Density Function | Random Number Generators[§] |
|---|---|---|
| Uniform | $f(x) = \dfrac{1}{b-a}, a \leq x \leq b$ | $x = a + (b-a)r$ |
| Exponential | $f(x) = \lambda e^{-\lambda x}, 0 < x < \infty$ | $x = -\dfrac{1}{\lambda}\ln r$ |
| Normal | $f(x) = \dfrac{1}{\alpha\sqrt{2\pi}}\exp\left[-\dfrac{1}{2}\left(\dfrac{x-\mu}{\sigma}\right)^2\right], -\infty < x < \infty$ | $x_1 = \left[\sqrt{-2\ln r_1}\cos(2\pi r_2)\right]\sigma + \mu$ <br> $x_2 = \left[\sqrt{-2\ln r_1}\sin(2\pi r_2)\right]\sigma + \mu$[†] |
| Lognormal | $f(x) = \dfrac{1}{\sigma x\sqrt{2\pi}}\exp\left[-\dfrac{1}{2}\left(\dfrac{\ln x - \mu}{\sigma}\right)^2\right], x > 0$ | $x_1 = \exp\left[\sqrt{-2\ln r_1}\cos(2\pi r_2)\right]\sigma + \mu$ <br> $x_2 = \left[\sqrt{-2\ln r_1}\sin(2\pi r_2)\right]\sigma + \mu$[†] |
| Weibull | $f(x) = \dfrac{\beta x^{\beta-1}}{\theta^\beta}\exp\left(\dfrac{x}{\theta}\right)^\beta, x > 0$ | $x = \theta(-\ln r)^{1/\beta}$ |
| Poisson | $f(x) = \dfrac{e^{-\lambda t}(\lambda t)^x}{x!}, x = 0, 1, 2, \ldots, \infty$ | $x = \begin{cases} 0, -\dfrac{1}{\lambda}\ln r > t^† \\ x, \sum_{i=1}^{x} -\dfrac{1}{\lambda}\ln r_i < t < \sum_{i=1}^{x+1} -\dfrac{1}{\lambda}\ln r_i \end{cases}$ |
| Chi-square | $f(x) = \dfrac{1}{2^{\nu/2}\Gamma(\nu/2)}x^{(\nu/2-1)}e^{-x/2}, x > 0$ | $x = \sum_{i=1}^{\nu} Z_i^2 \quad z_i$ is a standard normal random deviate. |
| Beta | $f(x) = \dfrac{1}{B(q,p)}x^{p-1}(1-x)^{q-1},$ <br> $0 \leq x \leq 1, p > 0, q > 0$ | $x = \dfrac{r^{1/p}}{r^{1/p} + r^{1/q}}$ |
| Gamma | $f(x) = \dfrac{\lambda^\eta}{\Gamma(\eta)}x^{(\eta-1)}e^{-\lambda x},$ <br> $x > 0, \eta > 0, \lambda > 0$ | 1. $\eta$ is a non-integer shape parameter. <br> 2. Let $\eta_1$ = the truncated integer root of $\eta$ <br> 3. Let $q = -\ln\prod_{j=1}^{\eta_1} r_j$. <br> 4. Let $A = \eta - \eta_1$, and $B = 1 - A$. <br> 5. Generate a random number and let $y_1 = r_j^{1/A}$. <br> 6. Generate a random number and let $y_2 = r_{j+1}^{1/B}$. <br> 7. If $y_1 + y_2 \leq 1$ go to 9. <br> 8. Let $j = j + 2$ and go to 5. <br> 9. Let $z = y_1/(y_1 + y_2)$. <br> 10. Generate a random number, $r_n$. <br> 11. Let $W = -\ln r_n$. <br> 12. $x = (q + zW)\lambda$. |
| Binomial | $P(x) = (n/x)p^x(1-p)^{n-x}, x = 0, 1, \ldots, n$ | $x = \sum_{i=1}^{n} y_i, y_i = \begin{cases} 0, r_i > p \\ 1, r_i \leq p \end{cases}$ |

*(continued on next page)*

**Table 11.11** Random Number Generators (*continued*)

| Distributors | Probability Density Function | Random Number Generators[§] |
|---|---|---|
| Geometric | $P(x) = p(1-p)^{x-1}, x = 1, 2, 3, \ldots$ | $\dfrac{\ln(1-r)}{\ln(1-p)} \leq x \leq \dfrac{\ln(1-r)}{\ln(1-p)} + 1$[‡] |
| Student's $t$ | $f(x) = \dfrac{\Gamma[(\nu-1)/2]}{\Gamma(\nu/2)\sqrt{\pi\nu}}\left(1 + \dfrac{x^2}{\nu}\right)^{-(\nu+1)/2}, -\infty < x < \infty$ | $x = \dfrac{z_t}{\left(\sum\limits_{i=2}^{\nu+1} \dfrac{z_i^2}{\nu}\right)^{1/2}}$<br>$z_i$ is a standard normal random deviate. |
| F | $f(x) = \left(\dfrac{\Gamma[(\nu_1+\nu_2)/2](\nu_1/\nu_2)^{\nu_1/2}}{\Gamma(\nu_1/2)\Gamma(\nu_2/2)}\right)$<br>$\times \left(\dfrac{x^{\nu_1/2-1}}{(1+\nu_1 x/\nu_2)^{(\nu_1+\nu_2)/2}}\right), x > 0$ | $x = \dfrac{\nu_2 \sum\limits_{i=1}^{\nu_1} z_i^2}{\nu_1 \sum\limits_{i=\nu_1+1}^{\nu_1+\nu_2} z_i^2}$<br>$z_i$ is a standard normal random deviate. |

[†]Two uniform random numbers must be generated, with the result being two normally distributed random numbers.
[‡]Increase the value of x until the inequality is satisfied.
[§]Statistical software, such as Minitab and MS Excel, have many of these functions built-in.

is experiencing a growing backlog of orders awaiting inspection. The backlog is having an adverse effect on production scheduling, including frequent missile assembly stoppages. A computer simulation will be conducted to answer the following questions:

- Is the backlog a chance event that will eventually correct itself without intervention, or is a permanent change of process required?
- If additional personnel are hired, will they be adequately utilized?
- Which types of job skills are required?
- Will additional automated or semi-automated inspection equipment alleviate the problem?

## *Model Development*

The first phase of the project is to develop an accurate model of the Receiving Inspection process. One element to evaluate is the distribution of arrivals of the various parts. Figure 11.15 compares the empirical distribution of the electrical lots with the predictions of an exponential arrival time model. Data were gathered from a recent work-month.

Similar "eyeball fits" were obtained from the arrival distributions of the other three part types. The exponential model seems to provide adequate representation of the data in each case (i.e., when we use the model parameters to simulate past performance, the results of the simulation are quite close to what actually happened). The parameter estimates (average arrival rates) used for the models are shown in Table 11.12.

**Figure 11.15** Electrical order arrivals predicted versus actual.

**Table 11.12** Average Arrival Rates

| Order Type | Mean Arrival Rate (Orders-per-Hour) |
| --- | --- |
| Electrical | 4.292 |
| Simple mechanical | 6.849 |
| Complex mechanical | 1.541 |
| Nondestructive test | 0.630 |

Another aspect of the model development is to describe the distribution of inspection time per order. Recent time studies conducted in Receiving Inspection provide data of actual inspection times for the four different parts. The exponential model proved to be adequate, passing a chi-square goodness-of-fit test as well as our "simulation of the past" check. The parameter estimates for the inspection times are given in Table 11.13.

Figure 11.16 shows the exponential curve, based on 228 orders, fitted to inspection times for electrical orders. Several studies showed that, on average, it takes four times longer to check a complex mechanical order using a manual surface plate layout than it takes on a coordinate measuring machine (CMM). (These interesting discoveries often result from simulation projects.)

**Table 11.13** Average Inspection Times

| Order Type | Average Inspection Time (Orders-per-Hour) |
|---|---|
| Electrical | 1.681 |
| Simple mechanical | 2.500 |
| Complex mechanical | 0.597 |
| Nondestructive test | 0.570 |

**Figure 11.16** Electrical order inspection times.

Time studies indicated that rejected lots required additional time to fill out rejected tags and complete the return authorization paperwork. The distribution of this process time is uniform on the interval [0.1 h, 0.5 h]. The proportion of lots rejected, by order type, was evaluated using conventional statistical process control techniques. The charts indicated that, with few exceptions, the proportion of lots rejected is in statistical control and the binomial distribution can be used as the model. The resulting estimated reject rates are given in Table 11.14. The evaluated lots were produced over a relatively short period, so the data in Table 11.14 should be regarded as tentative.

**Table 11.14** Average Reject Rates

| Order Type | % Lots Rejected | Lots Evaluated |
|---|---|---|
| Electrical | 2.2 | 762 |
| Simple mechanical | 1.1 | 936 |
| Complex mechanical | 25.0 | 188 |
| Nondestructive test | 0.5 | 410 |

## Management Constraints

A very important input in the model development process is a statement of constraints by management. With this project, the constraints involve a description of permissible job assignments for each inspector classification, the overtime policy, the queue discipline, and the priority discipline for assigning jobs to inspectors as well as to the CMMs. Table 11.15 summarizes the permissible job assignments. A "0" indicates that the assignment is not permitted, while a "1" indicates a permissible assignment.

**Table 11.15** Permissible Inspector Assignments

| | \multicolumn{4}{c}{Order Type} |
|---|---|---|---|---|
| Inspector Type | Electrical | Simple Mechanical | Complex Mechanical | NDT |
| Electrical | 1 | 0 | 0 | 0 |
| Grade 11 | 0 | 1 | 0 | 0 |
| Grade 19 | 0 | 1 | 1 | 0 |
| NDT | 0 | 0 | 0 | 1 |

The simulation will be run under the assumption that overtime is not permitted; 40 hours is considered a complete workweek. Preference is given to the lower paid Grade 11 inspectors when a simple mechanical order is assigned. The CMM is to be used for complex mechanical parts when it is available. Inspection is conducted using a first-in, first-out (FIFO) protocol.

## Backlog

The backlog is a composite of all part types. Information on when the particular order entered the backlog is not available. At the time the simulation was proposed, the backlog stood at 662 orders with the composition shown in Table 11.16.

**Table 11.16** Backlog Composition

| Order Type | Backlog | Percent |
|---|---|---|
| Electrical | 328 | 49 |
| Simple mechanical | 203 | 31 |
| Complex mechanical | 51 | 8 |
| NDT | 80 | 12 |

By the time the computer program was completed four weeks later, the backlog had dwindled to 200 orders. The assumption was made that the percentage of each order type remained constant and the simulation was run with a 200-order backlog.

## The Simulation

The first simulation mimics the current system so the decision maker can determine if the backlog is just a chance event that will work itself out. The simulation begins with the current staff, facilities, and backlog and runs four simulated regular weeks of 40 hours per week. This is done six times and the following statistics computed:

- Average delay awaiting inspection
- Maximum delay awaiting inspection
- Average backlog
- Maximum backlog
- Utilization of the inspectors
- Utilization of the CMM

Statistics 1 through 4 will be computed for each part type; statistic 5 will be computed for each inspector type.

*Modified Systems.* Simulations provide an ideal way of evaluating the impact of proposed management changes. Such changes might include inspection labor and the number of CMMs; therefore, these were programmed as input variables. In discussions with management, the following heuristic rules were established:

$$\text{If } U_i < (n_i - 1)/n_i, i = 1; 2; 3; 4, \text{ then let } n_i = n_i - 1$$

where $U_i$ = Utilization of inspector type $i$ and $n_i$ = Number of inspectors of type $i$.

For example, suppose there are three electrical inspectors (i.e., $n_i = 3$), and the utilization of electrical inspectors is 40% ($U_i = 0.4$). The heuristic rule would recommend eliminating an electrical inspector since $0.4 < (3 - 1)/3 = 0.67$.

A decision was made that the reductions would take place only if the backlog was under control for a given order type. The author interpreted this to mean that a two sigma interval about the average change in backlog should either contain zero backlog growth, or be entirely negative.

*Results of Simulations.* The first simulation was based on the existing system, coded 5–2–5–2–1, meaning:

- 5 electrical inspectors
- 2 grade 11 inspectors
- 5 grade 19 inspectors
- 2 NDT inspectors
- 1 CMM

The results of this simulation are shown in Table 11.17.

**Table 11.17** Current System 5–2–5–2–1 Simulation Results

*After 6 simulated weeks:*

| Type Inspection | Inspectors | Inspector Utilization | Backlog Avg. | Backlog Max. | CMM Number | CMM Utilization |
|---|---|---|---|---|---|---|
| **Run 1** | | | | | | |
| Electrical | 5 | 0.577 | 8.5 | 98 | | |
| Mech-simple | 2 | 0.704 | 1.6 | 61 | | |
| Mech-complex | 5 | 0.545 | 0.7 | 16 | 1 | 0.526 |
| NDT | 2 | 0.622 | 4.3 | 25 | | |
| **Run 2** | | | | | | |
| Electrical | 5 | 0.623 | 7.5 | 97 | | |
| Mech-simple | 2 | 0.752 | 1.9 | 68 | | |
| Mech-complex | 5 | 0.621 | 0.6 | 11 | 1 | 0.501 |
| NDT | 2 | 0.685 | 5.0 | 24 | | |
| **Run 3** | | | | | | |
| Electrical | 5 | 0.613 | 8.3 | 107 | | |
| Mech-simple | 2 | 0.732 | 1.5 | 51 | | |
| Mech-complex | 5 | 0.596 | 2.0 | 30 | 1 | 0.495 |
| NDT | 2 | 0.541 | 3.5 | 23 | | |

*(continued on next page)*

Table 11.17  Current System 5–2–5–2–1 Simulation Results (*continued*)

| Type Inspection | Inspectors | Inspector Utilization | Backlog Avg. | Backlog Max. | CMM Number | CMM Utilization |
|---|---|---|---|---|---|---|
| **Run 4** | | | | | | |
| Electrical | 5 | 0.608 | 4.9 | 93 | | |
| Mech-simple | 2 | 0.726 | 1.5 | 67 | | |
| Mech-complex | 5 | 0.551 | 0.8 | 14 | 1 | 0.413 |
| NDT | 2 | 0.665 | 3.5 | 28 | | |
| **Run 5** | | | | | | |
| Electrical | 5 | 0.567 | 6.8 | 91 | | |
| Mech-simple | 2 | 0.684 | 2.9 | 77 | | |
| Mech-complex | 5 | 0.554 | 0.6 | 13 | 1 | 0.506 |
| NDT | 2 | 0.592 | 2.1 | 21 | | |
| **Run 6** | | | | | | |
| Electrical | 5 | 0.598 | 6.6 | 96 | | |
| Mech-simple | 2 | 0.755 | 2.4 | 65 | | |
| Mech-complex | 5 | 0.584 | 1.6 | 19 | 1 | 0.493 |
| NDT | 2 | 0.735 | 5.0 | 22 | | |

| Job Type | Average Utilization | Average Change in Backlog | Std. Dev. of Change in Backlog |
|---|---|---|---|
| Electrical | 0.598 | −96.333 | 6.3140 |
| Mech-simple | 0.726 | −64.000 | 8.4617 |
| Mech-complex | 0.575 | −14.500 | 3.5637 |
| NDT | 0.640 | −22.500 | 3.7283 |

The heuristic rule describes the direction to go with staffing, but not how far. Based solely on the author's intuition, the following configuration was selected for the next simulation:

▲ 3 electrical inspectors
▲ 2 grade 11 inspectors
▲ 3 grade 19 inspectors
▲ 2 NDT inspectors
▲ 1 CMM

The results of simulating this 3–2–3–2–1 system are given in Table 11.18. All average utilization values pass the heuristic rule and the backlog growth is still, on the average, comfortably negative. However, the electrical order backlog reduction is considerably more erratic when the inspection staff is reduced.

**Table 11.18**  3–2–3–2–1 System Simulation Results

| Type Inspection | Inspectors | Inspector Utilization | Backlog Avg. | Backlog Max. | CMM Number | CMM Utilization |
|---|---|---|---|---|---|---|
| **Run 1** | | | | | | |
| Electrical | 3 | 0.935 | 49.4 | 101 | | |
| Mech-simple | 2 | 0.847 | 7.5 | 61 | | |
| Mech-complex | 3 | 0.811 | 2.0 | 16 | 1 | 0.595 |
| NDT | 2 | 0.637 | 8.2 | 28 | | |
| **Run 2** | | | | | | |
| Electrical | 3 | 0.998 | 81.7 | 114 | | |
| Mech-simple | 2 | 0.866 | 8.2 | 70 | | |
| Mech-complex | 3 | 0.863 | 2.5 | 16 | 1 | 0.629 |
| NDT | 2 | 0.631 | 3.5 | 22 | | |
| **Run 3** | | | | | | |
| Electrical | 3 | 0.994 | 74.3 | 109 | | |
| Mech-simple | 2 | 0.889 | 12.0 | 73 | | |
| Mech-complex | 3 | 0.891 | 6.2 | 32 | 1 | 0.623 |
| NDT | 2 | 0.679 | 6.4 | 27 | | |
| **Run 4** | | | | | | |
| Electrical | 3 | 0.879 | 31.2 | 109 | | |
| Mech-simple | 2 | 0.927 | 7.2 | 52 | | |
| Mech-complex | 3 | 0.924 | 5.6 | 26 | 1 | 0.632 |
| NDT | 2 | 0.715 | 3.8 | 25 | | |
| **Run 5** | | | | | | |
| Electrical | 3 | 0.992 | 45.6 | 117 | | |
| Mech-simple | 2 | 0.791 | 3.7 | 43 | | |
| Mech-complex | 3 | 0.761 | 1.8 | 18 | 1 | 0.537 |
| NDT | 2 | 0.673 | 2.3 | 24 | | |
| **Run 6** | | | | | | |
| Electrical | 3 | 0.990 | 39.9 | 95 | | |
| Mech-simple | 2 | 0.844 | 6.9 | 63 | | |
| Mech-complex | 3 | 0.800 | 1.7 | 18 | 1 | 0.606 |
| NDT | 2 | 0.716 | 4.2 | 24 | | |

**Table 11.18**  3–2–3–2–1 System Simulation Results (*continued*)

*After 6 simulations:*

| Job Type | Average Utilization | Average Change in Backlog | Std. Dev. of Change in Backlog |
|---|---|---|---|
| Electrical | 0.965 | −91.833 | 20.5856 |
| Mech-simple | 0.861 | −54.667 | 8.7331 |
| Mech-complex | 0.842 | −15.833 | 1.3292 |
| NDT | 0.676 | −23.500 | 1.3784 |

While this configuration was acceptable, the author believed that additional trials might allow replacement of one or more of the highly paid grade 19 inspectors with the lower paid grade 11 inspectors. A number of combinations were tried, resulting in the 3–3–1–2–1 system shown in Table 11.19.

**Table 11.19**  3–3–1–2–1 System Simulation Results

| Type Inspection | Inspectors | Inspector Utilization | Backlog Avg. | Max. | CMM Number | Utilization |
|---|---|---|---|---|---|---|
| **Run 1** | | | | | | |
| Electrical | 3 | 0.937 | 37.0 | 110 | | |
| Mech-simple | 3 | 0.885 | 13.1 | 61 | | |
| Mech-complex | 1 | 0.967 | 7.4 | 21 | 1 | 0.718 |
| NDT | 2 | 0.604 | 3.4 | 25 | | |
| **Run 2** | | | | | | |
| Electrical | 3 | 0.932 | 26.8 | 100 | | |
| Mech-simple | 3 | 0.888 | 7.9 | 58 | | |
| Mech-complex | 1 | 0.925 | 17.8 | 49 | 1 | 0.722 |
| NDT | 2 | 0.607 | 4.0 | 27 | | |
| **Run 3** | | | | | | |
| Electrical | 3 | 0.997 | 74.1 | 119 | | |
| Mech-simple | 3 | 0.915 | 14.6 | 58 | | |
| Mech-complex | 1 | 0.957 | 20.6 | 40 | 1 | 0.807 |
| NDT | 2 | 0.762 | 7.1 | 22 | | |
| **Run 4** | | | | | | |
| Electrical | 3 | 0.995 | 42.2 | 96 | | |
| Mech-simple | 3 | 0.976 | 38.4 | 79 | | |
| Mech-complex | 1 | 0.997 | 23.8 | 56 | 1 | 0.865 |
| NDT | 2 | 0.758 | 4.8 | 30 | | |

**Table 11.19** 3-3-1-2-1 System Simulation Results (*continued*)

| Type Inspection | Inspectors | Inspector Utilization | Backlog Avg. | Max. | CMM Number | Utilization |
|---|---|---|---|---|---|---|
| **Run 5** | | | | | | |
| Electrical | 3 | 0.996 | 61.3 | 121 | | |
| Mech-simple | 3 | 0.913 | 7.7 | 50 | | |
| Mech-complex | 1 | 0.996 | 21.7 | 52 | 1 | 0.909 |
| NDT | 2 | 0.820 | 7.4 | 30 | | |
| **Run 6** | | | | | | |
| Electrical | 3 | 0.933 | 35.3 | 101 | | |
| Mech-simple | 3 | 0.867 | 5.7 | 59 | | |
| Mech-complex | 1 | 0.938 | 17.8 | 49 | 1 | 0.736 |
| NDT | 2 | 0.674 | 8.8 | 33 | | |

*After 6 simulations:*

| Job Type | Average Utilization | Average Change in Backlog | Std. Dev. of Change in Backlog |
|---|---|---|---|
| Electrical | 0.965 | −93.667 | 6.9762 |
| Mech-simple | 0.908 | −57.500 | 5.8224 |
| Mech-complex | 0.963 | −5.500 | 18.1411 |
| NDT | 0.704 | −25.500 | 2.7386 |

The 3-3-1-2-1 system complies with all management constraints relating to resource utilization and backlog control. It is recommended to management with the caution that they carefully monitor the backlog of complex mechanical orders. For this type of order, the simulation indicates negative backlog growth on average, but with periods of positive backlog growth being possible

*Conclusion.* The simulation allowed the receiving inspection process to be "changed" without actually disrupting operations. In the computer, inspectors can be added, removed, or reassigned without the tremendous impact on morale and operations that would result from making these changes in the real world. It is a simple matter to add additional CMMs that would cost six figures in the real world. It is just as easy to try different job assignment protocols, examine the impact of a proposed new product line, look at new work area layouts, see if we can solve a temporary problem by working overtime or hiring temporary workers, etc. The effect of these changes can be evaluated in a few days, rather than waiting several months to learn that the problem was not resolved.

## Virtual DOE Using Simulation Software

Modern simulation software can interface with statistical analysis software to allow more detailed analysis of proposed new products and processes. In this section I'll demonstrate this capability with iGrafx Process for Six Sigma and Minitab. However, these capabilities are also incorporated into other software packages.

**Example**

A Six Sigma team has developed the process shown in Fig. 11.17. The CTQs for the process are the cycle times required for processing transactions for new and existing customers. They want to recommend staff levels that produce good CTQ results for both customer types during both normal and busy workload times. They will determine the recommended staff levels by performing simulations and analyzing the results using DOE techniques.

**Figure 11.17** Process to be evaluated using virtual DOE.

Figure 11.18 shows the dialog box for the iGrafx "RapiDOE" procedure. The top part of the box displays the available factors. The team wants to evaluate their two CTQs as the interarrival rate varies, and for different staff levels for six different types of workers. The middle of the box indicates that the experiment will be replicated four times. The bottom of the dialog box shows the CTQs the Black Belt has selected for evaluation.

**Figure 11.18** iGrafx Process for Six Sigma RapiDOE display.

Figure 11.19 shows the RapiDOE Minitab options dialog box. The Black Belt will use a fractional factorial design with 64 runs. Minitab's display of Available Factorial Designs (Fig. 11.20) indicates that this half-fraction seven-factor, 64-run design is resolution VII. This will allow the estimation of all main effects, two-factor interactions, and three-factor interactions.

**Figure 11.19** iGrafx process for Six Sigma RapiDOE Minitab options.

**Figure 11.20** Minitab display of available Factorial Designs.

In just a few minutes the 256 simulated experimental runs are completed (Fig. 11.21). The analysis of these results proceeds in exactly the same way as it would with the results from real-world experiments. Of course, the conclusions are not as trustworthy as real-world experimentation would provide. However, they are certainly a lot cheaper and faster to obtain and they provide a great deal of insight into the process bottlenecks, areas of potential improvement, and other important factors. Virtual DOE also allows trial-and-error without disrupting operations or impacting customers.

## Risk Assessment Tools

While reliability prediction is a valuable activity, it is even more important to design reliable systems in the first place. Proposed designs must be evaluated to detect potential failures prior to building the system. Some failures are more important than others, and the assessment should highlight those failures most deserving of attention and scarce resources. Once failures have been identified and prioritized, a system can be designed that is robust, that is, it is insensitive to most conditions that might lead to problems.

| ↓ | C4-T Generator1 | C5 Customer | C6 Sales Worker | C7 Credit | C8 Production | C9 Copying | C10 Assembly & | C11 ExistingCust | C12 NewCustCycle |
|---|---|---|---|---|---|---|---|---|---|
| 236 | Between(6,12) | 1 | 3 | 4 | 2 | 1 | 1 | 436.792 | 455.372 |
| 237 | Between(16,24) | 2 | 2 | 4 | 2 | 1 | 2 | 126.808 | 130.433 |
| 238 | Between(6,12) | 2 | 2 | 4 | 2 | 1 | 1 | 285.695 | 305.960 |
| 239 | Between(16,24) | 1 | 2 | 4 | 2 | 1 | 1 | 145.640 | 132.390 |
| 240 | Between(6,12) | 1 | 2 | 4 | 2 | 1 | 2 | 134.134 | 129.493 |
| 241 | Between(16,24) | 2 | 3 | 5 | 1 | 1 | 1 | 142.785 | 139.555 |
| 242 | Between(6,12) | 2 | 3 | 5 | 1 | 1 | 2 | 126.208 | 132.606 |
| 243 | Between(16,24) | 1 | 3 | 5 | 1 | 1 | 2 | 124.875 | 129.400 |
| 244 | Between(6,12) | 1 | 3 | 5 | 1 | 1 | 1 | 524.238 | 550.081 |
| 245 | Between(16,24) | 2 | 2 | 5 | 1 | 1 | 2 | 134.932 | 129.397 |
| 246 | Between(6,12) | 2 | 2 | 5 | 1 | 1 | 1 | 529.054 | 548.717 |
| 247 | Between(16,24) | 1 | 2 | 5 | 1 | 1 | 1 | 141.672 | 136.337 |
| 248 | Between(6,12) | 1 | 2 | 5 | 1 | 1 | 2 | 127.931 | 131.428 |
| 249 | Between(16,24) | 2 | 3 | 4 | 1 | 1 | 2 | 133.150 | 126.706 |
| 250 | Between(6,12) | 2 | 3 | 4 | 1 | 1 | 1 | 362.809 | 376.929 |
| 251 | Between(16,24) | 1 | 3 | 4 | 1 | 1 | 1 | 136.857 | 134.455 |
| 252 | Between(6,12) | 1 | 3 | 4 | 1 | 1 | 2 | 135.973 | 131.701 |
| 253 | Between(16,24) | 2 | 2 | 4 | 1 | 1 | 1 | 137.652 | 144.981 |
| 254 | Between(6,12) | 2 | 2 | 4 | 1 | 1 | 2 | 132.612 | 132.871 |
| 255 | Between(16,24) | 1 | 2 | 4 | 1 | 1 | 2 | 128.793 | 126.469 |
| 256 | Between(6,12) | 1 | 2 | 4 | 1 | 1 | 1 | 489.439 | 456.210 |

**Figure 11.21** Partial display of results from virtual DOE.

## Design Review

Design reviews are conducted by specialists, usually working on teams. Designs are, of course, reviewed on an ongoing basis as part of the routine work of a great number of people. However, the term as used here refers to the formal design review process. The purposes of formal design review are threefold:

▲ Determine if the product will actually work as desired and meet the customer's requirements.
▲ Determine if the new design is producible and inspectable.
▲ Determine if the new design is maintainable and repairable.

Design review should be conducted at various points in the design and production process. Review should take place on preliminary design sketches, after prototypes have been designed, and after prototypes have been built and tested, as developmental designs are released, etc. Designs subject to review should include parts, subassemblies, and assemblies.

## Fault-Tree Analysis

While FMEA (see section "Failure Mode and Effect Analysis") is a bottom-up approach to reliability analysis, FTA is a top-down approach. FTA provides a graphical representation of the events that might lead to failure. Some of the symbols used in construction of fault trees are shown in Table 11.20.

**Table 11.20** Fault-Tree Symbols

| Gate Symbol | Gate Name | Causal Relations |
|---|---|---|
|  | AND gate | Output event occurs if all the input events occur simultaneously |
|  | OR gate | Output event occurs if any one of the input events occurs |
|  | Inhibit gate | Input produces output when conditional event occurs |
|  | Priority AND gate | Output event occurs if all input events occur in the order from left to right |
|  | Exclusive OR gate | Output event occurs if one, but not both, of the input events occur |
|  | $m$-out-of-$n$ gate (voting or sample gate) | Output event occurs if $m$-out-of-$n$ input events occur |

| Event Symbol | Meaning |
|---|---|
| rectangle | Event represented by a gate |
| circle | Basic event with sufficient data |
| diamond | Undeveloped event |
| switch or house | Either occurring or not occurring |
| oval | Conditional event used with inhibit gate |
| triangles | Transfer symbol |

*Source:* Handbook of Reliability Engineering and Management, McGraw-Hill, reprinted with permission of the publisher.

In general, FTA follows these steps:

1. Define the top event, sometimes called the *primary event*. This is the failure condition under study.
2. Establish the boundaries of the FTA.
3. Examine the system to understand how the various elements relate to one another and to the top event.
4. Construct the fault tree, starting at the top event and working downward.
5. Analyze the fault tree to identify ways of eliminating events that lead to failure.
6. Prepare a corrective action plan for preventing failures and a contingency plan for dealing with failures when they occur.
7. Implement the plans.
8. Return to step 1 for the new design.

Figure 11.22 illustrates an FTA for an electric motor.

## *Safety Analysis*

Safety and reliability are closely related. A safety problem is created when a critical failure occurs, which reliability theory addresses explicitly with such tools as FMEA and FTA. The modern evaluation of safety/reliability takes into account the probabilistic nature of failures. With the traditional approach a safety factor would be defined using Eq. (11.7).

$$SF = \frac{\text{average strength}}{\text{worst expected stress}} \tag{11.7}$$

The problem with this approach is quite simple: it doesn't account for variation in either stress or strength. The fact of the matter is that both strength and stress will vary over time, and unless this variation is dealt with explicitly we have no idea what the "safety factor" really is. The modern view is that a safety factor is the difference between an improbably high stress (the maximum expected stress, or "reliability boundary") and an improbably low strength (the minimum expected strength). Figure 11.23 illustrates the modern view of safety factors. The figure shows two *distributions*, one for stress and one for strength.

Since any strength or stress is theoretically possible, the traditional concept of a safety factor becomes vague at best and misleading at worst. To deal intelligently with this situation, we must consider *probabilities* instead of *possibilities*. This is done by computing the probability that a stress/strength combination will occur such that the stress applied exceeds the strength. It is possible to do this since, if we have distributions of stress and strength, then the difference between the two distributions is also a distribution. In particular, if the distributions of stress and strength are normal, the distribution of the difference

**Figure 11.22** Fault tree for an electric motor.
(From Ireson, 1996. Reprinted with permission of the publisher.)

between stress and strength will also be normal. The average and standard distribution of the difference can be determined using statistical theory, and are shown in Eqs. (11.8) and (11.9).

$$\sigma^2_{SF} = \sigma^2_{STRENGTH} + \sigma^2_{STRESS} \tag{11.8}$$

$$\mu_{SF} = \mu_{STRENGTH} - \mu_{STRESS} \tag{11.9}$$

In Eqs. (11.8) and (11.9) the SF subscript refers to the safety factor.

**Figure 11.23** Modern view of safety factors.

## Example of Computing Probability of Failure

Assume that we have normally distributed strength and stress. Then the distribution of strength minus stress is also normally distributed with the mean and variance computed from Eqs. (11.8) and (11.9). Furthermore, the probability of a failure is the same as the probability that the difference of strength minus stress is less than zero. That is, a negative difference implies that stress exceeds strength, thus leading to a critical failure.

Assume that the strength of a steel rod is normally distributed with $\mu = 50{,}000^{\#}$ and $\sigma = 5{,}000^{\#}$. The steel rod is to be used as an undertruss on a conveyor system. The stress observed in the actual application was measured by strain gages and it was found to have a normal distribution with $\mu = 30{,}000^{\#}$ and $\sigma = 3{,}000^{\#}$. What is the expected reliability of this system?

*Solution.* The mean variance and standard deviation of the difference is first computed using Eqs. (11.8) and (11.9), giving:

$$\sigma^2_{\text{DIFFERENCE}} = \sigma^2_{\text{STRENGTH}} + \sigma^2_{\text{STRESS}} = 5{,}000^2 + 3{,}000^2 = 34{,}000{,}000$$

$$\sigma = \sqrt{34{,}000{,}000} = 5.831^{\#}$$

$$\mu_{\text{DIFFERENCE}} = \mu_{\text{STRENGTH}} - \mu_{\text{STRESS}} = 50{,}000^{\#} - 30{,}000^{\#} = 20{,}000$$

We now compute Z, which transforms this normal distribution to a standard normal distribution (see the section entitled "Normal Distribution" in Chap. 8).

$$z = \frac{0^{\#} - 20{,}000^3}{5{,}831^3} = -3.43$$

Using a normal table (Appendix 2), we now look up the probability associated with this Z value and find it is 0.0003. This is the probability of failure, about 3 chances in 10,000. The reliability is found by subtracting this probability from 1, giving 0.9997. Thus, the reliability of this system (and safety for this particular failure mode) is 99.97%. This example is summarized in Fig. 11.24.

**Figure 11.24** Distribution of strength minus stress.

## *Failure Mode and Effect Analysis*

Failure mode and effect analysis, or FMEA, is an attempt to delineate all possible failures, their effect on the system, the likelihood of occurrence, and the probability that the failure will go undetected. FMEA provides an excellent basis for classification of characteristics, that is, for identifying CTQs and other critical variables. As with Pareto analysis, one objective of FMEA is to direct the available resources toward the most promising opportunities. An extremely unlikely failure, even a failure with serious consequences, may not be the best place to concentrate preventative efforts. FMEA can be combined with decision analysis methods such as AHP and QFD to help guide preventive action planning.

FMEA came into existence on the space program in the 1960s. Later it was incorporated into military standards, in particular Mil-Std-1629A.[5] There are two primary approaches for accomplishing an FMEA:

▲ *The hardware approach*, which lists individual hardware items and analyzes their possible failure modes. This FMEA approach is sometimes used in product DFSS projects.
▲ *The functional approach*, which recognizes that every item is designed to perform a number of functions that can be classified as outputs. The outputs are listed and their failure modes analyzed. This approach to FMEA is most common on both DMAIC and DMADV projects involving improvement of processes or complex systems.

## FMEA Process

The FMEA is an integral part of the early design process, and it should take place during the improve phase of DMAIC or the design phase of DMADV. FMEAs are living documents, and they must be updated to reflect design changes, which makes them useful in the control or verify phases as well. The analysis is used to assess high risk items and the activities underway to provide corrective actions. The FMEA is also used to define special test considerations, quality inspection points, preventive maintenance actions, operational constraints, useful life, and other pertinent information and activities necessary to minimize failure risk. All recommended actions that result from the FMEA must be evaluated and formally dispositioned by appropriate implementation or documented rationale for no action. The following steps are used in performing an FMEA:

1. Define the system to be analyzed. Complete system definition includes identification of internal and interface functions, expected performance at all system levels, system restraints, and failure definitions. Functional narratives of the system should include descriptions of each goal in terms of functions that identify tasks to be performed for each goal and operational mode. Narratives should describe the environmental profiles, expected cycle times and equipment utilization, and the functions and outputs of each item.
2. Construct process maps that illustrate the operation, interrelationships, and interdependencies of functional entities.
3. Conduct SIPOC analysis for each subprocess in the system. All process and system interfaces should be indicated.
4. List the intended function of each step in the process or subprocess.
5. For each process step, identify all potential item and interface failure modes and define the effect on the immediate function or item, on the system, and on the mission to be performed for the customer.
6. Evaluate each failure mode in terms of the worst potential consequences that may result and assign a severity classification category, or SEV (see Table 11.21).
7. Determine the likelihood of occurrence of each failure mode and assign an occurrence risk category, or OCC (see Table 11.21).
8. Identify failure detection methods and assign a detectability risk category, or DET (see Table 11.21).
9. Calculate the risk priority number (RPN) for the current system. RPN = SEV × OCC × DET.
10. Determine compensating provisions for each failure mode.
11. Identify corrective design or other actions required to eliminate failure or control the risk. Assign responsibility and due dates for corrective actions.

**Table 11.21** FMEA Severity, Likelihood, Detectability Rating Guidelines

| Rating | Severity (SEV) | Occurrence (OCC) | Detectability (DET) |
|---|---|---|---|
| Rating | How significant is this failure's effect to the customer? | How likely is the cause of this failure to occur? | How likely is it that the existing system will detect the cause, if it occurs? Note: $p$ is the estimated probability of failure *not* being detected. |
| 1 | Minor. Customer won't notice the effect or will consider it insignificant. | Not likely. | Nearly certain to detect before reaching the customer. ($p \approx 0$) |
| 2 | Customer will notice the effect. | Documented low failure rate. | Extremely low probability of reaching the customer without detection. ($0 < p \leq 0.01$) |
| 3 | Customer will become irritated at reduced performance. | Undocumented low failure rate. | Low probability of reaching the customer without detection. ($0.01 < p \leq 0.05$) |
| 4 | Marginal. Customer dissatisfaction due to reduced performance. | Failures occur from time-to-time. | Likely to be detected before reaching the customer. ($0.05 < p \leq 0.20$) |
| 5 | Customer's productivity is reduced. | Documented moderate failure rate. | Might be detected before reaching the customer. ($0.20 < p \leq 0.50$) |
| 6 | Customer will complain. Repair or return likely. Increased internal costs (scrap, rework, etc.). | Undocumented moderate failure rate. | Unlikely to be detected before reaching the customer. ($0.50 < p \leq 0.70$) |
| 7 | Critical. Reduced customer loyalty. Internal operations adversely impacted. | Documented high failure rate. | Highly unlikely to detect before reaching the customer. ($0.70 < p \leq 0.90$) |
| 8 | Complete loss of customer goodwill. Internal operations disrupted. | Undocumented high failure rate. | Poor chance of detection. ($0.90 < p \leq 0.95$) |
| 9 | Customer or employee safety compromised. Regulatory compliance questionable. | Failures common. | Extremely poor chance of detection. ($0.95 < p \leq 0.99$) |
| 10 | Catastrophic. Customer or employee endangered without warning. Violation of law or regulation. | Failures nearly always occur. | Nearly certain that failure won't be detected. ($p \approx 1$) |

**12.** Identify effects of corrective actions on other system attributes.
**13.** Identify severity, occurrence, and detectability risks after the corrective action and calculate the "after" RPN.
**14.** Document the analysis and summarize the problems that could not be corrected and identify the special controls that are necessary to reduce failure risk.

RPNs are useful in setting priorities, with larger RPNs receiving greater attention than smaller RPNs. Some organizations have guidelines requiring action based on the absolute value of the RPN. For example, Boeing recommends that action be required if the RPN > 120.

A worksheet similar to worksheet 1 can be used to document and guide the team in conducting an FMEA. FMEA is incorporated into software packages, including some that perform QFD. There are numerous resources available on the web to assist you with FMEA, including spreadsheets, real-world examples of FMEA, and much more.[6]

## Defining New Performance Standards Using Statistical Tolerancing

For our discussion of statistical tolerancing we will use the definitions of limits proposed by Juran and Gryna (1993), which are shown in Table 11.22.

**Table 11.22** Definitions of Limits

| Name of Limit | Meaning |
|---|---|
| Tolerance | Set by the engineering design function to define the minimum and maximum values allowable for the product to work properly |
| Statistical tolerance | Calculated from process data to define the amount of variation that the process exhibits; these limits will contain a specified proportion of the total population |
| Prediction | Calculated from process data to define the limits that will contain all of $k$ future observations |
| Confidence | Calculated from data to define an interval within which a population parameter lies |
| Control | Calculated from process data to define the limits of chance (random) variation around some central value |

In manufacturing it is common that parts interact with one another. A pin fits through a hole, an assembly consists of several parts bonded together, etc. Figure 11.25 illustrates one example of interacting parts.

Suppose that all three layers of this assembly were manufactured to the specifications indicated in Fig. 11.25. A logical specification on the overall stack height would be found by adding the nominal dimensions and tolerances for each layer; for example, 0.175 inch

```
                    ┌─────────────────────────────────┐  ↑
                    │  Layer 1: 0.025" +/− 0.001"     │  │
                    ├─────────────────────────────────┤  │
                    │                                 │  │
                    │  Layer 2: 0.100" +/− 0.0005"    │  │  Stack
                    │                                 │  │  height
                    ├─────────────────────────────────┤  │
                    │                                 │  │
                    │  Layer 3: 0.050" +/− 0.002"     │  │
                    └─────────────────────────────────┘  ↓
```

**Figure 11.25**  A multilevel circuit board assembly.

± 0.0035 inch, giving limits of 0.1715 inch and 0.1785 inch. The lower specification is equivalent to a stack where all three layers are at their minimums, the upper specification is equivalent to a stack where all three layers are at their maximums, as shown in Table 11.23.

**Table 11.23**  Minimum and Maximum Multilayer Assemblies

| Minimum | Maximum |
|---------|---------|
| 0.0240  | 0.0260  |
| 0.0995  | 0.1005  |
| 0.0480  | 0.0520  |
| 0.1715  | 0.1785  |

Adding part tolerances is the usual way of arriving at assembly tolerances, but it is usually too conservative, especially when manufacturing processes are both capable and in a state of statistical control. For example, assume that the probability of getting any particular layer below its low specification was 1 in 100 (which is a conservative estimate for a controlled, capable process). Then the probability that a particular stack would be below the lower limit of 0.1715 inch is

$$\frac{1}{100} \times \frac{1}{100} \times \frac{1}{100} = \frac{1}{1,000,000}$$

Similarly, the probability of getting a stack that is too thick would be 1 in a million. Thus, setting component and assembly tolerances by simple addition is extremely conservative, and often costly.

The statistical approach to tolerancing is based on the relationship between the variances of a number of independent causes and the variance of the dependent or overall result. The equation is

$$\sigma_{result} = \sqrt{\sigma^2_{cause\,A} + \sigma^2_{cause\,B} + \sigma^2_{cause\,C} + \ldots} \qquad (11.10)$$

For our example, the equation is

$$\sigma_{stack} = \sqrt{\sigma^2_{layer\,1} + \sigma^2_{layer\,2} + \sigma^2_{layer\,3}} \qquad (11.11)$$

Of course, engineering tolerances are usually set without knowing which manufacturing process will be used to manufacture the part, so the actual variances are not known. However, a worst-case scenario would be where the process was just barely able to meet the engineering requirement. In Chap. 8 we learned that this situation occurs when the engineering tolerance is 6 standard deviations wide (±3 standard deviations). Thus, we can rewrite Eq. (11.11) as:

$$\frac{T_{stack}}{3} = \sqrt{\left(\frac{T_{layer\,1}}{3}\right)^2 + \left(\frac{T_{layer\,2}}{3}\right)^2 + \left(\frac{T_{layer\,3}}{3}\right)^2} \qquad (11.12)$$

or

$$T_{stack} = \sqrt{T^2_{layer\,1} + T^2_{layer\,2} + T^2_{layer\,3}}$$

In other words, instead of simple addition of tolerances, the squares of the tolerances are added to determine the square of the tolerance for the overall result.

The result of the statistical approach is a dramatic increase in the allowable tolerances for the individual piece-parts. For our example, allowing each layer a tolerance of ±0.002 inch would result in the same stack tolerance of 0.0035 inch. This amounts to doubling the tolerance for layer 1 and quadrupling the tolerance for layer 3, without changing the tolerance for the overall stack assembly. There are many other combinations of layer tolerances that would yield the same stack assembly result, which allows a great deal of flexibility for considering such factors as process capability and costs.

The penalty associated with this approach is a slight probability of an out-of-tolerance assembly. However, this probability can be set to as small a number as needed by adjusting the 3 sigma rule to a larger number. Another alternative is to measure the subassemblies prior to assembly and selecting different components in those rare instances where an out-of-tolerance combination results.

It is also possible to use this approach for internal dimensions of assemblies. For example, assume we had an assembly where a shaft was being assembled with a bearing as shown in Fig. 11.26.

**Figure 11.26** A bearing and shaft assembly.

The clearance between the bearing and the shaft can be computed as:

$$\text{Clearance} = \text{bearing inside diameter} - \text{shaft outside diameter} \quad (11.13)$$

The minimum clearance will exist when the bearing inside diameter is at its smallest allowed and the shaft outside diameter is at its largest allowed. Thus:

$$\text{Minimum clearance} = 0.999 \text{ in} - 0.998 \text{ in} = 0.001 \text{ in}$$

The maximum clearance will exist when the bearing inside diameter is at its largest allowed and the shaft outside diameter is at its smallest allowed:

$$\text{Maximum clearance} = 1.001 \text{ in} - 0.996 \text{ in} = 0.005 \text{ in}$$

Thus, the assembly tolerance can be computed as:

$$T_{\text{assembly}} = 0.005 \text{ in} - 0.001 \text{ in} = 0.004 \text{ in}$$

The statistical tolerancing approach is used here in the same way as it was used above. Namely:

$$\frac{T}{3} = \sqrt{\left(\frac{T_A}{3}\right)^2 + \left(\frac{T_B}{3}\right)^2} \quad (11.14)$$

or:

$$T_{\text{assembly}} = \sqrt{T_{\text{bearing}}^2 + T_{\text{shaft}}^2}$$

For our example we get

$$T_{assembly} = 0.004" = \sqrt{T^2_{bearing} + T^2_{shaft}}$$

If we assume equal tolerances for the bearing and the shaft the tolerance for each becomes

$$(0.004")^2 = T^2_{bearing} + T^2_{shaft} = 2T^2$$

$$T = \sqrt{\frac{(0.004)^2}{2}} = \pm 0.0028$$

which nearly triples the tolerance for each part.

## *Assumptions of Formula*

The formula is based on several assumptions:

- ▲ The component dimensions are independent and the components are assembled randomly. This assumption is usually met in practice.
- ▲ Each component dimension should be approximately normally distributed.
- ▲ The actual average for each component is equal to the nominal value stated in the specification. For the multilayer circuit board assembly example, the averages for layers 1, 2, and 3 must be 0.025 inch, 0.100 inch, and 0.050 inch, respectively. This condition can be met by applying SPC to the manufacturing processes.

Reasonable departures from these assumptions are acceptable. The author's experience suggests that few problems will appear as long as the subassembly manufacturing processes are kept in a state of statistical control.

## *Tolerance Intervals*

We have found that confidence limits may be determined so that the interval between these limits will cover a population parameter with a specified confidence, that is, a specified proportion of the time. Sometimes it is desirable to obtain an interval that will cover a fixed portion of the population distribution with a specified confidence. These intervals are called *tolerance intervals*, and the end points of such intervals are called *tolerance limits*. For example, a manufacturer may wish to estimate what proportion of product will have dimensions that meet the engineering requirement. In Six Sigma, tolerance intervals are typically of the form $\bar{X} \pm ks$, where $k$ is determined, so that the interval will cover a proportion $P$ of the population with confidence $\gamma$. Confidence limits for $\mu$ are also of the form $\bar{X} \pm ks$. However, we determine $k$ so that the confidence interval would cover the population mean $\mu$ a certain proportion of the time. It is obvious that the interval must be longer to cover a large portion of the distribution than to cover just the single value $\mu$.

Appendix 8 gives $k$ for $P = 0.90, 0.95, 0.99, 0.999$ and $\gamma = 0.90, 0.95, 0.99$ and for many different sample sizes $n$.

## Example of Calculating a Tolerance Interval

Assume that a sample of $n = 20$ from a stable process produced the following results: $\bar{X} = 20$, $s = 1.5$. We can estimate that the interval $\bar{X} \pm ks = 20 \pm 3.61(1.5) = 20 \pm 5.4225$; the interval from 14.5775 to 25.4225 will contain 99% of the population with confidence 95%. The $k$ values in the table assume normally distributed populations.

# CHAPTER 12

# Control/Verify Phase

The main objectives of the Control/Verify stage is to:

- Statistically validate that the new process or design meets the objectives and benefits sought through the project
- Develop and implement a control plan to institutionalize the new process or design
- Document lessons learned and project findings, as discussed in the "Tracking Six Sigma Project Results" section of Chap. 4

## Validating the New Process or Product Design

Although the design was thoroughly evaluated, there is no substitute for doing. The team should ensure that their operating procedures, operator training, materials, information systems, etc. actually produce the predicted results. The pilot run consists of a small-scale, limited time run of the new design under the careful watch of the process expert. Metrics are collected and analyzed using SPC analysis to determine if the CTQ predictions are reasonably accurate under real-world conditions Actual customers are served by the new design and their reactions closely monitored. Of course, the results of the pilot are analyzed bearing in mind that proficiency will improve with practice. Still, unanticipated problems are nearly always discovered during pilot runs and they should not be overlooked.

Full-scale operations are to the pilot run as the pilot run is to a simulation. The handoff should be gradual, with redesign options open until enough time has passed to ensure that the new design is stable. Process owners are the primary decision maker when it comes to declaring the handoff complete. The transition should be planned as a subproject, with tasks, due dates, and responsibility assigned.

# Business Process Control Planning

The project has finished successfully! Or has it? You've met the project's goals and the customer and sponsor have accepted the deliverables. But don't be too hasty to declare victory. The last battle is yet to be fought: the battle against creeping disorder, the battle against entropy. That battle to ensure that the gains you made are permanent.

## *Maintaining Gains*

All organizations have systems designed to ensure stability and to protect against undesirable change. Often these systems also make it more difficult to make beneficial change; perhaps you encountered an example or two while pursuing your Six Sigma project! Still, once you've created an improved business system these "anti-change" systems can be your friend. Here are some suggested ways to protect your hard-won gains.

- **Policy changes**—Which corporate policies should be changed as a result of the project? Have some policies been rendered obsolete? Are new policies needed?
- **New standards**—Did the project bring the organization into compliance with a standard (e.g., ISO 9000, environmental standards, product safety standards)? If so, having the company adopt the standard might prevent backsliding. Are there any industry standards which, if adopted, would help maintain the benefits of the project? Customer standards? ANSI, SAE, JCAHO, NCQA, ASTM, ASQ or any other standard-making organization standards? Government standards? Don't forget that compliance with accepted standards is often an effective marketing tool; ask your marketing people if this is the case and, if so, get their help in adopting the standard.
- **Modify procedures**—Procedures describe the way things are supposed to be done. Since the project produced better (different) results, presumably some things are being done differently. Be sure these differences are incorporated into formal procedures.
- **Modify quality appraisal and audit criteria**—The quality control activity in an organization exists to ensure conformance to requirements. This will work for you by ensuring that the changes made to documentation will result in changes in the way the work is done. In many cases, where the Six Sigma project results in fundamental improvements to a process, the existing control schemes may provide an unnecessarily stringent inspection or control regimen that may be eased given the improved performance. A project that built sufficient stakeholder buy-in with documented analysis will pay dividends in this case.
- **Update prices and contract bid models**—The way product is priced for sale is directly related to profit, loss, and business success. Because of this, project improvements that

are embedded in bid models and price models will be institutionalized by being indirectly integrated into an array of accounting and information systems.
- **Change engineering drawings**—Many Six Sigma projects create engineering change requests as part of their problem solution. For example, when a Six Sigma project evaluates process capability it is common to discover that the engineering requirements are excessively tight. Perhaps designers are using worst-case tolerancing instead of statistical tolerancing. The project team should ensure that these discoveries result in actual changes to engineering drawings.
- **Change manufacturing planning**—An organization's manufacturing plans describe in detail how product is to be processed and produced. Often the Six Sigma project team will discover better ways of doing things. If manufacturing plans are not changed the new and improved approach is likely to be lost due to personnel turnovers, etc. For those organizations that have no manufacturing plans, the Six Sigma project team should develop them, at least for products and processes developed as part of the project. Note: this should not be considered scope creep or scope drift because it is directly related to the team's goals. However, it will be better still if the team can obtain a permanent policy change to make manufacturing planning a matter of policy (see "Policy changes" above).
- **Revise accounting systems**—Six Sigma projects take a value stream perspective of business systems, that is, a global approach. However, many accounting systems (such as activity based costing) look at local activities in isolation from their place in the overall scheme of things. If kept in place, these accounting systems produce perverse incentives that will eventually undo all of the good the team has done by breaking the integrated value delivery process into a series of competing fiefdoms. Consider changing to throughput accounting or other accounting systems better aligned with a process and systems perspective.
- **Revise budgets**—Improvements mean that more can be done with less. Budgets should be adjusted accordingly. However, the general rule of free markets should also be kept in mind: capital flows to the most efficient.
- **Revise manpower forecasts**—Toyota's Taiichi Ohno says that he isn't interested in labor savings, only in manpower savings. In other words, if as a result of a Six Sigma project the same number of units can be produced with fewer people, this should be reflected in staffing requirements. I hasten to point out, however, that research shows that Six Sigma and total quality firms *increase* employment at roughly triple the rate of non–Six Sigma firms. Greater efficiency, higher quality, and faster cycle times allow firms to create more value for customers, thus generating more sales. Investors, employees, and other stakeholders benefit.

- ▲ **Modify training**—Personnel need to become familiar with the new way of doing things. Be sure all current employees are retrained, and new employees receive the proper indoctrination. Evaluate existing training materials and revise them as necessary.
- ▲ **Change information systems**—For example, Manufacturing Resource Planning MRP, inventory requirements, etc. Much of what occurs in the organization is not touched by humans. For example:
  - ▼ A purchase order might be issued automatically when inventories for a part reach a certain level. However, a Six Sigma project may have eliminated the need for safety stock.
  - ▼ MRP may generate a schedule based on cycle times rendered obsolete by improvements in cycle times.

When Six Sigma projects change the underlying relationships on which the automated information systems are based, programs should be modified to reflect this.

## *Tools and Techniques Useful for Control Planning*

- ▲ **Project planning**—Many of the Six Sigma tools and techniques used during the Define, Measure, Analyze, and Improve phases can also be used to develop a control plan. Perhaps most important is to keep in mind that control planning is a (sub) project. The deliverable is an effective and implemented control system. The activities, responsibilities, durations, and due dates necessary to produce the deliverable should be carefully listed. If the process changes are extensive, the control subproject may require another sponsor to take ownership of the control process after the team disbands and the main project sponsor accepts the new system. A detailed Business Process Change control plan should be prepared and kept up to date until the Black Belt, sponsor, and process owner are confident that the improvements are permanent.
- ▲ **Brainstorming**—The Six Sigma team should brainstorm to expand the list presented above with ideas from their own organization.
- ▲ **Force-field diagram**—A force-field diagram can be very useful at this point. Show the forces that will push to undo the changes, and create counterforces that will maintain them. The ideas obtained should be used to develop a process control plan that will ensure that the organization continues to enjoy the benefits of the Six Sigma project.
- ▲ **Process decision program chart**—The PDPC is a useful tool in developing a contingency plan.
- ▲ **Failure mode and effect analysis**—Using FMEA in the improve phase was discussed in detail in Chap. 11. Its output provides necessary input for the control plan.

## Using SPC for Ongoing Control

Assuming that the organization's leadership has created an environment where open and honest communication can flourish, SPC implementation becomes a matter of (1) selecting processes for applying the SPC approach and (2) selecting variables within each process as part of a detailed process control plan.

## *Preparing the Process Control Plan*

Process control plans should be prepared for each key process. The plans should be prepared by teams of people who understand the process. The team should begin by creating a flowchart of the process using the process elements determined in creating the house of quality (see the QFD discussion in Chap. 2). The flowchart will show how the process elements relate to one another, and it will help in the selection of control points. It will also show the point of delivery to the customer, which is usually an important control point. Note that the customer may be an internal customer.

For any given process there are a number of different types of process elements. Some process elements are internal to the process, others external. The rotation speed of a drill is an internal process element, while the humidity in the building is external. Some process elements, while important, are easy to hold constant at a given value so that they do not change unless deliberate action is taken. We will call these fixed elements. Other process elements vary of their own accord and must be watched; we call these variable elements. The drill rotation speed can be set in advance, but the line voltage for the drill press may vary, which causes the drill speed to change in spite of its initial setting (a good example of how a correlation matrix might be useful). Figure 12.1 provides a planning guide based on the internal/external and fixed/variable classification scheme. Of course, other classi-

|          | Internal | External |
|----------|----------|----------|
|          | I | II |
| **Fixed** | • Setup approval<br>• Periodic audits<br>• Preventive maintenance | • Audit<br>• Certification |
|          | III | IV |
| **Variable** | • Control charts<br>• Mistake-proof product<br>• Mistake-proof process<br>• Sort the output | • Supplier SPC<br>• Receiving inspection<br>• Supplier sorting<br>• Mistake-proof product |

**Figure 12.1** Guide to selecting and controlling process variables.

fication schemes may be more suitable on a given project and the analyst is encouraged to develop the approach that best serves his or her needs. For convenience, each class is identified with a Roman numeral; I = fixed–internal, II = fixed–external, III = variable–internal, and IV = variable–external.

In selecting the appropriate method of control for each process element, pay particular attention to those process elements which received high importance rankings in the house of quality analysis. In some cases an important process element is very expensive to control. When this happens, look at the QFD correlation matrix or the statistical correlation matrix for possible assistance. The process element may be correlated with other process elements that are less costly to control. Either correlation matrix will also help you to minimize the number of control charts. It is usually unnecessary to keep control charts on several variables that are correlated with one another. In these cases, it may be possible to select the process element that is least expensive (or most sensitive) to monitor as the control variable.

As Fig. 12.1 indicates, control charts are not always the best method of controlling a given process element. In fact, control charts are seldom the method of choice. When process elements are important we would prefer that they *not vary at all!* Only when this cannot be accomplished economically should the analyst resort to the use of control charts to monitor the element's variation. Control charts may be thought of as a control mechanism of last resort. Control charts are useful only when the element being monitored can be expected to exhibit measurable and "random-looking" variation when the process is properly controlled. A process element that always checks "10" if everything is okay is not a good candidate for control charting. Nor is one that checks "10" or "12," but never anything else. Ideally, the measurements being monitored with variables control charts will be capable of taking on any value, that is, the data will be continuous. Discrete measurement data can be used if it's not too discrete; indeed, all real-world data are somewhat discrete. As a rule of thumb, at least ten different values should appear in the data set and no one value should comprise more than 20% of the data set. When the measurement data become too discrete for SPC, monitor them with check sheets or simple time-ordered plots.

Of course, the above discussion applies to measurement data. Attribute control charts can be used to monitor process elements that are discrete counts.

Any process control plan must include instructions on the action to be taken if problems appear. This is particularly important where control charts are being used for process control. Unlike process control procedures such as audits or setup approvals, it is not always apparent just what is wrong when a control chart indicates a problem. The investigation of special causes of variation usually consists of a number of predetermined actions (such as checking the fixture or checking a cutting tool) followed by notifying someone if the items checked don't reveal the source of the problem. Also verify that the arithmetic was done correctly and that the point was plotted in the correct position on the control chart.

The reader may have noticed that Fig. 12.1 includes "sort the output" as part of the process control plan. Sorting the output implies that the process is not capable of meeting the customer's requirements, as determined by a process capability study and the application of Deming's all-or-none rules. However, even if sorting is taking place, SPC is still advisable. SPC will help ensure that things don't get any worse. SPC will also reveal improvements that may otherwise be overlooked. The improvements may result in a process that is good enough to eliminate the need for sorting.

## *Process Control Planning for Short and Small Runs*

A starting place for understanding SPC for short and small runs is to define our terms. The question "what is a short run?" will be answered for our purposes as an environment that has a large number of jobs per operator in a production cycle, each job involving different product. A production cycle is typically a week or a month. *A small run* is a situation where only a very few products of the same type are to be produced. An extreme case of a small run is the one-of-a-kind product, such as the Hubble Space Telescope. Short runs need not be small runs; a can manufacturing line can produce over 100,000 cans in an hour or two. Likewise, small runs are not necessarily short runs; the Hubble Space Telescope took over 15 years to get into orbit (and even longer to get into orbit and working properly)! However, it is possible to have runs that are both short and small. Programs such as Just-In-Time (JIT) inventory control are making this situation more common all of the time.

Process control for either small or short runs involves similar strategies. Both situations involve markedly different approaches than those used in the classical mass-production environment. Thus, this section will treat both the small run and the short run situations simultaneously. You should, however, select the SPC tool that best fits your particular situation.

### Strategies for Short and Small Runs

Juran's famous trilogy separates quality activities into three distinct phases (Juran and Gryna, 1988):

- Planning
- Control
- Improvement

Figure 12.2 provides a graphic portrayal of the Juran trilogy.

**Figure 12.2** Juran's trilogy.

When faced with small or short runs the emphasis should be placed in the planning phase. As much as possible needs to be done *before* any product is made, because it simply isn't possible to waste time or materials "learning from mistakes" made during production. It is also helpful to realize that the Juran trilogy is usually applied to *products*, while SPC applies to *processes*. It is quite possible that the element being monitored with SPC is a process element and not a product feature at all. In this case there really is no "short run," despite appearances to the contrary.

A common problem with application of SPC to short/small runs is that people fail to realize the limitations of SPC in this application. Even the use of SPC to *long production runs* will benefit from a greater emphasis on preproduction planning. In the best of all worlds, SPC will merely confirm that the correct process has been selected and controlled in such a way that it consistently produces well-designed parts at very close to the desired target values for every dimension.

### Preparing the Short Run Process Control Plan

Plans for short runs require a great deal of up-front attention. The objective is to create a list of as many potential sources of variation as possible and to take action to deal with them *before* going into production. One of the first steps to be taken is to identify which processes may be used to produce a given part; this is called the "Approved Process List." Analogously, parts that can be produced by a given process should also be identified; this is called the "Approved Parts List." These determinations are made based on process capability studies (Pyzdek, 1992a). The approach described in this guide uses process capability indices, specifically $C_{PK}$ (the number of standard deviations between the mean and the nearest specification limit). The use of this capability index depends on a number of assumptions, such as normality of the data etc. The proper use, and some common abuses, of capability indices is discussed in Chapter 8.

Because short runs usually involve less than the recommended number of pieces the acceptability criteria are usually modified. When less than 50 observations are used to determine the capability I recommend that the capability indices be modified by using a ±4σ minimum acceptable process width (instead of ±3σ) and a minimum acceptable $C_{PK}$ of 1.5 (instead of 1.33). Don't bother making formal capability estimates until you have at least 20 observations. (You can see in Chap. 8 that these observations need not always be from 20 separate parts.)

When preparing for short runs it often happens that actual production parts are not available in sufficient quantity for process capability studies. One way of dealing with this situation is to study process elements separately and to then sum the variances from all of the known elements to obtain an estimate of the best overall variance a given process will be able to produce.

For example, in an aerospace firm that produced conventional guided missiles, each missile contained thousands of different parts. In any given month only a small number of missiles were produced. Thus, the CNC machine shop (and the rest of the plant) was faced with a small/short run situation. However, it was not possible to do separate preproduction capability studies of each part separately. The approach used instead was to design a special test part that would provide estimates of the machine's ability to produce every basic type of characteristic (flatness, straightness, angularity, location, and so on.). Each CNC machine produced a number of these test parts under controlled conditions and the results were plotted on a short-run *R* chart (these are described in Chap. 8). The studies were repeated periodically for each machine.

These studies provided preproduction estimates of the machine's ability to produce different characteristics. However, these estimates were always *better* than the process would be able to do with actual production parts. Actual production would involve different operators, tooling, fixtures, materials, and other common and special causes not evaluated by the *machine capability study*. Preliminary Approved Parts lists and Preliminary Approved Process lists were created from the capability analysis using the more stringent acceptability criteria described earlier ($C_{PK}$ at least 1.5 based on a ±4σ process spread). When production commenced the actual results of the production runs were used instead of the estimates based on special runs. Once sufficient data were available, the parts were removed from the preliminary lists and placed on the appropriate permanent lists.

When creating Approved Parts and Approved Process lists, always use the most stringent product requirements to determine *the process requirement*. For example, if a process will be used to drill holes in 100 different parts with hole location tolerances ranging from 0.001 inch to 0.030 inch, the process requirement is 0.001 inch. The process capability estimate is based on its ability to hold the 0.001 inch tolerance.

The approach used is summarized as follows:

1. Get the process into statistical control.
2. Set the control limits *without regard to the requirement*.
3. Based on the calculated process capability, determine if the most stringent product requirement can be met.

## *Process Audits*

The requirements for all processes should be documented. A process audit checklist should be prepared and used to determine the condition of the process prior to production. The audit can be performed by the operator himself, but the results should be documented. The audit should cover known or suspected sources of variation. These include such things as the production plan, condition of fixtures, gage calibration, the resolution of the gaging being used, obvious problems with materials or equipment, operator changes, and so on.

SPC can be used to monitor the results of the process audits over time. For example, an audit score can be computed and tracked using an individual's control chart.

## *Selecting Process Control Elements*

Many short run SPC programs bog down because the number of control charts being used grows like Topsy. Before anyone knows what is happening they find the walls plastered with charts that few understand and no one uses. The operators and inspectors wind up spending more time filling out paperwork than they spend on true value-added work. Eventually the entire SPC program collapses under its own weight.

One reason for this is that people tend to focus their attention on the *product* rather than on the *process*. Control elements are erroneously selected because they are functionally important. A great fear is that an important product feature will be produced out of specification and that it will slip by unnoticed. This is a misunderstanding of the purpose of SPC, which is to provide a means of *process* control; SPC is not intended to be a substitute for inspection or testing. The guiding rule of selecting control items for SPC is

> SPC control items should be selected to provide a maximum amount of information regarding the state of the process at a minimum cost.

Fortunately, most process elements are correlated with one another. Because of this one process element may provide information not only about itself, but about several others as well. This means that a small number of process control elements will often explain a large portion of the process variance.

Although sophisticated statistical methods exist to help determine which groups of process elements explain the most variance, common sense and knowledge of the process can often do as well, if not better. The key is to think about the process carefully. What are the "generic process elements" that affect all parts? How do the process elements combine to affect the product? Do several process elements affect a single product feature? Do changes in one process element automatically cause changes in some other process elements? What process elements or product features are most sensitive to unplanned changes?

**Example One**

The CNC machines mentioned earlier were extremely complex. A typical machine had dozens of different tools and produced hundreds of different parts with thousands of characteristics. However, the SPC team reasoned that the machines themselves involved only a small number of "generic operations": select a tool, position the tool, remove metal, and so on. Further study revealed that nearly all of the problems encountered after the initial setup involved only the ability of the machine to position the tool precisely. A control plan was created that called for monitoring no more than one variable for each axis of movement. The features selected were those farthest from the machine's "home position" and involving the most difficult to control operations. Often a single feature provided control of more than one axis of movement, for example, the location of a single hole provides information on the location of the tool in both the $X$ and $Y$ directions.

As a result of this system no part had more than four features monitored with control charts, even though many parts had thousands of features. Subsequent sophisticated multivariate evaluation of the accumulated data by a statistician revealed that the choices made by the team explained over 90% of the process variance.

**Example Two**

A wave solder machine was used to solder printed circuit boards for a manufacturer of electronic test equipment. After several months of applying SPC the SPC team evaluated the data and decided that they needed only a single measure of product quality for SPC purposes: defects per 1,000 solder joints. A single control chart was used for dozens of different circuit boards. They also determined that most of the process variables being checked could be eliminated. The only process variables monitored in the future would be flux density, solder chemistry (provided by the vendor), solder temperature, and final rinse contamination. Historic data showed that one of these variables was nearly always out of control when process problems were encountered. Other variables were monitored with periodic audits using check sheets, but they were not charted.

Notice that in both of these examples all of the variables being monitored were related to the *process*, even though some of them were product features. The terms "short run" and "small run" refer to the product variables only; the process is in continuous operation so its run size and duration is neither small nor short.

**The Single Part Process**

The ultimate small run is the single part. A great deal can be learned by studying single pieces, even if your situation involves more than one part.

The application of SPC to single pieces may seem incongruous. Yet when we consider that the "P" in SPC stands for *process* and not product, perhaps it is possible after all. Even the company producing one-of-a-kind products usually does so with the same equipment, employees, facilities, etc. In other words, they use the same *process* to produce different *products*. Also, they usually produce products that are similar, even though not identical. This is also to be expected. It would be odd indeed to find a company fabricating microchips one day and baking bread the next. The processes are too dissimilar. The company assets are, at least to a degree, product-specific.

This discussion implies that the key to controlling the quality of single parts is to concentrate on the process elements rather than on the product features. This is the same rule we applied earlier to larger runs. In fact, it's a good rule to apply to all SPC applications, regardless of the number of parts being produced!

Consider a company manufacturing communications satellites. The company produces a satellite every year or two. The design and complexity of each satellite is quite different than any other. How can SPC be applied at this company?

A close look at a satellite will reveal immense complexity. The satellite will have thousands of terminals, silicon solar cells, solder joints, fasteners, and so on. Hundreds, even thousands of people are involved in the design, fabrication, testing, and assembly. In other words, there are *processes* that involve massive amounts of repetition. The processes include engineering (errors per engineering drawing); terminal manufacture (size, defect rates); solar cell manufacture (yields, electrical properties); soldering (defects per 1,000 joints, strength); fastener installation quality (torque) and so on.

Another example of a single-piece run is software development. The "part" in this case is the working copy of the software delivered to the customer. Only a singe unit of product is involved. How can we use SPC here?

Again, the answer comes when we direct our attention to the underlying process. Any marketable software product will consist of thousands, perhaps millions of bytes of finished machine code. This code will be compiled from thousands of lines of source code. The source code will be arranged in modules; the modules will contain procedures; the procedures will contain functions; and so on. Computer science has developed a number of ways of measuring the quality of computer code. The resulting numbers, called computer metrics, can be analyzed using SPC tools just like any other numbers. The processes that produced the code can thus be measured, controlled, and improved. If the process is in statistical control, the process elements, such as programmer selection and training, coding style, planning, procedures, etc. must be examined. If the process is not in statistical control, the special cause of the problem must be identified.

As discussed earlier, although the single part process is a small run, it isn't necessarily a short run. By examining the process rather than the part, improvement possibilities will begin to suggest themselves. The key is to find the process, to define its elements so they may be measured, controlled, and improved.

### *Other Elements of the Process Control Plan*

In addition to the selection of process control elements, the PCP should also provide information on the method of inspection, dates and results of measurement error studies, dates and results of process capability studies, subgroup sizes and methods of selecting subgroups, sampling frequency, required operator certifications, preproduction checklists, notes and suggestions regarding previous problems, etc. In short, the PCP provides a complete, detailed road map that describes how process integrity will be measured and maintained. By preparing a PCP the *inputs* to the process are controlled, thus assuring that the *outputs* from the process will be consistently acceptable.

## Multivariate Control Charts

Control charts are the primary tool for analysis of process data, overcoming the limitations of classical statistics to fixed populations by addition of the process-oriented time axis. They are thus essential for the various objectives of process analysis throughout DMAIC. In the Measure stage, control charts are necessary for baselining the process to quantify the current process performance; their ability to differentiate between common and special causes of variation provide essential information necessary for developing an improvement plan in the Analyze stage; defining the operational control limits of a process form the basis of the predictive abilities of process capability analysis used to verify the new process operating levels in the Improve stage. In the Control stage, they will provide essential monitoring of the key process indicators to ensure the project gains are continually held at the process level.

In most cases, the standard control charts presented in Chapter 8 will prove sufficient for the task. In other cases, a multivariate control chart approach will offer benefits. In any case, the Control stage will verify the changes implemented in the Improve stage have established a stable process, as evidenced by a suitable control chart, with the further objective to maintain this process stability by identifying instances of future process instability as rapidly as possible.

Multivariate control charts are developed to detect shifts in the means of several correlated variables (parameters) and/or the relationship (covariance) between these parameters. A multivariate control chart analysis may be useful in SPC whenever there is more than one process parameter whose effects are not independent or when one or more parameters are partial or complete measures of one or more other parameters (that is, they are correlated). In some cases, the true source of variation may not be recognized or may not be measurable. For example, you may seek to use control charts to monitor pressure and volumetric flow, but temperature may influence both, in addition to viscosity, which also impacts both pressure and flow. In manufacturing, feed rate, spindle speed, and material hardness in a grinding operation are interrelated.

An essential point is that almost all processes are multivariate but multivariate control chart analysis is often not necessary because there are only a few controlled variables, and these variables act sufficiently independent that they are best controlled individually. However, even when the variables do act independently, the use of a single control chart for each variable increases the number of charts to monitor, as well as the chance of randomly finding a variable out of control; the more variables there are the more likely it is that one of those charts will contain an out of control condition *even when the process has not shifted*. Thus, the false alarm rate (aka probability of Type 1 error) is increased if each variable is controlled separately. When two independent parameters are controlled separately, the resulting control region is a rectangle, as shown in Figure 12.3; the more precise control region for two parameters is an ellipse. When the parameters are jointly acting (i.e., dependent), the major axis of the ellipse will no longer be parallel to the x-axis, but sloped to reflect the correlation between the parameters, as shown by the slope of the ellipse in Figure 12.3.

Multivariate control charting provides a tool to identify shifts in the underlying relationship among k characteristics with a single set of charts. The Hotelling $T^2$ statistic may be used to monitor the means (or location) of the process variables, and a generalized variance used to monitor the process variability.

Consider the following example that Minitab provides for production of plastic sheeting. (The PlasticSheets.MTW file can be downloaded from their website, along with many other example data sets.) In this example, a quality engineer is monitoring three correlated characteristics of a plastic sheet: tear strength, glossiness, and opacity.

An initial multivariate control chart is constructed in Minitab as follows:

1. Select the *Control Charts > Multivariate Charts > T²-Generalized Variance* option from the *Stat* menu.
2. Select the columns C2 through C4 (i.e., Tear-Opacity) as the *Variables*; enter the subgroup size of 4 or use the *Subgroup* column as the subgroup size designator.
3. Select the *OK* button to create the analysis and show the completed control chart.

**Figure 12.3** Comparison of univariate and multivariate control regions.

Figure 12.4 shows the completed chart. The chart for Generalized Variance should be analyzed first to look for special causes impacting the covariance matrix, since the covariance matrix for each subgroup is used to calculate the subgroup value on the $T^2$ chart.

**Figure 12.4** Initial multivariate control charts for PlasticSheets.mtw (Green Belt XL).

The chart for Generalized Variance in Figure 12.4 has one subgroup (subgroup 6) exceeding the upper control limit. Each plotted point is the determinant of the covariance matrix for that subgroup, so an out of control group suggests that the variance of one or more of the parameters, or one or more of the covariances between two parameters, is significantly larger for this subgroup than the other subgroups. The presence of an instability is itself critical information; however, understanding the source of the instability is complicated by the number of parameters, the strength of the correlation between the parameters (which influences the likelihood that the special cause reflects a simultaneous change in multiple (correlated) parameters) and the relative simplicity of the determinant in quantifying the parameters variances and covariance.

One technique for troubleshooting the out of control condition on the Generalized Variance chart involves a review of the X-bar Sigma charts for each of the parameters. Although it was mentioned earlier that the use of multiple univariate control charts would increase the false alarm rate, here we have used the multivariate chart to establish an out of control condition, so the false alarm rate is not affected by this troubleshooting. Instead, we will use the univariate charts to investigate behavior only in the subgroup violating the multivariate chart limits.

Figure 12.5 shows the X-bar Sigma chart for the tear strength and opacity parameters, as well as a Moving Average Sigma chart of the covariance between the two parameters. A review of the X-bar Sigma chart for Glossiness, and separate Moving Average Sigma charts for the covariance between tear strength and glossiness and the covariance between opacity and glossiness, reveal nothing peculiar about subgroup 6. Indeed, the charts in Figure 12.5 show no special cause identified for subgroup 6, although the Sigma charts each show increased level of variation for subgroup 6. Based on this evidence, it seems probable that the out of control condition on the Generalized Variance chart was due to increased variation in Tear Strength and Opacity for that subgroup.

The $T^2$ chart in Figure 12.4 indicates that no subgroups are beyond the $T^2$ chart's upper control limit of 17.39. (Note that there is no lower control limit defined for the $T^2$ chart). If any subgroups were beyond the upper control limit, Minitab reports in its Session window which parameter(s) likely influenced the change in the parameter relationship.

Since there were no special causes identified on the $T^2$ chart, and we have some confidence the special cause on the Generalized Variance chart is understood, we can use the initial parameter means and the parameter covariance matrix to define new (sometimes called Phase II) control limits that may be applied to future data, once the effect of subgroup 6 has been removed. In Minitab:

1. Select the *Control Charts* > *Multivariate Charts* > *T²-Generalized Variance* option from the *Stat* menu (or the *Edit Last Dialog* from the *Edit* menu). The prior selection of the variables and subgroup sizes should remain intact.

**Figure 12.5** Univariate control charts for tear strength, opacity, and their covariance (Green Belt XL).

2. Select the *T²-GV Options* button from the *T²-Generalized Variance* dialog box.
3. In the *Estimate* tab, enter the value 6 in the field provided for omitting subgroups when estimating parameters.
4. In the *Storage* tab, select the *Means* and *Covariance Matrices* options in the *Store these estimates for each chart* group.
5. Select the *OK* button (two times) to save the settings and create a new T²-Generalized Variance chart window.
6. Select the *Display Data* option from the *Data* menu, select the *COVA* item from the left pane, then the *Select* button so the item appears in the right pane as a display column.
7. Select the *OK* button to close the dialog box and update the *Session Window* with the covariance matrix of the data (excluding subgroup 6).

Note the revised control limits on the T²-Generalized Variance charts shown in Figure 12.6. No additional subgroups were detected as out of control. Step 4 (above) created new

**604** | The Six Sigma Handbook

**T²-Generalized Variance Chart of Tear, ..., Opacity**

Subgroups omitted from the calculations: 6

**Figure 12.6** Multivariate control charts for PlasticSheets.mtw, excluding subgroup 6 (Minitab 18).

columns in the spreadsheet containing the mean of each of the parameters (*MEAN* in column C5 for the average of the *Tear* column; *MEAN_1* in column C6 for the average of the Gloss column; *MEAN_2* in column C7 for the average of the *Opacity* column).

The covariance matrix from the *Data Display* section in Minitab's *Session Window* is shown in Table 12.1. The main diagonal contains the variances for each parameter (excluding subgroup 6): the variance (i.e., standard deviation squared) of the tearing strength parameter is shown as 3.26296; the variance for glossiness as 21.8197; the variance of opacity as 5.07828. Note that the value shown is the average of the subgroup variances for each parameter, rather than the variance measured across all of the samples for each parameter. Likewise, the average of the subgroup covariances is shown in the table: the covariance between tear strength and glossiness is shown as –0.8793; the covariance of tear strength and opacity as –0.54561; the covariance of glossiness and opacity as -0.1853.

**Table 12.1** Covariance Matrix for Example Data (Minitab 18)

| Matrix COVA | | |
|---|---|---|
| 3.26296 | –0.8793 | –0.54561 |
| –0.87926 | 21.8197 | –0.18526 |
| –0.54561 | –0.1853 | 5.07828 |

Since the covariance of tear strength and glossiness equals the covariance of glossiness and tear strength (i.e., the order of the listed parameters makes no difference), the same values are shown in the upper and lower portions of the covariance matrix, and it's not uncommon to display only the upper or lower portion with the main diagonal (containing the variances of the univariate parameters).

The signs of the covariances can be used to understand the relationships between the parameters as a troubleshooting tool. In this case, each of the covariances is negative, implying that an increase in one parameter is accompanied by a decrease in the other. For example, an increase in tear strength is associated with a decrease in glossiness and/or opacity. The magnitude of the covariance cannot be compared, since the scale of each parameter is different, so we cannot make any assertions regarding the relative influence of the covariance of one pair of parameters to the covariance associated with another pair of parameters. (The correlation can be compared this way.)

The historical values can now be applied for calculation of Phase II control limits for future data, as follows:

1. Select the *Control Charts > Multivariate Chart > $T^2$-Generalized Variance* option from the *Stat* menu. The prior selection of the variables and subgroup sizes should remain intact.
2. Select the *$T^2$-GV Options* button from the *$T^2$-Generalized Variance* dialog box. In the *Parameters* tab:
   a. Place your cursor in the *Mean* field, then highlight the columns C5 through C7 containing the calculated mean for each parameter from the left pane. Use the *Select* button to move the selection to the *Mean* field.
   b. Place your cursor in the *Matrix* field, then select the *COVA* matrix containing the covariances from the left pane. Use the *Select* button to move the selection to the *Matrix* field.
   c. Enter the value 9 into the *Number of samples used to calculate covariances* field. The value of 9 reflects the initial 10 subgroups, minus the one subgroup omitted from the analysis when the historical means and covariance matrix were calculated.
3. In the *Storage* tab, deselect the *Means* and *Covariance Matrices* options in the *Store these estimates for each chart* group as these are unchanged.
4. Select the OK button (two times) to save the settings and create the analysis.

The revised multivariate control chart for future subgroups is shown in Figure 12.7.

Another example can be constructed using Minitab's LoanApplicant.MTW sample file. Although this file contains a sample of a population of loan applications, and (as such) would not be appropriate for the time-oriented analysis on a control chart, we will use the

**Figure 12.7** Phase II multivariate control charts for PlasticSheets.mtw (Minitab 18).

data on the premise that it is time-oriented, with each row representing a sample taken from a process at a later point in time than the preceding sample. In effect, we are using the data only for its value as example data, and not necessarily reflective of any known process, with a subgroup size of one for each parameter.

An initial multivariate control chart is constructed as shown in Figure 12.8, based on a subgroup size of one. Only subgroup 6 is out of control on the chart for Generalized Variance. The analysis is repeated, omitting subgroup 6, to obtain the multivariate charts shown in Figure 12.9. The $T^2$ chart indicates instability in subgroups 5 and 6. Minitab provides the information presented in Table 12.2 to diagnose the instability in the $T^2$ chart. The information suggests that, of the eight parameters included in the analysis, the instability in subgroup 5 is likely associated with changes in the parameters income, residence, employ and savings; for subgroup 6, parameters income, education, age, residence, savings and debt.

Univariate Moving Average Sigma charts were constructed (see Chapter 8) for each parameter, using the subgroup size of 1 and a moving cell width of 3, to investigate for disturbances in subgroups 5 and 6. Only the debt parameter showed excessive short-term variation in subgroup 6, while the income and education parameters had instability in their location for subgroups 5 and 6, as shown in Figure 12.10.

**Figure 12.8** Initial multivariate control charts for LoanApplication.mtw (Minitab 18).

**Figure 12.9** Multivariate control charts for LoanApplication.mtw, excluding subgroup 6 (Minitab 18).

**Table 12.2** Test Results for T² Chart of Income, …, Credit cards

|  | Point | Variable | P-Value |
|---|---|---|---|
| Greater Than UCL | 5 | Income | 0.0000 |
|  |  | Residence | 0.0003 |
|  |  | Employ | 0.0163 |
|  |  | Savings | 0.0076 |
|  | 6 | Income | 0.0000 |
|  |  | Education | 0.0000 |
|  |  | Age | 0.0105 |
|  |  | Residence | 0.0003 |
|  |  | Savings | 0.0010 |
|  |  | Debt | 0.0000 |

In this case, there are only eight parameters, yet the construction and analysis of charts to troubleshoot the out of control conditions on the multivariate charts increased rapidly from the earlier analysis of three parameters. With multivariate processes involving tens or hundreds of parameters, an alternate troubleshooting method is warranted.

## Principal Component Analysis

Principal components can be constructed as linear combinations of a set of related parameters. Unlike the process variables, which may be correlated, the principal components are constructed to be orthogonal (i.e., independent) of one-another. The number of principal components may never exceed the minimum of the number of process variables or the number of data. Often, an analysis with many process parameters can be simplified to a much smaller number of principal components that fully describe process variation

We can construct principal components for the loan application data in Minitab as follows:

1. Select the *Multivariate > Principal Components* option from the *Stat* menu.
2. Select the columns C1 through C8 (i.e., Income-Credit Cards) as the *Variables* and leave the other fields with their default values.
3. Using the *Graphs* button, select each of the graphs listed, then the OK button to save the selection.
4. Select the OK button to create the analysis in Minitab's *Session Window*, also shown in Tables 12.3 and 12.4.

**Figure 12.10** Univariate control charts for debt, income and education (Green Belt XL).

**Table 12.3** Eigenvectors

| Variable | PC1 | PC2 | PC3 | PC4 | PC5 | PC6 | PC7 | PC8 |
|---|---|---|---|---|---|---|---|---|
| Income | 0.314 | 0.145 | −0.676 | −0.347 | −0.241 | 0.494 | 0.018 | −0.030 |
| Education | 0.237 | 0.444 | −0.401 | 0.240 | 0.622 | −0.357 | 0.103 | 0.057 |
| Age | 0.484 | −0.135 | −0.004 | −0.212 | −0.175 | −0.487 | −0.657 | −0.052 |
| Residence | 0.466 | −0.277 | 0.091 | 0.116 | −0.035 | −0.085 | 0.487 | −0.662 |
| Employ | 0.459 | −0.304 | 0.122 | −0.017 | −0.014 | −0.023 | 0.368 | 0.739 |
| Savings | 0.404 | 0.219 | 0.366 | 0.436 | 0.143 | 0.568 | −0.348 | −0.017 |
| Debt | −0.067 | −0.585 | −0.078 | −0.281 | 0.681 | 0.245 | −0.196 | −0.075 |
| Credit cards | −0.123 | −0.452 | −0.468 | 0.703 | −0.195 | −0.022 | −0.158 | 0.058 |

**Table 12.4** Eigen Analysis of the Correlation Matrix

| Eigenvalue | 3.5476 | 2.1320 | 1.0447 | 0.5315 | 0.4112 | 0.1665 | 0.1254 | 0.0411 |
|---|---|---|---|---|---|---|---|---|
| Proportion | 0.443 | 0.266 | 0.131 | 0.066 | 0.051 | 0.021 | 0.016 | 0.005 |
| Cumulative | 0.443 | 0.710 | 0.841 | 0.907 | 0.958 | 0.979 | 0.995 | 1.000 |

The Eignevectors shown in Table 12.3 are the coefficients of the principal component's linear combination of process parameters. For example, the first principal component is defined as:

$$PC_1 = 0.314 x_{income} + 0.237 x_{education} + 0.484 x_{age} + 0.466 x_{residence} + 0.459 x_{employ} + 0.404 x_{savings} - 0.067 x_{debt} - 0.123 x_{creditcards}$$

These coefficients are also known as *loadings*, since they show the influence of a process parameter on each principal component. The Loading plot is shown in Figure 12.11, and is useful for visualizing the relative magnitudes of each parameter's loading. The magnitude of the coefficients provides insight into the dominant process parameters in each principal component, and the sign of each coefficient indicates a positive or negative correlation. In this case, age (0.484), residence (0.466), employ (0.459), and savings (0.404) are all important to the first principal component, and are also positively correlated with the first principal component: the first principal component increases in value when any of these parameters increase in value. Process parameters with approximately equal loading may be highly correlated to each other, which can be verified with a correlation matrix. When correlation is high between variables (close to one), it is recommended to remove the extraneous variables from the analysis (i.e., all but one of the highly correlated variables).

**Figure 12.11** Loading plot for LoanApplication.mtw (Minitab 18).

The Eigen analysis of the Correlation Matrix, presented in Table 12.4, shows the proportion of the total variation accounted for by each principal component. The first principal component explains more of the variation than any other possible linear combination of the original variables, and each subsequent principal component explains more than the next principal component (Jackson, 1980). In this example, the first principal component explains 44.3% of the total variance. The *Cumulative* line in Table 12.4 shows that the first four principal components explain 90.7% of the variation, so further analysis of this data will use only the first four principal components. The criterion is somewhat subjective, and often a cumulative probability approaching 90% or higher is desirable. One common determination uses the Scree Plot, as shown in Figure 12.12. Note the inflection (i.e., change in slope) after the fourth principal component, indicating the relative decline in information provided by additional principal components. Minitab's Help system references another method, known as the Kaiser method, which suggest including only the principal components whose eigenvalues exceed 1.0. This criterion would include only the first three principal components and account for 84% of the variance in this example.

We can further investigate the model by review of the outliers, as shown in Figure 12.13. Minitab 18 provides the description of each plotted point when you hover your mouse over the point. Curiously, subgroup six is sitting just beneath the reference line of 4.4 shown in the figure. We would generally exclude any group plotted above the reference line.

**612** | The Six Sigma Handbook

**Figure 12.12** Scree plot for LoanApplication.mtw (Minitab 18).

**Figure 12.13** Outlier plot for LoanApplication.mtw (Minitab 18).

Chapter 12 Control/Verify Phase | 613

We can now use the principal components to analyze the process for stability. First, we will calculate the first four principal components for each of the subgroups as follows:

1. Select the *Multivariate > Principal Components* option from the *Stat* menu (or the *Edit Last Dialog* from the *Edit* menu).
   a. Retain the prior selection of the variables and *Type of Matric* fields.
   b. Enter the value 4 into the *Number of components to compute* field.
2. Using the *Storage* button, enter "PC1 PC2 PC3 PC4" (without the quotes) into the *Scores* field. Select the OK button to save the selection.
3. Select the OK button to create the four new columns in Minitab's spreadsheet containing the calculated four principal components for each subgroup.

Since the principal components are constructed as independent variables, univariate Moving Average Sigma charts may be used for each principal component, using the subgroup size of 1 and a moving cell width of 3, to investigate process stability. When the process is stable, it has a stable set of principal components (Jackson, 1991).

In this case, the Moving Average Sigma charts show stability for the third and fourth principal components. The first and second principal components showed instability, as shown in Figures 12.14 and 12.15, respectively. Note that the out of control subgroup in the Principal Component 1 Sigma chart shown in Figure 12.14 is also approaching the upper limit on the Generalized Variance chart of Figure 12.9.

**Figure 12.14** Univariate control charts for first principal component (Green Belt XL).

**Figure 12.15** Univariate control charts for second principal component (Green Belt XL).

The control chart for the second principle component, shown in Figure 12.15, shows subgroup 6 as out of control, suggesting that the out of control situation on the multivariate chart in Figure 12.8 is explained by a special cause attributable to the second principal component. A review of the Score plot, generated earlier in Minitab and now shown in Figure 12.16, indicates a similar finding: note the lone point shown in the vicinity of $y = -4.5$ $x = 2.2$, corresponding to subgroup 6. Although the Score plot shows a fair amount of data of similar magnitude for the first component, there are few points whose second component score is as low as that of subgroup 6.

To understand which process parameter is influencing the second principal component, refer back to the Loading chart of Figure 12.11 and/or the Eigenvectors in Table 12.3, which indicated that the second principal component is heavily influenced by debt, followed by credit cards, then education. But which of these parameters changed, or did the relationship between the parameters change, at that subgroup?

Recall that the Sigma chart for debt in Figure 12.10 indicated an out of control condition for subgroup 6, while the Moving Average charts for education and income indicated the presence of special causes for subgroups 5 and 6. In this relatively simple case of only eight parameters., the univariate chart provided a possible solution to identify the process parameter responsible for out of control on the multivariate charts.

Another troubleshooting tool may be provided by a Contributions plot (Kourti and MacGregor, 1996), which looks for a significant shift at that subgroup for one of the process variables. To assess whether the parameter has significantly changed at that subgroup, each parameter's raw data at that subgroup is standardized by subtracting its mean and dividing by its standard deviation, so the signal indicates how far the current subgroup is from the mean, in standardized units for the parameter. An example is shown in Figure 12.17, indi-

Chapter 12  Control/Verify Phase | 615

**Figure 12.16**  Score plot for LoanApplication.mtw (Minitab 18).

**Figure 12.17**  Contribution plot of the standardized parameters for subgroup 6.

cating the highest contribution to the subgroup's value from the debt parameter. An alternative approach scales the subgroup's standardized values by the loading for that principal component, as shown in Figure 12.18, since the contribution to the principal component is a prime focus in this analysis. The loading weighted standardized value for debt is significantly larger than any other parameter (due to its loading on the second principal component), suggesting its primary role in influencing the out of control condition for the second principal component, and the resulting effect towards the special cause on the multivariate chart.

**Figure 12.18** Contribution plot of the standardized parameters weighted by loadings for subgroup 6.

# APPENDIX 1

# Glossary of Basic Statistical Terms

**Acceptable Quality Level (AQL)**—The maximum percentage or proportion of variant units in a lot or batch that, for the purposes of acceptance sampling, can be considered satisfactory as a process average.

**Analysis of Variance (ANOVA)**—A technique that subdivides the total variation of a set of data into meaningful component parts associated with specific sources of variation for the purpose of testing some hypothesis on the parameters of the model or estimating variance components.

**Assignable cause**—A factor that contributes to variation and that is feasible to detect and identify.

**Average Outgoing Quality (AOQ)**—The expected quality of outgoing product following the use of an acceptance sampling plan for a given value of incoming product quality.

**Average Outgoing Quality Limit (AOQL)**—For a given acceptance sampling plan, the maximum AOQ over all possible levels of incoming quality.

**Chance causes**—Factors, generally numerous and individually of relatively small importance, which contribute to variation, but which are not feasible to detect or identify.

**Coefficient of determination**—A measure of the part of the variance for one variable that can be explained by its linear relationship with a second variable. Designated by $\rho^2$ or $r^2$.

**Coefficient of multiple correlation**—A number between 0 and 1 that indicates the degree of the combined linear relationship of several predictor variables, $X_1$, $X_2, \ldots, X_p$ to the response variable $Y$. It is the simple correlation coefficient between predicted and observed values of the response variable.

---

From *Glossary & Tables for Statistical Quality Control*, prepared by the ASQ Statistics Division.
Copyright © 1983, ASQ Quality Press (800) 248-1946. Reprinted by permission of the publisher.

**Coefficient of variation**—A measure of relative dispersion that is the standard deviation divided by the mean and multiplied by 100 to give a percentage value. This measure cannot be used when the data take both negative and positive values or when it has been coded in such a way that the value $X = 0$ does not coincide with the origin.

**Confidence limits**—The end points of the interval about the sample statistic that is believed, with a specified confidence coefficient, to include the population parameter.

**Consumer's risk ($\beta$)**—For a given sampling plan, the probability of acceptance of a lot, the quality of which has a designated numerical value representing a level that it is seldom desired to accept. Usually the designated value will be the Limiting Quality Level (LQL).

**Correlation coefficient**—A number between –1 and 1 that indicates the degree of linear relationship between two sets of numbers:

$$r_{xy} = \frac{s_{xy}}{s_x s_y} = \frac{n\sum XY - \sum X \sum Y}{\sqrt{\left[n\sum X^2 - \left(\sum X\right)^2\right]\left[n\sum Y^2 - \left(\sum Y\right)^2\right]}}$$

**Defect**—A departure of a quality characteristic from its intended level or state that occurs with a severity sufficient to cause an associated product or service not to satisfy intended normal, or reasonably foreseeable, usage requirements.

**Defective**—A unit of product or service containing at least one defect, or having several imperfections that in combination cause the unit not to satisfy intended normal, or reasonably foreseeable, usage requirements. The word *defective* is appropriate for use when a unit of product or service is evaluated in terms of usage (as contrasted to conformance to specifications).

**Double sampling**—Sampling inspection in which the inspection of the first sample of size $n_1$ leads to a decision to accept a lot, not to accept it, or to take a second sample of size $n_2$, and the inspection of the second sample then leads to a decision to accept or not to accept the lot.

**Experiment design**—The arrangement in which an experimental program is to be conducted, and the selection of the versions (levels) of one or more factors or factor combinations to be included in the experiment.

**Factor**—An assignable cause that may affect the responses (test results) and of which different versions (levels) are included in the experiment.

**Factorial experiments**—Experiments in which all possible treatment combinations formed from two or more factors, each being studied at two or more versions (levels), are examined so that interactions (differential effects) as well as main effects can be estimated.

**Frequency distribution**—A set of all the various values that individual observations may have and the frequency of their occurrence in the sample or population.

**Histogram**—A plot of the frequency distribution in the form of rectangles whose bases are equal to the cell interval and whose areas are proportional to the frequencies.

**Hypothesis, alternative**—The hypothesis that is accepted if the null hypothesis is disproved. The choice of alternative hypothesis will determine whether "one-tail" or "two-tail" tests are appropriate.

**Hypothesis, null**—The hypothesis tested in tests of significance is that there is no difference (null) between the population of the sample and specified population (or between the populations associated with each sample). The null hypothesis can never be proved true. It can, however, be shown, with specified risks of error, to be untrue; that is, a difference can be shown to exist between the populations. If it is not disproved, one usually acts on the assumption that there is no adequate reason to doubt that it is true. (It may be that there is insufficient power to prove the existence of a difference rather than that there is no difference; that is, the sample size may be too small. By specifying the minimum difference that one wants to detect and $\beta$, the risk of failing to detect a difference of this size, the actual sample size required, however, can be determined.)

**In-control process**—A process in which the statistical measure(s) being evaluated are in a "state of statistical control."

**Kurtosis**—A measure of the shape of a distribution. A positive value indicates that the distribution has longer tails than the normal distribution (platykurtosis); while a negative value indicates that the distribution has shorter tails (leptokurtosis). For the normal distribution, the kurtosis is 0.

**Mean**—A measure of the location of a distribution. The centroid.

**Mean, standard error of**—The standard deviation of the average of a sample of size $n$.

$$s_{\bar{x}} = \frac{s_x}{\sqrt{n}}$$

**Median**—The middle measurement when an odd number of units are arranged in order of size; for an ordered set $X_1, X_2, \ldots, X_{2k-1}$

$$\text{Median} = X_k$$

When an even number are so arranged, the median is the average of the two middle units; for an ordered set $X_1, X_2, \ldots, X_{2k}$

$$\text{Median} = \frac{X_k + X_{k+1}}{2}$$

**Mode**—The most frequent value of the variable.

**Multiple sampling**—Sampling inspection in which, after each sample is inspected, the decision is made to accept a lot, not to accept it, or to take another sample to reach the decision. There may be a prescribed maximum number of samples, after which a decision to accept or not to accept must be reached.

**Operating Characteristics curve (OC curve)**—
1. For isolated or unique lots or a lot from an isolated sequence: a curve showing, for a given sampling plan, the probability of accepting a lot as a function of the lot quality. (Type A)
2. For a continuous stream of lots: a curve showing, for a given sampling plan, the probability of accepting a lot as a function of the process average. (Type B)
3. For continuous sampling plans: a curve showing the proportion of submitted product over the long run accepted during the sampling phases of the plan as a function of the product quality.
4. For special plans: a curve showing, for a given sampling plan, the probability of continuing to permit the process to continue without adjustment as a function of the process quality.

**Parameter**—A constant or coefficient that describes some characteristic of a population (e.g., standard deviation, average, regression coefficient).

**Population**—The totality of items or units of material under consideration.

NOTE: The items may be units or measurements, and the population may be real or conceptual. Thus *population* may refer to all the items actually produced in a given day or all that might be produced if the process were to continue *in-control*.

**Power curve**—The curve showing the relation between the probability $(1 - \beta)$ of rejecting the hypothesis that a sample belongs to a given population with a given characteristic(s) and the actual population value of that characteristic(s). NOTE: if $\beta$ is used instead of $(1 - \beta)$, the curve is called an operating characteristic curve (OC curve) (used mainly in sampling plans for quality control).

**Process capability**—The limits within which a tool or process operates based upon minimum variability as governed by the prevailing circumstances.

NOTE: The phrase "by the prevailing circumstances" indicates that the definition of inherent variability of a process involving only one operator, one source of raw material, etc., differs from one involving multiple operators, and many sources of raw material, etc. If the measure of inherent variability is made within very restricted circumstances, it is necessary to add components for frequently occurring assignable sources of variation that cannot economically be eliminated.

**Producer's risk** ($\alpha$)—For a given sampling plan, the probability of not accepting a lot the quality of which has a designated numerical value representing a level that it is generally desired to accept. Usually, the designated value will be the Acceptable Quality Level (AQL).

**Quality**—The totality of features and characteristics of a product or service that bear on its ability to satisfy given needs.

**Quality assurance**—All those planned or systematic actions necessary to provide adequate confidence that a product or service will satisfy given needs.

**Quality control**—The operational techniques and the activities that sustain a quality of product or service that will satisfy given needs; also the use of such techniques and activities.

**R (range)**—A measure of dispersion that is the difference between the largest observed value and the smallest observed value in a given sample. While the range is a measure of dispersion in its own right, it is sometimes used to estimate the population standard deviation, but is a biased estimator unless multiplied by the factor ($1/d_2$) appropriate to the sample size.

**Random sampling**—The process of selecting units for a sample of size *n* in such a manner that all combinations of *n* units under consideration have an equal or ascertainable chance of being selected as the sample.

**Replication**—The repetition of the set of all the treatment combinations to be compared in an experiment. Each of the repetitions is called a *replicate*.

**Sample**—A group of units, portion of material, or observations taken from a larger collection of units, quantity of material, or observations that serves to provide information that may be used as a basis for making a decision concerning the larger quantity.

**Single sampling**—Sampling inspection in which the decision to accept or not to accept a lot is based on the inspection of a single sample of size *n*.

**Skewness**—A measure of the symmetry of a distribution. A positive value indicates that the distribution has a greater tendency to tail to the right (positively skewed or skewed to the right), and a negative value indicates a greater tendency of the distribution to tail to the left (negatively skewed or skewed to the left). Skewness is 0 for a normal distribution.

**Standard deviation**—

1. $\sigma$—population standard deviation. A measure of variability (dispersion) of observations that is the positive square root of the population variance.

$$\sqrt{\frac{1}{n}\sum(X_i - \bar{X})^2}$$

2. *s*—sample standard deviation. A measure of variability (dispersion) that is the positive square root of the sample variance.

$$\sqrt{\frac{1}{n-1}\sum(X_i - \bar{X})^2}$$

**Statistic**—A quantity calculated from a sample of observations, most often to form an estimate of some population parameter.

**Type I error (acceptance control sense)**—The incorrect decision that a process is unacceptable when, in fact, perfect information would reveal that it is located within the "zone of acceptable processes."

**Type II error (acceptance control sense)**—The incorrect decision that a process is acceptable when, in fact, perfect information would reveal that it is located within the "zone of rejectable processes."

**Variance—**

1. $\sigma^2$—population variance. A measure of variability (dispersion) of observations based upon the mean of the squared deviation from the arithmetic mean.
2. $s^2$—sample variance. A measure of variability (dispersion) of observations in a sample based upon the squared deviations from the arithmetic average divided by the degrees of freedom.

# APPENDIX 2

# Area Under the Standard Normal Curve

| z | 0.00 | 0.01 | 0.02 | 0.03 | 0.04 | 0.05 | 0.06 | 0.07 | 0.08 | 0.09 |
|---|---|---|---|---|---|---|---|---|---|---|
| −3.4 | 0.0003 | 0.0003 | 0.0003 | 0.0003 | 0.0003 | 0.0003 | 0.0003 | 0.0003 | 0.0003 | 0.0002 |
| −3.3 | 0.0005 | 0.0005 | 0.0005 | 0.0004 | 0.0004 | 0.0004 | 0.0004 | 0.0004 | 0.0004 | 0.0003 |
| −3.2 | 0.0007 | 0.0007 | 0.0006 | 0.0006 | 0.0006 | 0.0006 | 0.0006 | 0.0005 | 0.0005 | 0.0005 |
| −3.1 | 0.0010 | 0.0009 | 0.0009 | 0.0009 | 0.0008 | 0.0008 | 0.0008 | 0.0008 | 0.0007 | 0.0007 |
| −3.0 | 0.0013 | 0.0013 | 0.0013 | 0.0012 | 0.0012 | 0.0011 | 0.0011 | 0.0011 | 0.0010 | 0.0010 |
| −2.9 | 0.0019 | 0.0018 | 0.0018 | 0.0017 | 0.0016 | 0.0016 | 0.0015 | 0.0015 | 0.0014 | 0.0014 |
| −2.8 | 0.0026 | 0.0025 | 0.0024 | 0.0023 | 0.0023 | 0.0022 | 0.0021 | 0.0021 | 0.0020 | 0.0019 |
| −2.7 | 0.0035 | 0.0034 | 0.0033 | 0.0032 | 0.0031 | 0.0030 | 0.0029 | 0.0028 | 0.0027 | 0.0026 |
| −2.6 | 0.0047 | 0.0045 | 0.0044 | 0.0043 | 0.0041 | 0.0040 | 0.0039 | 0.0038 | 0.0037 | 0.0036 |
| −2.5 | 0.0062 | 0.0060 | 0.0059 | 0.0057 | 0.0055 | 0.0054 | 0.0052 | 0.0051 | 0.0049 | 0.0048 |
| −2.4 | 0.0082 | 0.0080 | 0.0078 | 0.0075 | 0.0073 | 0.0071 | 0.0069 | 0.0068 | 0.0066 | 0.0064 |
| −2.3 | 0.0107 | 0.0104 | 0.0102 | 0.0099 | 0.0096 | 0.0094 | 0.0091 | 0.0089 | 0.0087 | 0.0084 |
| −2.2 | 0.0139 | 0.0136 | 0.0132 | 0.0129 | 0.0125 | 0.0122 | 0.0119 | 0.0116 | 0.0113 | 0.0110 |
| −2.1 | 0.0179 | 0.0174 | 0.0170 | 0.0166 | 0.0162 | 0.0158 | 0.0154 | 0.0150 | 0.0146 | 0.0143 |
| −2.0 | 0.0228 | 0.0222 | 0.0217 | 0.0212 | 0.0207 | 0.0202 | 0.0197 | 0.0192 | 0.0188 | 0.0183 |
| −1.9 | 0.0287 | 0.0281 | 0.0274 | 0.0268 | 0.0262 | 0.0256 | 0.0250 | 0.0244 | 0.0239 | 0.0233 |
| −1.8 | 0.0359 | 0.0351 | 0.0344 | 0.0336 | 0.0329 | 0.0322 | 0.0314 | 0.0307 | 0.0301 | 0.0294 |
| −1.7 | 0.0446 | 0.0436 | 0.0427 | 0.0418 | 0.0409 | 0.0401 | 0.0392 | 0.0384 | 0.0375 | 0.0367 |

| z | 0.00 | 0.01 | 0.02 | 0.03 | 0.04 | 0.05 | 0.06 | 0.07 | 0.08 | 0.09 |
|---|------|------|------|------|------|------|------|------|------|------|
| −1.6 | 0.0548 | 0.0537 | 0.0526 | 0.0516 | 0.0505 | 0.0495 | 0.0485 | 0.0475 | 0.0465 | 0.0455 |
| −1.5 | 0.0668 | 0.0655 | 0.0643 | 0.0630 | 0.0618 | 0.0606 | 0.0594 | 0.0582 | 0.0571 | 0.0559 |
| −1.4 | 0.0808 | 0.0793 | 0.0778 | 0.0764 | 0.0749 | 0.0735 | 0.0721 | 0.0708 | 0.0694 | 0.0681 |
| −1.3 | 0.0968 | 0.0951 | 0.0934 | 0.0918 | 0.0901 | 0.0885 | 0.0869 | 0.0853 | 0.0838 | 0.0823 |
| −1.2 | 0.1151 | 0.1131 | 0.1112 | 0.1093 | 0.1075 | 0.1056 | 0.1038 | 0.1020 | 0.1003 | 0.0985 |
| −1.1 | 0.1357 | 0.1335 | 0.1314 | 0.1292 | 0.1271 | 0.1251 | 0.1230 | 0.1210 | 0.1190 | 0.1170 |
| −1.0 | 0.1587 | 0.1562 | 0.1539 | 0.1515 | 0.1492 | 0.1469 | 0.1446 | 0.1423 | 0.1401 | 0.1379 |
| −0.9 | 0.1841 | 0.1814 | 0.1788 | 0.1762 | 0.1736 | 0.1711 | 0.1685 | 0.1660 | 0.1635 | 0.1611 |
| −0.8 | 0.2119 | 0.2090 | 0.2061 | 0.2033 | 0.2005 | 0.1977 | 0.1949 | 0.1922 | 0.1894 | 0.1867 |
| −0.7 | 0.2420 | 0.2389 | 0.2358 | 0.2327 | 0.2296 | 0.2266 | 0.2236 | 0.2206 | 0.2177 | 0.2148 |
| −0.6 | 0.2743 | 0.2709 | 0.2676 | 0.2643 | 0.2611 | 0.2578 | 0.2546 | 0.2514 | 0.2483 | 0.2451 |
| −0.5 | 0.3085 | 0.3050 | 0.3015 | 0.2981 | 0.2946 | 0.2912 | 0.2877 | 0.2843 | 0.2810 | 0.2776 |
| −0.4 | 0.3446 | 0.3409 | 0.3372 | 0.3336 | 0.3300 | 0.3264 | 0.3228 | 0.3192 | 0.3156 | 0.3121 |
| −0.3 | 0.3821 | 0.3783 | 0.3745 | 0.3707 | 0.3669 | 0.3632 | 0.3594 | 0.3557 | 0.3520 | 0.3483 |
| −0.2 | 0.4207 | 0.4168 | 0.4129 | 0.4090 | 0.4052 | 0.4013 | 0.3974 | 0.3936 | 0.3897 | 0.3859 |
| −0.1 | 0.4602 | 0.4562 | 0.4522 | 0.4483 | 0.4443 | 0.4404 | 0.4364 | 0.4325 | 0.4286 | 0.4247 |
| −0.0 | 0.5000 | 0.4960 | 0.4920 | 0.4880 | 0.4840 | 0.4801 | 0.4761 | 0.4721 | 0.4681 | 0.4641 |
| 0.0 | 0.5000 | 0.5040 | 0.5080 | 0.5120 | 0.5160 | 0.5199 | 0.5239 | 0.5279 | 0.5319 | 0.5359 |
| 0.1 | 0.5398 | 0.5438 | 0.5478 | 0.5517 | 0.5557 | 0.5596 | 0.5636 | 0.5675 | 0.5714 | 0.5753 |
| 0.2 | 0.5793 | 0.5832 | 0.5871 | 0.5910 | 0.5948 | 0.5987 | 0.6026 | 0.6064 | 0.6103 | 0.6141 |
| 0.3 | 0.6179 | 0.6217 | 0.6255 | 0.6293 | 0.6331 | 0.6368 | 0.6406 | 0.6443 | 0.6480 | 0.6517 |
| 0.4 | 0.6554 | 0.6591 | 0.6628 | 0.6664 | 0.6700 | 0.6736 | 0.6772 | 0.6808 | 0.6844 | 0.6879 |
| 0.5 | 0.6915 | 0.6950 | 0.6985 | 0.7019 | 0.7054 | 0.7088 | 0.7123 | 0.7157 | 0.7190 | 0.7224 |
| 0.6 | 0.7257 | 0.7291 | 0.7324 | 0.7357 | 0.7389 | 0.7422 | 0.7454 | 0.7486 | 0.7517 | 0.7549 |
| 0.7 | 0.7580 | 0.7611 | 0.7642 | 0.7673 | 0.7704 | 0.7734 | 0.7764 | 0.7794 | 0.7823 | 0.7852 |
| 0.8 | 0.7881 | 0.7910 | 0.7939 | 0.7967 | 0.7995 | 0.8023 | 0.8051 | 0.8078 | 0.8106 | 0.8133 |
| 0.9 | 0.8159 | 0.8186 | 0.8212 | 0.8238 | 0.8264 | 0.8289 | 0.8315 | 0.8340 | 0.8365 | 0.8389 |
| 1.0 | 0.8413 | 0.8438 | 0.8461 | 0.8485 | 0.8508 | 0.8531 | 0.8554 | 0.8577 | 0.8599 | 0.8621 |
| 1.1 | 0.8643 | 0.8665 | 0.8686 | 0.8708 | 0.8729 | 0.8749 | 0.8770 | 0.8790 | 0.8810 | 0.8830 |
| 1.2 | 0.8849 | 0.8869 | 0.8888 | 0.8907 | 0.8925 | 0.8944 | 0.8962 | 0.8980 | 0.8997 | 0.9015 |
| 1.3 | 0.9032 | 0.9049 | 0.9066 | 0.9082 | 0.9099 | 0.9115 | 0.9131 | 0.9147 | 0.9162 | 0.9177 |
| 1.4 | 0.9192 | 0.9207 | 0.9222 | 0.9236 | 0.9251 | 0.9265 | 0.9279 | 0.9292 | 0.9306 | 0.9319 |
| 1.5 | 0.9332 | 0.9345 | 0.9357 | 0.9370 | 0.9382 | 0.9394 | 0.9406 | 0.9418 | 0.9429 | 0.9441 |
| 1.6 | 0.9452 | 0.9463 | 0.9474 | 0.9484 | 0.9495 | 0.9505 | 0.9515 | 0.9525 | 0.9535 | 0.9545 |
| 1.7 | 0.9554 | 0.9564 | 0.9573 | 0.9582 | 0.9591 | 0.9599 | 0.9608 | 0.9616 | 0.9625 | 0.9633 |
| 1.8 | 0.9641 | 0.9649 | 0.9656 | 0.9664 | 0.9671 | 0.9678 | 0.9686 | 0.9693 | 0.9699 | 0.9706 |
| 1.9 | 0.9713 | 0.9719 | 0.9726 | 0.9732 | 0.9738 | 0.9744 | 0.9750 | 0.9756 | 0.9761 | 0.9767 |
| 2.0 | 0.9772 | 0.9778 | 0.9783 | 0.9788 | 0.9793 | 0.9798 | 0.9803 | 0.9808 | 0.9812 | 0.9817 |

## Appendix 2  Area Under the Standard Normal Curve

| z | 0.00 | 0.01 | 0.02 | 0.03 | 0.04 | 0.05 | 0.06 | 0.07 | 0.08 | 0.09 |
|---|------|------|------|------|------|------|------|------|------|------|
| 2.1 | 0.9821 | 0.9826 | 0.9830 | 0.9834 | 0.9838 | 0.9842 | 0.9846 | 0.9850 | 0.9854 | 0.9857 |
| 2.2 | 0.9861 | 0.9864 | 0.9868 | 0.9871 | 0.9875 | 0.9878 | 0.9881 | 0.9884 | 0.9887 | 0.9890 |
| 2.3 | 0.9893 | 0.9896 | 0.9898 | 0.9901 | 0.9904 | 0.9906 | 0.9909 | 0.9911 | 0.9913 | 0.9916 |
| 2.4 | 0.9918 | 0.9920 | 0.9922 | 0.9925 | 0.9927 | 0.9929 | 0.9931 | 0.9932 | 0.9934 | 0.9936 |
| 2.5 | 0.9938 | 0.9940 | 0.9941 | 0.9943 | 0.9945 | 0.9946 | 0.9948 | 0.9949 | 0.9951 | 0.9952 |
| 2.6 | 0.9953 | 0.9955 | 0.9956 | 0.9957 | 0.9959 | 0.9960 | 0.9961 | 0.9962 | 0.9963 | 0.9964 |
| 2.7 | 0.9965 | 0.9966 | 0.9967 | 0.9968 | 0.9969 | 0.9970 | 0.9971 | 0.9972 | 0.9973 | 0.9974 |
| 2.8 | 0.9974 | 0.9975 | 0.9976 | 0.9977 | 0.9977 | 0.9978 | 0.9979 | 0.9979 | 0.9980 | 0.9981 |
| 2.9 | 0.9981 | 0.9982 | 0.9982 | 0.9983 | 0.9984 | 0.9984 | 0.9985 | 0.9985 | 0.9986 | 0.9986 |
| 3.0 | 0.9987 | 0.9987 | 0.9987 | 0.9988 | 0.9988 | 0.9989 | 0.9989 | 0.9989 | 0.9990 | 0.9990 |
| 3.1 | 0.9990 | 0.9991 | 0.9991 | 0.9991 | 0.9992 | 0.9992 | 0.9992 | 0.9992 | 0.9993 | 0.9993 |
| 3.2 | 0.9993 | 0.9993 | 0.9994 | 0.9994 | 0.9994 | 0.9994 | 0.9994 | 0.9995 | 0.9995 | 0.9995 |
| 3.3 | 0.9995 | 0.9995 | 0.9995 | 0.9996 | 0.9996 | 0.9996 | 0.9996 | 0.9996 | 0.9996 | 0.9997 |
| 3.4 | 0.9997 | 0.9997 | 0.9997 | 0.9997 | 0.9997 | 0.9997 | 0.9997 | 0.9997 | 0.9997 | 0.9998 |

# APPENDIX 3

# Critical Values of the *t*-Distribution

|    | α     |       |        |        |        |
|----|-------|-------|--------|--------|--------|
| df | 0.1   | 0.05  | 0.025  | 0.01   | 0.005  |
| 1  | 3.078 | 6.314 | 12.706 | 31.821 | 63.657 |
| 2  | 1.886 | 2.920 | 4.303  | 6.965  | 9.925  |
| 3  | 1.638 | 2.353 | 3.182  | 4.541  | 5.841  |
| 4  | 1.533 | 2.132 | 2.776  | 3.747  | 4.604  |
| 5  | 1.476 | 2.015 | 2.571  | 3.365  | 4.032  |
| 6  | 1.440 | 1.943 | 2.447  | 3.143  | 3.707  |
| 7  | 1.415 | 1.895 | 2.365  | 2.998  | 3.499  |
| 8  | 1.397 | 1.860 | 2.306  | 2.896  | 3.355  |
| 9  | 1.383 | 1.833 | 2.262  | 2.821  | 3.250  |
| 10 | 1.372 | 1.812 | 2.228  | 2.764  | 3.169  |
| 11 | 1.363 | 1.796 | 2.201  | 2.718  | 3.106  |
| 12 | 1.356 | 1.782 | 2.179  | 2.681  | 3.055  |
| 13 | 1.350 | 1.771 | 2.160  | 2.650  | 3.012  |
| 14 | 1.345 | 1.761 | 2.145  | 2.624  | 2.977  |
| 15 | 1.341 | 1.753 | 2.131  | 2.602  | 2.947  |

| df | 0.1 | 0.05 | 0.025 | 0.01 | 0.005 |
|---|---|---|---|---|---|
| | | | $\alpha$ | | |
| 16 | 1.337 | 1.746 | 2.120 | 2.583 | 2.921 |
| 17 | 1.333 | 1.740 | 2.110 | 2.567 | 2.898 |
| 18 | 1.330 | 1.734 | 2.101 | 2.552 | 2.878 |
| 19 | 1.328 | 1.729 | 2.093 | 2.539 | 2.861 |
| 20 | 1.325 | 1.725 | 2.086 | 2.528 | 2.845 |
| 21 | 1.323 | 1.721 | 2.080 | 2.518 | 2.831 |
| 22 | 1.321 | 1.717 | 2.074 | 2.508 | 2.819 |
| 23 | 1.319 | 1.714 | 2.069 | 2.500 | 2.807 |
| 24 | 1.318 | 1.711 | 2.064 | 2.492 | 2.797 |
| 25 | 1.316 | 1.708 | 2.060 | 2.485 | 2.787 |
| 26 | 1.315 | 1.706 | 2.056 | 2.479 | 2.779 |
| 27 | 1.314 | 1.703 | 2.052 | 2.473 | 2.771 |
| 28 | 1.313 | 1.701 | 2.048 | 2.467 | 2.763 |
| 29 | 1.311 | 1.699 | 2.045 | 2.462 | 2.756 |
| $\infty$ | 1.282 | 1.645 | 1.960 | 2.326 | 2.576 |

# APPENDIX 4

# Chi-Square Distribution

| γ | 0.995 | 0.99 | 0.98 | 0.975 | 0.95 | 0.90 | 0.80 | 0.75 | 0.70 | 0.50 |
|---|---|---|---|---|---|---|---|---|---|---|
| 1 | 0.00004 | 0.000 | 0.001 | 0.001 | 0.004 | 0.016 | 0.064 | 0.102 | 0.148 | 0.455 |
| 2 | 0.0100 | 0.020 | 0.040 | 0.051 | 0.103 | 0.211 | 0.446 | 0.575 | 0.713 | 0.1386 |
| 3 | 0.0717 | 0.115 | 0.185 | 0.216 | 0.352 | 0.584 | 1.005 | 1.213 | 1.424 | 2.366 |
| 4 | 0.207 | 0.297 | 0.429 | 0.484 | 0.711 | 1.064 | 1.649 | 1.923 | 2.195 | 3.357 |
| 5 | 0.412 | 0.554 | 0.752 | 0.831 | 1.145 | 1.610 | 2.343 | 2.675 | 3.000 | 4.351 |
| 6 | 0.676 | 0.872 | 1.134 | 1.237 | 1.635 | 2.204 | 3.070 | 3.455 | 3.828 | 5.348 |
| 7 | 0.989 | 1.239 | 1.564 | 1.690 | 2.167 | 2.833 | 3.822 | 4.255 | 4.671 | 6.346 |
| 8 | 1.344 | 1.646 | 2.032 | 2.180 | 2.733 | 3.490 | 4.594 | 5.071 | 5.527 | 7.344 |
| 9 | 1.735 | 2.088 | 2.532 | 2.700 | 3.325 | 4.168 | 5.380 | 5.899 | 6.393 | 8.343 |
| 10 | 2.156 | 2.558 | 3.059 | 3.247 | 3.940 | 4.865 | 6.179 | 6.737 | 7.267 | 9.342 |
| 11 | 2.603 | 3.053 | 3.609 | 3.816 | 4.575 | 5.578 | 6.989 | 7.584 | 8.148 | 10.341 |
| 12 | 3.074 | 3.571 | 4.178 | 4.404 | 5.226 | 6.304 | 7.807 | 8.438 | 9.034 | 11.340 |
| 13 | 3.565 | 4.107 | 4.765 | 5.009 | 5.892 | 7.042 | 8.634 | 9.299 | 9.926 | 12.340 |
| 14 | 4.075 | 4.660 | 5.368 | 5.629 | 6.571 | 7.790 | 9.467 | 10.165 | 10.821 | 13.339 |
| 15 | 4.601 | 5.229 | 5.985 | 6.262 | 7.261 | 8.547 | 10.307 | 11.037 | 11.721 | 14.339 |
| 16 | 5.142 | 5.812 | 6.614 | 6.908 | 7.962 | 9.312 | 11.152 | 11.912 | 12.624 | 15.338 |
| 17 | 5.697 | 6.408 | 7.255 | 7.564 | 8.672 | 10.085 | 12.002 | 12.792 | 13.531 | 16.338 |
| 18 | 6.265 | 7.015 | 7.906 | 8.231 | 9.390 | 10.865 | 12.857 | 13.675 | 14.440 | 17.338 |
| 19 | 6.844 | 7.633 | 8.567 | 8.907 | 10.117 | 11.651 | 13.716 | 14.562 | 15.352 | 18.338 |
| 20 | 7.434 | 8.260 | 9.237 | 9.591 | 10.851 | 12.443 | 14.578 | 15.452 | 16.266 | 19.337 |
| 21 | 8.034 | 8.897 | 9.915 | 10.283 | 11.591 | 13.240 | 15.445 | 16.344 | 17.182 | 20.337 |
| 22 | 8.643 | 9.542 | 10.600 | 10.982 | 12.338 | 14.041 | 16.314 | 17.240 | 18.101 | 21.337 |

| | α | | | | | | | | | |
|---|---|---|---|---|---|---|---|---|---|---|
| γ | 0.995 | 0.99 | 0.98 | 0.975 | 0.95 | 0.90 | 0.80 | 0.75 | 0.70 | 0.50 |
| 23 | 9.260 | 10.196 | 11.293 | 11.689 | 13.091 | 14.848 | 17.187 | 18.137 | 19.021 | 22.337 |
| 24 | 9.886 | 10.856 | 11.992 | 12.401 | 13.848 | 15.659 | 18.062 | 19.037 | 19.943 | 23.337 |
| 25 | 10.520 | 11.524 | 12.697 | 13.120 | 14.611 | 16.473 | 18.940 | 19.939 | 20.867 | 24.337 |
| 26 | 11.160 | 12.198 | 13.409 | 13.844 | 15.379 | 17.292 | 19.820 | 20.843 | 21.792 | 25.336 |
| 27 | 11.808 | 12.879 | 14.125 | 14.573 | 16.151 | 18.114 | 20.703 | 21.749 | 22.719 | 26.336 |
| 28 | 12.461 | 13.565 | 14.847 | 15.308 | 16.928 | 18.939 | 21.588 | 22.657 | 23.647 | 27.336 |
| 29 | 13.121 | 14.256 | 15.574 | 16.047 | 17.708 | 19.768 | 22.475 | 23.567 | 24.577 | 28.336 |
| 30 | 13.787 | 14.953 | 16.306 | 16.791 | 18.493 | 20.599 | 23.364 | 24.478 | 25.508 | 29.336 |

| | α | | | | | | | | | |
|---|---|---|---|---|---|---|---|---|---|---|
| γ | 0.30 | 0.25 | 0.20 | 0.10 | 0.05 | 0.025 | 0.02 | 0.01 | 0.005 | 0.001 |
| 1 | 1.074 | 1.323 | 1.642 | 2.706 | 3.841 | 5.024 | 5.412 | 6.635 | 7.879 | 10.828 |
| 2 | 2.408 | 2.773 | 3.219 | 4.605 | 5.991 | 7.378 | 7.824 | 9.210 | 10.597 | 13.816 |
| 3 | 3.665 | 4.108 | 4.642 | 6.251 | 7.815 | 9.348 | 9.837 | 11.345 | 12.838 | 16.266 |
| 4 | 4.878 | 5.385 | 5.989 | 7.779 | 9.488 | 11.143 | 11.668 | 13.277 | 14.860 | 18.467 |
| 5 | 6.064 | 6.626 | 7.289 | 9.236 | 11.070 | 12.833 | 13.388 | 15.086 | 16.750 | 20.515 |
| 6 | 7.231 | 7.841 | 8.558 | 10.645 | 12.592 | 14.449 | 15.033 | 16.812 | 18.548 | 22.458 |
| 7 | 8.383 | 9.037 | 9.803 | 12.017 | 14.067 | 16.013 | 16.622 | 18.475 | 20.278 | 24.322 |
| 8 | 9.524 | 10.219 | 11.030 | 13.362 | 15.507 | 17.535 | 18.168 | 20.090 | 21.955 | 26.124 |
| 9 | 10.656 | 11.389 | 12.242 | 14.684 | 16.919 | 19.023 | 19.679 | 21.666 | 23.589 | 27.877 |
| 10 | 11.781 | 12.549 | 13.442 | 15.987 | 18.307 | 20.483 | 21.161 | 23.209 | 25.188 | 29.588 |
| 11 | 12.899 | 13.701 | 14.631 | 17.275 | 19.675 | 21.920 | 22.618 | 24.725 | 26.757 | 31.264 |
| 12 | 14.011 | 14.845 | 15.812 | 18.549 | 21.026 | 23.337 | 24.054 | 26.217 | 28.300 | 32.909 |
| 13 | 15.119 | 15.984 | 16.985 | 19.812 | 22.362 | 24.736 | 25.472 | 27.688 | 29.819 | 34.528 |
| 14 | 16.222 | 17.117 | 18.151 | 21.064 | 23.685 | 26.119 | 26.873 | 29.141 | 31.319 | 36.123 |
| 15 | 17.322 | 18.245 | 19.311 | 22.307 | 24.996 | 27.488 | 28.259 | 30.578 | 32.801 | 37.697 |
| 16 | 18.418 | 19.369 | 20.465 | 23.542 | 26.296 | 28.845 | 29.633 | 32.000 | 34.267 | 39.252 |
| 17 | 19.511 | 20.489 | 21.615 | 24.769 | 27.587 | 30.191 | 30.995 | 33.409 | 35.718 | 40.790 |
| 18 | 20.601 | 21.605 | 22.760 | 25.989 | 28.869 | 31.526 | 32.346 | 34.805 | 37.156 | 42.312 |
| 19 | 21.689 | 22.718 | 23.900 | 27.204 | 30.144 | 32.852 | 33.687 | 36.191 | 38.582 | 43.820 |
| 20 | 22.775 | 23.828 | 25.038 | 28.412 | 31.410 | 34.170 | 35.020 | 37.566 | 39.997 | 45.315 |
| 21 | 23.858 | 24.935 | 26.171 | 29.615 | 32.671 | 35.479 | 36.343 | 38.932 | 41.401 | 46.797 |
| 22 | 24.939 | 26.039 | 27.301 | 30.813 | 33.924 | 36.781 | 37.659 | 40.289 | 42.796 | 48.268 |
| 23 | 26.018 | 27.141 | 28.429 | 32.007 | 35.172 | 38.076 | 38.968 | 41.638 | 44.181 | 49.728 |
| 24 | 27.096 | 28.241 | 29.553 | 33.196 | 36.415 | 39.364 | 40.270 | 42.980 | 45.559 | 51.179 |
| 25 | 28.172 | 29.339 | 30.675 | 34.382 | 37.652 | 40.646 | 41.566 | 44.314 | 46.928 | 52.620 |
| 26 | 29.246 | 30.435 | 31.795 | 35.563 | 38.885 | 41.923 | 42.856 | 45.642 | 48.290 | 54.052 |
| 27 | 30.319 | 31.528 | 32.912 | 36.741 | 40.113 | 43.195 | 44.140 | 46.963 | 49.645 | 55.476 |
| 28 | 31.391 | 32.620 | 34.027 | 37.916 | 41.337 | 44.461 | 45.419 | 48.278 | 50.993 | 56.892 |
| 29 | 32.461 | 33.711 | 35.139 | 39.087 | 42.557 | 45.722 | 46.693 | 49.588 | 52.336 | 58.301 |
| 30 | 33.530 | 34.800 | 36.250 | 40.256 | 43.773 | 46.979 | 47.962 | 50.892 | 53.672 | 59.703 |

# APPENDIX 5

# F Distribution ($\alpha$ = 1%)

$$F_{.99}(n_1, n_2)$$

$n_1$ = degrees of freedom for numerator

$n_2$ = degrees of freedom for denominator

| $n_2$ \ $n_1$ | 1 | 2 | 3 | 4 | 5 | 6 | 7 | 8 | 9 | 10 |
|---|---|---|---|---|---|---|---|---|---|---|
| 1 | 4052 | 4999.5 | 5403 | 5625 | 5764 | 5859 | 5928 | 5982 | 6022 | 6056 |
| 2 | 98.50 | 99.00 | 99.17 | 99.25 | 99.30 | 99.33 | 99.36 | 99.37 | 99.39 | 99.40 |
| 3 | 34.12 | 30.82 | 29.46 | 28.71 | 28.24 | 27.91 | 27.67 | 27.49 | 27.35 | 27.23 |
| 4 | 21.20 | 18.00 | 16.69 | 15.98 | 15.52 | 15.21 | 14.98 | 14.80 | 14.66 | 14.55 |
| 5 | 16.26 | 13.27 | 12.06 | 11.39 | 10.97 | 10.67 | 10.46 | 10.29 | 10.16 | 10.05 |
| 6 | 13.75 | 10.92 | 9.78 | 9.15 | 8.75 | 8.47 | 8.26 | 8.10 | 7.98 | 7.87 |
| 7 | 12.25 | 9.55 | 8.45 | 7.85 | 7.46 | 7.19 | 6.99 | 6.84 | 6.72 | 6.62 |
| 8 | 11.26 | 8.65 | 7.59 | 7.01 | 6.63 | 6.37 | 6.18 | 6.03 | 5.91 | 5.81 |
| 9 | 10.56 | 8.02 | 6.99 | 6.42 | 6.06 | 5.80 | 5.61 | 5.47 | 5.35 | 5.26 |
| 10 | 10.04 | 7.56 | 6.55 | 5.99 | 5.64 | 5.39 | 5.20 | 5.06 | 4.94 | 4.85 |
| 11 | 9.65 | 7.21 | 6.22 | 5.67 | 5.32 | 5.07 | 4.89 | 4.74 | 4.63 | 4.54 |
| 12 | 9.33 | 6.93 | 5.95 | 5.41 | 5.06 | 4.82 | 4.64 | 4.50 | 4.39 | 4.30 |
| 13 | 9.07 | 6.70 | 5.74 | 5.21 | 4.86 | 4.62 | 4.44 | 4.30 | 4.19 | 4.10 |
| 14 | 8.86 | 6.51 | 5.56 | 5.04 | 4.69 | 4.46 | 4.28 | 4.14 | 4.03 | 3.94 |
| 15 | 8.68 | 6.36 | 5.42 | 4.89 | 4.56 | 4.32 | 4.14 | 4.00 | 3.89 | 3.80 |
| 16 | 8.53 | 6.23 | 5.29 | 4.77 | 4.44 | 4.20 | 4.03 | 3.89 | 3.78 | 3.69 |
| 17 | 8.40 | 6.11 | 5.18 | 4.67 | 4.34 | 4.10 | 3.93 | 3.79 | 3.68 | 3.59 |
| 18 | 8.29 | 6.01 | 5.09 | 4.58 | 4.25 | 4.01 | 3.84 | 3.71 | 3.60 | 3.51 |
| 19 | 8.18 | 5.93 | 5.01 | 4.50 | 4.17 | 3.94 | 3.77 | 3.63 | 3.52 | 3.43 |
| 20 | 8.10 | 5.85 | 4.94 | 4.43 | 4.10 | 3.87 | 3.70 | 3.56 | 3.46 | 3.37 |
| 21 | 8.02 | 5.78 | 4.87 | 4.37 | 4.04 | 3.81 | 3.64 | 3.51 | 3.40 | 3.31 |
| 22 | 7.95 | 5.72 | 4.82 | 4.31 | 3.99 | 3.76 | 3.59 | 3.45 | 3.35 | 3.26 |
| 23 | 7.88 | 5.66 | 4.76 | 4.26 | 3.94 | 3.71 | 3.54 | 3.41 | 3.30 | 3.21 |
| 24 | 7.82 | 5.61 | 4.72 | 4.22 | 3.90 | 3.67 | 3.50 | 3.36 | 3.26 | 3.17 |
| 25 | 7.77 | 5.57 | 4.68 | 4.18 | 3.85 | 3.63 | 3.46 | 3.32 | 3.22 | 3.13 |
| 26 | 7.72 | 5.53 | 4.64 | 4.14 | 3.82 | 3.59 | 3.42 | 3.29 | 3.18 | 3.09 |
| 27 | 7.68 | 5.49 | 4.60 | 4.11 | 3.78 | 3.56 | 3.39 | 3.26 | 3.15 | 3.06 |
| 28 | 7.64 | 5.45 | 4.57 | 4.07 | 3.75 | 3.53 | 3.36 | 3.23 | 3.12 | 3.03 |
| 29 | 7.60 | 5.42 | 4.54 | 4.04 | 3.73 | 3.50 | 3.33 | 3.20 | 3.09 | 3.00 |
| 30 | 7.56 | 5.39 | 4.51 | 4.02 | 3.70 | 3.47 | 3.30 | 3.17 | 3.07 | 2.98 |
| 40 | 7.31 | 5.18 | 4.31 | 3.83 | 3.51 | 3.29 | 3.12 | 2.99 | 2.89 | 2.80 |
| 60 | 7.08 | 4.98 | 4.13 | 3.65 | 3.34 | 3.12 | 2.95 | 2.82 | 2.72 | 2.63 |
| 120 | 6.85 | 4.79 | 3.95 | 3.48 | 3.17 | 2.96 | 2.79 | 2.66 | 2.56 | 2.47 |
| ∞ | 6.63 | 4.61 | 3.78 | 3.32 | 3.02 | 2.80 | 2.64 | 2.51 | 2.41 | 2.32 |

| $n_2$ \ $n_1$ | 12 | 15 | 20 | 24 | 30 | 40 | 60 | 120 | ∞ |
|---|---|---|---|---|---|---|---|---|---|
| 1 | 6106 | 6157 | 6209 | 6235 | 6261 | 6287 | 6313 | 6339 | 6366 |
| 2 | 99.42 | 99.43 | 99.45 | 99.46 | 99.47 | 99.47 | 99.48 | 99.49 | 99.50 |
| 3 | 27.05 | 26.87 | 26.69 | 26.60 | 26.50 | 26.41 | 26.32 | 26.22 | 26.13 |
| 4 | 14.37 | 14.20 | 14.02 | 13.93 | 13.84 | 13.75 | 13.65 | 13.56 | 13.46 |
| 5 | 9.89 | 9.72 | 9.55 | 9.47 | 9.38 | 9.29 | 9.20 | 9.11 | 9.02 |
| 6 | 7.72 | 7.56 | 7.40 | 7.31 | 7.23 | 7.14 | 7.06 | 6.97 | 6.88 |
| 7 | 6.47 | 6.31 | 6.16 | 6.07 | 5.99 | 5.91 | 5.82 | 5.74 | 5.65 |
| 8 | 5.67 | 5.52 | 5.36 | 5.28 | 5.20 | 5.12 | 5.03 | 4.95 | 4.86 |
| 9 | 5.11 | 4.96 | 4.81 | 4.73 | 4.65 | 4.57 | 4.48 | 4.40 | 4.31 |
| 10 | 4.71 | 4.56 | 4.41 | 4.33 | 4.25 | 4.17 | 4.08 | 4.00 | 3.91 |
| 11 | 4.40 | 4.25 | 4.10 | 4.02 | 3.94 | 3.86 | 3.78 | 3.69 | 3.60 |
| 12 | 4.16 | 4.01 | 3.86 | 3.78 | 3.70 | 3.62 | 3.54 | 3.45 | 3.36 |
| 13 | 3.96 | 3.82 | 3.66 | 3.59 | 3.51 | 3.43 | 3.34 | 3.25 | 3.17 |
| 14 | 3.80 | 3.66 | 3.51 | 3.43 | 3.35 | 3.27 | 3.18 | 3.09 | 3.00 |
| 15 | 3.67 | 3.52 | 3.37 | 3.29 | 3.21 | 3.13 | 3.05 | 2.96 | 2.87 |
| 16 | 3.55 | 3.41 | 3.26 | 3.18 | 3.10 | 3.02 | 2.93 | 2.84 | 2.75 |
| 17 | 3.46 | 3.31 | 3.16 | 3.08 | 3.00 | 2.92 | 2.83 | 2.75 | 2.65 |
| 18 | 3.37 | 3.23 | 3.08 | 3.00 | 2.92 | 2.84 | 2.75 | 2.66 | 2.57 |
| 19 | 3.30 | 3.15 | 3.00 | 2.92 | 2.84 | 2.76 | 2.67 | 2.58 | 2.49 |
| 20 | 3.23 | 3.09 | 2.94 | 2.86 | 2.78 | 2.69 | 2.61 | 2.52 | 2.42 |
| 21 | 3.17 | 3.03 | 2.88 | 2.80 | 2.72 | 2.64 | 2.55 | 2.46 | 2.36 |
| 22 | 3.12 | 2.98 | 2.83 | 2.75 | 2.67 | 2.58 | 2.50 | 2.40 | 2.31 |
| 23 | 3.07 | 2.93 | 2.78 | 2.70 | 2.62 | 2.54 | 2.45 | 2.35 | 2.26 |
| 24 | 3.03 | 2.89 | 2.74 | 2.66 | 2.58 | 2.49 | 2.40 | 2.31 | 2.21 |
| 25 | 2.99 | 2.85 | 2.70 | 2.62 | 2.54 | 2.45 | 2.36 | 2.27 | 2.17 |
| 26 | 2.96 | 2.81 | 2.66 | 2.58 | 2.50 | 2.42 | 2.33 | 2.23 | 2.13 |
| 27 | 2.93 | 2.78 | 2.63 | 2.55 | 2.47 | 2.38 | 2.29 | 2.20 | 2.10 |
| 28 | 2.90 | 2.75 | 2.60 | 2.52 | 2.44 | 2.35 | 2.26 | 2.17 | 2.06 |
| 29 | 2.87 | 2.73 | 2.57 | 2.49 | 2.41 | 2.33 | 2.23 | 2.14 | 2.03 |
| 30 | 2.84 | 2.70 | 2.55 | 2.47 | 2.39 | 2.30 | 2.21 | 2.11 | 2.01 |
| 40 | 2.66 | 2.52 | 2.37 | 2.29 | 2.20 | 2.11 | 2.02 | 1.92 | 1.80 |
| 60 | 2.50 | 2.35 | 2.20 | 2.12 | 2.03 | 1.94 | 1.84 | 1.73 | 1.60 |
| 120 | 2.34 | 2.19 | 2.03 | 1.95 | 1.86 | 1.76 | 1.66 | 1.53 | 1.38 |
| ∞ | 2.18 | 2.04 | 1.88 | 1.79 | 1.70 | 1.59 | 1.47 | 1.32 | 1.00 |

# APPENDIX 6

# F Distribution ($\alpha = 5\%$)

$$F_{.95}(n_1, n_2)$$
$n_1$ = degrees of freedom for numerator

$n_2$ = degrees of freedom for denominator

| $n_2$ \ $n_1$ | 1 | 2 | 3 | 4 | 5 | 6 | 7 | 8 | 9 | 10 |
|---|---|---|---|---|---|---|---|---|---|---|
| 1 | 161.4 | 199.5 | 215.7 | 224.6 | 230.2 | 234.0 | 236.8 | 238.9 | 240.5 | 241.9 |
| 2 | 18.51 | 19.00 | 19.16 | 19.25 | 19.30 | 19.33 | 19.35 | 19.37 | 19.38 | 19.40 |
| 3 | 10.13 | 9.55 | 9.28 | 9.12 | 9.01 | 8.94 | 8.89 | 8.85 | 8.81 | 8.79 |
| 4 | 7.71 | 6.94 | 6.59 | 6.39 | 6.26 | 6.16 | 6.09 | 6.04 | 6.00 | 5.96 |
| 5 | 6.61 | 5.79 | 5.41 | 5.19 | 5.05 | 4.95 | 4.88 | 4.82 | 4.77 | 4.74 |
| 6 | 5.99 | 5.14 | 4.76 | 4.53 | 4.39 | 4.28 | 4.21 | 4.15 | 4.10 | 4.06 |
| 7 | 5.59 | 4.47 | 4.35 | 4.12 | 3.97 | 3.87 | 3.79 | 3.73 | 3.68 | 3.64 |
| 8 | 5.32 | 4.46 | 4.07 | 3.84 | 3.69 | 3.58 | 3.50 | 3.44 | 3.39 | 3.35 |
| 9 | 5.12 | 4.26 | 3.86 | 3.63 | 3.48 | 3.37 | 3.29 | 3.23 | 3.18 | 3.14 |
| 10 | 4.96 | 4.10 | 3.71 | 3.48 | 3.33 | 3.22 | 3.14 | 3.07 | 3.02 | 2.98 |
| 11 | 4.84 | 3.98 | 3.59 | 3.36 | 3.20 | 3.09 | 3.01 | 2.95 | 2.90 | 2.85 |
| 12 | 4.75 | 3.89 | 3.49 | 3.26 | 3.11 | 3.00 | 2.91 | 2.85 | 2.80 | 2.75 |
| 13 | 4.67 | 3.81 | 3.41 | 3.18 | 3.03 | 2.92 | 2.83 | 2.77 | 2.71 | 2.67 |
| 14 | 4.60 | 3.74 | 3.34 | 3.11 | 2.96 | 2.85 | 2.76 | 2.70 | 2.65 | 2.60 |
| 15 | 4.54 | 3.68 | 3.29 | 3.06 | 2.90 | 2.79 | 2.71 | 2.64 | 2.59 | 2.54 |
| 16 | 4.49 | 3.63 | 3.24 | 3.01 | 2.85 | 2.74 | 2.66 | 2.59 | 2.54 | 2.49 |
| 17 | 4.45 | 3.59 | 3.20 | 2.96 | 2.81 | 2.70 | 2.61 | 2.55 | 2.49 | 2.45 |
| 18 | 4.41 | 3.55 | 3.16 | 2.93 | 2.77 | 2.66 | 2.58 | 2.51 | 2.46 | 2.41 |
| 19 | 4.38 | 3.52 | 3.13 | 2.90 | 2.74 | 2.63 | 2.54 | 2.48 | 2.42 | 2.38 |
| 20 | 4.35 | 3.49 | 3.10 | 2.87 | 2.71 | 2.60 | 2.51 | 2.45 | 2.39 | 2.35 |
| 21 | 4.32 | 3.47 | 3.07 | 2.84 | 2.68 | 2.57 | 2.49 | 2.42 | 2.37 | 2.32 |
| 22 | 4.30 | 3.44 | 3.05 | 2.82 | 2.66 | 2.55 | 2.46 | 2.40 | 2.34 | 2.30 |
| 23 | 4.28 | 3.42 | 3.03 | 2.80 | 2.64 | 2.53 | 2.44 | 2.37 | 2.32 | 2.27 |
| 24 | 4.26 | 3.40 | 3.01 | 2.78 | 2.62 | 2.51 | 2.42 | 2.36 | 2.30 | 2.25 |
| 25 | 4.24 | 3.39 | 2.99 | 2.76 | 2.60 | 2.49 | 2.40 | 2.34 | 2.28 | 2.24 |
| 26 | 4.23 | 3.37 | 2.98 | 2.74 | 2.59 | 2.47 | 2.39 | 2.32 | 2.27 | 2.22 |
| 27 | 4.21 | 3.35 | 2.96 | 2.73 | 2.57 | 2.46 | 2.37 | 2.31 | 2.25 | 2.20 |
| 28 | 4.20 | 3.34 | 2.95 | 2.71 | 2.56 | 2.45 | 2.36 | 2.29 | 2.24 | 2.19 |
| 29 | 4.18 | 3.33 | 2.93 | 2.70 | 2.55 | 2.43 | 2.35 | 2.28 | 2.22 | 2.18 |
| 30 | 4.17 | 3.32 | 2.92 | 2.69 | 2.53 | 2.42 | 2.33 | 2.27 | 2.21 | 2.16 |
| 40 | 4.08 | 3.23 | 2.84 | 2.61 | 2.45 | 2.34 | 2.25 | 2.18 | 2.12 | 2.08 |

| $n_2$ \ $n_1$ | 1 | 2 | 3 | 4 | 5 | 6 | 7 | 8 | 9 | 10 |
|---|---|---|---|---|---|---|---|---|---|---|
| 60 | 4.00 | 3.15 | 2.76 | 2.53 | 2.37 | 2.25 | 2.17 | 2.10 | 2.04 | 1.99 |
| 120 | 3.92 | 3.07 | 2.68 | 2.45 | 2.29 | 2.17 | 2.09 | 2.02 | 1.96 | 1.91 |
| ∞ | 3.84 | 3.00 | 2.60 | 2.37 | 2.21 | 2.10 | 2.01 | 1.94 | 1.88 | 1.83 |

$n_1$ = degrees of freedom for numerator

$n_2$ = degrees of freedom for denominator

| $n_2$ \ $n_1$ | 12 | 15 | 20 | 24 | 30 | 40 | 60 | 120 | ∞ |
|---|---|---|---|---|---|---|---|---|---|
| 1 | 243.9 | 245.9 | 248.0 | 249.1 | 250.1 | 251.1 | 252.2 | 253.3 | 254.3 |
| 2 | 19.41 | 19.43 | 19.45 | 19.45 | 19.46 | 19.47 | 19.48 | 19.49 | 19.50 |
| 3 | 8.74 | 8.70 | 8.66 | 8.64 | 8.62 | 8.59 | 8.57 | 8.55 | 8.53 |
| 4 | 5.91 | 5.86 | 5.80 | 5.77 | 5.75 | 5.72 | 5.69 | 5.66 | 5.63 |
| 5 | 4.68 | 4.62 | 4.56 | 4.53 | 4.50 | 4.46 | 4.43 | 4.40 | 4.36 |
| 6 | 4.00 | 3.94 | 3.87 | 3.84 | 3.81 | 3.77 | 3.74 | 3.70 | 3.67 |
| 7 | 3.57 | 3.51 | 3.44 | 3.41 | 3.38 | 3.34 | 3.30 | 3.27 | 3.23 |
| 8 | 3.28 | 3.22 | 3.15 | 3.12 | 3.08 | 3.04 | 3.01 | 2.97 | 2.93 |
| 9 | 3.07 | 3.01 | 2.94 | 2.90 | 2.86 | 2.83 | 2.79 | 2.75 | 2.71 |
| 10 | 2.91 | 2.85 | 2.77 | 2.74 | 2.70 | 2.66 | 2.62 | 2.58 | 2.54 |
| 11 | 2.79 | 2.72 | 2.65 | 2.61 | 2.57 | 2.53 | 2.49 | 2.45 | 2.40 |
| 12 | 2.69 | 2.62 | 2.54 | 2.51 | 2.47 | 2.43 | 2.38 | 2.34 | 2.30 |
| 13 | 2.60 | 2.53 | 2.46 | 2.42 | 2.38 | 2.34 | 2.30 | 2.25 | 2.21 |
| 14 | 2.53 | 2.46 | 2.39 | 2.35 | 2.31 | 2.27 | 2.22 | 2.18 | 2.13 |
| 15 | 2.48 | 2.40 | 2.33 | 2.29 | 2.25 | 2.20 | 2.16 | 2.11 | 2.07 |
| 16 | 2.42 | 2.35 | 2.28 | 2.24 | 2.19 | 2.15 | 2.11 | 2.06 | 2.01 |
| 17 | 2.38 | 2.31 | 2.23 | 2.19 | 2.15 | 2.10 | 2.06 | 2.01 | 1.96 |
| 18 | 2.34 | 2.27 | 2.19 | 2.15 | 2.11 | 2.06 | 2.02 | 1.97 | 1.92 |
| 19 | 2.31 | 2.23 | 2.16 | 2.11 | 2.07 | 2.03 | 1.98 | 1.93 | 1.88 |
| 20 | 2.28 | 2.20 | 2.12 | 2.08 | 2.04 | 1.99 | 1.95 | 1.90 | 1.84 |
| 21 | 2.25 | 2.18 | 2.10 | 2.05 | 2.01 | 1.96 | 1.92 | 1.87 | 1.81 |
| 22 | 2.23 | 2.15 | 2.07 | 2.03 | 1.98 | 1.94 | 1.89 | 1.84 | 1.78 |
| 23 | 2.20 | 2.13 | 2.05 | 2.01 | 1.96 | 1.91 | 1.86 | 1.81 | 1.76 |
| 24 | 2.18 | 2.11 | 2.03 | 1.98 | 1.94 | 1.89 | 1.84 | 1.79 | 1.73 |
| 25 | 2.16 | 2.09 | 2.01 | 1.96 | 1.92 | 1.87 | 1.82 | 1.77 | 1.71 |
| 26 | 2.15 | 2.07 | 1.99 | 1.95 | 1.90 | 1.85 | 1.80 | 1.75 | 1.69 |
| 27 | 2.13 | 2.06 | 1.97 | 1.93 | 1.88 | 1.84 | 1.79 | 1.73 | 1.67 |
| 28 | 2.12 | 2.04 | 1.96 | 1.91 | 1.87 | 1.82 | 1.77 | 1.71 | 1.65 |
| 29 | 2.10 | 2.03 | 1.94 | 1.90 | 1.85 | 1.81 | 1.75 | 1.70 | 1.64 |
| 30 | 2.09 | 2.01 | 1.93 | 1.89 | 1.84 | 1.79 | 1.74 | 1.68 | 1.62 |
| 40 | 2.00 | 1.92 | 1.84 | 1.79 | 1.74 | 1.69 | 1.64 | 1.58 | 1.51 |
| 60 | 1.92 | 1.84 | 1.75 | 1.70 | 1.65 | 1.59 | 1.53 | 1.47 | 1.39 |
| 120 | 1.83 | 1.75 | 1.66 | 1.61 | 1.55 | 1.50 | 1.43 | 1.35 | 1.25 |
| ∞ | 1.75 | 1.67 | 1.57 | 1.52 | 1.46 | 1.39 | 1.32 | 1.22 | 1.00 |

# APPENDIX 7

# Poisson Probability Sums

$$\sum_{x=0}^{r} p(x;\mu)$$

| | μ | | | | | | | | |
|---|---|---|---|---|---|---|---|---|---|
| r | 0.1 | 0.2 | 0.3 | 0.4 | 0.5 | 0.6 | 0.7 | 0.8 | 0.9 |
| 0 | 0.9048 | 0.8187 | 0.7408 | 0.6703 | 0.6065 | 0.5488 | 0.4966 | 0.4493 | 0.4066 |
| 1 | 0.9953 | 0.9825 | 0.9631 | 0.9384 | 0.9098 | 0.8781 | 0.8442 | 0.8088 | 0.7725 |
| 2 | 0.9998 | 0.9989 | 0.9964 | 0.9921 | 0.9856 | 0.9769 | 0.9659 | 0.9526 | 0.9371 |
| 3 | 1.0000 | 0.9999 | 0.9997 | 0.9992 | 0.9982 | 0.9966 | 0.9942 | 0.9909 | 0.9865 |
| 4 | 1.0000 | 1.0000 | 1.0000 | 0.9999 | 0.9998 | 0.9996 | 0.9992 | 0.9986 | 0.9977 |
| 5 | 1.0000 | 1.0000 | 1.0000 | 1.0000 | 1.0000 | 1.0000 | 0.9999 | 0.9998 | 0.9997 |
| 6 | 1.0000 | 1.0000 | 1.0000 | 1.0000 | 1.0000 | 1.0000 | 1.0000 | 1.0000 | 1.0000 |

| | μ | | | | | | | | |
|---|---|---|---|---|---|---|---|---|---|
| r | 1.0 | 1.5 | 2.0 | 2.5 | 3.0 | 3.5 | 4.0 | 4.5 | 5.0 |
| 0 | 0.3679 | 0.2231 | 0.1353 | 0.0821 | 0.0498 | 0.0302 | 0.0183 | 0.0111 | 0.0067 |
| 1 | 0.7358 | 0.5578 | 0.4060 | 0.2873 | 0.1991 | 0.1359 | 0.0916 | 0.0611 | 0.0404 |
| 2 | 0.9197 | 0.8088 | 0.6767 | 0.5438 | 0.4232 | 0.3208 | 0.2381 | 0.1736 | 0.1247 |
| 3 | 0.9810 | 0.9344 | 0.8571 | 0.7576 | 0.6472 | 0.5366 | 0.4335 | 0.3423 | 0.2650 |
| 4 | 0.9963 | 0.9814 | 0.9473 | 0.8912 | 0.8153 | 0.7254 | 0.6288 | 0.5321 | 0.4405 |
| 5 | 0.9994 | 0.9955 | 0.9834 | 0.9580 | 0.9161 | 0.8576 | 0.7851 | 0.7029 | 0.6160 |
| 6 | 0.9999 | 0.9991 | 0.9955 | 0.9858 | 0.9665 | 0.9347 | 0.8893 | 0.8311 | 0.7622 |
| 7 | 1.0000 | 0.9998 | 0.9989 | 0.9958 | 0.9881 | 0.9733 | 0.9489 | 0.9134 | 0.8666 |
| 8 | 1.0000 | 1.0000 | 0.9998 | 0.9989 | 0.9962 | 0.9901 | 0.9786 | 0.9597 | 0.9319 |
| 9 | 1.0000 | 1.0000 | 1.0000 | 0.9997 | 0.9989 | 0.9967 | 0.9919 | 0.9829 | 0.9682 |
| 10 | 1.0000 | 1.0000 | 1.0000 | 0.9999 | 0.9997 | 0.9990 | 0.9972 | 0.9933 | 0.9863 |
| 11 | 1.0000 | 1.0000 | 1.0000 | 1.0000 | 0.9999 | 0.9997 | 0.9991 | 0.9976 | 0.9945 |
| 12 | 1.0000 | 1.0000 | 1.0000 | 1.0000 | 1.0000 | 0.9999 | 0.9997 | 0.9992 | 0.9980 |
| 13 | 1.0000 | 1.0000 | 1.0000 | 1.0000 | 1.0000 | 1.0000 | 0.9999 | 0.9997 | 0.9993 |
| 14 | 1.0000 | 1.0000 | 1.0000 | 1.0000 | 1.0000 | 1.0000 | 1.0000 | 0.9999 | 0.9998 |
| 15 | 1.0000 | 1.0000 | 1.0000 | 1.0000 | 1.0000 | 1.0000 | 1.0000 | 1.0000 | 0.9999 |
| 16 | 1.0000 | 1.0000 | 1.0000 | 1.0000 | 1.0000 | 1.0000 | 1.0000 | 1.0000 | 1.0000 |

|   | μ | | | | | | | | |
|---|---|---|---|---|---|---|---|---|---|
| r | 5.5 | 6.0 | 6.5 | 7.0 | 7.5 | 8.0 | 8.5 | 9.0 | 9.5 |
| 0 | 0.0041 | 0.0025 | 0.0015 | 0.0009 | 0.0006 | 0.0003 | 0.0002 | 0.0001 | 0.0001 |
| 1 | 0.0266 | 0.0174 | 0.0113 | 0.0073 | 0.0047 | 0.0030 | 0.0019 | 0.0012 | 0.0008 |
| 2 | 0.0884 | 0.0620 | 0.0430 | 0.0296 | 0.0203 | 0.0138 | 0.0093 | 0.0062 | 0.0042 |
| 3 | 0.2017 | 0.1512 | 0.1118 | 0.0818 | 0.0591 | 0.0424 | 0.0301 | 0.0212 | 0.0149 |
| 4 | 0.3575 | 0.2851 | 0.2237 | 0.1730 | 0.1321 | 0.0996 | 0.0744 | 0.0550 | 0.0403 |
| 5 | 0.5289 | 0.4457 | 0.3690 | 0.3007 | 0.2414 | 0.1912 | 0.1496 | 0.1157 | 0.0885 |
| 6 | 0.6860 | 0.6063 | 0.5265 | 0.4497 | 0.3782 | 0.3134 | 0.2562 | 0.2068 | 0.1649 |
| 7 | 0.8095 | 0.7440 | 0.6728 | 0.5987 | 0.5246 | 0.4530 | 0.3856 | 0.3239 | 0.2687 |
| 8 | 0.8944 | 0.8472 | 0.7916 | 0.7291 | 0.6620 | 0.5925 | 0.5231 | 0.4557 | 0.3918 |
| 9 | 0.9462 | 0.9161 | 0.8774 | 0.8305 | 0.7764 | 0.7166 | 0.6530 | 0.5874 | 0.5218 |
| 10 | 0.9747 | 0.9574 | 0.9332 | 0.9015 | 0.8622 | 0.8159 | 0.7634 | 0.7060 | 0.6453 |
| 11 | 0.9890 | 0.9799 | 0.9661 | 0.9467 | 0.9208 | 0.8881 | 0.8487 | 0.8030 | 0.7520 |
| 12 | 0.9955 | 0.9912 | 0.9840 | 0.9730 | 0.9573 | 0.9362 | 0.9091 | 0.8758 | 0.8364 |
| 13 | 0.9983 | 0.9964 | 0.9929 | 0.9872 | 0.9784 | 0.9658 | 0.9486 | 0.9261 | 0.8981 |
| 14 | 0.9994 | 0.9986 | 0.9970 | 0.9943 | 0.9897 | 0.9827 | 0.9726 | 0.9585 | 0.9400 |
| 15 | 0.9998 | 0.9995 | 0.9988 | 0.9976 | 0.9954 | 0.9918 | 0.9862 | 0.9780 | 0.9665 |
| 16 | 0.9999 | 0.9998 | 0.9996 | 0.9990 | 0.9980 | 0.9963 | 0.9934 | 0.9889 | 0.9823 |
| 17 | 1.0000 | 0.9999 | 0.9998 | 0.9996 | 0.9992 | 0.9984 | 0.9970 | 0.9947 | 0.9911 |
| 18 | 1.0000 | 1.0000 | 0.9999 | 0.9999 | 0.9997 | 0.9993 | 0.9987 | 0.9976 | 0.9957 |
| 19 | 1.0000 | 1.0000 | 1.0000 | 1.0000 | 0.9999 | 0.9997 | 0.9995 | 0.9989 | 0.9980 |
| 20 | 1.0000 | 1.0000 | 1.0000 | 1.0000 | 1.0000 | 0.9999 | 0.9998 | 0.9996 | 0.9991 |
| 21 | 1.0000 | 1.0000 | 1.0000 | 1.0000 | 1.0000 | 1.0000 | 0.9999 | 0.9998 | 0.9996 |
| 22 | 1.0000 | 1.0000 | 1.0000 | 1.0000 | 1.0000 | 1.0000 | 1.0000 | 0.9999 | 0.9999 |
| 23 | 1.0000 | 1.0000 | 1.0000 | 1.0000 | 1.0000 | 1.0000 | 1.0000 | 1.0000 | 0.9999 |
| 24 | 1.0000 | 1.0000 | 1.0000 | 1.0000 | 1.0000 | 1.0000 | 1.0000 | 1.0000 | 1.0000 |

|   | μ | | | | | | | | |
|---|---|---|---|---|---|---|---|---|---|
| r | 10.0 | 11.0 | 12.0 | 13.0 | 14.0 | 15.0 | 16.0 | 17.0 | 18.0 |
| 0 | 0.0000 | 0.0000 | 0.0000 | 0.0000 | 0.0000 | 0.0000 | 0.0000 | 0.0000 | 0.0000 |
| 1 | 0.0005 | 0.0002 | 0.0001 | 0.0000 | 0.0000 | 0.0000 | 0.0000 | 0.0000 | 0.0000 |
| 2 | 0.0028 | 0.0012 | 0.0005 | 0.0002 | 0.0001 | 0.0000 | 0.0000 | 0.0000 | 0.0000 |
| 3 | 0.0103 | 0.0049 | 0.0023 | 0.0011 | 0.0005 | 0.0002 | 0.0001 | 0.0000 | 0.0000 |
| 4 | 0.0293 | 0.0151 | 0.0076 | 0.0037 | 0.0018 | 0.0009 | 0.0004 | 0.0002 | 0.0001 |
| 5 | 0.0671 | 0.0375 | 0.0203 | 0.0107 | 0.0055 | 0.0028 | 0.0014 | 0.0007 | 0.0003 |
| 6 | 0.1301 | 0.0786 | 0.0458 | 0.0259 | 0.0142 | 0.0076 | 0.0040 | 0.0021 | 0.0010 |
| 7 | 0.2202 | 0.1432 | 0.0895 | 0.0540 | 0.0316 | 0.0180 | 0.0100 | 0.0054 | 0.0029 |
| 8 | 0.3328 | 0.2320 | 0.1550 | 0.0998 | 0.0621 | 0.0374 | 0.0220 | 0.0126 | 0.0071 |
| 9 | 0.4579 | 0.3405 | 0.2424 | 0.1658 | 0.1094 | 0.0699 | 0.0433 | 0.0261 | 0.0154 |
| 10 | 0.5830 | 0.4599 | 0.3472 | 0.2517 | 0.1757 | 0.1185 | 0.0774 | 0.0491 | 0.0304 |
| 11 | 0.6968 | 0.5793 | 0.4616 | 0.3532 | 0.2600 | 0.1848 | 0.1270 | 0.0847 | 0.0549 |
| 12 | 0.7916 | 0.6887 | 0.5760 | 0.4631 | 0.3585 | 0.2676 | 0.1931 | 0.1350 | 0.0917 |
| 13 | 0.8645 | 0.7813 | 0.6815 | 0.5730 | 0.4644 | 0.3632 | 0.2745 | 0.2009 | 0.1426 |
| 14 | 0.9165 | 0.8540 | 0.7720 | 0.6751 | 0.5704 | 0.4657 | 0.3675 | 0.2808 | 0.2081 |

|  | μ | | | | | | | | |
|---|---|---|---|---|---|---|---|---|---|
| r | 10.0 | 11.0 | 12.0 | 13.0 | 14.0 | 15.0 | 16.0 | 17.0 | 18.0 |
| 15 | 0.9513 | 0.9074 | 0.8444 | 0.7636 | 0.6694 | 0.5681 | 0.4667 | 0.3715 | 0.2867 |
| 16 | 0.9730 | 0.9441 | 0.8987 | 0.8355 | 0.7559 | 0.6641 | 0.5660 | 0.4677 | 0.3751 |
| 17 | 0.9857 | 0.9678 | 0.9370 | 0.8905 | 0.8272 | 0.7489 | 0.6593 | 0.5640 | 0.4686 |
| 18 | 0.9928 | 0.9823 | 0.9626 | 0.9302 | 0.8826 | 0.8195 | 0.7423 | 0.6550 | 0.5622 |
| 19 | 0.9965 | 0.9907 | 0.9787 | 0.9573 | 0.9235 | 0.8752 | 0.8122 | 0.7363 | 0.6509 |
| 20 | 0.9984 | 0.9953 | 0.9884 | 0.9750 | 0.9521 | 0.9170 | 0.8682 | 0.8055 | 0.7307 |
| 21 | 0.9993 | 0.9977 | 0.9939 | 0.9859 | 0.9712 | 0.9469 | 0.9108 | 0.8615 | 0.7991 |
| 22 | 0.9997 | 0.9990 | 0.9970 | 0.9924 | 0.9833 | 0.9673 | 0.9418 | 0.9047 | 0.8551 |
| 23 | 0.9999 | 0.9995 | 0.9985 | 0.9960 | 0.9907 | 0.9805 | 0.9633 | 0.9637 | 0.8989 |
| 24 | 1.0000 | 0.9998 | 0.9993 | 0.9980 | 0.9950 | 0.9888 | 0.9777 | 0.9594 | 0.9317 |
| 25 | 1.0000 | 0.9999 | 0.9997 | 0.9990 | 0.9974 | 0.9938 | 0.9869 | 0.9748 | 0.9554 |
| 26 | 1.0000 | 1.0000 | 0.9999 | 0.9995 | 0.9987 | 0.9967 | 0.9925 | 0.9848 | 0.9718 |
| 27 | 1.0000 | 1.0000 | 0.9999 | 0.9998 | 0.9994 | 0.9983 | 0.9959 | 0.9912 | 0.9827 |
| 28 | 1.0000 | 1.0000 | 1.0000 | 0.9999 | 0.9997 | 0.9991 | 0.9978 | 0.9950 | 0.9897 |
| 29 | 1.0000 | 1.0000 | 1.0000 | 1.0000 | 0.9999 | 0.9996 | 0.9989 | 0.9973 | 0.9941 |
| 30 | 1.0000 | 1.0000 | 1.0000 | 1.0000 | 0.9999 | 0.9998 | 0.9994 | 0.9986 | 0.9967 |
| 31 | 1.0000 | 1.0000 | 1.0000 | 1.0000 | 1.0000 | 0.9999 | 0.9997 | 0.9993 | 0.9982 |
| 32 | 1.0000 | 1.0000 | 1.0000 | 1.0000 | 1.0000 | 1.0000 | 0.9999 | 0.9996 | 0.9990 |
| 33 | 1.0000 | 1.0000 | 1.0000 | 1.0000 | 1.0000 | 1.0000 | 0.9999 | 0.9998 | 0.9995 |
| 34 | 1.0000 | 1.0000 | 1.0000 | 1.0000 | 1.0000 | 1.0000 | 1.0000 | 0.9999 | 0.9998 |
| 35 | 1.0000 | 1.0000 | 1.0000 | 1.0000 | 1.0000 | 1.0000 | 1.0000 | 1.0000 | 0.9999 |
| 36 | 1.0000 | 1.0000 | 1.0000 | 1.0000 | 1.0000 | 1.0000 | 1.0000 | 1.0000 | 0.9999 |
| 37 | 1.0000 | 1.0000 | 1.0000 | 1.0000 | 1.0000 | 1.0000 | 1.0000 | 1.0000 | 1.0000 |

# APPENDIX 8

# Tolerance Interval Factors

Values of *k* for Two-Sided Limits

| n | γ = 0.90 p = 0.90 | p = 0.95 | p = 0.99 | p = 0.999 | γ = 0.95 p = 0.90 | p = 0.95 | p = 0.99 | p = 0.999 | γ = 0.99 p = 0.90 | p = 0.95 | p = 0.99 | p = 0.999 |
|---|---|---|---|---|---|---|---|---|---|---|---|---|
| 2 | 15.978 | 18.800 | 24.167 | 30.227 | 32.019 | 37.674 | 48.430 | 60.573 | 160.193 | 188.491 | 242.300 | 303.054 |
| 3 | 5.847 | 6.919 | 8.974 | 11.309 | 8.380 | 9.916 | 12.861 | 16.208 | 18.930 | 22.401 | 29.055 | 36.616 |
| 4 | 4.166 | 4.943 | 6.440 | 8.149 | 5.369 | 6.370 | 8.299 | 10.502 | 9.398 | 11.150 | 14.527 | 18.383 |
| 5 | 3.494 | 4.152 | 5.423 | 6.879 | 4.275 | 5.079 | 6.634 | 8.415 | 6.612 | 7.855 | 10.260 | 13.015 |
| 6 | 3.131 | 3.723 | 4.870 | 6.188 | 3.712 | 4.414 | 5.775 | 7.337 | 5.337 | 6.345 | 8.301 | 10.548 |
| 7 | 2.902 | 3.452 | 4.521 | 5.750 | 3.369 | 4.007 | 5.248 | 6.676 | 4.613 | 5.488 | 7.187 | 9.142 |
| 8 | 2.743 | 3.264 | 4.278 | 5.446 | 3.136 | 3.732 | 4.891 | 6.226 | 4.147 | 4.936 | 6.468 | 8.234 |
| 9 | 2.626 | 3.125 | 4.098 | 5.220 | 2.967 | 3.532 | 4.631 | 5.899 | 3.822 | 4.550 | 5.966 | 7.600 |
| 10 | 2.535 | 3.018 | 3.959 | 5.046 | 2.839 | 3.379 | 4.433 | 5.649 | 3.582 | 4.265 | 5.594 | 7.129 |
| 11 | 2.463 | 2.933 | 3.849 | 4.906 | 2.737 | 3.259 | 4.277 | 5.452 | 3.397 | 4.045 | 5.308 | 6.766 |
| 12 | 2.404 | 2.863 | 3.758 | 4.792 | 2.655 | 3.162 | 4.150 | 5.291 | 3.250 | 3.870 | 5.079 | 6.477 |
| 13 | 2.355 | 2.805 | 3.682 | 4.697 | 2.587 | 3.081 | 4.044 | 5.158 | 3.130 | 3.727 | 4.893 | 6.240 |
| 14 | 2.314 | 2.756 | 3.618 | 4.615 | 2.529 | 3.012 | 3.955 | 5.045 | 3.029 | 3.608 | 4.737 | 6.043 |
| 15 | 2.278 | 2.713 | 3.562 | 4.545 | 2.480 | 2.954 | 3.878 | 4.949 | 2.945 | 3.507 | 4.605 | 5.876 |
| 16 | 2.246 | 2.676 | 3.514 | 4.484 | 2.437 | 2.903 | 3.812 | 4.865 | 2.872 | 3.421 | 4.492 | 5.732 |
| 17 | 2.219 | 2.643 | 3.471 | 4.430 | 2.400 | 2.858 | 3.754 | 4.791 | 2.808 | 3.345 | 4.393 | 5.607 |
| 18 | 2.194 | 2.614 | 3.433 | 4.382 | 2.366 | 2.819 | 3.702 | 4.725 | 2.753 | 3.279 | 4.307 | 5.497 |
| 19 | 2.172 | 2.588 | 3.399 | 4.339 | 2.337 | 2.784 | 3.656 | 4.667 | 2.703 | 3.221 | 4.230 | 5.399 |
| 20 | 2.152 | 2.564 | 3.368 | 4.300 | 2.310 | 2.752 | 3.615 | 4.614 | 2.659 | 3.168 | 4.161 | 5.312 |
| 21 | 2.135 | 2.543 | 3.340 | 4.264 | 2.286 | 2.723 | 3.577 | 4.567 | 2.620 | 3.121 | 4.100 | 5.234 |
| 22 | 2.118 | 2.524 | 3.315 | 4.232 | 2.264 | 2.697 | 3.543 | 4.523 | 2.584 | 3.078 | 4.044 | 5.163 |
| 23 | 2.103 | 2.506 | 3.292 | 4.203 | 2.244 | 2.673 | 3.512 | 4.484 | 2.551 | 3.040 | 3.993 | 5.098 |
| 24 | 2.089 | 2.480 | 3.270 | 4.176 | 2.225 | 2.651 | 3.483 | 4.447 | 2.522 | 3.004 | 3.947 | 5.039 |
| 25 | 2.077 | 2.474 | 3.251 | 4.151 | 2.208 | 2.631 | 3.457 | 4.413 | 2.494 | 2.972 | 3.904 | 4.985 |
| 30 | 2.025 | 2.413 | 3.170 | 4.049 | 2.140 | 2.549 | 3.350 | 4.278 | 2.385 | 2.841 | 3.733 | 4.768 |
| 35 | 1.988 | 2.368 | 3.112 | 3.974 | 2.090 | 2.490 | 3.272 | 4.179 | 2.306 | 2.748 | 3.611 | 4.611 |
| 40 | 1.959 | 2.334 | 3.066 | 3.917 | 2.052 | 2.445 | 3.213 | 4.104 | 2.247 | 2.677 | 3.518 | 4.493 |
| 45 | 1.935 | 2.306 | 3.030 | 3.871 | 2.021 | 2.408 | 3.165 | 4.042 | 2.200 | 2.621 | 3.444 | 4.399 |
| 50 | 1.916 | 2.284 | 3.001 | 3.833 | 1.996 | 2.379 | 3.126 | 3.993 | 2.162 | 2.576 | 3.385 | 4.323 |

## Values of k for One-Sided Limits

| n | γ = 0.90 p = 0.90 | p = 0.95 | p = 0.99 | p = 0.999 | γ = 0.95 p = 0.90 | p = 0.95 | p = 0.99 | p = 0.999 | γ = 0.99 p = 0.90 | p = 0.95 | p = 0.99 | p = 0.999 |
|---|---|---|---|---|---|---|---|---|---|---|---|---|
| 3 | 4.258 | 5.310 | 7.340 | 9.651 | 6.158 | 7.655 | 10.552 | 13.857 | – | – | – | – |
| 4 | 3.187 | 3.957 | 5.437 | 7.128 | 4.163 | 5.145 | 7.042 | 9.215 | – | – | – | – |
| 5 | 2.742 | 3.400 | 4.666 | 6.112 | 3.407 | 4.202 | 5.741 | 7.501 | – | – | – | – |
| 6 | 2.494 | 3.091 | 4.242 | 5.556 | 3.006 | 3.707 | 50.62 | 6.612 | 4.408 | 5.409 | 7.334 | 9.540 |
| 7 | 2.333 | 2.894 | 3.972 | 5.201 | 2.755 | 3.399 | 4.641 | 6.061 | 3.856 | 4.730 | 6.411 | 8.348 |
| 8 | 2.219 | 2.755 | 3.783 | 4.955 | 2.582 | 3.188 | 4.353 | 5.686 | 3.496 | 4.287 | 5.811 | 7.566 |
| 9 | 2.133 | 2.649 | 3.641 | 4.772 | 2.454 | 3.031 | 4.143 | 5.414 | 3.242 | 3.971 | 5.389 | 7.014 |
| 10 | 2.065 | 2.568 | 3.532 | 4.629 | 2.355 | 2.911 | 3.981 | 5.203 | 3.048 | 3.739 | 5.075 | 6.603 |
| 11 | 2.012 | 2.503 | 3.444 | 4.515 | 2.275 | 2.815 | 3.852 | 5.036 | 2.897 | 3.557 | 4.828 | 6.284 |
| 12 | 1.966 | 2.448 | 3.371 | 4.420 | 2.210 | 2.736 | 3.747 | 4.900 | 2.773 | 3.410 | 4.633 | 6.032 |
| 13 | 1.928 | 2.403 | 3.310 | 4.341 | 2.155 | 2.670 | 3.659 | 4.787 | 2.677 | 3.290 | 4.472 | 5.826 |
| 14 | 1.895 | 2.363 | 3.257 | 4.274 | 2.108 | 2.614 | 3.585 | 4.690 | 2.592 | 3.189 | 4.336 | 5.651 |
| 15 | 1.866 | 2.329 | 3.212 | 4.215 | 2.068 | 2.566 | 3.520 | 4.607 | 2.521 | 3.102 | 4.224 | 5.507 |
| 16 | 1.842 | 2.299 | 3.172 | 4.146 | 2.032 | 2.523 | 3.463 | 4.534 | 2.458 | 3.028 | 4.124 | 5.374 |
| 17 | 1.820 | 2.272 | 3.136 | 4.118 | 2.001 | 2.468 | 3.415 | 4.471 | 2.405 | 2.962 | 4.038 | 5.268 |
| 18 | 1.800 | 2.249 | 3.106 | 4.078 | 1.974 | 2.453 | 3.370 | 4.415 | 2.357 | 2.906 | 3.961 | 5.167 |
| 19 | 1.781 | 2.228 | 3.078 | 4.041 | 1.949 | 2.423 | 3.331 | 4.364 | 2.315 | 2.855 | 3.893 | 5.078 |
| 20 | 1.765 | 2.208 | 3.052 | 4.009 | 1.926 | 2.396 | 3.295 | 4.319 | 2.275 | 2.807 | 3.832 | 5.003 |
| 21 | 1.750 | 2.190 | 3.028 | 3.979 | 1.905 | 2.371 | 3.262 | 4.276 | 2.241 | 2.768 | 3.776 | 4.932 |
| 22 | 1.736 | 2.174 | 3.007 | 3.952 | 1.887 | 2.350 | 3.233 | 4.238 | 2.208 | 2.729 | 3.727 | 4.866 |
| 23 | 1.724 | 2.159 | 2.987 | 3.927 | 1.869 | 2.329 | 3.206 | 4.204 | 2.179 | 2.693 | 3.680 | 4.806 |
| 24 | 1.712 | 2.145 | 2.969 | 3.904 | 1.853 | 2.309 | 3.181 | 4.171 | 2.154 | 2.663 | 3.638 | 4.755 |
| 25 | 1.702 | 2.132 | 2.952 | 3.882 | 1.838 | 2.292 | 3.158 | 4.143 | 2.129 | 2.632 | 3.601 | 4.706 |
| 30 | 1.657 | 2.080 | 2.884 | 3.794 | 1.778 | 2.220 | 3.064 | 4.022 | 2.029 | 2.516 | 3.446 | 4.508 |
| 35 | 1.623 | 2.041 | 2.833 | 3.730 | 1.732 | 2.166 | 2.994 | 3.934 | 1.957 | 2.431 | 3.334 | 4.364 |
| 40 | 1.598 | 2.010 | 2.793 | 3.679 | 1.697 | 2.126 | 2.941 | 3.866 | 1.902 | 2.365 | 3.250 | 4.255 |
| 45 | 1.577 | 1.986 | 2.762 | 3.638 | 1.669 | 2.092 | 2.897 | 3.811 | 1.857 | 2.313 | 3.181 | 4.168 |
| 50 | 1.560 | 1.965 | 2.735 | 3.604 | 1.646 | 2.065 | 2.963 | 3.766 | 1.821 | 2.296 | 3.124 | 4.096 |

## Proportion of Population Covered with $\gamma$ % Confidence and Sample Size $n$

| $n$ | $\gamma = 0.90$ | $\gamma = 0.95$ | $\gamma = 0.99$ | $\gamma = 0.995$ |
|---|---|---|---|---|
| 2 | 0.052 | 0.026 | 0.006 | 0.003 |
| 4 | 0.321 | 0.249 | 0.141 | 0.111 |
| 6 | 0.490 | 0.419 | 0.295 | 0.254 |
| 10 | 0.664 | 0.606 | 0.496 | 0.456 |
| 20 | 0.820 | 0.784 | 0.712 | 0.683 |
| 40 | 0.907 | 0.887 | 0.846 | 0.829 |
| 60 | 0.937 | 0.924 | 0.895 | 0.883 |
| 80 | 0.953 | 0.943 | 0.920 | 0.911 |
| 100 | 0.962 | 0.954 | 0.936 | 0.929 |
| 150 | 0.975 | 0.969 | 0.957 | 0.952 |
| 200 | 0.981 | 0.977 | 0.968 | 0.961 |
| 500 | 0.993 | 0.991 | 0.987 | 0.986 |
| 1000 | 0.997 | 0.996 | 0.994 | 0.993 |

## Sample Size Required to Cover $(1 - \alpha)$% of the Population with $\gamma$ % Confidence

| $\alpha$ | $\gamma = 0.90$ | $\gamma = 0.95$ | $\gamma = 0.99$ | $\gamma = 0.995$ |
|---|---|---|---|---|
| 0.005 | 777 | 947 | 1325 | 1483 |
| 0.01 | 388 | 473 | 662 | 740 |
| 0.05 | 77 | 93 | 130 | 146 |
| 0.01 | 38 | 46 | 64 | 72 |
| 0.15 | 25 | 30 | 42 | 47 |
| 0.20 | 18 | 22 | 31 | 34 |
| 0.25 | 15 | 18 | 24 | 27 |
| 0.30 | 12 | 14 | 20 | 22 |
| 0.40 | 6 | 10 | 14 | 16 |
| 0.50 | 7 | 8 | 11 | 12 |

# APPENDIX 9

# Control Chart Constants

| | Chart for Averages | | | Chart for Standard Deviations | | | | | |
|---|---|---|---|---|---|---|---|---|---|
| | Factors for Control Limits | | | Factors for Central Line | | Factors for Control Limits | | | |
| Observations in Sample, $n$ | $A$ | $A_2$ | $A_3$ | $c_4$ | $1/c_4$ | $B_3$ | $B_4$ | $B_5$ | $B_6$ |
| 2 | 2.121 | 1.880 | 2.659 | 0.7979 | 1.2533 | 0 | 3.267 | 0 | 2.606 |
| 3 | 1.732 | 1.023 | 1.954 | 0.8862 | 1.1284 | 0 | 2.568 | 0 | 2.276 |
| 4 | 1.500 | 0.729 | 1.628 | 0.9213 | 1.0854 | 0 | 2.266 | 0 | 2.088 |
| 5 | 1.342 | 0.577 | 1.427 | 0.9400 | 1.0638 | 0 | 2.089 | 0 | 1.964 |
| 6 | 1.225 | 0.483 | 1.287 | 0.9515 | 1.0510 | 0.030 | 1.970 | 0.029 | 1.874 |
| 7 | 1.134 | 0.419 | 1.182 | 0.9594 | 1.0423 | 0.118 | 1.882 | 0.113 | 1.806 |
| 8 | 1.061 | 0.373 | 1.099 | 0.9650 | 1.0363 | 0.185 | 1.815 | 0.179 | 1.751 |
| 9 | 1.000 | 0.337 | 1.032 | 0.9693 | 1.0317 | 0.239 | 1.761 | 0.232 | 1.707 |
| 10 | 0.949 | 0.308 | 0.975 | 0.9727 | 1.0281 | 0.284 | 1.716 | 0.276 | 1.669 |
| 11 | 0.905 | 0.285 | 0.927 | 0.9754 | 1.0252 | 0.321 | 1.679 | 0.313 | 1.637 |
| 12 | 0.866 | 0.266 | 0.886 | 0.9776 | 1.0229 | 0.354 | 1.646 | 0.346 | 1.610 |
| 13 | 0.832 | 0.249 | 0.850 | 0.9794 | 1.0210 | 0.382 | 1.618 | 0.374 | 1.585 |
| 14 | 0.802 | 0.235 | 0.817 | 0.9810 | 1.0194 | 0.406 | 1.594 | 0.399 | 1.563 |
| 15 | 0.775 | 0.223 | 0.789 | 0.9823 | 1.0180 | 0.428 | 1.572 | 0.421 | 1.544 |
| 16 | 0.750 | 0.212 | 0.763 | 0.9835 | 1.0168 | 0.448 | 1.552 | 0.440 | 1.526 |
| 17 | 0.728 | 0.203 | 0.739 | 0.9845 | 1.0157 | 0.466 | 1.534 | 0.458 | 1.511 |
| 18 | 0.707 | 0.194 | 0.718 | 0.9854 | 1.0148 | 0.482 | 1.518 | 0.475 | 1.496 |
| 19 | 0.688 | 0.187 | 0.698 | 0.9862 | 1.0140 | 0.497 | 1.503 | 0.490 | 1.483 |
| 20 | 0.671 | 0.180 | 0.680 | 0.9869 | 1.0133 | 0.510 | 1.490 | 0.504 | 1.470 |
| 21 | 0.655 | 0.173 | 0.663 | 0.9876 | 1.0126 | 0.523 | 1.477 | 0.516 | 1.459 |
| 22 | 0.640 | 0.167 | 0.647 | 0.9882 | 1.0119 | 0.534 | 1.466 | 0.528 | 1.448 |
| 23 | 0.626 | 0.162 | 0.633 | 0.9887 | 1.0114 | 0.545 | 1.455 | 0.539 | 1.438 |
| 24 | 0.612 | 0.157 | 0.619 | 0.9892 | 1.0109 | 0.555 | 1.445 | 0.549 | 1.429 |
| 25 | 0.600 | 0.153 | 0.606 | 0.9896 | 1.0105 | 0.565 | 1.435 | 0.559 | 1.420 |

| Observations in Sample, $n$ | Chart for Ranges ||||||| $\bar{x}$ Charts |
| | Factors for Central Line ||| Factors for Control Limits |||| |
| | $d_2$ | $1/d_2$ | $d_3$ | $D_1$ | $D_2$ | $D_3$ | $D_4$ | $E_2$ |
|---|---|---|---|---|---|---|---|---|
| 2 | 1.128 | 0.8865 | 0.853 | 0 | 3.686 | 0 | 3.267 | 2.660 |
| 3 | 1.693 | 0.5907 | 0.888 | 0 | 4.358 | 0 | 2.574 | 1.772 |
| 4 | 2.059 | 0.4857 | 0.880 | 0 | 4.698 | 0 | 2.282 | 1.457 |
| 5 | 2.326 | 0.4299 | 0.864 | 0 | 4.918 | 0 | 2.114 | 1.290 |
| 6 | 2.534 | 0.3946 | 0.848 | 0 | 5.078 | 0 | 2.004 | 1.184 |
| 7 | 2.704 | 0.3698 | 0.833 | 0.204 | 5.204 | 0.076 | 1.924 | 1.109 |
| 8 | 2.847 | 0.3512 | 0.820 | 0.388 | 5.306 | 0.136 | 1.864 | 1.054 |
| 9 | 2.970 | 0.3367 | 0.808 | 0.547 | 5.393 | 0.184 | 1.816 | 1.010 |
| 10 | 3.078 | 0.3249 | 0.797 | 0.687 | 5.469 | 0.223 | 1.777 | 0.975 |
| 11 | 3.173 | 0.3152 | 0.787 | 0.811 | 5.535 | 0.256 | 1.744 | 0.945 |
| 12 | 3.258 | 0.3069 | 0.778 | 0.922 | 5.594 | 0.283 | 1.717 | 0.921 |
| 13 | 3.336 | 0.2998 | 0.770 | 1.025 | 5.647 | 0.307 | 1.693 | 0.899 |
| 14 | 3.407 | 0.2935 | 0.763 | 1.118 | 5.696 | 0.328 | 1.672 | 0.881 |
| 15 | 3.472 | 0.2880 | 0.756 | 1.203 | 5.741 | 0.347 | 1.653 | 0.864 |
| 16 | 3.532 | 0.2831 | 0.750 | 1.282 | 5.782 | 0.363 | 1.637 | 0.849 |
| 17 | 3.588 | 0.2787 | 0.744 | 1.356 | 5.820 | 0.378 | 1.622 | 0.836 |
| 18 | 3.640 | 0.2747 | 0.739 | 1.424 | 5.856 | 0.391 | 1.608 | 0.824 |
| 19 | 3.689 | 0.2711 | 0.734 | 1.487 | 5.891 | 0.403 | 1.597 | 0.813 |
| 20 | 3.735 | 0.2677 | 0.729 | 1.549 | 5.921 | 0.415 | 1.585 | 0.803 |
| 21 | 3.778 | 0.2647 | 0.724 | 1.605 | 5.951 | 0.425 | 1.575 | 0.794 |
| 22 | 3.819 | 0.2618 | 0.720 | 1.659 | 5.979 | 0.434 | 1.566 | 0.786 |
| 23 | 3.858 | 0.2592 | 0.716 | 1.710 | 6.006 | 0.443 | 1.557 | 0.778 |
| 24 | 3.895 | 0.2567 | 0.712 | 1.759 | 6.031 | 0.451 | 1.548 | 0.770 |
| 25 | 3.931 | 0.2544 | 0.708 | 1.806 | 6.056 | 0.459 | 1.541 | 0.763 |

# APPENDIX 10

# Control Chart Equations

| | **np Chart** | **p Chart** |
|---|---|---|
| **LCL** | $LCL = n\bar{p} - 3\sqrt{n\bar{p}\left(1 - \frac{n\bar{p}}{n}\right)}$<br>or 0 if LCL is negative | $LCL = \bar{p} - 3\sqrt{\frac{\bar{p}(1-\bar{p})}{n}}$<br>or 0 if LCL is negative |
| **Center Line** | $n\bar{p} = \frac{\text{Sum of items with problems}}{\text{Number of subgroups}}$ | $\bar{p} = \frac{\text{Sum of items with problems}}{\text{Number of items in all subgroups}}$ |
| **UCL** | $UCL = n\bar{p} + 3\sqrt{n\bar{p}\left(1 - \frac{n\bar{p}}{n}\right)}$<br>or n if UCL is greater than n | $UCL = \bar{p} + 3\sqrt{\frac{\bar{p}(1-\bar{p})}{n}}$<br>or 1 if UCL is greater than 1 |

| | **c Chart** | **u Chart** |
|---|---|---|
| **LCL** | $LCL = \bar{c} - 3\sqrt{\bar{c}}$ or 0 if LCL is negative | $LCL = \bar{u} - 3\sqrt{\frac{\bar{u}}{n}}$ or 0 if LCL is negative |
| **Center Line** | $\bar{c} = \frac{\text{Sum of problems}}{\text{Number of subgroups}}$ | $\bar{u} = \frac{\text{Sum of problems}}{\text{Number of units in all subgroups}}$ |
| **UCL** | $UCL = \bar{c} + 3\sqrt{\bar{c}}$ | $UCL = \bar{u} + 3\sqrt{\frac{\bar{u}}{n}}$ |

|  | x Chart | $\bar{X}$ Chart |
|---|---|---|
| LCL | LCL = $\bar{X}$ − 2.66($M\bar{R}$) | LCL = $\bar{\bar{X}}$ − $A_2\bar{R}$ |
| Center Line | $\bar{X} = \dfrac{\text{Sum of measurements}}{\text{Number of measurements}}$ | $\bar{X} = \dfrac{\text{Sum of subgroup averages}}{\text{Number of averages}}$ |
| UCL | LCL = $\bar{X}$ + 2.66($M\bar{R}$) | UCL = $\bar{\bar{X}}$ + $A_2\bar{R}$ |

|  | R Chart |
|---|---|
| LCL | LCL = $D_3\bar{R}$ |
| Center Line | $\bar{R} = \dfrac{\text{Sum of ranges}}{\text{Number of ranges}}$ |
| UCL | UCL = $D_4\bar{R}$ |

# APPENDIX 11

# Table of $d_2^*$ Values

|   | \multicolumn{7}{c}{$m$ = # measurements used to calculate range value} |
|---|---|---|---|---|---|---|---|
| | 2 | 3 | 4 | 5 | 6 | 7 | 8 |
| 1 | 1.41 | 1.91 | 2.24 | 2.48 | 2.67 | 2.83 | 2.96 |
| 2 | 1.28 | 1.81 | 2.15 | 2.40 | 2.60 | 2.77 | 2.91 |
| 3 | 1.23 | 1.77 | 2.12 | 2.38 | 2.58 | 2.75 | 2.89 |
| 4 | 1.21 | 1.75 | 2.11 | 2.37 | 2.57 | 2.74 | 2.88 |
| 5 | 1.19 | 1.74 | 2.10 | 2.36 | 2.56 | 2.73 | 2.87 |
| 6 | 1.18 | 1.73 | 2.09 | 2.35 | 2.56 | 2.73 | 2.87 |
| 7 | 1.17 | 1.73 | 2.09 | 2.35 | 2.55 | 2.72 | 2.87 |
| 8 | 1.17 | 1.72 | 2.08 | 2.35 | 2.55 | 2.72 | 2.87 |
| 9 | 1.16 | 1.72 | 2.08 | 2.34 | 2.55 | 2.72 | 2.86 |
| 10 | 1.16 | 1.72 | 2.08 | 2.34 | 2.55 | 2.72 | 2.86 |
| 11 | 1.16 | 1.71 | 2.08 | 2.34 | 2.55 | 2.72 | 2.86 |
| 12 | 1.15 | 1.71 | 2.07 | 2.34 | 2.55 | 2.72 | 2.85 |
| 13 | 1.15 | 1.71 | 2.07 | 2.34 | 2.55 | 2.71 | 2.85 |
| 14 | 1.15 | 1.71 | 2.07 | 2.34 | 2.54 | 2.71 | 2.85 |
| 15 | 1.15 | 1.71 | 2.07 | 2.34 | 2.54 | 2.71 | 2.85 |
| >15 | 1.128 | 1.693 | 2.059 | 2.326 | 2.534 | 2.704 | 2.847 |

$g$ = # range estimates

|  | | \multicolumn{7}{c}{$m$ = Repeat Readings Taken} |
|---|---|---|---|---|---|---|---|---|
|  | | 9 | 10 | 11 | 12 | 13 | 14 | 15 |
| | 1 | 3.08 | 3.18 | 3.27 | 3.35 | 3.42 | 3.49 | 3.55 |
| | 2 | 3.02 | 3.13 | 3.22 | 3.30 | 3.38 | 3.45 | 3.51 |
| | 3 | 3.01 | 3.11 | 3.21 | 3.29 | 3.37 | 3.43 | 3.50 |
| | 4 | 3.00 | 3.10 | 3.20 | 3.28 | 3.36 | 3.43 | 3.49 |
| | 5 | 2.99 | 3.10 | 3.19 | 3.28 | 3.35 | 3.42 | 3.49 |
| $g$ = # parts × # inspectors | 6 | 2.99 | 3.10 | 3.19 | 3.27 | 3.35 | 3.42 | 3.49 |
| | 7 | 2.99 | 3.10 | 3.19 | 3.27 | 3.35 | 3.42 | 3.48 |
| | 8 | 2.98 | 3.09 | 3.19 | 3.27 | 3.35 | 3.42 | 3.48 |
| | 9 | 2.98 | 3.09 | 3.18 | 3.27 | 3.35 | 3.42 | 3.48 |
| | 10 | 2.98 | 3.09 | 3.18 | 3.27 | 3.34 | 3.42 | 3.48 |
| | 11 | 2.98 | 3.09 | 3.18 | 3.27 | 3.34 | 3.41 | 3.48 |
| | 12 | 2.98 | 3.09 | 3.18 | 3.27 | 3.34 | 3.41 | 3.48 |
| | 13 | 2.98 | 3.09 | 3.18 | 3.27 | 3.34 | 3.41 | 3.48 |
| | 14 | 2.98 | 3.08 | 3.18 | 3.27 | 3.34 | 3.41 | 3.48 |
| | 15 | 2.98 | 3.08 | 3.18 | 3.26 | 3.34 | 3.41 | 3.48 |
| | >15 | 2.970 | 3.078 | 3.173 | 3.258 | 3.336 | 3.407 | 3.472 |

# APPENDIX 12

# Factors for Short Run Control Charts for Individuals, $\bar{X}$, and $R$ Charts

|   | \multicolumn{4}{c}{1 (R Based on Moving Range of 2)} | \multicolumn{4}{c}{Subgroup Size 2} | \multicolumn{4}{c}{3} |
|---|---|---|---|---|---|---|---|---|---|---|---|---|
| g | $A_{2F}$ | $D_{4F}$ | $A_{2S}$ | $D_{4S}$ | $A_{2F}$ | $D_{4F}$ | $A_{2S}$ | $D_{4S}$ | $A_{2F}$ | $D_{4F}$ | $A_{2S}$ | $D_{4S}$ |
| 1 | NA | NA | 236.5 | 128 | NA | NA | 167 | 128 | NA | NA | 8.21 | 14 |
| 2 | 12.0 | 2.0 | 20.8 | 16.0 | 8.49 | 2.0 | 15.70 | 15.6 | 1.57 | 1.9 | 2.72 | 7.1 |
| 3 | 6.8 | 2.7 | 9.6 | 15.0 | 4.78 | 2.7 | 6.76 | 14.7 | 1.35 | 2.3 | 1.90 | 4.5 |
| 4 | 5.1 | 3.3 | 6.6 | 8.1 | 3.62 | 3.3 | 4.68 | 8.1 | 1.26 | 2.4 | 1.62 | 3.7 |
| 5 | 4.4 | 3.3 | 5.4 | 6.3 | 3.12 | 3.3 | 3.82 | 6.3 | 1.20 | 2.4 | 1.47 | 3.4 |
| 6 | 4.0 | 3.3 | 4.7 | 5.4 | 2.83 | 3.3 | 3.34 | 5.4 | 1.17 | 2.5 | 1.39 | 3.3 |
| 7 | 3.7 | 3.3 | 4.3 | 5.0 | 2.65 | 3.3 | 3.06 | 5.0 | 1.14 | 2.5 | 1.32 | 3.2 |
| 8 | 3.6 | 3.3 | 4.1 | 4.7 | 2.53 | 3.3 | 2.87 | 4.7 | 1.13 | 2.5 | 1.28 | 3.1 |
| 9 | 3.5 | 3.3 | 3.9 | 4.5 | 2.45 | 3.3 | 2.74 | 4.5 | 1.12 | 2.5 | 1.25 | 3.0 |
| 10 | 3.3 | 3.3 | 3.7 | 4.5 | 2.37 | 3.3 | 2.62 | 4.5 | 1.10 | 2.5 | 1.22 | 3.0 |
| 15 | 3.1 | 3.5 | 3.3 | 4.1 | 2.18 | 3.5 | 2.33 | 4.1 | 1.08 | 2.5 | 1.15 | 2.9 |
| 20 | 3.0 | 3.5 | 3.1 | 4.0 | 2.11 | 3.5 | 2.21 | 4.0 | 1.07 | 2.6 | 1.12 | 2.8 |
| 25 | 2.9 | 3.5 | 3.0 | 3.8 | 2.05 | 3.5 | 2.14 | 3.8 | 1.06 | 2.6 | 1.10 | 2.7 |

*Numbers enclosed in bold boxes represent the recommended minimum number of subgroups for starting a control chart.*

|   | Subgroup Size ||||||||
|---|---|---|---|---|---|---|---|---|
|   | 4 |||| 5 ||||
| g | $A_{2F}$ | $D_{4F}$ | $A_{2s}$ | $D_{4S}$ | $A_{2F}$ | $D_{4F}$ | $A_{2s}$ | $D_{4S}$ |
| 1 | NA | NA | 3.05 | 13 | NA | NA | 1.8 | 5.1 |
| 2 | 0.83 | 1.9 | 1.44 | 3.5 | 0.58 | 1.7 | 1.0 | 3.2 |
| 3 | 0.81 | 1.9 | 1.14 | 3.2 | 0.59 | 1.8 | 0.83 | 2.8 |
| 4 | 0.79 | 2.1 | 1.01 | 2.9 | 0.59 | 1.9 | 0.76 | 2.6 |
| 5 | 0.78 | 2.1 | 0.95 | 2.8 | 0.59 | 2.0 | 0.72 | 2.5 |
| 6 | 0.77 | 2.2 | 0.91 | 2.7 | **0.59** | **2.0** | 0.70 | 2.4 |
| 7 | **0.76** | **2.2** | 0.88 | 2.6 | 0.59 | 2.0 | 0.68 | 2.4 |
| 8 | 0.76 | 2.2 | 0.86 | 2.6 | 0.59 | 2.0 | 0.66 | 2.3 |
| 9 | 0.76 | 2.2 | 0.85 | 2.5 | 0.59 | 2.0 | 0.65 | 2.3 |
| 10 | 0.75 | 2.2 | 0.83 | 2.5 | 0.58 | 2.0 | 0.65 | 2.3 |
| 15 | 0.75 | 2.3 | 0.80 | 2.4 | 0.58 | 2.1 | 0.62 | 2.2 |
| 20 | 0.74 | 2.3 | 0.78 | 2.4 | 0.58 | 2.1 | 0.61 | 2.2 |
| 25 | 0.74 | 2.3 | 0.77 | 2.4 | 0.58 | 2.1 | 0.60 | 2.2 |

*Numbers enclosed in bold boxes represent the recommended minimum number of subgroups for starting a control chart.*

# APPENDIX 13

# Sample Customer Survey

Taken from *How did we do?*, a patient satisfaction survey for the XXX Community Hospital.

| For each of the following statements, please check the appropriate box. Mark the NA box if you had no opportunity to judge that aspect of care during your stay at XXX Community Hospital. | Strongly agree | Agree | Neither agree nor disagree | Disagree | Strongly disagree | NA |
|---|---|---|---|---|---|---|
| I received my medication on time | | | | | | |
| The menu offered foods I liked | | | | | | |
| My doctor kept me informed | | | | | | |
| My room was clean | | | | | | |
| The discharge process was smooth | | | | | | |
| My doctor was available | | | | | | |
| The hospital was well supplied | | | | | | |
| I received the foods I selected from the menu | | | | | | |
| The staff answered my call light quickly | | | | | | |
| The food looked good | | | | | | |
| I was informed of what I should do after discharge | | | | | | |
| My bed was comfortable | | | | | | |
| The hospital staff took good care of me | | | | | | |
| I knew my doctor's name | | | | | | |
| The staff treated one another with respect | | | | | | |
| The hospital was well maintained | | | | | | |
| The food tasted good | | | | | | |
| My medications were ready when I was ready to go | | | | | | |
| The billing procedures were explained to me | | | | | | |
| I was served the right amount of food | | | | | | |
| The nurse checked on me frequently | | | | | | |
| I had assistance making plans to leave the hospital | | | | | | |
| My doctor told me when I was going home | | | | | | |
| The food servers were pleasant | | | | | | |

| For each of the following statements, please check the appropriate box.<br>Mark the NA box if you had no opportunity to judge that aspect of care during your stay at XXX Community Hospital. | Strongly agree | Agree | Neither agree nor disagree | Disagree | Strongly disagree | NA |
|---|---|---|---|---|---|---|
| The hospital was clean | | | | | | |
| Overall, the hospital staff treated me with respect | | | | | | |
| My room was quiet | | | | | | |
| The staff met my special needs | | | | | | |
| The attitude of the staff was nice | | | | | | |
| I was escorted out of the hospital at discharge | | | | | | |
| My room was comfortable | | | | | | |
| My diet was what the doctor ordered | | | | | | |
| The staff kept me informed about my care | | | | | | |
| I was satisfied with my doctor(s) | | | | | | |
| Meals were served on time | | | | | | |
| The staff were helpful | | | | | | |
| The discharge process was speedy | | | | | | |
| My doctor knew who I was | | | | | | |
| My medications/wound care/equipment were explained to me | | | | | | |
| I was treated well | | | | | | |
| I was prepared to go home | | | | | | |
| The staff were attentive to my needs | | | | | | |
| I had the same doctor(s) throughout my hospitalization | | | | | | |
| The nurses acted in a professional manner | | | | | | |
| The staff knew what care I needed | | | | | | |
| I would refer a family member to XXX Community Hospital | | | | | | |
| I would choose to come back to XXX Community Hospital | | | | | | |

Were there any incidents you remember from your stay that were especially PLEASANT?

Were there any incidents you remember from your stay that were especially UNPLEASANT?

We welcome any other suggestions you have to offer.

Thank you for your assistance!

# APPENDIX 14

# Process σ Levels and Equivalent PPM Quality Levels

Based on the assumption that in the long term the process could drift by plus or minus 1.5σ.

| Process σ Level | Process PPM |
|---|---|
| 6.27 | 1 |
| 6.12 | 2 |
| 6.0 | 3.4 |
| 5.97 | 4 |
| 5.91 | 5 |
| 5.88 | 6 |
| 5.84 | 7 |
| 5.82 | 8 |
| 5.78 | 9 |
| 5.77 | 10 |
| 5.61 | 20 |
| 5.51 | 30 |
| 5.44 | 40 |
| 5.39 | 50 |
| 5.35 | 60 |
| 5.31 | 70 |
| 5.27 | 80 |

| Process σ Level | Process PPM |
|---|---|
| 5.25 | 90 |
| 5.22 | 100 |
| 5.04 | 200 |
| 4.93 | 300 |
| 4.85 | 400 |
| 4.79 | 500 |
| 4.74 | 600 |
| 4.69 | 700 |
| 4.66 | 800 |
| 4.62 | 900 |
| 4.59 | 1,000 |
| 4.38 | 2,000 |
| 4.25 | 3,000 |
| 4.15 | 4,000 |
| 4.08 | 5,000 |
| 4.01 | 6,000 |
| 3.96 | 7,000 |

| Process σ Level | Process PPM |
|---|---|
| 3.91 | 8,000 |
| 3.87 | 9,000 |
| 3.83 | 10,000 |
| 3.55 | 20,000 |
| 3.38 | 30,000 |
| 3.25 | 40,000 |
| 3.14 | 50,000 |
| 3.05 | 60,000 |
| 2.98 | 70,000 |
| 2.91 | 80,000 |
| 2.84 | 90,000 |
| 2.78 | 100,000 |
| 2.34 | 200,000 |
| 2.02 | 300,000 |
| 1.75 | 400,000 |
| 1.50 | 500,000 |

# APPENDIX 15

# Black Belt Effectiveness Certification

**Black Belt Certification Recommendation**

Name _____ (as it will appear on the certificate)
Address _____
City _____ State _____ Zip _____
Social Security Number _____
We the undersigned, on behalf of _____, the Six Sigma organization, certify the above named individual as a Six Sigma Black Belt within [COMPANY].

| Printed or typed Board member name | Signature | Date Signed |
|---|---|---|
|  |  |  |
|  |  |  |
|  |  |  |
|  |  |  |
|  |  |  |

## [COMPANY] Black Belt Skill Set Certification Process

### *Introduction*

This document describes the process and provides the minimum acceptable criteria for certifying an individual as a [COMPANY] Six Sigma Black Belt. [COMPANY] certification involves recognition by the [COMPANY] and his or her peers, and should not be construed as a professional license.

## Process

The [COMPANY] determines recognition as a [COMPANY] Six Sigma Black Belt. [COMPANY] certification requires that the applicant pass the [COMPANY]'s Black Belt examination. The exam covers the core skill set of the Black Belt Body of Knowledge (BOK) as defined by the [COMPANY]. The [COMPANY] will score the candidate and determine if his or her score meets the [COMPANY]'s minimum passing score for each section of the BOK, as well as for the overall score. The [COMPANY] also provides criteria for assessing the candidate's effectiveness by evaluating his or her

- Ability to achieve significant, tangible results by applying the Six Sigma approach.
- Ability to lead organizational change as demonstrated by the candidate's leadership, teamwork, project management, and communication skills.

The exam will be administered by the Six Sigma organization. The Six Sigma organization is responsible for assuring the integrity of the exam, verifying the identity of the candidate sitting for the exam, and enforcing time limits. The Six Sigma organization will evaluate the candidate's effectiveness using the [COMPANY] requirements and will notify the [COMPANY] when a candidate who has passed the [COMPANY] BOK exam has met the effectiveness requirements.

## [COMPANY] Black Belt Effectiveness Certification Criteria

This section describes the criteria for certifying that a [COMPANY] Black Belt candidate is "effective" in applying the Six Sigma approach. Effectiveness means that the candidate has demonstrated the ability to lead the change process by successfully applying Six Sigma methodologies on more than one significant project. Success is demonstrated by achieving documented substantial, sustained, and tangible results. Examples of results are cost savings or cost avoidance validated by finance and accounting experts, improved customer satisfaction, reduced cycle time, increased revenues and profits, reduced accident rates, improved morale, reduction of critical to customer defects, etc. Merely demonstrating the use of Six Sigma tools is not sufficient. Nor is the delivery of intermediate "products" such as Pareto diagrams or process maps.

In addition on passing the [COMPANY] BOK exam, certification requires the following:

1. Acceptable completion of a Black Belt training curriculum approved by the Six Sigma organization.

2. Demonstration of clear and rational thought process.
   a. Ability to analyze a problem following a logical sequence.
   b. Usage of facts and data to guide decisions and action.
3. Ability to clearly explain Six Sigma and the DMAIC project cycle in layman's terms.
4. Ability to achieve tangible results, for example,
   a. Completed two or more projects which employed the Six Sigma approach (DMAIC or equivalent).
      i. Projects reviewed by appropriate personnel.
      ii. Deliverables accepted by the project sponsor.
      iii. Projects documented in the manner prescribed by the Six Sigma organization.
      iv. Projects used the Six Sigma approach and correctly employed a significant subset of basic, intermediate, and advanced Six Sigma tools and techniques (see later in this appendix for a listing).
   b. Ability to perform benefit/cost analysis.
   c. Ability to quantify deliverables in terms meaningful to the organization, for example, cost, quality, cycle time, safety improvement, etc.
   d. Ability to identify and overcome obstacles to progress.
   e. Ability to work within time, budget, and operational constraints.
5. Demonstrated ability to explain the tools of Six Sigma to others.
6. Demonstrated interpersonal and leadership skills necessary to be an effective change agent within the organization.

## [COMPANY] Black Belt Certification Board

The [COMPANY] recommends that each area of effectiveness be rated by at least two qualified individuals. Table A15.1 provides guidelines for identifying members of the [COMPANY] Black Belt Certification Board.

**Table A15.1** [COMPANY] Black Belt Certification Board Member Selection Guide

| Assessment Subject Area | Board Member |
| --- | --- |
| Change agent skills | Supervisor, project sponsor(s), Six Sigma champion, mentor, process owner, Green Belt |
| Application of tools and techniques | Black Belt instructor, Master Black Belt, [COMPANY] Certified Master Black Belt consultant |
| Ability to achieve results | Project sponsor, process owner, team members, Green Belt, Six Sigma champion, [COMPANY] Certified Master Black Belt consultant |

## Effectiveness Questionnaire

The [COMPANY] provides questionnaires to assist [COMPANY] Certification Board members with their assessment. It is strongly recommended that the candidate perform a self-assessment using the [COMPANY]'s questionnaire prior to applying for certification. The candidate should provide the Six Sigma champion with a list of potential members of his or her Certification Board.

The effectiveness questionnaire includes a set of assessment questions for each subject area. The results of the questionnaires can be summarized and used as input into the Six Sigma organization's certification process. A form for this is provided below. The scoring summary sheet summarizes the evaluator's scores by category. Worksheet items scored in the top 3 boxes are considered to be acceptable. Particular attention should be directed to any worksheet item scored in the lower 4 boxes. Since there are 10 choices for each item, any score below 5 indicates that the evaluator disagreed with the survey item. Survey items are worded in such a way that evaluators should agree with them for qualified Black Belt candidates. Disagreement indicates an area for improvement. The scores are, of course, not the only input. The [COMPANY] Certification Board must also consider any other relevant factors before reaching their decision.

The Scoring Summary and Assessment Worksheets may be reproduced as necessary.

## [COMPANY] Black Belt Notebook and Oral Review

[COMPANY] Black Belt candidates should provide Certification Board members with written documentation of their on-the-job applications of the Six Sigma approach. These "notebooks" should include all relevant information, including project charters, demonstrations of tool usage, samples of data used, excerpts of presentations to sponsors or leaders, team member names, project schedules and performance to these schedules, financial and other business results, etc. The notebooks can be distributed to Certification Board members as either soft copies or hard copies, at their discretion.

Even with the best documentation, it is difficult to assess effectiveness properly without providing the candidate the opportunity to present his or her work and respond to questions. Six Sigma organizations should require that [COMPANY] Black Belt candidates deliver an oral presentation to the Certification Board. The oral review will also provide the Certification Board with a firsthand demonstration of the candidate's communication skills.

| **Change Agent Skills**  **Assessment Worksheet** |||||
|---|---|---|---|
| Black Belt Candidate | | Date of Assessment | |
| Certification Board Member | | Role | |

**1.** The candidate effectively identifies and recruits Six Sigma team members.

Strongly Disagree                                                                           Strongly Agree

☐    ☐    ☐    ☐    ☐    ☐    ☐    ☐    ☐    ☐

**2.** The candidate effectively develops Six Sigma team dynamics and motivates participants.

Strongly Disagree                                                                           Strongly Agree

☐    ☐    ☐    ☐    ☐    ☐    ☐    ☐    ☐    ☐

**3.** The candidate is able to apply conflict resolution techniques.

Strongly Disagree                                                                           Strongly Agree

☐    ☐    ☐    ☐    ☐    ☐    ☐    ☐    ☐    ☐

**4.** The candidate is able to overcome obstacles to change.

Strongly Disagree                                                                           Strongly Agree

☐    ☐    ☐    ☐    ☐    ☐    ☐    ☐    ☐    ☐

**5.** The candidate utilizes a logical approach to problem solving.

Strongly Disagree                                                                           Strongly Agree

☐    ☐    ☐    ☐    ☐    ☐    ☐    ☐    ☐    ☐

**6.** The candidate effectively facilitates group discussions and meetings.

Strongly Disagree                                                                           Strongly Agree

☐    ☐    ☐    ☐    ☐    ☐    ☐    ☐    ☐    ☐

**7.** The candidate's presentations are well organized and easy to understand.

Strongly Disagree                                                                           Strongly Agree

☐    ☐    ☐    ☐    ☐    ☐    ☐    ☐    ☐    ☐

8. The candidate identifies and mobilizes sponsors for change.

    Strongly Disagree                                              Strongly Agree
    ❏   ❏   ❏   ❏   ❏   ❏   ❏   ❏   ❏   ❏

9. The candidate builds a shared vision of the desired state with champions and sponsors.

    Strongly Disagree                                              Strongly Agree
    ❏   ❏   ❏   ❏   ❏   ❏   ❏   ❏   ❏   ❏

10. The candidate effectively communicates with and obtains support from all levels of management.

    Strongly Disagree                                              Strongly Agree
    ❏   ❏   ❏   ❏   ❏   ❏   ❏   ❏   ❏   ❏

11. The candidate identifies gaps between as-is and desired performance.

    Strongly Disagree                                              Strongly Agree
    ❏   ❏   ❏   ❏   ❏   ❏   ❏   ❏   ❏   ❏

12. The candidate identifies and obtains support from all key stakeholders.

    Strongly Disagree                                              Strongly Agree
    ❏   ❏   ❏   ❏   ❏   ❏   ❏   ❏   ❏   ❏

| Application of Tools and Techniques Assessment Worksheet ||||
|---|---|---|---|
| Black Belt Candidate | | Date of Assessment | |
| Certification Board Member | | Role | |

1. The candidate uses an appropriate mix of basic, intermediate, and advanced Six Sigma tools.[1]

    Strongly Disagree                                              Strongly Agree
    ❏   ❏   ❏   ❏   ❏   ❏   ❏   ❏   ❏   ❏

2. The candidate uses the tools of Six Sigma properly.

    Strongly Disagree                                              Strongly Agree
    ❏   ❏   ❏   ❏   ❏   ❏   ❏   ❏   ❏   ❏

---

[1] See later in this appendix for examples of these tools.

**3.** The candidate applies the correct Six Sigma tools at the proper point in the project.

Strongly Disagree                                               Strongly Agree

❒     ❒     ❒     ❒     ❒     ❒     ❒     ❒     ❒     ❒

**4.** The candidate asks for help with Six Sigma tools when necessary.

Strongly Disagree                                               Strongly Agree

❒     ❒     ❒     ❒     ❒     ❒     ❒     ❒     ❒     ❒

**5.** The candidate has a working knowledge of word processors, spreadsheets, and presentation software.

Strongly Disagree                                               Strongly Agree

❒     ❒     ❒     ❒     ❒     ❒     ❒     ❒     ❒     ❒

**6.** The candidate has a working knowledge of a full-featured statistical software package.

Strongly Disagree                                               Strongly Agree

❒     ❒     ❒     ❒     ❒     ❒     ❒     ❒     ❒     ❒

**7.** The candidate understands the limitations as well as the strengths of quantitative methods.

Strongly Disagree                                               Strongly Agree

❒     ❒     ❒     ❒     ❒     ❒     ❒     ❒     ❒     ❒

| \multicolumn{4}{c}{**Ability to Achieve Results Assessment Worksheet**} |||||
|---|---|---|---|
| Black Belt Candidate | | Date of Assessment | |
| Certification Board Member | | Role | |

**1.** The candidate has completed more than one Six Sigma project which produced tangible results.

Strongly Disagree                                               Strongly Agree

❒     ❒     ❒     ❒     ❒     ❒     ❒     ❒     ❒     ❒

**2.** The candidate's projects had an acceptable project charter, including sponsorship, problem statement, business case, etc.

Strongly Disagree                                               Strongly Agree

❒     ❒     ❒     ❒     ❒     ❒     ❒     ❒     ❒     ❒

**3.** The projects employed the Six Sigma approach (DMAIC or equivalent).

Strongly Disagree                                                    Strongly Agree

☐     ☐     ☐     ☐     ☐     ☐     ☐     ☐     ☐     ☐

**4.** The projects' deliverables were clearly defined in tangible terms.

Strongly Disagree                                                    Strongly Agree

☐     ☐     ☐     ☐     ☐     ☐     ☐     ☐     ☐     ☐

**5.** The projects produced significant improvements to an important business process.

Strongly Disagree                                                    Strongly Agree

☐     ☐     ☐     ☐     ☐     ☐     ☐     ☐     ☐     ☐

**6.** The current baseline sigma level was determined using valid data.

Strongly Disagree                                                    Strongly Agree

☐     ☐     ☐     ☐     ☐     ☐     ☐     ☐     ☐     ☐

**7.** The final sigma level was calculated using valid data and showed improvements that were both statistically significant and important to the organization.

Strongly Disagree                                                    Strongly Agree

☐     ☐     ☐     ☐     ☐     ☐     ☐     ☐     ☐     ☐

**8.** An acceptable control plan has been implemented to assure that improvements are maintained.

Strongly Disagree                                                    Strongly Agree

☐     ☐     ☐     ☐     ☐     ☐     ☐     ☐     ☐     ☐

**9.** The projects' financial benefits were validated by experts in accounting or finance.

Strongly Disagree                                                    Strongly Agree

☐     ☐     ☐     ☐     ☐     ☐     ☐     ☐     ☐     ☐

**10.** Key customers were identified and their critical requirements defined.

Strongly Disagree                                                    Strongly Agree

☐     ☐     ☐     ☐     ☐     ☐     ☐     ☐     ☐     ☐

**11.** Project sponsors are satisfied with their project's deliverables.

Strongly Disagree                                                    Strongly Agree

☐     ☐     ☐     ☐     ☐     ☐     ☐     ☐     ☐     ☐

12. Projects identified and corrected root causes, not symptoms.

    Strongly Disagree ❏ ❏ ❏ ❏ ❏ ❏ ❏ ❏ ❏ ❏ Strongly Agree

13. All key stakeholders were kept informed of project status and are aware of final outcomes.

    Strongly Disagree ❏ ❏ ❏ ❏ ❏ ❏ ❏ ❏ ❏ ❏ Strongly Agree

14. Projects were completed on time.

    Strongly Disagree ❏ ❏ ❏ ❏ ❏ ❏ ❏ ❏ ❏ ❏ Strongly Agree

15. Projects were completed within budget.

    Strongly Disagree ❏ ❏ ❏ ❏ ❏ ❏ ❏ ❏ ❏ ❏ Strongly Agree

16. Projects were conducted in a manner that minimized disruptions to normal work.

    Strongly Disagree ❏ ❏ ❏ ❏ ❏ ❏ ❏ ❏ ❏ ❏ Strongly Agree

| Assessment Comments ||
|---|---|
| **Assessment Subject Area** | **Comments** |
| Change agent skills | |
| Applications of tools and techniques | |
| Ability to achieve results | |

## Scoring Summary

| Evaluator | Subject Area | Items Scored 4 or Less | % in Top 3 Boxes | Comment |
|---|---|---|---|---|
| | Change agent skills | | | |
| | Application of tools and techniques | | | |
| | Ability to achieve results | | | |
| | Change agent skills | | | |
| | Application of tools and techniques | | | |
| | Ability to achieve results | | | |
| | Change agent skills | | | |
| | Application of tools and techniques | | | |
| | Ability to achieve results | | | |
| | Change agent skills | | | |
| | Application of tools and techniques | | | |
| | Ability to achieve results | | | |
| | Change agent skills | | | |
| | Application of tools and techniques | | | |
| | Ability to achieve results | | | |

| Examples of Six Sigma Tools and Analytical Concepts |||
|---|---|---|
| **Basic** | **Intermediate** | **Advanced** |
| • DMAIC<br>• SIPOC<br>• DPMO<br>• Computer skills<br>• Scales of measurement<br>• Pareto analysis<br>• Process mapping, flowcharts<br>• Check sheets<br>• Cause-and-effect diagrams<br>• Scatter plots<br>• Run charts<br>• Histograms<br>• Ogives<br>• Descriptive statistics (e.g., mean, standard deviation, skewness)<br>• Enumerative vs. analytic statistics<br>• Stem-and-leaf, boxplots<br>• Basic probability concepts<br>• Discrete probability distributions (binomial, Poisson, hypergeometric)<br>• Continuous probability distributions (normal, exponential, etc.)<br>• 7M tools<br>• FMEA<br>• Sampling<br>• CTx identification | • Control charts for measurements<br>• Control charts for attributes<br>• Process capability<br>• Yield analysis (e.g., first pass yield, rolled throughput yield)<br>• Measurement error analysis (gage R&R)<br>• Correlation analysis<br>• Simple linear regression<br>• Chi-square<br>• Type I and Type II errors<br>• Confidence interval interpretation<br>• Hypothesis tests<br>• Normality assessment and transformations<br>• Z transformations<br>• Process sigma calculations | • Exponentially weighted moving average control charts<br>• Short run SPC<br>• Design and analysis of experiments<br>• ANOVA, MANOVA, and other general linear models<br>• Multiple linear regression<br>• Basic reliability analysis<br>• Design for Six Sigma<br>• Simulation and modeling<br>• Statistical tolerancing<br>• Response surface methods<br>• Robust design concepts<br>• Design, validation, and analysis of customer surveys<br>• Logistic regression |

# APPENDIX 16

# Green Belt Effectiveness Certification

Green Belt Certification Recommendation

Name _____ (as it will appear on the Certificate)

Payroll Number _____

Org Code _____ Date _____

We the undersigned, on behalf of [COMPANY] certify the above named individual as a Six Sigma Green Belt.

| Printed or Type Board Member Name | Signature | Date Signed |
|---|---|---|
|  |  |  |
|  |  |  |
|  |  |  |
|  |  |  |
|  |  |  |
|  |  |  |

## Green Belt Skill Set Certification Process

### Introduction

This document describes the process and provides the minimum criteria for certifying an individual as a Six Sigma Green Belt. Certification involves recognition by [COMPANY], and should not be construed as a professional license.

## Green Belt Effectiveness Certification Criteria

To become a Certified Green Belt, the candidate must demonstrate:

1. Ability to lead organizational change as demonstrated by the candidate's leadership, teamwork, project management, communication, and technical skills.
2. Ability to achieve tangible results that have a significant impact by applying the Six Sigma approach.

This section describes the criteria for certifying that a Green Belt candidate is "effective" in applying the Six Sigma approach. Effectiveness means that the candidate has demonstrated the ability to lead the change process by successfully applying Six Sigma methodologies on a significant project. Success is demonstrated by achieving documented substantial, tangible, and sustained results. Examples of results are cost savings or cost avoidance validated by finance and accounting experts, improved customer satisfaction, reduced cycle time, increased revenues and profits, reduced accident rates, improved employee morale, reduction of critical to customer defects, etc. Merely demonstrating the use of Six Sigma tools is *not* sufficient. Nor is the delivery of intermediate "products" such as Pareto diagrams or process maps.

Certification as a Green Belt requires the following:

1. Acceptable completion of a Green Belt training curriculum approved by the Six Sigma organization.
2. Demonstration of clear and rational thought process.
    a. Ability to analyze a problem following a logical sequence.
    b. Usage of facts and data to guide decisions and action.
3. Ability to clearly explain Six Sigma and the DMAIC project cycle in layman's terms.
4. Ability to achieve tangible results, for example,
    a. Completed one or more projects that employed the Six Sigma approach (DMAIC or equivalent).
        i. Projects reviewed by appropriate personnel.
        ii. Deliverables accepted by the project sponsor.
        iii. Projects documented in a Green Belt notebook arranged in the DMAIC or equivalent format.
        iv. Projects used the Six Sigma approach and correctly employed a significant subset of basic tools and at least some intermediate Six Sigma tools and techniques (see later in this appendix for a listing).
    b. Ability to perform benefit/cost analysis.
    c. Ability to quantify deliverables in terms meaningful to the organization, for example, cost, quality, cycle time, safety improvement, etc.

d. Ability to identify and overcome obstacles to progress.
e. Ability to work within time, budget, and operational constraints.
5. Demonstrated ability to explain the tools of Six Sigma to others in ordinary language.
6. Demonstrated interpersonal and leadership skills necessary to be an effective change agent within the organization.

## *Green Belt Certification Board*

Effectiveness must be determined by qualified individuals familiar with the candidate's performance in the given effectiveness area. Table A16.1 provides guidelines for identifying prospective members of the Green Belt Certification Board. It is the Green Belt's responsibility to assist with the selection of their Certification Board.

## *Effectiveness Questionnaire*

It is strongly recommended that the candidate perform a self-assessment prior to applying for certification.

Certification Board members are encouraged to use the following questionnaires to assist them with their assessment. The candidate should provide the Six Sigma champion with a list of potential members of his or her Certification Board. When questionnaires are completed by someone other than a Certification Board member, they should be sent directly to a Certification Board member.

## *Scoring Guidelines*

The effectiveness questionnaire includes a set of assessment questions for each subject area. The results of the questionnaires can be summarized and used as input into the certification process. A form for this is provided below. The scoring summary sheet sum-

**Table A16.1** Green Belt Certification Board Member Selection Guide

| Assessment Subject Area | Board Member |
| --- | --- |
| Change agent skills | Supervisor, project sponsor(s), Six Sigma champion, mentor, process owner, Black Belt |
| Application of tools and techniques | Green Belt instructor, Master Black Belt, qualified Certified Master Black Belt Six Sigma consultant |
| Ability to achieve results | Project sponsor, process owner, team members, Green Belt, Six Sigma champion, Certified Master Black Belt, qualified Six Sigma consultant |

marizes the evaluator's scores by category. Worksheet items scored in the top 3 boxes are considered to be acceptable. Particular attention should be directed to any worksheet item scored in the lower 4 boxes. Since there are 10 choices for each item, any score below 5 indicates that the evaluator disagreed with the survey item. Survey items are worded in such a way that evaluators should agree with them for qualified Green Belt candidates; that is, higher scores are always better. Disagreement (low scores) in a few areas does not necessarily disqualify a candidate for certification. However, it indicates areas which need improvement and it is recommended that certification be granted only if the candidate agrees to a program for addressing these areas. The scores are, of course, not the only input. Ultimately each Certification Board member must exercise his or her own judgment and consider any other relevant factors before reaching a decision.

The Scoring Summary and Assessment Worksheets may be reproduced as necessary.

## Green Belt Notebook

Green Belt candidates should provide Certification Board members with written documentation of their on-the-job applications of the Six Sigma approach. These "Green Belt notebooks" should include all relevant information, including project characters, demonstrations of tool usage, samples of data used, excerpts of presentations to sponsors or leaders, team member names, project schedules and performance to these schedules, financial and other business results, etc. The notebooks can be distributed to Certification Board members as either soft copies or hard copies, at the candidate's discretion.

| | Change Agent Skills<br>Assessment Worksheet | | |
|---|---|---|---|
| Green Belt Candidate | | Date of Assessment | |
| Certification Board Member | | Role | |

**1.** The candidate effectively identifies and recruits Six Sigma team members.

Strongly Disagree                                                                 Strongly Agree

❏   ❏   ❏   ❏   ❏   ❏   ❏   ❏   ❏   ❏

**2.** The candidate effectively develops Six Sigma team dynamics and motivates participants.

Strongly Disagree                                                                 Strongly Agree

❏   ❏   ❏   ❏   ❏   ❏   ❏   ❏   ❏   ❏

**3.** The candidate is able to apply conflict resolution techniques.

Strongly Disagree                                          Strongly Agree

❐    ❐    ❐    ❐    ❐    ❐    ❐    ❐    ❐    ❐

**4.** The candidate is able to overcome obstacles to change.

Strongly Disagree                                          Strongly Agree

❐    ❐    ❐    ❐    ❐    ❐    ❐    ❐    ❐    ❐

**5.** The candidate utilizes a logical approach to problem solving.

Strongly Disagree                                          Strongly Agree

❐    ❐    ❐    ❐    ❐    ❐    ❐    ❐    ❐    ❐

**6.** The candidate effectively facilitates group discussions and meetings.

Strongly Disagree                                          Strongly Agree

❐    ❐    ❐    ❐    ❐    ❐    ❐    ❐    ❐    ❐

**7.** The candidate's presentations are well organized and easy to understand.

Strongly Disagree                                          Strongly Agree

❐    ❐    ❐    ❐    ❐    ❐    ❐    ❐    ❐    ❐

**8.** The candidate identifies and mobilizes sponsors for change.

Strongly Disagree                                          Strongly Agree

❐    ❐    ❐    ❐    ❐    ❐    ❐    ❐    ❐    ❐

**9.** The candidate builds a shared vision of the desired state with champions and sponsors.

Strongly Disagree                                          Strongly Agree

❐    ❐    ❐    ❐    ❐    ❐    ❐    ❐    ❐    ❐

**10.** The candidate identifies gaps between as-is and desired performance.

Strongly Disagree                                          Strongly Agree

❐    ❐    ❐    ❐    ❐    ❐    ❐    ❐    ❐    ❐

**11.** The candidate identifies all key stakeholders and obtains support for the project.

Strongly Disagree                                          Strongly Agree

❐    ❐    ❐    ❐    ❐    ❐    ❐    ❐    ❐    ❐

|  | Application of Tools and Techniques<br>Assessment Worksheet |  |  |
|---|---|---|---|
| Green Belt Candidate |  | Date of Assessment |  |
| Certification Board Member |  | Role |  |

**1.** The candidate uses an appropriate mix of basic and intermediate Six Sigma tools.[2]

Strongly Disagree                                                Strongly Agree

❐ ❐ ❐ ❐ ❐ ❐ ❐ ❐ ❐ ❐

**2.** The candidate uses the tools of Six Sigma properly.

Strongly Disagree                                                Strongly Agree

❐ ❐ ❐ ❐ ❐ ❐ ❐ ❐ ❐ ❐

**3.** The candidate applies the correct Six Sigma tools at the proper point in the project.

Strongly Disagree                                                Strongly Agree

❐ ❐ ❐ ❐ ❐ ❐ ❐ ❐ ❐ ❐

**4.** The candidate asks for help with Six Sigma tools when necessary.

Strongly Disagree                                                Strongly Agree

❐ ❐ ❐ ❐ ❐ ❐ ❐ ❐ ❐ ❐

**5.** The candidate can clearly explain all of the Six Sigma tools used on his or her projects in ordinary language. Note: candidates are not required to be able to *perform* all of the analyses without assistance, but they are required to understand basic or intermediate tools used for their projects.

Strongly Disagree                                                Strongly Agree

❐ ❐ ❐ ❐ ❐ ❐ ❐ ❐ ❐ ❐

**6.** The candidate understands the limitations as well as the strengths of quantitative methods.

Strongly Disagree                                                Strongly Agree

❐ ❐ ❐ ❐ ❐ ❐ ❐ ❐ ❐ ❐

---

[2] See later in this appendix for examples of these tools.

| Ability to Achieve Results<br>Assessment Worksheet |||||
|---|---|---|---|
| Green Belt Candidate | | Date of Assessment | |
| Certification Board Member | | Role | |

1. The candidate has successfully completed at least one Six Sigma project which produced tangible results.

   Strongly Disagree □ □ □ □ □ □ □ □ □ □ Strongly Agree

2. The candidate's project(s) had an acceptable project charter, including sponsorship, problem statement, business case, etc.

   Strongly Disagree □ □ □ □ □ □ □ □ □ □ Strongly Agree

3. The projects employed the Six Sigma approach (DMAIC or equivalent).

   Strongly Disagree □ □ □ □ □ □ □ □ □ □ Strongly Agree

4. The projects' deliverables were clearly defined in tangible terms.

   Strongly Disagree □ □ □ □ □ □ □ □ □ □ Strongly Agree

5. The projects produced significant improvements to an important business process.

   Strongly Disagree □ □ □ □ □ □ □ □ □ □ Strongly Agree

6. The baseline performance level was determined using valid data.

   Strongly Disagree □ □ □ □ □ □ □ □ □ □ Strongly Agree

7. The final performance level was calculated using valid data and showed improvements that were both statistically significant and important to the organization.

   Strongly Disagree □ □ □ □ □ □ □ □ □ □ Strongly Agree

8. An acceptable control plan has been implemented to assure that improvements are maintained.

   Strongly Disagree                                                                 Strongly Agree
   ❏    ❏    ❏    ❏    ❏    ❏    ❏    ❏    ❏    ❏    ❏

9. The projects' financial benefits were validated by experts in accounting or finance.

   Strongly Disagree                                                                 Strongly Agree
   ❏    ❏    ❏    ❏    ❏    ❏    ❏    ❏    ❏    ❏    ❏

10. Key customers were identified and their critical requirements defined.

    Strongly Disagree                                                                Strongly Agree
    ❏    ❏    ❏    ❏    ❏    ❏    ❏    ❏    ❏    ❏    ❏

11. Project sponsors are satisfied with their project's deliverables.

    Strongly Disagree                                                                Strongly Agree
    ❏    ❏    ❏    ❏    ❏    ❏    ❏    ❏    ❏    ❏    ❏

12. Project identified and corrected root causes, not symptoms.

    Strongly Disagree                                                                Strongly Agree
    ❏    ❏    ❏    ❏    ❏    ❏    ❏    ❏    ❏    ❏    ❏

13. All key stakeholders were kept informed of project status and are aware of final outcomes.

    Strongly Disagree                                                                Strongly Agree
    ❏    ❏    ❏    ❏    ❏    ❏    ❏    ❏    ❏    ❏    ❏

14. Projects were completed on time.

    Strongly Disagree                                                                Strongly Agree
    ❏    ❏    ❏    ❏    ❏    ❏    ❏    ❏    ❏    ❏    ❏

15. Projects were completed within budget.

    Strongly Disagree                                                                Strongly Agree
    ❏    ❏    ❏    ❏    ❏    ❏    ❏    ❏    ❏    ❏    ❏

16. Projects were conducted in a manner that minimized disruptions to normal work.

    Strongly Disagree                                                                Strongly Agree
    ❏    ❏    ❏    ❏    ❏    ❏    ❏    ❏    ❏    ❏    ❏

| Assessment Comments ||
|---|---|
| **Assessment Subject Area** | **Comments** |
| Change agent skills | |
| Application of tools and techniques | |
| Ability to achieve results | |

| Evaluator | Subject Area | Items scored 4 or less | % in top 3 boxes | Comment |
|---|---|---|---|---|
| | **Scoring Summary** | | | |
| | Change agent skills | | | |
| | Application of tools and techniques | | | |
| | Ability to achieve results | | | |
| | Change agent skills | | | |
| | Application of tools and techniques | | | |
| | Ability to achieve results | | | |
| | Change agent skills | | | |
| | Application of tools and techniques | | | |
| | Ability to achieve results | | | |
| | Change agent skills | | | |
| | Application of tools and techniques | | | |
| | Ability to achieve results | | | |
| | Change agent skills | | | |
| | Application of tools and techniques | | | |
| | Ability to achieve results | | | |

## Examples of Six Sigma Tools and Analytical Concepts

| Basic | Intermediate |
|---|---|
| • DMAIC<br>• SIPOC<br>• DPMO<br>• Computer skills<br>• Scales of measurement<br>• Pareto analysis<br>• Process mapping, flowcharts<br>• Check sheets<br>• Cause-and-effect diagrams<br>• Scatter plots<br>• Run charts<br>• Histograms<br>• Ogives<br>• Descriptive statistics (e.g., mean, standard deviation, skewness)<br>• Enumerative vs. analytic statistics<br>• Stem-and-leaf, boxplots<br>• Basic probability concepts<br>• Discrete probability distributions (binominal, Poisson, hypergeometric)<br>• Continuous probability distributions (normal, exponential, etc.)<br>• 7M tools<br>• FMEA<br>• Sampling<br>• CTx identification | • Control charts for measurements<br>• Control charts for attributes<br>• Process capability<br>• Yield analysis (e.g., first pass yield, rolled throughput yield)<br>• Measurement error analysis (gage R&R)<br>• Correlation analysis<br>• Simple linear regression<br>• Chi-square<br>• Type I and Type II errors<br>• Confidence interval interpretation<br>• Hypothesis tests<br>• Normality assessment and transformations<br>• Z-transformations<br>• Process sigma calculations |

# APPENDIX 17

# AHP Using Microsoft Excel

The analytic hierarchical process (AHP) is a powerful technique for decision making. It is also quite elaborate, and if you wish to obtain exact results you will probably want to acquire specialized software, such as *Expert Choice 2000* (www.expertchoice.com). However, if all you need is a good approximation, and if you are willing to forgo some of the bells and whistles, you can use a spreadsheet to perform the analysis. To demonstrate this, we will use Microsoft Excel to repeat the analysis we performed in Chap. 2.

## Example

In Chap. 2 we analyzed the high-level requirements for a software development process and obtained this matrix of pairwise comparisons from our customers.

|  | Easy to Learn | Easy to Use | Internet Connectivity | Interacts Well | Easy to Maintain |
|---|---|---|---|---|---|
| Easy to learn |  | 4.0 | 1.0 | 3.0 | 1.0 |
| Easy to use |  |  | 0.20 | 0.33 | 0.25 |
| Internet connectivity |  |  |  | 3.0 | 3.0 |
| Interacts well |  |  |  |  | 0.33 |
| Easy to maintain |  |  |  |  |  |

The meaning of the numbers is described in Chap. 2. The Excel equivalent of this is

|   | A | B | C | D | E | F |
|---|---|---|---|---|---|---|
| 1 | Attribute | A | B | C | D | E |
| 2 | A-Easy to learn | 0.00 | 4.00 | 1.00 | 3.00 | 1.00 |
| 3 | B-Easy to use | 0.25 | 0.00 | 0.20 | 0.33 | 0.25 |
| 4 | C-Connectivity | 1.00 | 5.00 | 0.00 | 3.00 | 3.00 |
| 5 | D-Compatible | 0.33 | 3.00 | 0.33 | 0.00 | 0.33 |
| 6 | E-Easy to maintain | 1.00 | 4.00 | 0.33 | 3.00 | 0.00 |

Note that the numbers on the diagonal are zeros, that is, the comparison of an attribute with itself has no meaning. Also note that the numbers below the diagonals are the reciprocals of the corresponding comparison above the diagonal. For example, the cell C2 has a 4.00, indicating that attribute A is preferred over attribute B; so the cell B3 must contain 1/4 = 0.25 to show the same thing.

To calculate the weight for each item, we must obtain the grand total for the entire matrix, then divide the row totals by the grand total. This is shown below:

|   | A | B | C | D | E | F | Total | Weight |
|---|---|---|---|---|---|---|---|---|
| 1 | Attribute | A | B | C | D | E | Total | Weight |
| 2 | A-Easy to learn | 0.00 | 4.00 | 1.00 | 3.00 | 1.00 | 9.00 | 26.2% |
| 3 | B-Easy to use | 0.25 | 0.00 | 0.20 | 0.33 | 0.25 | 1.03 | 3.0% |
| 4 | C-Connectivity | 1.00 | 5.00 | 0.00 | 3.00 | 3.00 | 12.00 | 34.9% |
| 5 | D-Compatible | 0.33 | 3.00 | 0.33 | 0.00 | 0.33 | 4.00 | 11.6% |
| 6 | E-Easy to maintain | 1.00 | 4.00 | 0.33 | 3.00 | 0.00 | 8.33 | 24.2% |
| 7 | Grand total |  |  |  |  |  | 34.36 |  |

These results are shown in the next illustration.

## Category Importance Weights

- A-Easy to learn: 26.2%
- B-Easy to use: 3.0%
- C-Connectivity: 34.9%
- D-Compatible: 11.6%
- E-Easy to maintain: 24.2%

Compare these weights to those obtained by the exact analysis obtained using Expert Choice 2000.

| Category | Exact Weight | Spreadsheet Weight |
| --- | --- | --- |
| Easy to learn | 26.4% | 26.2% |
| Easy to use quickly after I've learned it | 5.4% | 3.0% |
| Internet connectivity | 35.8% | 34.9% |
| Works well with other software I own | 10.5% | 11.6% |
| Easy to maintain | 21.8% | 24.2% |

The conclusions are essentially the same for both analyses.

# NOTES

## Chapter 1

1. Statistician's note: The area under the normal curve beyond Six Sigma is 2 parts-per-billion. In calculating failure rates for Six Sigma purposes, we assume that performance experienced by customers over the life of the product or process will be much worse than internal short-term estimates predict. To compensate, a "shift" of 1.5 sigma from the mean is added before calculating estimated long-term failures. Thus, you will find 3.4 parts-per-million as the area beyond 4.5 sigma on the normal curve.
2. This is about twice the rate of improvement reported by companies using TQM. For example, Baldrige winner Milliken & Co. implemented a "ten-four" improvement program requiring reductions in key adverse measures by a factor of ten every four years.
3. Although Green Belts are also highly trained change agents, they are not full-time change agents and we will not discuss their compensation here.
4. This section (up to but not including The Six Sigma Process Enterprise) is an excerpt from *The Handbook for Quality Management* (McGraw-Hill, 2013).
5. There are exceptions to this. Many large customers, such as the Department of Defense or automobile or aircraft manufacturers, take a very active interest in the internal operations of their key suppliers.

## Chapter 2

1. All detailed quality cost descriptions are from John T. Hagan, ed., *Principles of Quality Costs*, Milwaukee, WI: ASQ Quality Press, Appendix B.
2. See Appendix 17 for an example of how to derive approximate importance weights using Microsoft Excel.
3. Excerpt from *The Handbook for Quality Management* (McGraw-Hill, 2013), by permission.
4. Thanks to H. William Dettmer from Goal Systems International for writing this section.

## Chapter 3

1. The key requirements probably won't require explicit support plans. However, if they do, QFD can be used to evaluate the plans. Key requirements QFD should be handled separately.
2. www.attar.com/tutor/deploy.htm.

## Chapter 8

1. There is some debate over the value of moving $R$ charts. Academic researchers have failed to show statistical value in their usage. However, some practitioners contend that moving $R$ charts provide valuable additional information useful in troubleshooting.
2. We aren't saying this situation actually exists, we are just saying that the $\bar{X}$ treats the process as if this were true. Studying the patterns of variation will often reveal clues to making the process more consistent, even if the process variation remains within the control limits.
3. The null hypothesis is that the data are normally distributed. The $P$-value is the probability of obtaining the observed results if the null hypothesis were true. In this case, the probability is 0.

## Chapter 9

1. A problem with this analysis is that the datum for each part is an average of 12 measurements, not individual measurements. If we could obtain the 60 actual measurements the $P$-value would probably be different because the standard error would be based on 60 measurements rather than 5. On the other hand, the individual measurements would also be more variable, so the exact magnitude of the difference is impossible to determine without the raw data.

## Chapter 10

1. See Chap. 6.
2. It is possible to evaluate the association of $x$ and $y$ by removing the effect of the lurking variable. This can be done using regression analysis and computing partial correlation coefficients. This advanced procedure is described in most texts on regression analysis.
3. Referring to the critical value is actually unnecessary; the $P$-value of 0.000 indicates that the probability of getting an $F$ value as large as that computed is less than 1 in 1,000.

## Chapter 11

1. Technically, a Poisson model would be the correct choice here. However, use of a normal model, which the analysis assumes, is reasonably accurate for defect counts of this magnitude. The team also evaluated the variance, more specifically, the log of the variance. The variances at each factor combination did not differ significantly and are not shown here.
2. It is possible, however, to include various transformed variables to "help" the neural net if one has a model in mind. For example, in addition to feeding the neural net X1 and X2 raw data, one could include higher-order polynomial and interaction terms as inputs to the neural network.
3. Crystal Ball Professional, www.decisioneering.com.
4. Pyzdek, 1992b.
5. Mil-Std-1629A actually calls the approach FMECA, which stands for Failure mode, effect, and criticality analysis, but over time the "C" has been dropped from common usage. However, criticality analysis is still very much a part of FMEA.
6. www.fmeainfocentre.com/.

# REFERENCES

Abraham, B., and Whitney, J. B. (1990). "Applications of EWMA Charts to Data from Continuous Processess," *Annual Quality Congress Transactions*, Milwaukee, WI: ASQ Quality Press.

Akao, Y., Editor (1990). *Quality Function Deployment: Integrating Customer Requirements into Product Design*, Cambridge, MA: Productivity Press.

Alwan, L. C., and Roberts, H. V. (1989). "Time Series Modeling for Statistical Process Control," in Keats, J. B., and Hubel, N. F., Editors, *Statistical Process Control in Automated Manufacturing*, New York: Marcel Dekker.

Argyris, C. (1991). "Teaching Smart People How to Learn," *Harvard Business Review,* May–June.

ASQ Statistics Division. (1983). *Glossary and Tables for Statistical Quality Control*, Milwaukee, WI: ASQ Quality Press.

Athey, Thomas H. (1982). *The Systematic Systems Approach*, NJ: Prentice-Hall, Inc..

Aubrey, C. A., and Felkins, P. K. (1988). *Teamwork: Involving People in Quality and Productivity Improvement*, Milwaukee, WI: ASQ Quality Press.

Automotive Industry Action Group (2003). *Measurement Systems Analysis Reference Manual Third Edition*, Detroit, MI: AIAG.

Berry, Michael J. A., and Linoff, Gordon. (1997). *Data Mining Techniques for Marketing, Sales, and Customer Support*, New York: John Wiley & Sons, pp. 369–370.

Blanchard, Kenneth, and Johnson, Spencer. (1988). *The One-Minute Manager*, NY: William Morrow & Co.

Blauth, Chris (AchieveGlobal). (2008). Web presentation: *Managing Change: Keys to Success,* June 12.

Box, G. E. P., and Draper, N. R. (1969). *Evolutionary Operation: A Statistical Method for Process Improvement*, New York: John Wiley & Sons.

Box, G .E. P., and Draper, N. R. (1987). *Empirical Model-Building and Response Surfaces*, New York: John Wiley & Sons.

Box, G. E. P., Hunter, W. G., and Hunter, J. S. (1978). *Statistics for Experimenters*, New York: John Wiley & Sons.

Brassard, M. (1989). *The Memory Jogger Plus +*, Methuen, MA: GOAL/QPC.

Brown, B., and Anthony, S. D. (2011). "How P&G Tripled Its Innovation Success Rate," http://hbr.org/2011/06/how-pg-tripled-its-innovation-success-rate/ar/pr.

Buffett, Warren E. (1996). *An Owners Manual*, Omaha, NE: Berkshire Hathaway.

Burke, E., Kloeber, J. M., and Deckro, R. F. (2002). "Using and Abusing QFD Scores," *Quality Engineering*, vol. 15, no. l, pp. 23–36.

Burr, I. W. (1976). *Statistical Quality Control Methods, Statistics: Textbooks and Monographs*, vol. 16, New York: Marcel-Dekker, Inc.

Camp, R. C. (1989). *Benchmarking: The Search for Industry Best Practices That Lead to Superior Performance*, Milwaukee, WI: ASQ Quality Press and White Plains, NY: Quality Resources.

Campanella, J., Editor. (1990). *Principles of Quality Costs*, 2nd Edition, Milwaukee, WI: ASQ Quality Press.

Campanella, J., Editor. (1999). *Principles of Quality Costs*, 3rd Edition, Milwaukee, WI: ASQ Quality Press.

Carder, B., and Clark, J. D. (1992). "The Theory and Practice of Employee Recognition," *Quality Progress*, December.

Carlzon, J. (1987). *Moments of Truth*, New York: Harper and Row.

Christensen, C. (1997). *The Innovator's Dilemma*, Boston, MA: Harvard Business School Press.

Christensen, C., and Overdorf, M. (2000). "Meeting the Challenge of Disruptive Change," *Harvard Business Review*, March 2000.

Corbett, Thomas. (1998). *Throughput Accounting*, Great Barrington, MA: North River Press.

Cornell, John A. (1990). *How to Apply Response Surface Methodology*, Milwaukee, WI: ASQ Quality Press.

Deming, W. E. (1975). "On Probability as a Basis for Action," *The American Statistician*, vol. 29, no. 4, pp. 146–152.

Deming, W. E. (1986). *Out of the Crisis*, Cambridge, MA: MIT Center for Advanced Engineering Study.

Deming, W. E. (1993). *The New Economics for Industry, Government, Education*, Cambridge, MA: MIT Center for Advanced Engineering Study.

DeToro, I. (1995). "The 10 Pitfalls of Benchmarking," *Quality Progress*, January, pp. 61–63.

Dettmer, H. William. (1997). *Goldratt's Theory of Constraints: A Systems Approach to Continuous Improvement*, Milwaukee, WI: ASQ Quality Press.

Dettmer, H. William. (1998). *Breaking the Constraints to World-Class Performance*, Milwaukee, WI: ASQ Quality Press.

Dillman, D. A. (1983). "Mail and Other Self-Administered Questionnaires," in Rossi, P., Wright, J., and Anderson, A., Editors, *Handbook of Survey Research*, New York: Academic Press, Inc., pp. 359–377.

Draper, N., and Smith, H. (1981). *Applied Regression Analysis*, 2nd Edition, New York: John Wiley & Sons.

Drucker, P. F. (1974). *Management: Tasks, Responsibilities, Practices*, New York: Harper and Row.

Duncan, A. J. (1974). *Quality Control and Industrial Statistics*, 4th Edition, Homewood, IL: Irwin.

Eades, Keith M. (2004). *Solution Selling*, New York: McGraw-Hill.

Edosomwan, J. A. (1993). *Customer and Market-Driven Quality Management*, Milwaukee, WI: ASQ Quality Press.

Efron, B. (1982). *The Jackknife, the Bootstrap, and Other Resampling Plans*, Philadelphia, PA: Society for Industrial and Applied Mathematics.

Endler, M. (2013). "Elon Musk: I Thought Tesla Was Dying," www.informationweek.com/strategic-cio/executive-insights-and-innovation/elon-musk-i-thought-tesla-was-dying/d/d-id/1113102.

Fivars, G. (1980). "The Critical Incident Technique: a Bibliography," Research and Publication Service, American Institutes for Research ED 195 681.

Flanagan, J. C. (1954). "The Critical Incident Technique," *Psychological Bulletin*, vol. 51, no. 4, July, pp. 327–358.

Forsha, H. J. (1992). *The Pursuit of Quality through Personal Change*, Milwaukee, WI: ASQ Quality Press.

Foster, R., and Kaplan, S. (2001). *Creative Destruction: Why Companies That Are Built to Last Underperform the Market—And How to Successfully Transform Them*, New York: Doubleday.

Forum Corporation. (1996). Annual Report.

Galloway, D. (1994). *Mapping Work Processes*, Milwaukee, WI: ASQ Quality Press.

GAO (1986). "Developing and Using Questionnaires—Transfer Paper 7," Washington, DC: United States General Accounting Office.

George, M. L. (2002). *Lean Six Sigma*, New York: McGraw-Hill.

Gibbons, Jean Dickinson. (1993). *Nonparametric Statistics: An Introduction*, New York: Sage Publications, p. 63.

Goldratt, Eliyahu M. (1986). *The Goal*, Croton-on-Hudson, New York: The North River Press.

Goldratt, Eliyahu M. (1990). *The Haystack Syndrome: Sifting Information Out of the Data Ocean*, Great Barrington, MA: North River Press, pp. 59–63.

Goldratt, Elyahu M., and Fox, Robert E. (1986). *The Race*, Croton-on-Hudson, New York: The North River Press.

Goldratt, E.M. (1997). *Critical Chain*, Great Barrington, MA: The North River Press.

Hagan, John T., Editor. (1999). *Principles of Quality Costs*, Milwaukee: ASQ Quality Press.

Hammer, M., and Champy, J. (1993). *Reengineering the Corporation: A Manifesto for Business Revolution*, New York: HarperCollins Publishers.

Hammer, Michael, and Stanton, Steve. (1999). "How Process Enterprises *Really* Work," *Harvard Business Review*, November–December, pp. 108–118.

Harrington, H. J. (1992). "Probability and Statistics," in Pyzdek, T., and Berger, R. W., Editors, *Quality Engineering Handbook*, Milwaukee, WI: ASQ Quality Press, pp. 513–577.

Harry, M., and Schroeder, R. (2000). *Six Sigma: The Breakthrough Management Strategy Revolutionizing the World's Top Corporations*, New York: Doubleday.

Harry, M. Contributor. (2013). "Another Controversy to Debate," SixSigma Network Discussion on LinkedIn, www.linkedin.com/groupItem?view=&srchtype=discussedNews&gid=42396&item=5821675147173572612&type=member&trk=eml-anet_dig-b_pd-ttl-cn&fromEmail=&ut=3w4n8j_xpbTS41.

Hayes, Bob E. (1992). *Measuring Customer Satisfaction: Development and Use of Questionnaires*, Milwaukee, WI: ASQ Quality Press.

Hicks, Charles R. (1973). *Fundamental Concepts in the Design of Experiments*, 2nd Edition, New York: Holt, Rinehart, Winston, pp. 31–38.

Hicks, Charles R. (1993). *Fundamental Concepts in the Design of Experiments*, 3rd Edition, New York: Holt, Rinehart, Winston.

Hillier, F. S. (1969). "X-bar and R-chart Control Limits Based on a Small Number of Subgroups," *Journal of Quality Technology*, vol. 1, no. 1, January, pp. 17–26.

Hoerl, R. W. (2001). "Six Sigma Black Belts: What Do They Need to Know?" *Journal of Quality Technology*, vol. 33, no. 4, October, pp. 391–406.

Holland, John H. (1996). *Hidden Order: How Adaptation Builds Complexity*, Reading, MA: Perseus Books, p. 56.

Hunter, J. S. (1986). "The Exponentially Weighted Moving Average," *Journal of Quality Technology*, vol. 18, pp. 203–210.

Hunter, J. S. (1989). "A One Point Plot Equivalent to the Shewhart Chart with Western Electric Rules," *Quality Engineering*, vol. 2, pp. 13–19.

Hurley, H., and Loew, C. (1996). "A Quality Change for New Product Development," *Quality Observer*, January, pp. 10–13.

Hutton, D. W. (1994). *The Change Agent's Handbook: A Survival Guide for Quality Improvement Champions*, Milwaukee, WI: ASQ Quality Press.

Imai, M. (1986). *Kaizen*, New York: Random House.

Innosight Executive Briefing Winter 2012: "Creative Destruction Whips Through Corporate America," www.innosight.com.

Ireson, W., Coombs, C., Moss, R. (1996). *Handbook of Reliability Engineering and Management*, New York: McGraw-Hill.

Ishikawa, K. (1985). *What is Total Quality Control the Japanese Way?*, Englewood Cliffs, NJ: Prentice-Hall, Inc.

Jackson, J.E. (1980). "Principal Components & Factor Analysis: Part 1-Principal Components" *Journal of Quality Technology*, vol. 12, No. 4.

Jackson, J.E. (1991). *A User's Guide to Principal Components*, New York: Wiley and Sons.

Johnson, D. W., and Johnson, F. P. (1999). *Joining Together: Group Theory and Group Skills*, Englewood Cliffs, NJ: Prentice-Hall.

Johnson, R. S. (1993). *TQM: The Mechanics of Quality Processes*, Milwaukee, WI: ASQ Quality Press.

Joiner, B. L. (1994). *Fourth Generation Management: The New Business Consciousness*, New York: McGraw-Hill, Inc.

Juran, J. M., and DeFeo, J. A. (2010). *Juran's Quality Handbook*, 6th Edition, New York: McGraw-Hill.

Juran, J. M., and Gryna, F. M. (1988). *Juran's Quality Control Handbook*, 4th Edition, New York: McGraw-Hill.

Juran, J. M., and Gryna, F. M. (1993). *Quality Planning and Analysis*, 3rd Edition, New York: McGraw-Hill.

Kackar, R. N. (1985). "Off-line Quality Control, Parameter Design, and the Taguchi Method" *Journal of Quality Technology*, vol. 17, pp. 176–188.

Kano, N. (1993). "A Perspective on Quality Activities in American Firms." *California Management Review*, 35(3), Spring 1993, pp. 12–31.

Kanter, R. M. (2006). "Innovation: The Classic Traps," *Harvard Business Review*, November 2006.

Kaplan, Robert S., and Norton, David P. (1992). "The Balanced Scorecard—Measures that Drive Performance," *Harvard Business Review*, January–February, pp. 71–79.

Kaplan, Robert S., and Norton, David P. (1996). "Using the Balanced Scorecard as a Strategic Management System," *Harvard Business Review*, January.

Keller, Paul A. (2005). *Six Sigma Demystified*, New York: McGraw-Hill, Inc.

Keller, Paul (2011). *Statistical Process Control Demystified*, New York: McGraw-Hill, Inc.

King, B. (1987). *Better Designs in Half the Time: Implementing QFD in America*, Methuen, MA: Goal/QPC.

Kirkpatrick, D. L. (1996). "Evaluation," in Craig, R. L., Editor-in-Chief, *The ASTD Training and Development Handbook: A Guide to Human Resources Development*, New York: McGraw-Hill, pp. 294–312.

Koch, G. G., and Gillings, D. B. (1983). "Statistical Inference, Part I," in Kotz, Samuel, and Johnson, Norman L., Editors-in-Chief, *Encyclopedia of Statistical Sciences*, New York: John Wiley & Sons, vol. 4, pp. 84–88.

Koetzier, W., and Alon, A. (2013). "Why Low-Risk Innovation Is Costly," www.accenture.com/us-en/Pages/insight-low-risk-innovation-costly.aspx?c=mc_prposts_10000040&n=otc_1013.

Kohn, A. (1993). *Punished by Rewards: The Trouble with Gold Stars, Incentive Plans, A's, Praise and Other Bribes*, New York: Houghton Mifflin Company.

Kotler, P. (1991). *Marketing Management: Analysis, Planning, Implementation, and Control*, 7th Edition, Englewood Cliffs, NJ: Prentice-Hall.

Kotter, J. (2012). www.forbes.com/sites/johnkotter/2012/05/02/barriers-to-change-the-real-reason-behind-the-kodak-downfall/.

Kourti, Theodora and MacGregor, John F. (1996). "Multivariate SPC Methods for Process and Product Monitoring," *Journal of Quality Technology*, Vol 28, No. 4, pp. 409-428.

Krisnamoorthi, K. S. (1991). "On Assignable Causes that Cannot be Eliminated—An Example from a Foundry," *Quality Engineering*, vol. 3, pp. 41–47.

Lafley, A. G., Martin, R. L., Rivkin, J.W., Siggelkow, N. (2012). "Bringing Science to the Art of Strategy," *Harvard Business Review*. (September) pp. 57–66.

Leach, L. P. (2000). *Critical Chain Project Management*, Boston, MA: Artech House, Inc.

Levinson, Jay Conrad; Godin, Seth (Contributor); and Rubin, Charles. (1995). *The Guerrilla Marketing Handbook*, Boston, MA: Houghton Mifflin Co.

Main, Jeremy. (1994). *Quality Wars: The Triumphs and Defeats of American Business*, New York: The Free Press, p. 173.

Meyers, Raymond H., and Montgomery, Douglas C. (1995). *Response Surface Methodology: Process and Product Optimization Using Designed Experiments*, New York: John Wiley & Sons.

Mintzberg, H. (1994). *The Rise and Fall of Strategic Planning*, New York: The Free Press.

Mintzberg, H., and Quinn, J. B. (1991). *The Strategy Process—Concepts, Contexts, Cases*, 2nd edition, Englewood Cliffs, NJ: Prentice-Hall, Inc.

Mizuno, S., Editor. (1988). *Management for Quality Improvement: The 7 New QC Tools*, Cambridge, MA: Productivity Press.

Montgomery, D. C. (1984). *Design and Analysis of Experiments*, 2nd Edition, New York: John Wiley & Sons.

Montgomery, Douglas C. (1996). *Design and Analysis of Experiments*, New York: John Wiley & Sons.

Natrella, M. G. (1963). *Experimental Statistics: NBS Handbook 91*, Washington, DC: U.S. Government Printing Office.

Nelson, L. S. (1984). "The Shewhart Control Chart—Tests for Special Causes," *Journal of Quality Technology*, vol. 16, no. 4, October 1984, pp. 237–239.

Newbold, Robert C. (1998). *Project Management in the Fast Lane: Applying the Theory of Constraints*, Boca Raton, FL: St. Lucie Press.

Noreen, Eric, Debra Smith, and James T. Mackey. (1995). *The Theory of Constraints and Its Implications for Management Accounting*, Great Barrington, MA: North River Press.

Ohno, Taiichi. (1988). *Toyota Production System: Beyond Large-Scale Production*, Portland, OR: Productivity Press.

Ohno, Taiichi (2013). *Taiichi Ohno's Workplace Management, Special 100th Birthday Edition*, New York: McGraw-Hill.

Palm, A. C. (1990). "SPC Versus Automatic Process Control," *Annual Quality Congress Transactions*, Milwaukee, WI: ASQ Quality Press.

Phillips, J. J. (1996). "Measuring the Results of Training," in Craig, R. L., Editor-in-chief, *The ASTD Training and Development Handbook: A Guide to Human Resources Development*, New York: McGraw-Hill, pp. 313–341.

Porter, M. E. (1996). "What Is Strategy?" *Harvard Business Review*, November 1996.

Proschan, F., and Savage, I. R. (1960). "Starting a Control Chart," *Industrial Quality Control*, 17(3), September, pp. 12–13.

Provost, L. P. (1988). "Interpretation of Results of Analytic Studies," paper presented at 1989 NYU Deming Seminar for Statisticians, March 13, New York: New York University.

Pyzdek, T. (1990). *Pyzdek's Guide to SPC Volume One*, Tucson, AZ: Quality Publishing, Inc.

Pyzdek, T. (1992a). *Pyzdek's Guide to SPC Volume Two—Applications and Special Topics*, Tucson, AZ: Quality Publishing, Inc.

Pyzdek, T. (1992b). "A Simulation of Receiving Inspection," *Quality Engineering*, 4(1), pp. 9–19.

Pyzdek, T., and Keller, P. (2013). *The Handbook for Quality Management: A Complete Guide to Operational Excellence*, New York: McGraw-Hill.

Reeves, M., Love, C., Tillmanns, P. (2012). "Your Strategy Needs a Strategy." *Harvard Business Review*, September, pp. 76–83.

Reichheld, F. F. (1996). *The Loyalty Effect: The Hidden Force Behind Growth, Profits, and Lasting Value*, New York: McGraw-Hill.

Rencher, (1995). *Methods of Multivariate Analysis*, New York: Wiley & Sons.

Rose, K. H. (1995). "A Performance Measurement Model," *Quality Progress*, February, pp. 63–66.

Ruskin, A. M., and Estes, W. E. (1995). *What Every Engineer Should Know About Project Management*, 2nd Edition, New York: Marcel-Dekker.

Saaty, T. L. (1988). *Decision Making for Leaders: The Analytic Hierarchy Process for Decisions in a Complex World*, Pittsburgh, PA: RWS Publications.

Scholtes, P. R. (1988). *The Team Handbook: How to Use Teams to Improve Quality*, Madison, WI: Joiner Associates, Inc.

Schragenheim, Eli, and Dettmer, H. William. (2000). *Manufacturing at Warp Speed*, Boca Raton, FL: St. Lucie Press.

Schuman, S. P. (1996). "What to Look for in a Group Facilitator," *Quality Progress*, June, pp. 69–72.

Senge, Peter M. (1990). *The Fifth Discipline: The Art and Practice of the Learning Organization*, New York: Doubleday.

Shainin, D., and Shainin, P. D. (1988). "Statistical Process Control," in Juran, J. M., and Gryna, F. M., *Juran's Quality Control Handbook*, 4th Edition, New York: McGraw-Hill.

Sheridan, B. M. (1993). *Policy Deployment: The TQM Approach to Long-Range Planning*, Milwaukee, WI: ASQ Quality Press.

Shewhart, W. A. (1939, 1986). *Statistical Method from the Viewpoint of Quality Control*, New York: Dover Publications.

Shewhart, W. A. (1931, 1980). *Economic Control of Quality of Manufacturing*, Milwaukee, WI: ASQ Quality Press.

Silberzahn, P., and Jones, M. (6/9/2012). "That's All Folks: Why the Writing Is on the Wall at Microsoft," www.forbes.com/sites/frederickallen/2012/06/19/thats-all-folks-why-the-writing-ison-the-wall-at-microsoft/.

Simon, J. L. (1992). *Resampling: the New Statistics*, Arlington, VA: Resampling Stats, Inc.

Smith, Debra. (2000). *The Measurement Nightmare: How the Theory of Constraints Can Resolve Conflicting Strategies, Policies, and Measures*, Boca Raton, FL: The St. Lucie Press.

Spencer, Mary. (1999). "DFE and Lean Principles," *The Monitor*, 6(2), Fall, pp. 15–16.

Stewart, T. A. (1995). "After All You've Done for Your Customers, Why Are They Still NOT HAPPY?" *Fortune*, December 11.

Suminski, L. T., Jr. (1994). "Measuring the Cost of Quality," *Quality Digest*, March, pp. 26–32.

Taguchi, G. (1986). *Introduction to Quality Engineering: Designing Quality into Products and Processes*, White Plains, NY: Quality Resources.

Taylor, B. (2013). http://blogs.hbr.org/2013/12/stop-me-before-i-innovate-again/.

Taylor, Frederick W. (1947). *Scientific Management*, New York: Harper & Brothers.

Tuckman, B. W. (1965). "Development Sequence in Small Groups," *Psychological Bulletin*, 63, June, pp. 384–399.

Tufte, E. R. (2001). *The Visual Display of Quantitative Information*, Cheshire, CT: Graphics Press.

Tukey, J. W. (1977). *Exploratory Data Analysis*, Reading, MA: Addison-Wesley.

Vaziri, H. K. (1992). "Using Competitive Benchmarking to Set Goals," *Quality Progress*, November, pp. 81–85.

Waddington, P. (1996). *Dying for Information: An Investigation of Information Overload in the UK and World-wide*, London: Reuters Business Information.

Wahl, P. R., and Bersbach, P. L. (1991). "TQM Applied—Cradle to Grave," *Annual Quality Congress Transactions*, Milwaukee, WI: ASQ Quality Press.

*Webster's New Universal Unabridged Dictionary*. (1989). New York: Barnes & Noble Books.

Wheeler, D. J. (1991). "Shewhart's Charts: Myths, Facts, and Competitors," *Annual Quality Congress Transactions*, Milwaukee, WI: ASQ Quality Press.

Wheeler, D.J. (2010). "Problems With Gauge R&R Studies," https://www.qualitydigest.com/inside/twitter-ed/problems-gauge-rr-studies.html#

Wilson, L. (2010). *How to Implement Lean Manufacturing*, New York: McGraw-Hill.

Womack, J. P., and Jones, D. T. (1996). *Lean Thinking: Banish Waste and Create Wealth in Your Corporation*, New York: Simon & Shuster.

Zaltman, Berald. (2002). "Hidden Minds," *Harvard Business Review*, June, pp. 26–27.

# INDEX

*References to figures are in italics.*

$\chi^2$. *See* chi-square

Accenture, 103
Acceptable Quality Level (AQL), 617
accuracy, 416
adaptive strategy, 110–111
adjusted *R* squared, 469
AHP. *See* analytic hierarchical process (AHP)
Akao, Yoji, 91–93
alternative hypotheses. *See* hypotheses, alternative
Analysis of Variance. *See* ANOVA
analytic hierarchical process (AHP), 96, 98
    using Microsoft Excel, 679–681
analytic studies, 280–282
Analyze phase, 56
    binary logistic regression, 503–506
    correlation analysis, 463, 470–471
    designed experiments, 472
    hypothesis testing, 452–459
    Lean, 423–426
    least-squares fit, 466–470
    linear models, 463–466
    logistic regression, 501–503
    making comparisons using chi-square tests, 499–500
    non-parametric methods, 513–516
    objectives, 423
    point and interval estimation, 444–452
    regression analysis, 460, 463, 468–470
    resampling, 459–460
    scatter diagrams, 460–463
    spaghetti charts, 433–434
    value stream analysis, 423–426
    value stream mapping, 427–433
    *See also* Define-Measure-Analyze-Improve-Control (DMAIC)
ANOVA, 469, 617
    one-factor, 477–479
    two-way with no replicates, 479–480
    two-way with replicates, 480–482
ANOVA method, 403–404
Anthony, Scott, 104
AOQ. *See* Average Outgoing Quality (AOQ)
AOQL. *See* Average Outgoing Quality Limit (AOQL)
APC. *See* automatic process control (APC)
appraisal costs, 79–80
AQL. *See* Acceptable Quality Level (AQL)
area under the standard normal curve, 623–625
assignable causes, 285
    defined, 617
assumptions
    equal-variance, 497
    independence, 493–494
    linear model, 498–499
    normality, 494–497
    testing, 491–499
*AT&T Statistical Quality Control Handbook*, 376
Athey, Thomas H., 117
attribute data, 276
    control charts for, 323–336
attribute measurement
    accuracy, 416
    attribute gage R&R, 417–421
    bias, 416
    concepts, 411
    evaluating attribute inspection, 413
    example of attribute inspection error analysis, 414
    individual inspector accuracy, 414–415
    operational definitions, 412–413
    overview, 410
    repeatability, 416
    repeatability and pairwise reproducibility, 415–416
    reproducibility, 416
    stability, 416, 417

autocorrelation, 339–341, 493
automatic process control (APC)
  and statistical process control, 379–380
Average Outgoing Quality (AOQ), 617
Average Outgoing Quality Limit (AOQL), 617
averages charts, 310–317
awareness
  building through communications, 54–60
  implementing Six Sigma, 17

background variables, 473
backlog, 563–564
balanced scorecards, 150–151
  customer perspective, 151–153
  financial perspective, 155–157
  innovation and learning perspective, 154–155
  internal process perspective, 153–154
  strategy deployment plan, 157–159
Batch Means chart, 338–339
batch processing, 429, 432
benchmarking
  benefits of, 86
  causes of failure, 85–86
  dangers of, 87
  exchange services, 84
  getting started with, 83–85
  management commitment, 85
  overview, 82
  steps, 82–83
  teams, 85
benefit-cost analysis, 179–180
bias, 396–397, 416
  defined, 391
  illustrated, *392, 398*
billing-by-the-minute (BBTM), 273, *274, 275*
binary logistic regression, 503–506
binomial distribution, 295–296
binomial probability distribution, 295–296
Black Belts, 3, 19–20
  alternatives to, 34–35
  candidate selection process and criteria, *33*
  centralized reporting, 21–22
  compensation and retention, 37–38
  effectiveness certification, 655–666
  sample curriculum for finance, *29*
  sample curriculum for manufacturing, *30*
  success factors and importance weights, 32
  and supply chain management, 53–54
  training, 28–35
  *See also* training
Blanchard, Ken, 117
boards of directors, support from, 47
bootstrapping, 459–460

Box-Cox transformation, 322
boxplots, 437–439
brainstorming, 590
break-even points, 261
Brown, Bruce, 104
budget reviews, 229
budgeting, 230–231
buffer management, 138
business model capabilities, 113
business planning
  deploying differentiators to operations, 163–165
  deploying operations plans to projects, 165–166
  overview, 162–163
business process control planning
  maintaining gains, 588–590
  overview, 588
  preparing the process control plan, 591–593
  tools and techniques, 590–591
Business Process Executives (BPEs), 48–49, 50–51
business value added (BVA) tasks, 428
buy-in, using DMAIC methodology, 54–60

$c$ charts, 332–336
calibration, 391, 392
capability analysis, Minitab, 385–390
capacity-constrained resources (CCRs), 134
cash flow streams, 263
category importance weights, 98–99
category labels, 276
causal correlation, 345
cause and effect diagrams, 435–437
  *See also* CEDAC
causes and effects, measuring, 147–149
CCRs. *See* capacity-constrained resources (CCRs)
CEDAC, 437
central limit theorem, 284
centralized reporting, 21–22
champions, 27
chance causes, 285
  defined, 617
change
  the change imperative, 12–14
  goals of, 14
  managing, 14–15
  process of personal change, *198*
change agents. *See* Black Belts; Green Belts; Master Black Belts
channel, 113
chemical processes, 340
chi-square
  distributions, 440–443, 629–630
  making comparisons using chi-square tests, 499–500

Christensen, Clayton, 87, 101
CIT. *See* Critical Incident Technique (CIT)
classical strategy, 110
$C_M$ indices, 382–383, 384–385
coefficient of determination, defined, 617
coefficient of multiple correlation, defined, 617
coefficient of variation, defined, 618
command-and-control organizational structure, 42–43, 196
common bosses, 200
common cause variation, 380
common causes, 285
    charts, 370–371
    EWMA common cause charts, 371–378
communication
    and awareness, 54–60
    communicating a vision, 108–109
    communications plan and requirement matrix, 58
    effective, 46
    facilitating, 242
    implementing Six Sigma, 17
    by leadership, 47–48
    and linking, 159–162
competitors, 113
completely randomized design, 475
composite design phase, 530
compounding, 260, 262–265
confidence intervals, 444, 445, 446–447, 452
confidence limits, 446
    defined, 618
conflict resolution, 233–234
conjoint analysis, 96
consensus technique, 234
consolidating improvements, 16
constraint management, 113–114
    assumptions, 119
    Five Focusing Steps, 121–126
    goals and necessary conditions, 119–121
    logical thinking process, 126–133
    measurements, 138–143
    overview, 117–118
    tools of, 126–138
    *See also* Theory of Constraints (TOC)
Constraint Theory. *See* Theory of Constraints (TOC)
constraints
    defined, 118
    elevation, 123–125
    exploitation, 122
    identification, 122
    inertia, 125–126
    nonconstraints, 122
    subordination, 122–123
    types of, 118–119
    using constraint information to focus Six Sigma projects, 191–192
    *See also* Theory of Constraints (TOC)
consumers, 113
consumer's risk (b), defined, 618
continuous, 278
contrasts, 478
Contributions plot, 614, *615, 616*
control charts, 310
    for average occurrences-per-unit, 329–332
    averages and standard deviation (sigma) charts, 314–317
    averages charts, 310–317
    *c* charts, 332–336
    constants, 643–644
    for count of items, 327–329
    for counts of occurrences-per-unit, 332–336
    cycles, 343
    demerit control charts, 366–368
    discrete data patterns, *344*
    drift, 341–343, 370–371
    equations, 645–646
    EWMA control charts, 322, 323, 378–380
    factors for short run control charts for individuals, *X*, and *R* charts, 649–650
    freaks, 341, *342*
    individuals control charts, 317–323, 378–380
    interpreting, 341–348
    mixture, *346*
    moving range charts, 317–323
    multivariate control charts, 599–608
    *np* charts, 327–329
    *p* charts, 323–327
    planned changes, 345
    for proportion, 323–327
    ranges charts, 310–314
    repeating patterns, *344*
    selecting, 336–341
    special causes, 378–379
    stabilized attribute control charts, 362–366
    suspected differences, *346*
    *u* charts, 329–332
    *X* charts, 317–323
    zones, 347–348
control limits
    for averages and ranges charts, 311–312
    for averages and sigma charts, 315–316
    for *c* charts, 333–334
    EWMA, 375–376
    for *np* charts, 327
    for *p* charts, 324

Control phase, 60
　　control planning tools and techniques, 590–591
　　maintaining gains, 588–590
　　multivariate control charts, 599–608
　　objectives, 587
　　preparing the process control plan, 591–593
　　principal component analysis, 608
　　process audits, 596
　　process control planning for short and small runs, 593–596
　　selecting process control elements, 596–599
　　validating the new process or product design, 587
　　See also Define-Measure-Analyze-Improve-Control (DMAIC)
COPQ. See cost of poor quality (COPQ)
correlation analysis, 470–471
correlation coefficient, defined, 618
cost of poor quality (COPQ), 5, 75–78
　　bases, 81–82
　　examples, 78–81
cost of quality, 5
　　See also cost of poor quality (COPQ)
costs, 113
$C_p$ indices, 381–382, 384–385
$C_{PK}$ indices, 383–385
CPM, 218–222
$C_R$ indices, 382, 384–385
"create, operate, and trade" model, 103
creative destruction, 101–106
critical chain, 135–138
　　project portfolio management, 193
　　scheduling illustration, 195
Critical Incident Technique (CIT), 65–71
critical mass, 12
critical path, 222
Critical Path Method. See CPM
Critical to Cost (CTC), 259–265
　　break-even points, 261
　　financial analysis of benefit and cost, 260
Critical to Quality (CTQ)
　　defects per million opportunities (DPMO) and sigma level, 251–254
　　metrics, 250–251
　　normalized yield, 255–257
　　performance, 552–553
　　rolled throughput yield, 255, 256
　　throughput yield, 254–255
Critical to Schedule (CTS)
　　Cycle Time, 257
　　Overall Equipment Efficiency (OEE), 259
　　Process Cycle Efficiency, 257–258
　　Process Lead Time, 258
　　Velocity, 258

cross-functional collaboration, 200–201
CRT. See Current Reality Tree (CRT)
CTQ. See Critical to Quality (CTQ)
cubes, 173
cumulative distributions, 293
cumulative frequency, 293
Current Reality Tree (CRT), 127, 133
curvilinear relationships, 498
customer audits, 229
customer data collection methods
　　focus groups, 71–72
　　operational feedback systems, 72–75
　　surveys, 62–71
customer demands, 517–518
　　model, 97
　　translating, 93–101
customer feedback, 72–75
customer panels, 72
　　See also focus groups
customer retention, calculating the value of, 180–182
customer satisfaction, and quality, 88–89
customer survey, sample, 651–652
customer value, 61
　　projects, 178
customer-driven organizational structure, 44–46
Cycle Time, 257

$d_2^*$, table of values, 647–648
dashboard metrics, 150
　　dashboard design, 167–169
　　flowdown of strategies to drivers and projects, 150
data, continuous, 278, 492–493
data, discrete, 278, 492–493
data mining, 171, 173–175, 549–550
　　and Six Sigma, 175
data space, 464, 465
data warehousing, 171–172
　　and Six Sigma, 175
DBR. See Drum-Buffer-Rope (DBR)
Decision Rights Matrix, 49
deductive thinking, 281
defective, defined, 618
defects, 251
　　defined, 618
defects per million opportunities (DPMO), 18, 19, 251–254
defects per unit (DPU), 254
Define phase, 54
　　assembling the team, 267–268
　　Critical to Cost (CTC) metrics, 259–265
　　Critical to Quality (CTQ) metrics, 250–257

Critical to Schedule (CTS) metrics, 257–259
deliverables, 248–249
key objectives, 243
Pareto analysis, 247–248, *249*
process maps, 266
Project Charters, 243–245
project decomposition, 246–247
top-level process definition, 265–266
*See also* Define-Measure-Analyze-Improve-Control (DMAIC)
Define-Measure-Analyze-Design-Verify (DMADV)
vs. DMAIC, 214
overview, 211, *214*, 215–216
*See also specific phases*
Define-Measure-Analyze-Improve-Control (DMAIC), 3–4, 54–60
activities, durations, and critical path calculations, *221, 224*
vs. DMADV, *214*
overview, 211–212, 215–216
project network diagram, *222*
schedule cost for activities, *227*
Six Sigma tools commonly used in each phase, *213*
and strategy development, 109
typical project tasks and responsibilities, 217–218
using on a Six Sigma project, *212*
*See also specific phases*
deliverables, Define phase, 248–249
demerit control charts, 366–368
Deming, W. Edwards, 285
constraint management, 117
dependent variables, 460, 463–464, 473
deployment timetable, 18–21
Design phase, 517
data mining, 549–550
empirical model building, 529–548
Lean techniques for optimizing flow, 518–529
neural networks, 548–552
optimization using simulation, 552–572
Pugh concept selection method, 517–518
risk assessment tools, 572–581
statistical tolerancing, 581–586
virtual process mapping, 550
design review, 573
designed experiments, 472
ANOVA, 477–482
design characteristics, 474–475
full and fractional factorial, 482–491
power and sample size, 491
terminology, 472–474
testing common assumptions, 491–499
types of design, 475–477

differentiators
deploying differentiators to operations, 163–165
metrics, 157–159
direct labor budgets, 230
discrete data, 278
discrete part manufacturing, 340
discrimination, 395, *396*
disruptive-innovation, 87
distributions, 291
binomial distribution, 295–296
chi-square distribution, 440–443
cumulative distributions, 293
exponential distribution, 307–309
*F* distribution, 443–444, *445*
frequency distributions, 293
Gaussian distribution, 300
hypergeometric distribution, 298–299
lognormal distribution, 306–307
methods of enumeration, 291–293
normal distribution, 299–306
Poisson distribution, 297–298
sampling, 294–295
student's *t* distribution, 443, *444*
Weibull distribution, 309–310
DMADV. *See* Define-Measure-Analyze-Design-Verify (DMADV)
DMAIC. *See* Define-Measure-Analyze-Improve-Control (DMAIC)
double sampling, 618
DPMO. *See* defects per million opportunities (DPMO)
drift, 341–343, 370–371
drum, 138
Drum-Buffer-Rope (DBR), 133–135
dynamics management, 233–234

earliest finish (EF), 220
earliest start (ES), 220
EC. *See* Evaporating Cloud (EC)
education, 22
effect, 469
effectiveness, defined, 40
efficiency, defined, 40
Eigenvectors, 610
empirical model building
factorial experiments, 538–539
overview, 529–530
Phase 0, 531
Phase I, 532–536
Phase II, 536
Phase III, 538–539
Phase IV, 540–544
Phase V, 544–548

employees, rewarded, 47–48
enumeration, methods of, 291–293
enumerative studies, 280–282
errors, 251
Evaporating Cloud (EC), 128
Evolutionary Operation (EVOP), 544
EWMA, 371
    choosing the value of λ, 376
    common cause charts, 371–378
    control charts, 322, 323, 378–380
    Minitab example, 377–378
exchange services, 84
experiment design, defined, 618
experimental errors, 473–474
exponential distribution, 307–309
exponentially weighted moving averages (EWMA). *See* EWMA
external failure costs, 80–81
external setup activities, 433
external validity, 439
extrapolation, 461, 464

*F* distribution, 443–444, *445*, 631–634
facilitation
    of the group maintenance process, 242
    of the group task process, 241–242
    principles of team leadership and, 240–241
    selecting a facilitator, 239
    using an outside facilitator, 238–239
factorial experiments, 530, 538–539
    defined, 618–619
factors, defined, 618
failure mode and effect analysis (FMEA), 578–581, 590
Fault Tree Analysis. *See* process decision program chart (PDPC)
fault-tree analysis, 574–575
feasibility, 179
feedback
    customer, 72–75
    and learning, 166–175
feedback systems
    implementing Six Sigma, 17
    operational, 72–75
feeding buffers, 136–138
*The Fifth Discipline* (Senge), 115–116, 117
fill-in-the-blank questions, 63
financial analysis of benefit and cost, 260
Five Focusing Steps, 121–126
fixed-effects model, 475
flowcharts, 270, *271*
FMEA. *See* failure mode and effect analysis (FMEA)

focus groups, 71–72
focused brainstorming, 272
force-field diagrams, 590
Ford, Henry, 52, 87, 107, 423
Ford River Rouge Complex, 52
forecasting, vs. strategic positioning, 107
fractional factorial, 482–491
freaks, 341, *342*
frequency distributions, 293
    defined, 619
FRT. *See* Future Reality Tree (FRT)
FTA. *See* fault-tree analysis
Fukuda, Ryuji, 437
full factorial, 482–491
Future Reality Tree (FRT), 129, 133
future value, 262

gage R&R analysis, 404–406
    attribute gage R&R, 417–421
gains, maintaining, 588–590
Galvin, Bob, 4
Gantt charts, 216–218
Gaussian distribution, 300
    *See also* normal distribution
General Electric, 19, *20*
global impact, 100
global importance weights, 100–101, 102
GM, 87
*The Goal* (Goldratt), 114, 133
goals, 46, 119–121
Goldratt, Eliyahu M., 114, 116–117, 120
    critical chain, 135–138
    Drum-Buffer-Rope (DBR), 133–135
    financial measures, 139
    Five Focusing Steps, 121–126
    logical thinking process, 126–133
Gosset, W.S., 443
Green Belt XL software, 403, 404
Green Belts
    candidate selection process and criteria, 36
    compensation and retention, 38
    effectiveness certification, 667–677
    overview, 35–36
    and supply chain management, 53
    *See also* training
grouped frequency distributions, 293
groups
    counterproductive group roles, 237–238
    development stages, 234–236
    facilitation techniques, 238–242
    productive group roles, 236–237
    *See also* teams
Guttman format, 64

handle time, 385
hard savings, 180
Harry, Mikel, 253–254
*Harvard Business Review*, 87
hierarchies, flattened, 46
histograms, defined, 619
*Hoshin Kanri*, 150
hypergeometric distribution, 298–299
hypotheses, alternative, 440
    defined, 619
hypotheses, null, 440, 454
    defined, 619
hypothesis testing, 452–459

*I* (inventory), 139–143
implementing Six Sigma
    activities and systems required, 17–18
    infrastructure, 21–38
    timetable for deployment, 18–21
importance weights
    Black Belt success factors and, 32
    category, 98–99
    global, 100–101, 102
    local, 100
    subcategory, 99
imprecision, defined, 392
Improve phase, 56–59, 517
    data mining, 549–550
    empirical model building, 529–548
    Lean techniques for optimizing flow, 518–529
    neural networks, 548–552
    optimization using simulation, 552–572
    Pugh concept selection method, 517–518
    risk assessment tools, 572–581
    statistical tolerancing, 581–586
    virtual process mapping, 550
    *See also* Define-Measure-Analyze-Improve-Control (DMAIC)
Improvement Opportunities (IO), 74
in-control process, defined, 619
independent variables, 460, 463–464
individuals control charts, 317–323
    vs. EWMA control charts, 378–380
inductive thinking, 281
industry segmentation, 112
industry structure, 113
inferences
    defined, 439
    model-based, 439
    statistical, 439–440
information overload, 10–12
information systems requirements, 170

infrastructure
    centralized reporting, 21–22
    design-based, 439
    implementing Six Sigma, 17
    overview, 21–27
Innosight, 101, 103, 104
innovation
    creative destruction, 101–106
    Kano model, 88–89
    key areas for success, 105–106
    overview, 87–88
    Quality Function Deployment (QFD), 89–93
    spurring growth through, 104–105
    translating customer demands, 93–101
*The Innovator's Dilemma* (Christensen), 87
institutionalizing new approaches, 16
intensity scales, 64–65
interaction, 474
intercept, 469
internal failure costs, 80
internal marketing, 46
Internal Rate of Return (IRR), 259–260, 264–265
internal setup activities, 433
interval estimation, 444–452
interval scales, 277
inventory (*I*), 139–143
IO. *See* Improvement Opportunities (IO)
IRR. *See* Internal Rate of Return (IRR)
Ishikawa, Kaoru, 435

*Joining Together* (Johnson and Johnson), 240–241

Kano, Noritaki, 88–89
Kano model, 88–89
Kellogg's, 87
knowledge discovery, 529
Kodak, 87
Kolmogorov-Smirnov (K-S) statistic, 305–306
kurtosis, defined, 619

lambda, 308–309
latest finish (LF), 220
latest start (LS), 220
Latin-squares design, 476–477
lazy man's approach, 112
leadership
    champions and sponsors, 27
    communication, 47–48
    implementing Six Sigma, 17
Lean, 423–426
    CLOSEDMITTS, 426
    flow charts vs. spaghetti charts, *434*
    techniques for optimizing flow, 518–529

*Lean Thinking* (Womack and Jones), 424–425
least-squares fit, 466–470
level loading, 429
Likert format, 64–65
linear models, 463–466
    assumptions, 498–499
linearity, 407–408
    analysis using Minitab, 409–410
    defined, 392
    illustrated, *394*
linking, and communicating, 159–162
loadings, 610
local importance weights, 100
location, 283
logical thinking process, 126–133
logistic regression, 501–503
    binary, 503–506
    nominal, 510–512
    ordinal, 506–509
logit, 502
lognormal distribution, 306–307
loss, 546–547
lurking variables, 471

machine capability studies, 595
Malcolm Baldrige National Quality Award, 4
management constraints, 563
management reviews, 229
management roles, 238
management support, 195
    effective strategies, 199–200
    external roadblocks, 197
    individual barriers to change, 197–198
    ineffective strategies, 198–199
    internal roadblocks, 196
marketing, internal, 46
Master Black Belts, 19–20, 31
    compensation and retention, 37–38
    overview, 36–37
    *See also* Black Belts
Matsushita, 107
mean
    defined, 619
    estimates of $\mu$ when $\sigma$ is unknown, 447–449
    estimates of the mean when $\sigma$ is known, 445–447
mean, standard error of, defined, 619
Measure phase, 55–56
    metric definition, 275–278
    objectives, 269
    process baselines, 278–290
    process definition, 269–275
    *See also* Define-Measure-Analyze-Improve-Control (DMAIC)

measurement scales, 276–278
measurement systems
    accuracy, 416
    bias, 391, *392*, 396–397, *398*, 416
    discrimination, 395, *396*
    gage R&R analysis, 404–406
    linearity, 392, *394*, 407–410
    overview, 391
    pairwise reproducibility, 415–416
    part-to-part variation, 403
    repeatability, 391–392, 397–400, 415–416
    reproducibility, 392, *393*, 400–403, 416
    stability, 395–396, 416, 417
    summary reporting, 403–404
measurements, 47
    constraint management, 138–143
    defined, 276
    *See also* metrics
median, defined, 619–620
meeting management, facilitating, 242
mentoring, 199
metrics
    attributes of good metrics, 145–147
    Critical to Cost (CTC), 259–265
    Critical to Quality (CTQ), 250–257
    Critical to Schedule (CTS), 257–259
    differentiator metrics, 157–159
    measuring causes and effects, 147–149
    metric definition, 275–278
    organizational performance metrics, 146–147, *148*
    *See also* measurements
Microsoft, 87
milestone charts. *See* Gantt charts
Minitab
    ANOVA, 404
    attribute gage R&R, 417–421
    binary logistic regression, 503–506
    capability analysis, 385–390
    nominal logistic regression, 510–512
    non-parametric tests, 513–516
    ordinal logistic regression, 506–509
mission statements, and organizational vision, 107–109
mixed model, 475
mode, defined, 620
moments of truth, 75
monuments, 432
Motorola, 4, 8–9, 89
moving range charts, 317–323
*muda*, 424
    *See also* waste
multi-bossed individuals, 200–201

multiple bosses, 200
multiple linear regression, 468
multiple $R$, 469
multiple sampling, defined, 620
multiple server systems, 527
multiplication principle, 291, *292*
multitasking, 135, 136–138, 192–193
multivariate control charts, 599–608
Musk, Elon, 105

NB. *See* Negative Branch (NB)
necessary conditions, 119–121
Negative Branch (NB), 130, 133
Net Present Value (NPV), 259–260, 263–264
neural networks, 548–549
    example of neural net models, 549–552
    neural net model for solder defects, *551*
    neural net process map for solder defects, *551*
*The New Economics for Industry, Government,*
    *Education* (Deming), 116
Newton, Isaac, 6
Nietzsche, Friedrich, 119
nominal logistic regression, 510–512
nominal scale, 276
nominal variables, 276
non-parametric methods, 513–516
normal distribution, 299–306
normalized yield, 255–257
norming stage, 235
*np* charts, 327–329
NPV. *See* Net Present Value (NPV)
null hypotheses. *See* hypotheses, null

observations, 469
OC curve. *See* Operating Characteristics curve
    (OC curve)
occurrence intervals, estimates of, 450–452
odds ratios, 502–503
*OE* (operating expense), 139–143
*The Office* (TV show), 105
ogives, 293
Ohno, Taiichi, 156, 423, 424, 589
OLAP, 170, 171, 173
    and Six Sigma, 175
*One-Minute Manager* (Blanchard), 117
online analytic processing. *See* OLAP
open-ended questions, 63
Operating Characteristics curve (OC curve), 620
operating expense (*OE*), 139–143
operational control limits, 289
operational definitions, 412–413
operational feedback systems, 72–75
ordinal logistic regression, 506–509

ordinal variables, 276–277
organizational performance metrics, 146–147, *148*
organizational structure, 41–48
    command-and-control, 42–43, 196
    "correct" view of the company organizational
        chart, 44
    cross-functional collaboration, 200–201
    customer-driven, 44–46
organizational transformation, steps of, 15–16
organizational vision, 107–109
outliers, 611, *612*
Overall Equipment Efficiency (OEE), 259

*p* charts, 323–327
padding time, 135
pairwise reproducibility, 415–416
parameter design, 545
parameters, defined, 620
Pareto principle, 117
    analysis, 247–248
    analyzing Six Sigma project candidates, 187–189
    example Pareto diagram, *249*
Pareto Priority Index (PPI), 187–189
Parkinson's law, 135
partial autocorrelation function (PACF), 340
part-to-part variation, 403
PDPC. *See* process decision program chart (PDPC)
Pearson's product-moment correlation, 470
performance evaluations, 204–205
performance measurement model, 146–147, *148*
performing stage, 235
permutation, 292
PERT, 218–219, 222–223
PERT-CPM, 218–223
PID control equation, 380
point estimators, 444–452
Poisson
    distribution, 297–298
    probability sums, 635–637
pooled standard deviation, 312, 315–316
Pop Tarts, 87
population, defined, 620
possibilities-based strategic decisions, 111–113
power, 491
power curve, defined, 620
PPI. *See* Pareto Priority Index (PPI)
precedence relationship, 219
precision, defined, 391
predictor variables, 463–464
Prerequisite Tree (PRT), 131
prescriptive theories, 117
present value, 262
prevention costs, 78–79

primary variables, 473
principal component analysis, 608–616
prioritization matrix, 96
probability theory, applying to discrete variables, 292
process, 105
process audits, 596
process baselines, 278–280
    enumerative and analytic studies, 280–282
    estimating using process capability analysis, 289–290
    principles of statistical process control, 283–289
process capability
    defined, 620
    indices, 380–390
    using process capability analysis to estimate process baselines, 289–290
process control plan
    preparing, 591–593
    selecting process control elements, 596–599
    for short and small runs, 593–596
Process Cycle Efficiency, 257–258
process decision program chart (PDPC), 225, 590
process definition
    flowcharts, 270, *271*
    overview, 269–270
    process maps, 270, *273*
    SIPOC, 270, 271–275
Process Enterprises, 48–49
    roles and responsibilities, *50*
    *See also* Six Sigma Process Enterprise
Process Excellence Leadership Teams (PELTs), 49
Process Excellence (PEX), 49, 50–51
process feedback systems, implementing Six Sigma, 17
process focus, 41
process improvement, vs. system optimization, 116
Process Lead Time, 258
process maps, 41, 266, 270, *273*
    virtual process mapping, 550
Process Owners. *See* Business Process Executives (BPEs)
process performance indices, 381
process σ levels, and equivalent PPM quality levels, 653
process standard deviation, 320
    formulas for estimating, 380–381
processes, 40–41
    examples of high-level processes and subprocesses, 42
Procter and Gamble (P&G), 103–106, 112–113
producer's risk (α), defined, 621

production scheduling, Drum-Buffer-Rope (DBR), 133–135
profitability, 120
Program Evaluation and Review Technique. *See* PERT
project budgets, 230–231
project buffers, 136
Project Charter, 228, 243–245
project decomposition, 246
    work breakdown structures (WBS), 246–247
project deployment, implementing Six Sigma, 18
project launch synchronization, 194
project planning, 590
project records, 231–232
project reporting, 228–229
project scheduling, 192–193
    control and prevention of schedule slippage, 223–225
    cost considerations, 226–228
    Gantt charts, 216–218
    PERT-CPM, 218–223
    typical DMAIC project tasks and responsibilities, 217–218
project scope, 246
project selection, implementing Six Sigma, 17–18
projects
    analyzing project candidates, 179–187
    analyzing Six Sigma project candidates using the Pareto principle, 187–189
    assessing Six Sigma projects, 182–186
    benefit-cost analysis, 179–180
    calculating the value of customer retention, 180–182
    choosing the right projects, 177–178
    customer value projects, 178
    shareholder value projects, 178
    throughput-based project selection, 189–195
    types of savings, 180
    using constraint information to focus Six Sigma projects, 191–192
propagation of error, 36
prospective studies, 175
PRT. *See* Prerequisite Tree (PRT)
Pugh concept selection method, 517–518
purchased items budgets, 230
*P*-value, 470, 539
Pyzdek's law, 7

QFD. *See* Quality Function Deployment (QFD)
QMS. *See* Quality Management System (QMS)
quadratic loss function, *546*
quality
    and customer satisfaction, 88–89
    defined, 621

quality assurance, 621
quality circles, 34
quality control, 39
    defined, 621
    off-line, 545–546
quality function, 38–39
Quality Function Deployment (QFD)
    data collection and review of customer expectations, needs, requirements, and specifications, 91–93
    Macabe approach, *91*
    overview, 89–91
    sample matrix, *92*
quality improvement, 39
Quality Management System (QMS), 74
quality planning, 39
queuing theory, 526

*R* (range), 621
*R* square, 469
*The Race* (Goldratt and Fox), 133–134
random number generators, 557–560
random sampling, defined, 621
random-effects model, 475
randomization, 474, 475
randomized-block design, 475–476
range. *See R* (range)
ranges charts, 310–314
ranking questions, 63–64
RapiDOE, 570–572
rating questions, 64
ratio scales, 278
rational subgroups, 311
    sampling, 336–341
recognition and reward, 205–207
reengineering, 40–41
regression analysis
    example, 468–470
    scatter diagrams, 460–463
Reichheld, Frederick, 180
relative strength, 122
relative weakness, 122
repeatability, 397–400, 415–416
    defined, 391–392
    illustrated, *393*
replication, 474–475, 621
reports, 228–229
reproducibility, 400–403, 416
    defined, 392
    illustrated, *393*
    pairwise reproducibility, 415–416
resampling, 459–460
resolution, 395

resource redirection, 229
response variables, 463–464, 473
retrospective studies, 175
rewarded employees, 47–48
risk assessment tools
    design review, 573
    failure mode and effect analysis (FMEA), 578–581
    fault-tree analysis, 574–575
    overview, 572
    safety analysis, 575–578
risk-taking, 46
robust product and process design, 530, 540–544
roles and responsibilities, 23–26
rolled throughput yield (RTY), 255, *256*
run tests, 347–348

safety analysis, 575–578
sample, defined, 621
sample mean, 294
    calculating, 301–303
sample size, 491
sample standard deviation, 294
sampling distributions, 294–295
savings, types of, 180
scatter diagrams, 370, 460–463, *465*
scatter plots. *See* scatter diagrams
scientific management, 114
SCM. *See* supply chain management (SCM)
Score plot, 614, *615*
Scree Plot, 611, *612*
screening experiments, 530, 532–536
semantic differential format, 65
Senge, Peter, 115–116, 117
serial correlation. *See* autocorrelation
service processes, 339–340
shape, 283
shaping strategy, 111
shareholder value projects, 178
short-term wins, 16
sigma ($\sigma$), 3
sigma charts, 314–317
sigma levels, 4–6, 251–254
silos, 196
simple linear regression, 468
simulation
    backlog, 563–564
    management constraints, 563
    model development, 560–569
    overview, 552
    predicting CTQ performance, 552–553
    random number generators, 557–560
    tools, 553–557

simulation (*continued*)
    virtual DOE using simulation software, 570–572, *573*
single sampling, defined, 621
SIPOC, 153, 270, 271–275
Six Sigma
    activities and systems required to implement, 17–18
    defined, 3–4
    focusing on important metrics, 10–12
    history of, 4
    impact on human behavior, 11–12
    organizations, 10
    overview, 4–6
    philosophy, 6–8
    project evaluation, 182–186
    vs. traditional three sigma performance, 8–10
Six Sigma Process Enterprise, 40–51
*Six Sigma: The Breakthrough Management Strategy Revolutionizing the World's Top Corporations* (Harry), 253–254
skewness, 621
skills, 105–106
slack time, 220
Sloan, Alfred P., 42
smokestacks, 196
soft savings, 180
solution selling, 74
SpaceX, 105
spaghetti charts, 433–434
    vs. Lean flow, *434*
SPC. *See* statistical process control (SPC)
special causes, 285, 341
    charts, 370–371, 378–379
sponsors, 27
    cascading of sponsorship, *60*
    lack of sponsorship, 85
spread, 283
stability, 395–396, 416, 417
    defined, 392
    illustrated, *394*
stabilized attribute control charts, 362–366
stakeholder feedback systems, implementing Six Sigma, 17
stakeholders, 266
standard deviation
    calculating, 301–303
    defined, 621–622
    estimates of, 449–450
standard deviation charts, 314–317
standard error, 466, 469, 470
standard error of the mean, 294–295
standard normal curve, area under, 623–625

standard operating procedures (SOPs), 196
statistical association, 345
statistical inference, 439–440, 452
statistical process control (SPC), 249, 282
    attribute SPC for small and short runs, 361–368
    and automatic process control, 379–380
    principles of, 283–289
    using for ongoing control, 591
    *See also* control charts
statistical process control (SPC) for automated manufacturing
    overview, 369
    problems with traditional SPC techniques, 369–370
    special and common cause charts, 370–371
statistical process control (SPC) for short runs
    attribute SPC for small and short runs, 361–368
    overview, 348–349
    variables data, 350–361
statistical stability, 395
statistical test of hypotheses. *See* hypothesis testing
statistical tolerancing, 581–586
statistics, defined, 622
status reports, 229
steepest ascent, 530, 536–538
storming stage, 235
strategic development
    system optimization versus process improvement, 116
    systems as chains, 116–117
    systems management, 114–115
    systems thinking, 115–116
    using Constraint Theory, 113–114
strategic positioning
    vs. forecasting, 107
    organizational vision, 107–109
    overview, 106–107
    possibilities-based strategic decisions, 111–113
    strategic styles, 110–111
    strategy development, 109–110
strategic vision, translating, 151–159
strategy, 105
strategy deployment matrix, 160–162
strategy deployment plan, 157–159
Strengths and Weaknesses to external Opportunities and Threats (SWOT), 109, *110*
structure, 105
student syndrome, 135
student's $t$ distribution, 443, *444*
subcategory importance weights, 99
summary reporting, 403–404
supply chain, deployment to, 52–54

supply chain management (SCM), 52–54
support services budgets, 230
surveys
    case study using the Critical Incident Technique, 65–71
    guidelines for developing questions, 62–63
    overview, 62
    response types, 63–65
SWOT. *See* Strengths and Weaknesses to external Opportunities and Threats (SWOT)
synchronizer resources, 194
system constraints, 114
system design, 545
system optimization, vs. process improvement, 116
systems
    as chains, 116–117
    defined, 117
systems management, 114–115
systems thinking, 115–116

$T$ (throughput), 139–143
$t$ stat, 470
Taguchi, Genichi, 472, 545–548
Taguchi method, 547–548
tampering, 290
    effects and diagnosis, 288–289
Taylor, Frederick, 114
$t$-distribution, critical values of, 627–628
TDM. *See* total design method (TDM)
teams, 15
    assembling, 267–268
    common problems and remedial action, 235–236
    conflict resolution, 233–234
    counterproductive group roles, 237–238
    dynamics management, 233–234
    facilitation techniques, 238–242
    group development stages, 234–236
    membership, 233
    overview, 232–233
    performance evaluation, 204–205
    productive group roles, 236–237
    recognition and reward, 205–207
terminology, 472–474, 617–622
Tesla Motors, 105
Texas Instruments, 41
Theory of Constraints (TOC), 189–190
    comparison of TOC with traditional approaches, 191
    defined, 114
    *See also* constraint management
throughput ($T$), 139–143
    throughput-based project selection, 189–195
  yield, 254

time value of money, 261–265
TOC. *See* Theory of Constraints (TOC)
tolerance design, 545–546
tolerance interval factors, 585–586, 639–641
tolerance limits, 585
total design method (TDM), 70
Total Quality Management (TQM), 3
Toyota, 423
tracking results, 201–202
    financial results validation, 204
    lessons-learned capture and replication, 207–208
    possible information to be captured, 202
    team performance evaluation, 204–205
    team recognition and reward, 205–207
    typical view of Six Sigma projects, 203
trade unions, partnering with, 47
tragedy of the commons, 44
training, 22, 27
    Black Belts, 28–35
transfer functions, 148–149
transformation process, steps of, 15–16
Transition Tree (TT), 132, 133
TT. *See* Transition Tree (TT)
Type I error, 452, 622
Type II error, 452, 622

$u$ charts, 329–332
Udry, J. Richard, 7
unbiased sample variance, 294
uncaused variation, 345
underspending, 231
ungrouped frequency distributions, 293
universe, defined, 280
updating plans and timetables, 229
urgency, 15
*Using the Balanced Scorecard as a Strategic Management System* (Kaplan and Norton), 151

value, Lean and, 424
value stream analysis, 423–426
value stream mapping, 427–433
    common symbols, *428*
    example, 430–431
values, 46
    and organizational vision, 107–109
variables
    background, 473
    dependent, 460, 463–464, 473
    independent, 460, 463–464
    predictor, 463–464
    primary, 473
    response, 463–464, 473

variables data, 350–361
variance, 230–231
    analysis of, 469
    defined, 622
    equal-variance assumptions, 497
    estimates of, 449–450
    unbiased sample variance, 294
variation
    common causes, 285–287, 380
    part-to-part, 403
    special causes, 285–287
    types of, *286*
    uncaused, 345
variation, sources of, 434–435
    boxplots, 437–439
    cause and effect diagrams, 435–437
    statistical inference, 439–440
Velocity, 258
Verify phase
    control planning tools and techniques, 590–591
    maintaining gains, 588–590
    multivariate control charts, 599–608
    objectives, 587
    preparing the process control plan, 591–593
    principal component analysis, 608
    process audits, 596
    process control planning for short and small runs, 593–596
    selecting process control elements, 596–599
    validating the new process or product design, 587
virtual process mapping, solder process data for, 550
vision, 15–16, 46
    communicating, 108–109
    organizational, 107–109
visionary strategy, 111
visioning, 27

waiting line systems, 526
waste, 4, 424–426
Weibull distribution, 309–310
Welch, Jack, 177
Western Electric Rules, 376
work breakdown structures (WBS), 246–247
work in progress (WIP) inventories, 429

$X$ charts, 317–323

yes/no questions, 63

$Z$ transformation, 303
zero slack, *222*
$Z_{MIN}$ indices, 383, 384–385

CPSIA information can be obtained
at www.ICGtesting.com
Printed in the USA
JSHW042131100622
26832JS00003B/7

9 781260 121827